LACTIC ACID BACTERIA: GENETICS, METABOLISM AND APPLICATIONS

Lactic Acid Bacteria: Genetics, Metabolism and Applications

Proceedings of the Fifth Symposium held in Veldhoven, The Netherlands, 8–12 September 1996

Edited by

G. VENEMA

University of Groningen, The Netherlands

J.H.J. HUIS IN 'T VELD

University of Utrecht, The Netherlands

and

J. HUGENHOLTZ

Netherlands Institute for Dairy Research (NIZO), Ede, The Netherlands

Reprinted from *Antonie van Leeuwenhoek*, Volume 70, Nos. 2–4

Kluwer Academic Publishers

Dordrecht / Boston / London

A C.I.P. Catalogue record for this book is available from
the Library of Congress.

ISBN-13: 978-94-010-7292-2 e-ISBN-13: 978-94-009-1774-3
DOI: 10.1007/978-94-009-1774-3

Published by Kluwer Academic Publishers,
P.O. Box 17, 3300 AA Dordrecht, The Netherlands.

Kluwer Academic Publishers incorporates
the publishing programmes of
D. Reidel, Martinus Nijhoff, Dr W. Junk and MTP Press.

Sold and distributed in the U.S.A. and Canada
by Kluwer Academic Publishers,
101 Philip Drive, Norwell, MA 02061, U.S.A.

In all other countries, sold and distributed
by Kluwer Academic Publishers Group,
P.O. Box 322, 3300 AH Dordrecht, The Netherlands.

Printed on acid-free paper

Softcover reprint of the hardcover 1st edition 1996

Table of Contents

Editorial

Lactic acid bacteria constitute a heterogeneous group of organisms of considerable importance for industrial applications such as the production of wholesome foods and feeds. In addition, there is a growing interest in lactic acid bacteria with respect to health benefits and other medical applications. The importance of lactic acid bacteria is reflected in an increasingly growing body of research directed towards their biochemistry, genetics and biophysics as well as industrial and medical applications. In all of these research sectors impressive progress has been made in recent years, which is documented in this special issue of *Antonie van Leeuwenhoek*. These advances will be discussed at this lustrum symposium at Veldhoven, the Netherlands, from September 8–12, 1996, the fifth symposium in succession of the series that was started in 1983.

Collectively, the 14 contributions, all from experts in the field, provide an excellent summary of the present state of our knowledge of lactic acid bacteria both with respect to fundamental biology, application and their potential in promoting human and animal health and nutrition.

Gerard Venema — *University of Groningen, the Netherlands*
Jos H.J. Huis in't Veld — *Utrecht University, the Netherlands*
Jeroen Hugenholtz — *Netherlands Institute for Dairy Research (NIZO), Ede, the Netherlands*

Antonie van Leeuwenhoek **70**: 99–110, 1996.

Biotechnology of lactic acid bacteria with special reference to bacteriophage resistance

Charles Daly, Gerald F. Fitzgerald & Ruth Davis
Department of Microbiology and the National Food Biotechnology Centre, University College, Cork, Ireland

Abstract

Lactic acid bacteria play an important role in many food and feed fermentations. In recent years major advances have been made in unravelling the genetic and molecular basis of significant industrial traits of lactic acid bacteria. Bacteriophages which can infect and destroy lactic acid bacteria pose a particularly serious threat to dairy fermentations that can result in serious economic losses. Consequently, these organisms and the mechanisms by which they interact with their hosts have received much research attention. This paper reviews some of the key discoveries over the years that have led us to our current understanding of bacteriophages themselves and the means by which their disruptive influence may be minimized.

Introduction

The ability to preserve various perishable foods by means of fermentation is an ancient and well-accepted form of biotechnology. One has only to view the wide range of fermented cheeses, milks, breads, wines, pickles and meats available to the consumer to recognise their valuable contribution to our daily diet. Members of the lactic acid bacteria (LAB) which include the genera *Lactococcus*, *Leuconostoc*, *Lactobacillus*, *Streptococcus* and *Pediococcus*, play an essential role in the majority of these fermentations. In addition to providing an effective form of 'natural' preservation, they determine the flavour, texture and frequently, the nutritional attributes of the products as well as enhancing the variety of foods available to us. Some key technological traits of the LAB include carbohydrate and citrate metabolism, proteolysis, production of exopolysaccharides and antimicrobials and bacteriophage resistance (McKay & Baldwin, 1990; Cogan & Accolas, 1990).

Of the many factors that can contribute to poor starter culture performance, attack by lytic bacteriophages (phages) is surely the most significant and difficult to overcome. The problem has been exacerbated by the current trends within the industry towards more efficient manufacturing practices that depend on fewer, more finely-tuned starter cultures and by the requirement for consistently high quality end- products. In recent years, tremendous strides have been made in unravelling the genetics and molecular biology of LAB phages and their complex interactions with host strains. These developments have been reviewed in detail by a number of authors (Hill, 1993; Klaenhammer & Fitzgerald, 1994; Garvey et al., 1995b; Chopin & Fitzgerald, in preparation). Despite this increased knowledge, the control of phage proliferation in commercial practice continues to challenge both scientists and manufacturers alike. In particular, the extraordinary capacity of phages to respond to and overcome an array of host- mediated defence mechanisms has required that researchers devise novel and innovative strategies to effectively protect starter cultures.

The purpose of this paper is to provide an overview of the area of bacteriophage resistance in LAB but particularly in the lactococci, the hosts in which the most spectacular advances have been gained and to highlight some of the achievements in that research that have brought us to our current level of understanding of this phenomenon.

European unions' biotechnology frameworks for LAB research

The economic importance of LAB is well recognised both by governments and industries worldwide who have invested considerable financial resources into furthering scientific research of these bacteria. In particular, the exceptional support of the European Union (EU) for these endeavours must be acknowledged. Over the last twenty years the Commission of European Communities (CEC) has sponsored an integrated long-term research programme to advance fundamental and applied knowledge of LAB in an effort to identify and construct improved starter strains which are better able to withstand the rigours of increasingly demanding manufacturing conditions. Individual and transnational collaborative research efforts have been funded and coordinated through the successive frameworks of BEP (Biomolecular Engineering Programme), BAP (Biotechnology Action Programme), BRIDGE T (Biotechnology Research for Innovation Development) and the current programme BIOTECH G (Biotechnology Programme) as well as through the applied programmes of ECLAIR (European Collaborative Linkage of Agriculture and Industry through Research), FLAIR (Food-Linked Agro-Industrial Research) and FAIR (Food and Agro-Industrial Research). BIOTECH-G involves the varied research activities of 36 laboratories throughout Europe and actively disseminates its results to almost 40 companies that are either directly involved with BIOTECH or are affiliated to LABIP (Lactic Acid Bacteria Industrial Platform).

Considering the devastating nature of phage in the industry it is not surprising that a core activity of the above programmes has been dedicated to improving the resistance of starter cultures to infection by these agents.

Starter culture management and phage control

Disruption of dairy fermentations by phage has been recognised since the mid 1930's (Whitehead & Cox, 1935). Around the same time these researchers outlined a series of Good Manufacturing Practices aimed at reducing the level of phage buildup in factories (Whitehead & Cox, 1936). The effectiveness of these measures is reflected by their continued use in current manufacturing facilities. These include segregation of the starter culture room from the cheese production area, heat treatment of bulk starter media, the use of well designed starter tanks and effective sanitation procedures (Mullan, 1986; Limsowtin et al., 1996). The use of phage inhibitory medium (PIM) for bulk starter propagation (Reiter, 1956; Hargrove et al., 1961) is another measure that offers a further degree of protection against phage. PIM's contain phosphate and/or citrate which are chelating agents that sequester Ca^{++} ions essential for the adsorption of most LAB phage to host cell surfaces and these media are widely used in US cheese plants and to a lesser extent in Europe (Cogan & Hill, 1994). Despite external precautions however, phage have not been entirely eliminated. Ultimately, the most effective approach to preventing phage build up lies in combining these exclusion measures with the development of cultures which themselves are better able to resist attack by phage.

Traditionally, starter cultures have tended to be mixed strain varieties consisting of an unknown number of uncharacterized strains, or even species, of LAB. The complex composition of mixed strain cultures ensured that a certain level of resistance to phage attack was maintained. However, repeated subculturing of mixed strain starters leads to unpredictable changes in species composition and eventual strain dominance (Limsowtin et al., 1996). This in turn leads to increased susceptibility to phage attack and fermentation failures. Many fermentations of this type are still performed. In particular, producers of artisanal or 'farmhouse' cheeses rely on mixed strain starters to deliver the subtle flavour notes that define various cheeses. The production of mixed strain starters by commercial starter culture houses has to some extent alleviated problems of strain dominance by providing a more constant source of culture preparations and by avoiding unnecessary culture transfers in factories.

Although mixed strain starters may be used more or less continuously in small production units, they are generally unable to provide the necessary control and consistent performance required for the intensive, high volume cheese production more commonly encountered nowadays. One notable and highly successful exception to this is the so-called P- or practice cultures used in the manufacture of Gouda cheese in the Netherlands (for reviews see, Cogan & Accolas, 1990; Limsowtin et al., 1996). P-starters are derived from cultures that were transferred daily under factory conditions and normally contain high numbers of 'own' phage that do not affect the acid producing ability of the culture. When a new disturbing phage arises in the plant, P-cultures suffer some reduction in activity as the sensitive components of the starter are attacked; however, the overall culture recovers quickly (Stadhouders & Leenders, 1984). To overcome such fluctuations, the Netherlands Institute of Dairy Research (NIZO) founded a central culture collection in 1986 comprised of reliable phage resistant P-starters collected from the factories (Stadhouders, 1986). These cultures are then grown under carefully controlled conditions and distributed to the cheese manufacturers where the bulk starter is propagated under aseptic conditions to prevent phage contamination. In this manner mixed strain starters are able to perform consistently within very stringent manufacturing parameters without sacrificing the nuances of traditional Gouda flavour.

As research into the roles of starter bacteria evolved during the 20th century, it became clear that successful dairy fermentations could be achieved using limited number of specific LAB species, usually *Lactococcus lactis* ssp. *lactis*, *L. lactis* ssp. *cremoris* and occasionally, *L. lactis* ssp. *lactis* biovar. *diacetylactis*. As a result, Cheddar cheese manufacturers began to favour the use of a defined starter culture system comprised of one (rarely) or a combination of 2–6 well-characterized lactococcal strains. This starter culture system was pioneered in New Zealand (Lawrence & Pearce, 1972) and has since been adopted by the industries in Australia (Czulak et al., 1979; Hull, 1983), US (Richardson et al., 1980; Daniell & Sandine, 1981; Thunell et al., 1981) and Ireland (Timmons et al., 1988). The Irish defined strain starter system is founded on 35 strains of *L. lactis* ssp. *lactis* and *L. lactis* ssp. *cremoris*. These strains or their descendants have been used successfully by all of the Irish Cheddar cheese plants since 1986 producing in excess of 60,000 tonnes of cheese per annum.

Components of defined strain starters are chosen for their fast acid producing ability, high phage insensitivity and differing phage-host susceptibility profiles. Specific strain combinations may be used continuously until they become phage sensitive or, alternatively, they may be rotated in order to prevent phage build-up by limiting the exposure of any one combination to the phage environment. In either case, daily monitoring of factory phage levels is essential to preempt high phage titres developing against individual starter components. When significant levels of lytic phage are detected for a particular strain, it is replaced by a phage unrelated or phage resistant derivative. As the choice of genuine phage unrelated strains is quite limited, spontaneous phage insensitive mutants (BIMs) have provided an alternative source of replacement strains. Such mutants have proved successful in some instances; however, they frequently exhibit slow acid producing ability and/or revert to their original phage sensitivity which limits their usefulness in practice (Klaenhammer, 1987).

The benefits of using defined strains lie in improved fermentation reliability and in the achievement of consistently high quality end- products. Furthermore, the use of defined culture systems has allowed the development of specialized strains which exhibit enhanced functionality in cheese making. Unfortunately, as manufacturers come to rely on an increasingly small number of strains, the potential for fermentation failures due to phage attack also becomes greater.

Phage classification and genomic analysis

The designation of reliable and effective criteria by which phage may be classified has been an important objective of researchers from the points of view of determining the evolutionary origin of these organisms and of identifying the most prevalent phage species disturbing commercial starter cultures. Differentiation of phages by host susceptibility profiles is essential to the cheese maker in order to devise effective strain rotation programmes and to identify suitable phage- unrelated replacement strains. Unfortunately, phage-host range data do not adequately reflect phylogenetic relationships between various phages and are also subject to changes arising from minor DNA mutations, various genetic recombination/rearrangement events or even simply host mediated modification of phage DNA.

Early classification schemes relied on morphological, physiological and serological characteristics to group related phage (Jarvis, 1977; 1989). However, in recent years, advances in the areas of phage genetics and molecular biology have provided a more sound evolutionary basis for classification. As a result of a major international collaborative effort by members of the Lactococcal and Streptococcal Phage Study Group, Jarvis *et al.* (1991) segregated phages into 12 categories (with corresponding type phages) according to differences in morphology, DNA-DNA homology and protein profiles. Under this scheme, the most commonly encountered phages in commercial practice belong to the c2, 936 and P335 phage species. In New Zealand, US, and Ireland, the majority of phage isolated from factory wheys tend to be small isometric-headed with only a small minority of isolates being prolate- or large isometric-headed (Jarvis, 1977; Jarvis & Klaenhammer, 1986; Prévots et al., 1990; Casey et al., 1993) while in Continental Europe and parts of Canada a more even distribution of phage types is encountered (Lembke et al., 1980; Relano et al., 1987; Saxelin et al., 1986; Moineau et al., 1992). It has been speculated that these observations may relate to differences in starter culture practices (i.e. defined vs. mixed strain starters) and the wider range of cheese types produced on the Continent (Jarvis, 1988; Klaenhammer & Fitzgerald, 1994). Interestingly, phages isolated from widely separated geographical areas frequently exhibit high levels of genetic homology (Jarvis, 1995).

There has been tremendous progress in the recent past in elucidating the molecular structure and genetic organisation of phage genomes. The complete nucleotide sequences of four lactococcal phage genomes are known; two temperate small isometric representatives, r1t (van Sinderen et al., 1996) and Tuc 2009 (Garvey et al., 1995b; D. van Sinderen & G.F. Fitzgerald, personal communication), and two lytic prolate-headed phages, bIL67 (Schouler et al., 1994) and c2 (Lubbers et al., 1995). In addition, the sequence of approximately half of the genome of the temperate small isometric lactococcal phage, BK5-T has been reported by Boyce et al. (1995) while the entire sequence of a lytic phage infecting *Lactobacillus delbrueckii* has been determined (T. Alatossava, personal communication). Moreover, many individual genes and expression signals from a variety of LAB phage sources have been cloned and in some cases sequenced (Johnsen et al., 1996 & reviews, Hill, 1993; Klaenhammer & Fitzgerald, 1994; Garvey et al., 1995b) which have proved to be very rewarding in terms of obtaining a greater understanding of the molecular basis of lytic and lysogenic phage-host interactions as well as in identifying points in the phage lytic cycle that may be targeted by various defence mechanisms. More importantly, it has provided invaluable information as to how phages evolve and overcome various host-encoded resistance mechanisms. Comparisons between different phages has shown that their genomes are composed of exchangeable functional units (e.g. involved in lysis, integration and replication). Although the composition and nucleotide sequence may be different, the overall organisation of the genomes of many phages appears to be similar. This feature, together with their remarkable recombinogenic capacity, allows phages to exchange genetic modules, originating either from other phages or from the host, quickly and efficiently in order to escape imminent elimination by a host-encoded resistance mechanism. Two examples demonstrating such a recombination event have been described (Hill et al., 1991b; Moineau et al., 1994). This genetic flexibility is probably the major cause of the emergence of novel phages and must be taken into account when devising strategies to protect industrially used strains.

Host-mediated bacteriophage insensitivity

It is to be expected that lactic cultures exposed to bacteriophage contamination over extended periods of time are under intense selective pressure to evolve a variety of tactics by which productive phage attack may be prevented completely or at least minimised. Early in the study of lytic phage-host relationships it was recognised that specific strains existed that were inherently unsusceptible to the effects of phage under commercial manufacturing conditions (Zehran & Whitehead, 1954; Chopin et al., 1976; Lawrence et al., 1976; 1978). This was soon followed by the recognition that cell-mediated bacteriophage resistance, like many of the industrially important traits, was an unstable characteristic (Collins, 1958; Limsowtin et al., 1978). Limsowtin et al. (1978) conjectured that the instability of the phage resistance phenotype might be caused by its association with plasmid DNA elements and Sanders & Klaenhammer (1981) provided the first definitive evidence to support this theory when they correlated loss of phage resistance in *L. lactis* ssp. *cremoris* KH with the absence of a 17.5 kb plasmid pME100. A decade and a half later, more than 40 different lactococcal plasmids have been identified that confer phage resistance. It should be noted that some, relatively rare, chromosomal phage resistance loci have also been reported. Significantly, many of the phage resistance plasmids have conjugative functions (Chopin et al., 1984; McKay & Baldwin, 1984; Klaenhammer & Sanozky, 1985; Baumgartner et al., 1986; Daly & Fitzgerald, 1987; Jarvis, 1988; Coffey et al., 1989) a feature which has proved to be invaluable in the development of lactococcal starters with enhanced resistance properties (Sanders et al., 1986; Coffey et al., 1989; Jarvis et al., 1989; Klaenhammer, 1991; Harrington & Hill, 1991).

Currently four distinct types of bacteriophage insensitivity phenotypes have been described - adsorption inhibition, prevention of phage DNA penetration, restriction - modification (R-M) and abortive infection.

Adsorption inhibition

It is generally recognised that prevention of productive contact between phage and the bacterial cell greatly reduces the ability of the phage to proliferate. Ironically, although most of the phage resistance mechanisms in lactococci are now known to be plasmid encoded, preliminary reports of adsorption inhibition related to the generation of BIMs in which chromosomal mutations may have altered phage cell- surface receptors.

Little research has been directed at these mutants as the inhibition is highly specific and they tend to revert to phage sensitivity quite frequently. A novel means of BIM formation was recently described by Harrington & Hill (1992) who observed that a plasmid cointegration event was responsible for the reduction in phage adsorption to variants of *L. lactis* ssp. *lactis* biovar. *diacetylactis* DPC220. However, despite the relative lack of information, BIMs play a central role in many defined culture rotation programmes.

At least seven different plasmids encoding adsorption inhibition have been identified in lactococci. The first of these to be described was pME0030, a 48 kb plasmid from *L. lactis* ssp. *lactis* ME2 which was reported by Sanders & Klaenhammer (1983) to reduce the efficiency of phage adsorption from 80–90% to 20–40% when present in its host strain. More recently, considerable research work has focused on two plasmids pSK11 (54 kb; de Vos et al., 1984) and pCI528 (46 kb; Costello, 1988; Coffey et al., 1989) from *L. lactis* ssp. *lactis* SK110 and *L. lactis* ssp. *cremoris* UC503, respectively. Both these plasmids appear to prevent phage adsorption by masking the phage receptor sites. In the case of pSK11, a galactose-containing lipotechoic acid moiety was implicated in the shielding process (Sijtsma et al., 1988, 1990) while for pCI528, a cell wall associated hydrophilic polymer containing rhamnose and galactose was demonstrated to be responsible for the phage inhibition (Lucey et al., 1992). Two other plasmids pKC50 and P2520L have been shown to direct the synthesis of cell surface antigens; however, the nature by which they interfere with phage adsorption remains unclear (Tortorello et al., 1990; Akcelik & Tunail, 1992). Unfortunately, despite extensive efforts by some researchers (Lucey et al., 1992) none of the genetic determinants for these systems have been characterized to date.

Phage DNA injection blocking

Evidence that adsorption of phage to the host cell surface is not always sufficient for translocation of the phage DNA into the cell came first from studies performed by Watanabe et al. (1984) on *Lactobacillus casei* mutants which adsorbed ϕPL–1 efficiently, but did not permit injection of the phage DNA. Very recently exciting new evidence for an early acting mechanism encoded by the lactococcal plasmid pNP40 that prevented phage DNA penetration was presented (Garvey, 1995; Garvey et al., 1995a, 1996). Two groups have demonstrated the involvement of specific host cell membrane proteins in productive phage infections (Valyasevi et al., 1991; Monteville et al., 1994) and it is tempting to speculate that such proteins may represent potential targets for the operation of the pNP40 defence mechanism.

Restriction-modification

Given that the main biological role of R-M is the protection of the host cell from viral attack, one would expect that these systems should be widespread among LAB. Indeed since the first reported incidence by Collins as far back as 1956, numerous descriptions of R-M activities in *Lactococcus* spp. and other LAB have been cited in the literature, clearly confirming that this is the case. As with other LAB phage defence mechanisms, the majority of the R-Ms described have plasmid associated genotypes. However, *Scr*FI, a type II R-M system identified in the commercial starter strain *L. lactis* ssp. *cremoris* UC503 (formerly *Streptococcus cremoris* F; Schleifer & Kilpper-Bälz, 1987) is a notable exception that has been shown to be chromosomally determined (Davis et al., 1993; Twomey et al., 1993; Twomey, 1996) and there is evidence that this may be true for a number of other lactococcal R-Ms also (Davies & Gasson, 1984; Gautier & Chopin, 1987; Mayo et al., 1991). Interestingly, Harrington & Hill (1992) reported the activation of a silent R-M system following a plasmid cointegration event in a derivative of *L. lactis* ssp. *lactis* biovar. *diacetylactis* DPC220.

During the 1980's in particular, analysis of lactococcal R-Ms focused mainly on description of host-dependent variations of phage plaquing ability, the effects of temperature on R-M activity and assessments of the synergism observed between multiple R-M systems of different specificities and/or combinations of R-M with other phage defence mechanisms (e.g. adsorption inhibition and abortive infection). This latter information is especially relevant to the construction of commercially viable starter cultures with improved phage resistance properties. Throughout this period, there was only one report in the literature concerning the biochemical nature of LAB R-M reactions. Fitzgerald et al. (1982) isolated the first lactococcal restriction endonuclease (ENase) *Scr*FI and determined the DNA sequence of its target cleavage site. More recently, however, similar information has become available for four more lactococcal ENases (Mayo et al., 1991; Nyengaard et al., 1993; Moineau et al., 1995a) and a further four ENases from other LAB genera (Solaiman & Somkuti, 1990, 1991; Takano et al, 1990; Benbadis et al., 1991).

Major breakthroughs in the understanding of lactococcal R-Ms came from the laboratories of Klaenhammer and Daly/Fitzgerald who undertook the genetic characterization of the *Lla*I R-M system from pTR2030 (as opposed to the *Lla*I ENase activity described by Mayo et al., 1991) and the *Scr*FI R-M system, respectively. DNA sequencing revealed that at least six open reading frames (ORFs) are required for the expression of the pTR2030 *Lla*I R-M. The largest ORF specified a protein, M.*Lla*I, capable of performing two distinct N^6A methylase (MTase) activities. Analysis of the deduced amino acid (AA) sequence indicated that M.*Lla*I is homologous to *Fok*I and *Sts*I type IIs MTases and, by analogy, it was suggested that the dual MTase action of M.*Lla*I was necessary to methylate both strands of an asymmetric target DNA recognition sequence. Surprisingly, *in vivo* phage restriction by *Lla*I is mediated by three ORFs immediately following M.*Lla*I. Thus, R.*Lla*I would appear to be a complex multisubunit enzyme and may in fact represent the first of a new, fourth class of ENases (O'Sullivan et al., 1995). A further ORF was shown to encode a regulatory protein C.*Lla*I which promotes the activity of the *Lla*I ENase while concomitantly reducing the level of transcription of the *Lla*I operon. This complex interaction highlights the fine control of the cell over the expression of these potentially lethal genes.

The *Scr*FI R-M has also been completely sequenced (Davis et al., 1993; Twomey et al., 1995; Twomey, 1996). This system comprises of three ORFs, two of which encode functional 5^mC MTases separated by a third ORF that encodes the *Scr*FI ENase. The AA sequences of both MTases revealed the 10 consensus motifs characteristic of all 5^mC MTases. In addition, M.*Scr*FIA was found to share strong sequence homology with isomethylomeric MTases *Eco*RI, *Dcm* and *Nla*X which extended throughout the so-called variable or target-recognition domain (TRD). In contrast, M.*Scr*FIB MTase differs significantly from these proteins particularly throughout the TRD suggesting that it recognises its target sequence in a manner fundamentally different to M. *Scr*FIA. The sequence information for R.*Scr*FI did not reveal any homologies with known restriction ENases. The requirement for two *Scr*FI MTases is not yet clearly understood. However, only M.*Scr*FIA was found to protect phage from *in vivo Scr*FI restriction and was essential for stable maintenance of the ENase gene. Three other R-M systems from lactococci have been cloned and sequenced: *Lla*DCHI (formerly designated *Lla*II; Moineau et al., 1995a) and *Lla*AI (Nyangaard et al., 1993, 1995) which share common sequence specificities (↓GATC) and genetic organization with the *Dpn*II R-M system from *Streptococcus pneumoniae* and *Lla*B1 which recognizes C↓TRYAG. In addition, three other systems have been identified but have not yet been cloned (Nyangaard et al., 1995). Recently it was shown by Moineau et al. (1995b) that the *Lla*DCHI system is also expressed in *Streptococcus thermophilus* where it conferred a high level of phage resistance.

Abortive infection

Among lactococci, the term abortive infection encompasses a broad range of phage defence mechanisms that suppress intracellular phage development but which are not the result of adsorption inhibition, blocking of phage DNA injection or restriction-modification. Phenotypically, they are all characterized by either a total absence of plaques or by severely reduced plaquing efficiencies and plaque sizes.

The first Abi plasmid to be identified was the 64 kb nisin resistance plasmid pNP40 by McKay & Baldwin (1984). These researchers noted that total insensitivity to the prolate-headed phage c2 was acquired following conjugal transfer of pNP40 to a plasmid-free recipient *L. lactis* ssp. *lactis* LM2301. Similar approaches were employed to isolate numerous other conjugative or mobilizable plasmids encoding these phenotypes. Several non-conjugative Abi plasmids have also been reported (Klaenhammer & Fitzgerald, 1994).

The molecular, genetic and biochemical characterization of the 46 kb phage resistance plasmid pTR2030 has consistently represented the forefront of our understanding not only of host-mediated phage resistance in lactococci but also of phage counter responses to these resistance systems. In addition, this plasmid was the first to be exploited to generate more desirable starter cultures for the dairy industry. pTR2030 is a self-transmissible plasmid harboured by *L. lactis* ssp. *lactis* ME2. Two separate phage resistance phenotypes have been attributed to this plasmid - *Lla*I R-M (described earlier) - and a heat sensitive, abortive infection mechanism AbiA (also known as Hsp). AbiA was the first such system to be analysed at the DNA sequence level. It is comprised of a single 1887 bp ORF capable of specifying a 74 kDa gene product. No homology was observed between it and other known protein sequences (Hill et al., 1990a). Shortly afterwards, Coffey et al. (1991) presented sequence data for the pCI829 encoded Abi from *L. lactis* ssp. *lactis* UC811 which proved to be identical to the pTR2030 system. Examination of the rate of phage DNA accumulation in infected cells harbouring pTR2030 indicate that AbiA inhibits replication of specific phage types (Hill et al. 1991a).

More recently, the Abi plasmid, pNP40, has been the focus of extensive research efforts. The presence of pNP40 in *L. lactis* ssp. *lactis* MG1363, completely prevented plaque formation of prolate- and small isometric-headed phages. Three independent phage insensitivity traits were governed by this plasmid. Two of the systems operate via abortive infection (AbiE and AbiF) while the third, which was discussed in an earlier section, interrupts phage DNA injection. Both Abi loci have been cloned and sequenced (Garvey et al., 1995a,b). Two ORFs, *abi*Ei and *abi*Eii were required for the operation of AbiE. This system which affects small isometric- headed phage only has little, if any, effect on phage replication and it is assumed, therefore, to act at a later stage of the phage lytic cycle possibly during viral gene transcription/translation, packaging or release. The second Abi system (AbiF) is encoded by a single ORF which specifies a 41.2 kDa protein. In contrast to AbiE, AbiF inhibits both prolate and small isometric phage development. It also differs from AbiE in that it operates at the level of phage replication. The coordinated action of the two pNP40 abortive infection mechanisms directed at consecutive stages of the phage lytic cycle in addition to the contribution of the infection-blocking mechanism permitted only 3% of the infected cells to release progeny phage (Garvey et al., 1995b).

The DNA sequences of several other Abi genes have also been reported: AbiB (*L. lactis* ssp. *lactis* IL416, pIL415, Cluzel et al., 1991); AbiC (*L. lactis* ssp. *lactis* ME2, pTN20, Durmaz et al., 1992); AbiD (*L. lactis* ssp. *lactis* KR5, pBF61, Mc Landsborough et al., 1995) AbiD1 (*L. lactis* ssp. *cremoris* IL964, pIL105, Anba et al., 1995) and the two gene operon AbiGi-ii (*L. lactis* ssp. *cremoris* UC653, pCI750, Fitzgerald, G.F., personal communication). It is notable that in all cases the genetic loci have lower % G/C contents (approx. 26–29%) than the lactococcal average (approx. 37–38%). Unfortunately, these data have not provided the much anticipated clues that would signal the mechanisms of the various systems. In general, none of the predicted Abi products relate to proteins of known functionality. However, AbiD, AbiD1 and AbiF share 26–47% amino acid identity with one another which may prove significant in identifying common functional domains for these proteins (Garvey et al., 1995b).

Preliminary investigations by researchers in the laboratory of M.-C. Chopin have provided important insights as to the mode of action of AbiB and AbiD1 systems. Parreira et al. (1996) revealed a dramatic decrease in phage-specific mRNA transcripts ten minutes after infection of cells with AbiB. Thus AbiB may itself have RNase activity or, alternatively, it may stimulate the synthesis/activity of an existing bacterial or phage RNase. A phage operon consisting of four ORFs was implicated in the action of AbiD1. In this instance Bidnenko et al. (1995) proposed that the interaction of *orf*1 and *abi*D1 gene products prevented translation of *orf*3 RNA, expression of which appeared to be essential for phage multiplication.

Novel phage resistance mechanisms

The increased understanding of the structure and organization of lactococcal phage genomes and of the manner in which phages interact with their hosts have placed researchers in the dynamic position of being able to design and implement new strategies of phage resistance. Two such specifically designed measures have been reported to date. Per (*p*hage *e*ncoded *r*esistance) mimics the effects of the pTR2030 AbiA system by cloning of a phage origin of replication on a high copy number vector (Hill et al., 1990b). This artificial origin out competes the true phage origin for essential replication factors thereby promoting an increase of vector rather than phage DNA. In the second scheme, a promoter is directed to transcribe part of the antisense strand of phage DNA and the resulting RNA transcripts interact with the phage mRNA preventing its further translation into phage protein products (Kim & Batt, 1991; Batt et al., 1995).

Maximizing host-encoded bacteriophage resistance in industrial starter cultures

Over the past 15–20 years, starter cultures have evolved from a heterogeneous mix of undefined strains into the small number of highly characterized, uniform strains that support today's intensive cheese-making practices. This trend is likely to become even more pronounced in future years as specialized cultures with advantageous fermentative properties become available to the industry. As a result, it is increasingly difficult to maintain traditional rotation programmes that rely on the selection of naturally occurring phage-unrelated strains and/or the isolation of BIMs in order to protect cultures from phage attack for extended periods of time.

From the start it was recognised that the conjugative properties of many phage resistance plasmids could be exploited to develop cultures for the industry with improved resistance to phage (McKay & Baldwin, 1984). Sanders et al. (1986) reported the first practical application of this strategy when pTR2030 was introduced into an industrial starter strain. Significantly, selection of the transconjugants was on the basis of lactose fermenting ability and increased phage resistance thereby maintaining the food-grade status of the manipulation. Several researchers have expanded on this work using both pTR2030 (Sing & Klaenhammer, 1986; Klaenhammer, 1991) and a variety of other phage resistance plasmids (Jarvis et al., 1989; Coffey et al., 1989; Kelly et al., 1990). In some of these studies, higher levels of phage insensitivity were achieved by 'stacking' two or more phage resistance mechanisms in a single strain that target different points of the phage lytic development (Coffey et al., 1989; Klaenhammer, 1989). In addition, several plasmids can each naturally encode more than one mechanism. For example, R-M and abortive infection systems, a particularly effective combination, were genetically linked on plasmids such as pTR2030 (Hill et al., 1989), pBF61 (Froseth et al., 1988) and pKR223 (Laible et al., 1987). As described above, pNP40, the plasmid chosen by Harrington and Hill (1991) for strain development programmes, encoded not just one but two abortive infection loci as well as a third gene specifying the blocking of phage DNA injection (Garvey, 1995; Garvey et al., 1996).

Successful applications of such strategies in culture rotation programmes have been reported in the US (Sanders, 1988; Klaenhammer, 1991) and in Ireland (Kelly et al., 1990; Klaenhammer & Fitzgerald, 1994). However over time, phages have evolved that can defeat even the multi-layer defences of such 'enhanced' strains. ϕ50, which is completely resistant to pT2030 (*Lla*I$^+$, *abi*A$^+$), has been shown to have acquired part of the*Lla*I MTese gene, while resistance to AbiA is believed to be due to its nonsusceptible *ori* (Hill et al., 1991b). In a separate incident, recombination with a particular host chromosomal DNA segment enabled ϕul37 to overcome the action of AbiC. Interestingly, development of ϕul37 was prevented by disruption of the host DNA sequence (Moineau et al., 1994).

The manner in which these 'new' phages evolved alerted scientists to the remarkable genetic flexibility and adaptability of these organisms. In response, Sing & Klaenhammer (1993) devised an alternative form of starter rotation aimed at minimizing the development of new resistant phages arising from such DNA recombination and mutation events. The strategy employs several derivatives of a single starter strain constructed such that they each harbour a different set of defence mechanisms and which are then used for fermentations in a specific sequence. As a result, the phage population is only briefly exposed to any single set of host defences making it more difficult for modified variants to emerge that can disrupt the fermentation. Durmaz & Klaenhammer (1995) further developed this concept by constructing three derivatives of a lactococcal strain, each with a different R-M or Abi phage defence mechanism, which were rotated in repeated cycles of a milk starter activity test (Heap & Lawrence, 1976). The result was to create a highly effective, designed culture scheme with uniform fermentation characteristics that, on the basis of laboratory tests, could be predicted to withstand phage attack in commercial circumstances for extended periods of time. As an added advantage, the phage defence rotation strategy (PDRS) is particularly suited to protect the new specialized strains that are currently being developed to meet specific product and process requirements.

An alternative approach to increasing the range of lactococcal cultures available for cheese fermentation processes is being developed at Oregon State University and is based on the use of molecular probes to screen for new *Lactococcus lactis* strains from natural environments (Salama et al., 1991, 1993, 1995). Employing species and subspecies specific rRNA probes, *L. lactis* subsp. *lactis* strains were readily recovered from a range of plant and milk- derived materials but more significantly, it was also possible to isolate *L. lactis* subsp. *cremoris* strains from raw milk and cottage cheese samples from regions as diverse as The Peoples Republic of China, Morocco, Yugoslavia and The Ukraine. Studies examining the growth performance of the subsp. *cremoris* isolates, in particular in milk, indicated that a number were capable of rapid acid production and yielded an acceptable flavour which suggested that they may be suitable for use as starter cultures in milk fermentations. In addition, the *cremoris* isolates were resistant to a range of phages commonly encountered in the cheese industry. Indeed, it has been proposed that unlike most commercially available strains which are ultimately derived from a relatively limited number of starter cultures, these new isolates may potentially represent an important source of genetic diversity for the dairy industry (Salama et al., 1995). It is notable that four of these 'new' *cremoris* strains have been successfully used either singly or in various combinations in cheese making trials under commercial conditions and that no phages were detected (W. Sandine, personal communication). Thus, it can be anticipated that this novel approach will provide an extended bank of commercially useful strains whose genetic diversity will confer, for a time at least, some measure of protection against phage.

Concluding remarks

This paper has attempted to acknowledge some of the major research achievements that have contributed to our present comprehension of dairy phages and the various means by which they may be controlled in the industry. Over the last 10 years especially, there have been tremendous advances made in elucidating the molecular basis of phage infection and the cell directed interference of this process. It is particularly gratifying to note that some of the phage resistance plasmids mentioned here are already in commercial use. In addition, the ability to devise novel defence mechanisms and to design sophisticated rationales such as the PDRS emphasizes the fact that the ultimate goal of strong, long-lasting phage protection is within our grasp. Clearly, recombinant DNA technology must figure in future strategies. With this in mind, ongoing developments in 'self-cloning' techniques is very pertinent (de Vos & Simons, 1994). Such methods facilitate *in vitro* DNA manipulation in a food-grade manner. Significantly, as self-cloning only utilizes the DNA of the GRAS organism in question, it does not require regulatory approval as a genetically engineered microorganism (GEM) from the Commission of the European Communities. It is likely that other regulatory agencies will adopt similar approaches to this issue also.

Unfortunately the adaptability of phages is such that the prevention of their dominance in dairy fermentations will remain an ongoing battle for many years to come and scientists will have to be ever vigilant against the emergence of new 'super phage' capable of overcoming even the elaborate measures currently being developed.

References

Akcelik M & Tunail N (1992) A 30Kd cell wall protein produced by plasmid DNA which encodes inhibition of phage adsorption in *Lactococcus lactis* subsp. *lactis* P25. Milchwissenschaft 47: 215–217

Anba J, Bidnenko E, Hillier A, Ehrlich D & Chopin MC (1995) Characterization of the lactococcal *abiD1* gene coding for phage abortive infection. J. Bacteriol. 177: 3818–3823

Batt CA, Erlandson K & Bsat N (1995) Design and implementation of a strategy to reduce bacteriophage infection of dairy starter cultures. Int. Dairy J. 5: 949–962

Baumgartner A, Murphy M, Daly C & Fitzgerald GF (1986). Conjugative co-transfer of lactose and bacteriophage resistance plasmids from *Streptococcus cremoris* UC653. FEMS Microbiol. Lett. 35: 233–237

Benbadis L, Garel JR & Hartley DL (1991) Purification, properties and sequence specificity of *SslI* a new type II restriction endonuclease from *Streptococcus salivarius* subsp. *thermophilus*. Appl. Environ. Microbiol. 57: 3677–3678

Bidnenko E, Ehrlich D & Chopin MC (1995) Phage operon involved in sensitivity to the *Lactococcus lactis* abortive infection mechanism AbiD1. J. Bacteriol. 177: 3824–3829

Boyce JD, Davidson BE & Hillier AJ (1995) Sequence and analysis of the *Lactococcus lactis* temperate bacteriophage BK5-T and demonstration that the phage DNA has cohesive ends. Appl. Environ. Microbiol. 61: 4089–4098

Casey CN, Morgan E, Daly C & Fitzgerald GF (1993) Characterization and classification of virulent lactococcal bacteriophages isolated from a Cheddar cheese plant. J. Appl. Bacteriol. 74: 268–275

Chopin MC & Fitzgerald GF (1996) Molecular analysis of bacteriophage and bacteriophage resistance mechanisms of lactic acid bacteria. In preparation

Chopin MC, Chopin A & Roux C (1976). Definition of bacteriophage groups according to their action on mesophilic lactic streptococci. Appl. Environ. Microbiol. 32: 741–746

Chopin A, Chopin MC, Mollio-Batt A & Langella P (1984) Two plasmid determined restriction and modification systems in *Streptococcus lactis*. Plasmid 11: 260–263

Cluzel PJ, Chopin A, Ehrlich DS & Chopin MC (1991) Phage abortive infection mechanism from *Lactococcus lactis* subsp. *lactis*, expression of which is mediated by an ISO-ISSI element. Appl. Environ. Microbiol. 57: 3547–3551

Coffey AG, Fitzgerald GF & Daly C (1989) Identification and characterization of a plasmid encoding abortive infection from *Lactococcus lactis* ssp. *lactis*UC811. Neth. Milk Dairy J. 43: 229–244

Coffey AG, Fitzgerald GF & Daly C (1991) Cloning and characterisation of the determinant for abortive infection of bacteriophage from lactococcal plasmid pC1829. J. Gen. Microbiol. 137: 1355–1362

Cogan TM & Accolas JP (1990) Starter cultures: types, metabolism and bacteriophage. In Dairy Microbiology. Vol. 1, The Microbiology of Milk. (Robinson, R.K. Ed.) pp 77–114 Elsevier Applied Science, London

Cogan TM & Hill C (1994) Cheese starter cultures. In: Cheese Chemistry, Physics and Microbiology, Vol. 1. (Fox, P.F. Ed.) pp 193–255 Chapman and Hall, London

Collins EB (1956). Host-controlled variations in bacteriophages active against lactic streptococci. Virology 2: 261–271

— (1958) Changes in the bacteriophage sensitivity of lactic streptococci. J. Dairy Sci. 41: 41–48

Costello VA (1988) Characterisation of bacteriophage-host interactions in *Streptococcus cremoris* UC503 and related lactic streptococci. PhD Thesis. The National University of Ireland, University College, Cork

Czulak J, Bant DJ, Blyth SC & Crace JB (1979) A new cheese starter system. Dairy Industries Int. 44: 17–19

Daly C & Fitzgerald GF (1987) Mechanisms of bacteriophage insensitivity in the lactic streptococci. In Streptococcal Genetics. (Ferretti, J. and Curtiss, R. Eds.) pp 259–268 American Society for Microbiology. Washington DC

Daniell SD & Sandine WE (1981) Development and commercial use of a multiple strain starter. J. Dairy Sci. 64: 407–415

Davies FL & Gasson MJ (1984) Bacteriophages of lactic acid bacteria. In: Advances in the Microbiology and Biochemistry of Cheese and Fermented Milk (Davies, F.L. and Law. B.A., Eds.) pp 127–151 Elsevier Applied Science Publisher, London

Davis R, van der Lelie D, Mercenier A, Daly C & Fitzgerald GF (1993) *Scr*FI restriction-modification system of *Lactococcus lactis* subsp. *cremoris* UC503: cloning and characterization of two *Scr*FI methylase genes. Appl. Environ. Microbiol. 59: 777–785

de Vos WM & Simons G (1994) Gene cloning and expression systems in lactococci. In: Genetics and Biotechnology of Lactic Acid Bacteria (Gasson, M.J. and de Vos, W.M. Eds.) pp 52–105

de Vos WM, Underwood HM & Davies FL (1984) Plasmid encoded bacteriophage resistance in *Streptococcus cremoris* SK11. FEMS Microbiol. Lett. 23: 175–178

Durmaz E & Klaenhammer TR (1995) A starter culture rotation strategy incorporating paired restriction/modification and abortive infection bacteriophage defences in a single *Lactococcus lactis* strain. Appl. Environ. Microbiol. 61: 1266–1273

Durmaz E, Higgins DL & Klaenhammer TR (1992) Molecular characterization of a second abortive phage resistance gene present in *Lactococcus lactis* subsp. *lactis* ME2. J. Bacteriol. 174: 7463–7469

Fitzgerald GF, Daly C, Brown LR & Gingeras TR (1982) *Scr*FI: a new sequence - specific endonuclease from *Streptococcuscremoris*. Nuc. Acid Res. 10: 8171–8179

Froseth BR, Harlander SK & Mc Kay LL (1988) Plasmid mediated reduced phage sensitivity in *Streptococcus lactis* KR5. J. Dairy Sci. 71: 275–284

Garvey P (1995) Analysis of three phage resistance mechanisms and a *rec*A homologue encoded by the lactococcal plasmid pNP40. PhD Thesis. The National University of Ireland, University College, Cork

Garvey P, Fitzgerald GF & Hill C (1995a) Cloning and DNA sequence analysis of two abortive infection phage resistance determinants from the lactococcal plasmid pNP40. Appl. Environ. Microbiol. 61: 4321–4328

Garvey P, van Sinderen D, Twomey DP, Hill C & Fitzgerald GF (1995b) Molecular genetics of bacteriophage and natural phage defence systems in the genus *Lactococcus*. Int. Dairy J. 5: 905–947

Garvey P, Hill C & Fitzgerald GF (1996) The lactococcal plasmid pNP40 encodes a third bacteriophage resistance mechanism, one which affect phage DNA penetration. Appl. Environ. Microbiol. 62: 676–679

Gautier M & Chopin MC (1987) Plasmid determined systems for restriction and modification activity and abortive infection in *Streptococcus cremoris*. Appl. Environ. Microbiol. 53: 923–297

Hargrove RE, McDonough FE & Tittsler RP (1961). Phosphate heat treatment of milk to prevent phage proliferation in lactic cultures. J. Dairy Sci. 44: 1977–1810

Harrington A & Hill C (1991) Construction of a bacteriophage resistance derivative of *Lactococcus lactis* subsp. *lactis* 425A by using the conjugal plasmid pNP40. Appl. Environ. Microbiol. 57: 3405–3409

— (1992) Plasmid involvement in the formation of a spontaneous bacteriophage insensitive mutant of *Lactococcus lactis*. FEMS Microbiol. Lett. 96: 135–142

Heap HA & Lawrence RC (1976) The selection of starter strains for cheesemaking. N.Z.J. Dairy Sci. Technol. 11: 16–20

Hill C (1993) Bacteriophage and bacteriophage resistance in lactic acid bacteria. FEMS Microbiol. Rev. 12: 87–108

Hill C, Miller LA & Klaenhammer TR (1990a). Nucleotide sequence and distribution of the pTR2030 resistance determinant (*hsp*) which aborts bacteriophage infection in lactococci. Appl. Environ. Microbiol. 56: 2255–2258

— (1990b) Cloning, expression and sequence determination of a bacteriophage fragment encoding bacteriophage resistance in *Lactococcus lactis*. J. Bacteriol. 172: 6419–6426

Hill C, Massey IJ & Klaenhammer TR (1991a) Rapid method to characterize lactococcal bacteriophage genomes. Appl. Environ. Microbiol. 57: 283–288

Hill C, Miller LA & Klaenhammer TR (1991b) *In vivo* genetic exchange of a functional domain from a type IIA methylase between lactococcal plasmid pTR2030 and a virulent bacteriophage. J. Bacteriol. 172: 6419–6426

Hill C, Pierce K & Klaenhammer TR (1989) The conjugative plasmid pTR2030 encodes two bacteriophage defence mechanisms in lactococci, restriction-modification (R+/M+) and abortive infection (Hsp+). Appl. Environ. Microbiol. 55: 2416–2419

Hull RR (1983) Factory-derived starter cultures for the control of bacteriophage in cheese manufacture. Austr. J. Dairy Technol. 38: 149–154

Jarvis AW (1977) The serological differentiation of lactic streptococcal bacteriophage. N.Z.J. Dairy Sci. Technol. 12: 176–181

— (1989) Bacteriophages of lactic acid bacteria. J. Dairy Sci. 72: 3406–3428

— (1988) Conjugal transfer in lactic streptococci of plasmid- encoded insensitivity to prolate and small isometric-headed bacteriophages. Appl. Environ. Microbiol. 54: 777–783

— (1995) Relationships by DNA homology between lactococcal phages, 7–9, P335 and New Zealand industrial lactococcal phages. Int. Dairy J. 5: 335–366

Jarvis AW & Klaenhammer TR (1986) Bacteriophage resistance conferred on lactic streptococci by the conjugative plasmid pTR2030: effects on small isometric-, large isometric- and prolate-headed phages. Appl. Environ. Microbiol. 51: 1272–1277

Jarvis AW, Heap HA & Limsowtin GKY (1989) Resistance against industrial bacteriophages conferred on lactococci by plasmid pAJ1106 and related plasmids. Appl. Environ. Microbiol. 55: 1537–1543

Jarvis AW, Fitzgerald GF, Mata M, Mercenier A, Neve H, Powell IB, Ronda C, Saxelin M & Tueber M (1991) Species and types phages of lactococcal bacteriophages. Intervirol. 32: 2–9

Johnsen MG, Appel KF, Madsen PL, Vogensen FK, Hammer K & Arnau J (1996) A genome region of lactococcal temperate bacteriophage TP901–1 encoding major virion proteins. Virol. In press

Kelly W, Dobson J, Jorck-Ramberg D, Fitzgerald GF & Daly C (1990) Introduction of bacteriophage resistance plasmids into commercial *Lactococcus* starter cultures. FEMS Microbiol. Rev. 87: Abst. C20

Kim SG & Batt CA (1991) Antisense mRNA mediated bacteriophage resistance in *Lactococcus lactis* ssp.*lactis* Appl. Environ. Microbiol. 57: 1109–1113

Klaenhammer TR (1984) Interactions of bacteriophage with lactic streptococci. Advances in Appl. Microbiol. 30: 1–29

— (1987) Plasmid directed mechanisms for bacteriophage defence in lactic streptococci. FEMS Microbiol. Rev. 46: 313–325

— (1989) Genetic characterization of multiple mechanisms of phage defence from a prototype phage-insensitive strain, *Lactococcus lactis* ME2. J. Dairy Sci. 72: 3429–3442

— (1991) Development of bacteriophage-resistant strains of lactic acid bacteria. Biochemical Soc. Transactions 19: 675–681

Klaenhammer TR & Fitzgerald GF (1994) Bacteriophage and bacteriophage resistance. In: Genetics and Biotechnology of Lactic Acid Bacteria (Gasson, M.J. and de Vos., W.M. Eds.) pp 106–168 Blackie Academic and Professional, Glasgow

Klaenhammer TR & Sanozky RB (1985) Conjugal transfer from *Streptococcus lactis* ME2 of plasmids encoding phage resistance, nisin resistance and lactose-fermenting ability: evidence for a high- frequency conjugative plasmid responsible for abortive infection of virulent bacteriophage. J. Gen. Microbiol. 181: 1531–1541

Laible NJ, Rule PL, Harlander SK & Mc Kay LL (1987) Identification and cloning of plasmid deoxyribonucleic acid coding for abortive infection from *Streptococcus lactis* ssp. *diacetylactis* KR2. J. Dairy Sci. 70: 2211–2219

Lawrence RC & Pearce LE (1972) Cheese starters under control. Dairy Industries Int. 37: 73

Lawrence RC, Thomas TD & Terzaghi BE (1976) Reviews of the progress of dairy science: cheese starters. J. Dairy Res. 43: 141–193

Lawrence RC, Heap HA, Limsowtin G & Jarvis AW (1978) Cheddar cheese starters: current knowledge and practices of phage characteristics and strain selection. J. Dairy Sci. 61: 1181–1191

Lembke J, Krusch U, Lompe A & Teuber M (1980) Isolation and ultrastructure of bacteriophages of group N (lactic) streptococci. Zbl. Bakt. 1: 79–91

Limsowtin GKY, Heap HA & Lawrence RC (1978) Heterogeneity among strains of lactic streptococci. N.Z.J. Dairy Sci. Technol.. 13: 1–8

Limsowtin GKY, Powell IB & Parente E (1996) Types of starters. In: Dairy Starter Cultures (Cogan, T.M. and Accolas, J.P. Eds.) pp 101–129 VCH Publishers, Cambridge

Lubbers MW, Waterfield NR, Beresford TPJ, Le Page RWF & Jarvis AW (1995) Sequencing and analysis of the prolate-headed lactococcal bacteriophage c2 genome and identification of the structural genes. Appl. Environ. Microbiol. 61: 4348–4356

Lucey M, Daly C & Fitzgerald GF (1992) Cell surface characteristics of *Lactococcus lactis* harbouring pC1528, a 46 kb plasmid encoding inhibition of bacteriophage adsorption. J. Gen. Microbiol. 138: 2137–2143

Mayo B, Hardisson C & Braña AF (1991) Nucleolytic activities in*Lactococcus lactis* subsp. *lactis*, NCDO497. FEMS Microbiol. Lett. 79: 195–198

McKay LL & Baldwin KA (1984) Conjugative 40 - mega dalton plasmid in *Streptococcus lactis* subsp. *diacetylactis* DRC3 is associated with resistance to nisin and bacteriophage. Appl. Environ. Microbiol. 47: 68–74

— (1990) Applications for biotechnology: present and future improvements in lactic acid bacteria. FEMS Microbiol. Rev. 87: 3–14

McLandsborough LA, Kolaetis KM, Requena T & McKay LL (1995) Cloning and characterization of the abortive infection genetic determinant*abiD* isolated from pBF61 of *Lactococcus lactis* subsp. *lactis* KR5. Appl. Environ. Microbiol. 61: 2023–2026

Moineau S, Fortier J, Ackermann HW & Pandian S (1992) Characterization of lactococcal bacteriophages from Quebec cheese plants. Can. J. Microbiol. 38: 875–882

Moineau S, Pandian S & Klaenhammer TR (1994) Evolution of a lytic bacteriophage via DNA acquisition from the *Lactococcus lactis* chromosome. Appl. Environ. Microbiol. 60: 1832–1841

Moineau S, Walker SA, Vedamuthu ER & Vandenbergh PA (1995a) Cloning and sequencing of *Lla*DCHI restriction/modification genes from *Lactococcus lactis* and relatedness of this system to the *Streptococcus pneumoniae DpnII* system. Appl. Environ. Microbiol. 61: 2193–2202

— (1995b) Expression of a *Lactococcus lactis* phage resistance mechanism by *Streptococcus thermophilus*. Appl. Environ. Microbiol. 60: 2461–2466

Monteville MR, Ardestani B & Gellar BR (1994). Lactococcal phages require a host cell wall carbohydrate and a plasma membrane protein for adsorption and ejection of DNA. Appl. Environ. Microbiol. 60: 3204–3211

Mullan WMA (1986) Bacteriophage induced starter problems. Dairy Industries Int. 51: 39- 42

Nyengaard N, Vogensen FK & Josephsen J (1993) *Lla*AI and *Lla*BI, two type II restriction endonucleases from *Lactococcus lactis* subsp. *cremoris* W9 and W56 recognizing respectively, 5'-/GATC-3' and 5'-C/TRYAG-3'. Gene 136: 371–372

— (1995) Restriction- modification systems in *Lactococcus lactis*. Gene 157: 13–18

O'Sullivan DJ, Zagula K & Klaenhammer TR (1995) *In vivo* restriction by *LlaI* is encoded by three genes, arranged in an operon with *llaIM*, on the conjugative plasmid pTR2030. J. Bacteriol. 177: 134–143

Parreira R, Ehrlich SD & Chopin MC (1996) Dramatic decay of phage transcripts in lactococcal cells carrying the abortive infection determinant AbiB. Molecular Microbiol. 19: 221–230

Prévots F, Mata M & Ritzenthaler P (1990) Taxonomic differentiation of 101 lactococcal bacteriophages and characterization of bacteriophages with unusually large genomes. Appl. Environ. Microbiol. 56: 2180–2185

Reiter B (1956) Inhibition of lactic *Streptococcus* bacteriophage. Dairy Industries Int. 21: 877

Relano P, Mata M, Bonneau M & Ritzenthaler P (1987) Molecular characterization and comparison of 38 virulent and temperate bacteriophages of *Streptococcus lactis*. J. Gen. Microbiol. 133: 3053–3063

Richardson GH, Hong GL & Ernstrom CA (1980) Defined single strains of lactic streptococci in bulk culture for Cheddar and Monteray cheese manufacture. J. Dairy Sci. 63: 1981–1986

Salama MS, Sandine WE & Giovannoni SJ (1991) Development and application of oligonucleotide probes for identification of *Lactococcus lactis* ssp. *cremoris*. Appl. Environ. Microbiol. 57: 1313–1318

— (1993) Isolation of lactococci and *Lactococcus lactis* ssp. *cremoris* from nature by colony hybridization with ribosomal RNA. Appl. Environ. Microbiol. 59: 3941–3945

Salama MS, Musafija-Jeknic T, Sandine WE & Giovannoni SJ (1995) An ecological study of lactic acid bacteria: Isolation of new strains of *Lactococcus* including *Lactococcus lactis* subsp. *cremoris*. J. Dairy Sci. 78: 1004–1017

Sanders ME (1988) Phage resistance in lactic acid bacteria. Biochimie 70: 411–421

Sanders ME & Klaenhammer TR (1981) Evidence for plasmid linkage of restriction and modification in *Streptococcus cremoris* KH. Appl. Environ. Microbiol. 42: 944–950

— (1983) Characterization of phage-insensitive mutants from a phage-sensitive strain of *Streptococcus* lactis: Evidence for a plasmid determinant that prevents phage adsorption. Appl. Environ. Microbiol. 46: 1125–1133

Sanders ME, Leonhard PJ, Sing WE & Klaenhammer TR (1986) Conjugal strategy for construction of fast acid-producing, bacteriophage - resistance lactic streptococci for use in dairy fermentations. Appl. Environ. Microbiol. 52: 1001–1007

Saxelin ML, Nurmiaho-Lassila EL, Merilainen VT & Forse RI (1986) Ultrastructure and host specificity of bacteriophages of *Streptococcus cremoris*, *Streptococcus lactis* subsp. *diacetylactis* and *Leuconostoc cremoris* from Finnish fermented milk villi. Appl. Environ. Microbiol. 52: 771–777

Schleifer KH & Kilpper-Bälz R (1987) Molecular and chemotaxonomic approaches to the classification of streptococci, enterococci and lactococci: a review. Syst. Appl. Microbiol. 10: 1–9

Schouler C, Ehrlich SD & Chopin MC (1994) Sequence and organisation of the lactococcal prolate-headed bIL67 phage genome. Microbiology 140: 3061–3069

Sijtsma L, Sterkenburg A & Wouters JTM (1988) Properties of the cell walls of *Lactococcus lactis* subsp. *cremoris* SK110 and SK112 and their relation to bacteriophage resistance. Appl. Environ. Microbiol. 54: 2808–2811

Sijtsma L, Wouters JTM & Hellingwerf KJ (1990) Isolation and characterisation of lipoteichoic acid, a cell envelope component involved in preventing phage adsorption from *Lactococcus lactis* ssp. *cremoris* SK110. J. Bacteriol. 172: 7126–7130

Sing WD & Klaenhammer TR (1986) Conjugal transfer of bacteriophage resistance determinants on pTR2030 to *Streptococcus cremoris* strains. Appl. Environ. Microbiol. 51: 1264–1271

— (1993) A strategy for rotation of different bacteriophage defences in a lactococcal single strain starter culture system. Appl. Environ. Microbiol. 59: 365–372

Solaiman DKY & Somkuti GA (1990) Isolation and characterization of a type II restriction endonuclease from *Streptococcus thermophilus*. FEMS Microbiol. Lett. 67: 261–266

— (1991) A type II restriction endonuclease of *Streptococcus thermophilus* ST117. FEMS Microbiol. Lett. 80: 75–80

Stadhouders J (1986) The control of cheese starter activity. Neth. Milk Dairy J. 40: 155–173

Stadhouders J & Leenders GJM (1984) Spontaneously developed mixed strain cheese starters. Their behaviour towards phages and their use in the Dutch cheese industry. Neth. Milk Dairy J. 38: 157–181

Takano T, Ochi A & Yamamoto N (1990) Restriction enzyme from*Lactobacillus fermentum*. FEMS Microbiol. Rev. 87: ABST-C64

Thunnell RK, Sandine WE & Bodyfelt FW (1981) Phage insensitive multiple strain starter approach to Cheddar cheese making. J. Dairy Sci. 64: 2270–2279

Timmons P, Hurley M, Drinan FD, Daly C & Cogan T (1988) Development and use of a defined strain starter system for Cheddar cheese. J. Soc. Dairy Technol. 41: 49–53

Tortorello ML, Chang PK, Ladford RA & Dunny GM (1990) Plasmid associated antigens associated with resistance to phage adsorption in *Lactococcus lactis*. In: Abstracts of 3rd International ASM Conference on Streptococcal Genetics. P50

Twomey DP (1996) Molecular characterization of the *Scr*FI restriction-modification system from *Lactococcus lactis* ssp. *cremoris* UC503. PhD Thesis, The National University of Ireland, University College, Cork

Twomey DP, Davis R, Daly C & Fitzgerald GF (1993) Sequence of the gene encoding a second *Scr*FI m5C methyltransferase of *Lactococcus lactis*. Gene 136: 205–209

Valyasevi R, Sandine WE & Geller BL (1991) A membrane protein is required for bacteriophage c2 infection of *Lactococcuslactis* subsp. *lactis* C2. J. Bacteriol. 173: 6095–6100

van Sinderen D, Karsens H, Kok J, Terpstra P, Ruiters MHJ, Venema G & Nauta A (1996) Sequence analysis and molecular characterization of the temperated lactococcal bacteriophage r1t. Mol. Microbiol. 19: 1343–135

Watanabe K, Ishibashi K, Nakashima Y & Sakurai T (1984). A phage resistance mutant of *Lactobacillus casei* which permits phage adsorption but not genome injection. J. Gen. Virol. 65: 981–986

Whitehead HR & Cox GA (1935) The occurrence of bacteriophage in lactic streptococci. N.Z.J. Dairy Sci. Technol.. 16: 319–320

— (1936) Phage phenomena in cultures of lactic streptococci. J. Dairy Res. 7: 55–62

Zehern VL & Whitehead MR (1954) Growth characteristics of streptococcal phages in relation to cheese manufacture. J. Dairy Sci. 37: 209–219

GENETICS

Antonie van Leeuwenhoek **70:** 113–128, 1996.

Biosynthesis of bacteriocins in lactic acid bacteria

Ingolf F. Nes*, Dzung Bao Diep, Leiv Sigve Håvarstein, May Bente Brurberg, Vincent Eijsink & Helge Holo
Laboratory of Microbial Gene Technology, Department of Biotechnological Sciences, Agricultural University of Norway, P.O. Box 5051, N–1432 ås, Norway (author for correspondence)*

Summary

A large number of new bacteriocins in lactic acid bacteria (LAB) has been characterized in recent years. Most of the new bacteriocins belong to the class II bacteriocins which are small (30–100 amino acids) heat- stable and commonly not post-translationally modified. While most bacteriocin producers synthesize only one bacteriocin, it has been shown that several LAB produce multiple bacteriocins (2–3 bacteriocins).

Based on common features, *some* of the class II bacteriocins can be divided into separate groups such as the pediocin-like and strong anti-listeria bacteriocins, the two-peptide bacteriocins, and bacteriocins with a *sec*-dependent signal sequence. With the exception of the very few bacteriocins containing a *sec*-dependent signal sequence, class II bacteriocins are synthesized in a preform containing an N-terminal double-glycine leader. The double-glycine leader-containing bacteriocins are processed concomitant with externalization by a dedicated ABC-transporter which has been shown to possess an N-terminal proteolytic domain. The production of some class II bacteriocins (plantaricins of *Lactobacillus plantarum* C11 and sakacin P of *Lactobacillus sake*) have been shown to be transcriptionally regulated through a signal transduction system which consists of three components: an induction factor (IF), histidine protein kinase (HK) and a response regulator (RR). An identical regulatory system is probably regulating the transcription of the sakacin A and carnobacteriocin B2 operons. The regulation of bacteriocin production is unique, since the IF is a bacteriocin-like peptide with a double-glycine leader processed and externalized most probably by the dedicated ABC-transporter associated with the bacteriocin. However, IF is *not* constituting the bacteriocin activity of the bacterium, IF is only activating the transcripion of the regulated class II bacteriocin gene(s).

The present review discusses recent findings concerning biosynthesis, genetics, and regulation of class II bacteriocins.

Introduction

The antimicrobial effect of lactic acid bacteria (LAB) has been appreciated by man for more than 10 000 years and has enabled him to extend the shelf life of many foods through fermentation processes. The major preservative effect of LAB is due to their production of lactic acid which results in a concomitant lowering of pH. For a long time it has been known that many LAB also produce additional antimicrobial compounds and among these the antimicrobial ribosomally synthesized peptides, generally termed bacteriocins, have received special attention, from both scientific and the food-industrial communities.

Since the late 1920s and early 1930s, when the discovery of nisin initiated the investigation of proteinaceous antimicrobial compounds from LAB, a large number of chemically diverse bacteriocins have been identified and characterized, particularly in recent years. Nonetheless, we can observe common traits which justify their classification into just a few classes (Klaenhammer, 1993). On a sound scientific basis, three defined classes of bacteriocins in LAB have been established, class I: the lantibiotics; class II: the small heat-stable non-lantibiotics and class III: large heat-labile bacteriocins (Table 1). A fourth class of bacteriocins has also been defined, which contains bacteriocins composed of an undefined mixture of proteins,

Table 1. Classification of LAB Bacteriocins

Class I.	*Lantibiotics*
Class II	*Small heat stable non-lantibiotcs*
	IIa: Pediocin-like bacteriocins with strong antilisterial effect
	IIb: Two-peptide bacteriocins
	IIc: *sec*-dependent secreted bacteriocins
Class III.	*Large heat-labile proteins*

lipids, and carbohydrates. The existence of the fourth class was supported mainly by the observation that some bacteriocin activities obtained in cell-free supernatant, exemplified by the activity of *Lactobacillus plantarum* LPCO10, were abolished not only by protease treatments, but also by glycolytic and lipolytic enzymes (Jiménez-Diaz et al., 1993). However, such bacteriocins have not yet been characterized adequately at the biochemical level and the recognition of this separate class therefore seems premature. Indeed, the experimental data suggest that these complex bacteriocinogenic activities may be artifacts caused by interaction between constituents from the cells or the growth medium and the undefined bacteriocin activities are likely to be regular peptide bacteriocins. This view is strongly supported by experiments showing that proper purification of such activities indeed leads to the isolation of regular peptide bacteriocins (Jiménez-Diaz et al., 1995). Bacteriocin activity is frequently found associated with large aggregates in cell free extracts. These aggregates include not only proteinaceous material but, most probably, also lipids and other macromolecules which could affect the bacteriocin activity. Several bacteriocin containing aggregates are resolved into simple peptide bacteriocins by purification. Various enzymatic treatments may affect the bacteriocin activity of these crude complexes.

LAB bacteriocins of class I and II are by far the most studied because they are both the most abundant ones and the most prominent candidates for industrial application. Members of the two classes are generally clearly different, both in terms of the structure of the bacteriocin itself and in terms of the machinery involved in production and processing. A few intermediate cases have been observed, namely a few lantibiotics (class I) that are secreted by mechanisms characteristic for class II bacteriocins (Piard et al., 1993; Ross et al., 1993; Håvarstein et al., 1994, 1995). The present review focuses on class II bacteriocins only. Class I bacteriocins have recently been reviewed extensively by others (De Vos et al., 1995; Jack et al., 1995; Sahl et al., 1995; Konings & Hilbers, 1996).

Diversity of class II bacteriocins

A large number of class II bacteriocins has now been biochemically characterized mainly due to the development of efficient and standardized protocols for purification of these hydrophobic and cationic peptides. The availability of biochemical characteristics of a large number of class II bacteriocins now permits a subgrouping of many of these compounds. However, one should keep in mind that the subgrouping of bacteriocins is just a way to organize our present knowledge in a functional way, and future research will certainly change our present concepts of the class II bacteriocins.

One major subgroup of bacteriocins shows very strong antilisterial activity. Members of this subgroup (class IIa) are found in a wide variety of LAB including *Pediocicoccus* (Henderson et al., 1992; Marug et al., 1992; Motlag et al., 1992; Nieto Lozano et al., 1992; Ray, 1992), *Leuconostoc* (Hastings et al., 1991; Hechard et al., 1992), *Lactobacillus* (Holck et al., 1992; Larsen et al., 1993; Tichaczeck et al., 1992; Kanatani et al., 1995) and *Enterococcus* (Aymerich et al., 1996) . The antilisterial bacteriocins share strong amino acid sequence homology (between 38–55% identity) which is most pronounced in the N-terminal part of the peptides (Aymerich et al. 1996). This subclass of bacteriocins has been termed the pediocin-family after the first and most extensively studied example of this class, pediocin PA–1.

A second subgroup (class IIb) contains bacteriocins whose activity depends on the complementary action of two peptides and several examples of such bacteriocins have been studied (van Belkum et al., 1991, Nissen-Meyer et al., 1992, 1993b; Allison et al., 1994; Jiménez-Diaz et al., 1995; Diep et al., 1996). It should also be mentioned that one example of a two-peptide lantibiotic has been characterized (Gillmore et al., 1994).

All bacteriocins are synthesized with an N-terminal leader sequence and, until recently, only the double-glycine type of leader was found in class II bacteriocins (see below) (Holo et al., 1991, Muriana & Klaenhammer, 1991; Klaenhammer, 1993; Håvarstein et al., 1994). However, it has now been disclosed that some small, heat stable, and non-modified bacteriocins are translated with *sec*-dependent leaders (Leer et al., 1995, Worobo et al., 1995). Due to their similarity to the class II bacteriocins they should be included as a separate subgroup, the class IIc (Table 1).

It has been suggested that a subgroup of thiol-activated bacteriocins (lactococcin B) should be included (Venema et al. 1994), but recent findings suggest that this subgroup should be excluded since oxidation of the sulphydryl group with other chemicals did not interfere with its activity (Venema et al. 1995). It was also shown that when this cysteine residue was replaced by all other amino acids, only the positive charged amino acids were reducing/abolishing the bacteriocin activity which suggests that the cysteine is note essential for the biological activity (Venema et al. 1995).

The synthesis of the cationic peptide-bacteriocins rest upon a general genetic structure encompassing four different genes which encode the basic functions required for production of the extracellular antimicrobial activity (Nes et al. 1995). These four genes are: 1) the structural gene encoding the prebacteriocin, 2) a dedicated immunity gene always located next to the bacteriocin gene and on the same transcription unit, 3) a gene encoding a dedicated ABC-transporter which externalizes the bacteriocin concomitant with processing of the leader, and 4) a gene encoding an accessory protein essential for the externalization of the bacteriocin, the specific role of which is not known. The four basic genes are organized either in one or two operons. In the lactococcin A system two operons are found (Holo et al, 1991, van Belkum et al. 1991, Stoddard et al., 1992) while the pediocin PA–1 system possesses one operon comprising the four genes involved in production of the active bacteriocin molecule (Marugg et al. 1991). In addition to the four basic genes, regulatory genes have been found associated with the genetic determinants of some class II bacteriocins (Diep et al., 1994, 1996; Axelsson & Holck, 1995; Quadri et al., 1995a; Huehne et al., 1996; Brurberg et al., 1996). These findings are discussed separately below.

The works of van Belkum et al., (1991, 1992) revealed that one *Lactococcus lactic* strain can produce more than one bacteriocin. This particular *Lactococcus* strain produces three plasmid-encoded bacteriocins, two one-peptide bacteriocins and one two-peptide bacteriocin. Recent studies have shown that production of multiple bacteriocins by one organism is quite common. In *Carnobacterium piscicola* LV17B it has been shown that at least two different bacteriocins are produced, one is plasmid-encoded while the other is found on the chromosome (Quadri et al., 1994; 1995a,b). The bacteriocin secretion system which was physically located on the plasmid next to the bacteriocin gene, was also used by the chromosomally encoded bacteriocin. *Lactobacillus plantarum* C11 also produces multiple bacteriocins including two two-peptide bacteriocins and one one-peptide bacteriocin (Diep et al., 1996). In *Lactobacillus plantarum* C11 the bacteriocin genes and their accessory genes are clustered on the bacterial chromosome.

The bacteriocin and its gene

The structural bacteriocin gene encodes a preform of the bacteriocin containing an N-terminal leader sequence (termed double-glycine leader) whose function seems a) to prevent the bacteriocin from being biologically active while detained inside the producer, and b) to provide the recognition signal for the transporter system (see below). The double-glycine leader varies in length from 14 residues up to approximately 30 residues (Klaenhammer, 1993, Håvarstein et al., 1994). The consensus elements found in the double-glycine leader include the two glycine residues at the C-terminus of the cleavage site, conserved hydrophobic and hydrophilic residues separated by defined distance between the conserved residues. In addition the minimum length of the double-glycine leader seems to be 14 amino acids (Table 2A). Of the consensus residues in the leader, only the glycine at position –2 residue is fully conserved. The mature bacteriocins identified so far vary in size from less than 30 residues to more than 100 residues in some cases. It should also be mentioned that colicin V (88 residues) of *E. coli* can be formally classified as a class II bacteriocin (Fath et al., 1995, Håvarstein et al., 1995). Recently, a second *E. coli* bacteriocin, microcin 24, can also be classified as a class II bacteriocin according to the nomenclature (O'Brian et al., 1996).

The class II bacteriocins share a number of common features. They have a high content of small amino acids like glycine. They are strongly cationic, with pI's usually varying from 8 to 11, and they possess a hydrophobic domain and/or amphiphilic region, which may relate to their activity on membranes (Abee, 1995).

As referred to above, a number of bacteriocins consists of two peptides. Both peptides possess a double-glycine leader and are encoded by individual, contiguous genes in the same operon. Both peptides are structurally indistinguishable from the one-peptide bacteriocins, however, both peptides are apparently required for activity, or for obtaining optimal activity.

Table 2. Leader sequences found in Class II bacteriocins

A	
ABC-transporter-dependent leaders (consensus sequence)	
Double glycine leaders: (14–30 residues)	- - LS - - EL - - I - GG
Consensus:	- -*¤ - ¤¤* - -* - GG
**Hydrophobic residues, ¤Hydrophilic residues*	
B	
sec-dependent signal sequences	
Divergicin A: MKKQILKGLVIVVCLSGATFFSTPQASA	
Acidicin B: MVTKYGRNLGLSKKVELFAIWAVLVVALLATA	

The antimicrobial activity of lactococcin G and probably lactococcin MN is completely dependent on both peptides (Nissen-Meyer et al., 1992, van Belkum et al., 1991). On the other hand, one of the peptides of the two-peptide bacteriocins plantaricin S and lactacin F possesses apparently some antimicrobial activity. However, the second peptide has a dramatic effect by enhancing and/or modifying the activity (Allison et al., 1994, Jiménez-Diaz et al., 1995). In the plantaricin S system the (α-peptide does not show any detectable bacteriocin activity while the bacteriocin activity of β-peptide is enhanced approximately 50-fold by the α-peptide. Lactacin F activity is defined by the two functional *lafA* and *lafX* genes located next to each other. LafA is a bacteriocinogenic peptide by itself but upon addition of LafX the activity of LafA increases and the inhibitory spectrum expands.

While most bacteriocins appear to be secreted by the *sec-* independent universal ABC-transporter system (see below), it has recently been shown that some bacteriocins do not possess a double-glycine leader sequence but are, instead, synthesized with a typical N-terminal leader sequence of the *sec*-type. This is a new and very interesting feature, which demonstrates that bacteriocins can be secreted/processed by two different pathways. So far only two LAB bacteriocins containing the *sec*-dependent signal sequences are known (Table 2B) (Leer et al., 1995; Worobo et al., 1995).

The immunity protein and its gene

Bacteriocin producers have developed a protection system against their own bacteriocin. This system is referred to as immunity. Each bacteriocin has its own dedicated protein conferring immunity, which is expressed concomitantly with the bacteriocin. In all bacteriocin operons studied so far, potential immunity genes have been identified next to and downstream of the bacteriocin structural genes. While synthesis of extracellular bacteriocin requires a dedicated secretion/processing system, the immunity protein is functionally expressed in the absence of transport and processing. This has been proven convincingly for a number of immunity factors including LciA and PedC of lactococcin A and pediocin PA–1, respectively (Holo et al. 1991, Venema et al. 1994, 1995a). When lactococcin A and its immunity gene (*lciA*) were cloned into lactococcin A sensitive strains, only LciA was functionally expressed while LcnA activity was not detected because the transporter/processing system was missing in the new host.

The immunity proteins are fairly small, ranging from 51 to approximately 150 amino acids and the homology between various immunity proteins is surprisingly low when considering the similarity found between several bacteriocins (Aymerich et al. 1996, Holo 1996). This lack of similarity is particularly strong between the immunity proteins of the two identical bacteriocins, sakacin A and curvacin A (Axelsson et al. 1993, Tichaczek et al. 1993), where the putative immunity proteins are 90 amino acids and 51, respectively. Most of the immunity proteins of the bacteriocins belonging to the pediocin-family do not share significant homology while these bacteriocins share 38–55 % identity (Aymerich et al. 1996). This observation may suggest that no direct interaction occurs between bacteriocins and their immunity proteins.

The putative immunity proteins of curvacin A and acidocin A (51 amino 55 amino acids respectively) are smaller than other immunity factors and computer-assisted amino acid sequencing analyses indicates that the N-terminal regions of both immunity proteins can form membrane spanning helices. Hydropatic profile analyses of some immunity proteins have revealed up to four putative transmembrane segments, which suggests that these immunity factors may integrate in the membrane of the bacteriocin producer (Fremaux et al. 1993).

The immunity protein of lactococcin A (LciA) has been purified to homogeneity and *in vitro* experiments suggest that LciA does not interact directly with lactococcin A, although such experiments, of course, do not exclude that direct interaction can take place *in vivo* (Nissen-Meyer et al. 1993). It has also been shown that the immunity protein of lactococcin A is intracellular-

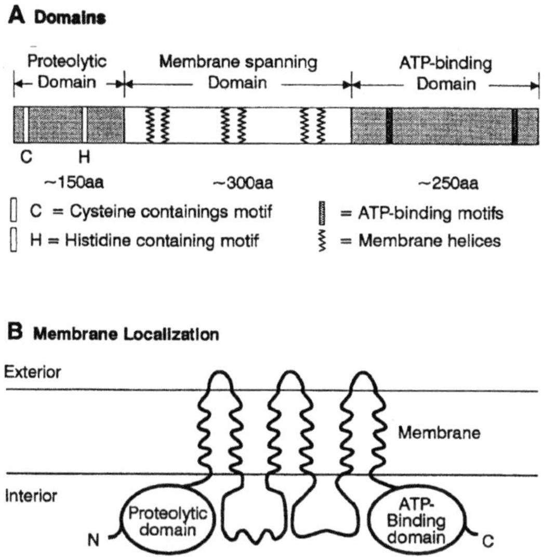

Figure 1. The ABC-transporter of class II bacteriocins with double-glycine leaders.
A: The organization of the domains of the transporter.
B: The presumed localization of the domains in relation to the cytoplasmatic membrane.

ly located and between 50 and 90% of the immunity protein is found free in the cytoplasm (Nissen- Meyer et al. 1993, Venema et al. 1994). The work of Venema et al. (1994) also suggests that the amphiphilic α-helix domain (residue 29 to 47) of the immunity protein of lactococcin A is embedded in the membrane with the C-terminal end directed towards the exterior of the cell. Free intracellular LciA is considered a reservoir to be used for the defense when needed. It has also been proposed that LciA is closely associated with a postulated bacteriocin receptor (Venema et al. 1994), however, the presence of such a receptor is still speculative and the molecular entity has yet to be identified. The most intriguing and challenging problems in the field of bacteriocin research today have to do with the molecular mechanism behind the immunity of bacteriocins and questions related to the existence and identity of bacteriocin receptors.

The ABC-transporter

As mentioned above, most bacteriocins are synthesized in a preform containing an N-terminal extension, the so called double- glycine leader (Table 2A), first identified in lactacin F and lactococcin A (Holo et al., 1991, Muriana & Klaenhammer, 1991). The strong conservation of the cleavage site of the leaders strongly suggests a common processing mechanism for these peptides. The first evidence suggesting that ABC-transporters are required for extracellular activation of class II bacteriocins was presented by Stoddard et al. (1992), during studies of the lactococcin A, B and MN systems. In this study the genes of a dedicated ABC-transporter (LcnC) and its accessory protein (LcnD) were identified to be needed for production of active extracellular lactococcin A. Today it is well established that secretion of the double-glycine leader containing bacteriocins is mediated by a dedicated transmembrane translocator belonging to the ATP-binding cassette (ABC) transporter superfamily (Gilson et al., 1987, 1990; Gillmore et al., 1990, 1994, Stoddard et al., 1992; Marugg et al., 1992; Håvarstein et al., 1995). The bacteriocin ABC- transporter gene is usually either part of the bacteriocin operon or found on a separate operon in the vicinity of the bacteriocin-containing operon. It has been generally recognized that all ABC-transporters contain two domains, a hydrophobic integral membrane domain and a cytoplasmatic ATP-binding domain (Figure 1). By comparing the amino acid sequence of seven ABC-transporters dedicated to the translocation of bacteriocins containing the double-glycine leader type with other ABC-transporters, a unique feature in the N-terminal part of the bacteriocin transporters has become apparent (Figure 1). It was noticed that the bacteriocin transporters carried an N-terminal extension of approximately 150 amino acids which was also found in a few other systems including the α-haemolysin transporter of *Escherichia coli*, that of the competence system of *Streptococcus pneumonia* and in the transporters of some lantibiotics (Håvarstein et al., 1995). However, the ABC-transporter of lantibiotics such as nisin, with a leader sequence different from the double-glycine type, does not have this extension. For some time the unanswered questions were: where, when and by which protease is the double-glycine leader removed? The observation suggested that the N-terminal extension of the transporter could be involved in the processing of the bacteriocins. Functional studies of the N-terminal region of the ABC-transporter of lactococccin G were performed. The N-terminal extension of the transporter was cloned and expressed, and enzymatic studies were performed on the cloned polypeptide fragment in order to determine its enzymatic role in the transport process. It was convincingly demonstrated that the N-terminal polypeptide was able to cleave off the leader of the lactococcin G α-peptide specifically at the C-terminus of the double-glycine motif, as observed in the *in vivo* process (Håvarstein et al., 1995). In the unique N-terminal extension of the bacteriocin transporters two conserved motifs (the cysteine motif; $QX_4D/ECX_2AX_3MX_4Y/FGX_4I/L$ and histidine motif; $HY/FY/VVX_{10}I/LXDP$) were identified. These two motifs were not present in the N-terminal extension of haemolysin transporter, (Figure 1A). Replacement of the conserved cysteine residue in the cycteine motif by an alanine residue abolished the leader-specific proteolytic cleavage. Based on this work Håvarstein et al. (1995) proposed the following hypothesis for the export of bacteriocins containing the double-glycine leader: The proteolytic domain of the ABC-transporter binds a bacteriocin precursor. The hydrolysis of ATP induces conformational changes in the transporter which result in an intimately integrated process of removing the leader sequence and transport of the bacteriocin molecule across the cytoplasmatic membrane. It has been shown that mutation of the glycine residue in position –2 in both colicin V and lactacin F abolished bacteriocin secretion (Gilson et al. 1990, Fremaux et al. 1993). In the colV mutants bacteriocin activity was detected *inside* the cells thus explaining the export deficiency of these mutants. Combined with the discovery that the processing site is part of the transporter, this strongly indicates that the two processes of cleavage and translocation are integrated and, also, that the leader peptide serves as a recognition signal for the transmembrane transport process of the bacteriocin.

In addition to the biochemical work demonstrating the proteolytic role of the N-terminal domain of the ABC-transporters, an *in vivo* study on the pediocin PA–1 ABC-transporter also suggested that the N- terminal part of this bacteriocin transporter takes part in the proteolytic removal of the bacteriocin leader. By cloning the N-terminal part of PedD (transporter) together with the structural gene of prepediocin PA–1, a proteolytic activation of bacteriocin was observed intracellularly which strongly suggested that the N-terminal part of the ABC- transporter is required for the cleavage and that this process can be uncoupled from secretion (Venema et al. 1995).

The accessory protein

Transposon mutagenesis combined with cloning and DNA sequencing of the lactococcin system, Stoddard et al. (1992) identified a second gene, termed *lcnD*, which was needed for production of *extracellular* bacteriocin activity. The translation product LcnD shares some homology with another group of proteins which is known to be involved in ATP-dependent translocation processes. These gene products (in general about 470 amino acids), often designated as accessory proteins, are apparently required in the ABC-transporter-dependent translocation process.

Computer-assisted amino acid sequence analysis of LcnD predicted the presence of an N-terminal transmembrane (TM) sequence. The localization of the TM structure of LcnD in the membrane has now been confirmed by topological studies using various in-frame translation fusions to reporter proteins. The present model suggests that the N- terminus of LcnD is intracellularly located, while the transmembrane spanning helix (from residue 22 to 43) directs the C-terminal part of the protein towards the outside (Franke et al., 1996). Also work on the pediocin PA–1 system has shown that the analogous *pedD* gene encoding an accessory protein is required for successful externalization of pediocin PA–1 (Venema et al., 1995). The specific role of the accessory protein in the tranlocation process has not yet been resolved.

Regulation of class II bacteriocin synthesis -A three-component regulatory system

It has been occasionally observed that bacteriocinogenic LAB loose their ability to produce bacteriocin. Some of these observations have been attributed to plasmid loss or to transposition-mediated inactivation. Recently, it has been shown that production of some bacteriocins can be transcriptionally regulated and that bacteriocins can not been produced in the absence of an induction factor (IF). This phenomenon may also explain why loss of bacteriocin production is occasionally observed as demonstrated in the induction experiment shown in Figure 2.

During the past several years, it has been become evident that prokaryotic organisms process much of their sensory information through families of proteins which make up the two-component regulatory system (Stock et al., 1989; Bourret et al., 1991; Parkinson & Kofoid, 1992). The genetic elements common for two-component regulatory system, namely a response regulator (RR) gene and a sensor histidine protein kinase (HK) gene, have been identified and related to bacteriocin production. Nucleotide sequencing of class II bac-

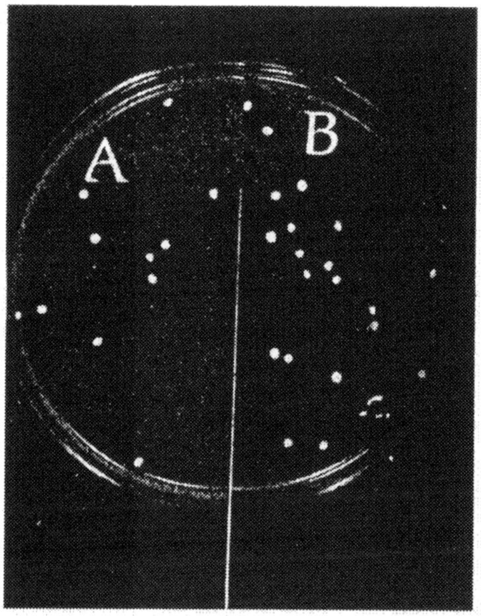

Figure 2. Growth of *L. plantarum* C11 demonstrating the effect of induction factor on bacteriocin production.
A: In the absence of induction factor and B: In the presence of induction factor (40 ng/ml synthetic plantaricin A). The indicator organism used was *Lactobacillus plantarum* 965.

teriocin operons has uncovered a regulatory systems associated with the gene clusters of several LAB bacteriocins which include the plantaricin system of *Lactobacillus plantarum* C11 (Diep et al. 1994, 1995, 1996), sakacin P of *Lactobacillus sake* LTH673 and Lb674 (Brurberg et al. 1996, Huehne et al. 1996) , carnobacteriocin(s) of *Carnobacterium piscicola* LV17 (Quadri et al., 1995a,b) and sakacin A of *Lactobacillus sake* Lb706 (Axelsson et al., 1994, 1995) (Figure 3A).

Among the non-lantibiotics it has recently been shown that bacteriocin production of *L. plantarum* C11 was indeed regulated by a two-component-like regulatory system and that the regulatory signal was a bacteriocin-like peptide encoded by *pln*A which is the first ORF of the regulatory operon (*plnABCD*) (Diep et al., 1994, 1995). The regulatory peptide, termed plantaricin A, shared most of the physico- chemical properties of a regular class II bacteriocin and this peptide was originally considered to be responsible for the bacteriocin activity of *L. plantarum* C11 (Nissen-Meyer et al., 1993b). However, subsequently it has been shown that neither synthetic plantaricin A, nor the product of the cloned *plnA* gene exhibited significant bacteriocin activity (Diep et al., 1995). In addition, no gene encoding a potential immunity protein was found in the *plnABCD* operon. It has now become apparent that there are structural differences between regulator peptides and bacteriocins, the former normally being considerably shorter (see also below). Thus, it is now clear that *plnABCD* operon encodes a regulatory system composed of three elements : the induction factor, the histidine protein kinase and the response regulator.

Sakacin P is another regulated class II bacteriocin of which the induction factor (IF) has been identified. Eijsink et al. (1996) purified a non-bactericidal 19-mer peptide (Figure 3B) from the bacterial growth medium and showed that this peptide was able to induce sakacin P production at a concentration of 0.2 ng/ml (0.1 nM) in liquid medium. Nucleotide sequencing has identified the structural gene of this induction factor (IF or OrfY in *spp* system, Figure 3A) as the first gene in the regulatory operon (Figure 3A). The IF, which is a bacteriocin- like peptide with a double-glycine leader, is located next to the histidine kinase as found in *plnA* (Brurberg et al., 1996). The sakacin P gene cluster has been sequenced by Huehne et al. (1996) and their data confirm the organization of a regulatory operon consisting of the three genes that encode a bacteriocin-like peptide followed by histidine protein kinase and a response regulator protein.

Axelsson et al. (1995) identified a two-component regulatory system associated with sakacin A production, and Diep et al.(1995) suggested that the peptide translated from *orf4* (previously termed *orf45)* in the *sap* system (Figure 3), preceding the regulatory genes in the sakacin A gene cluster, is the most likely factor regulating the bacteriocin production. *Orf4* encodes a putative protein (23 residues, Figure 3B) containing a double glycine leader and, again, three components apparently determine the transcriptional regulation of sakacin A production.

Although lantibiotics are not treated in this review, the recent work on transcriptional regulation of nisin is very relevant to the regulation of group II bacteriocins. Eleven genes organized in a single locus are responsible for nisin production. Two of the gene products, a response regulator and a sensor histidine protein kinase, have been identified to participate in the regulation of nisin biosynthesis (van der Meer et al., 1993; Engelke et al., 1994). It has now been convincingly demonstrated that nisin can regulate its own production, via signal transduction, by acting as an extracellular signal leading to transcriptional activation of the genes which are required for nisin biosynthesis (Kuipers et al., 1995). While the class II bacteriocins

A:

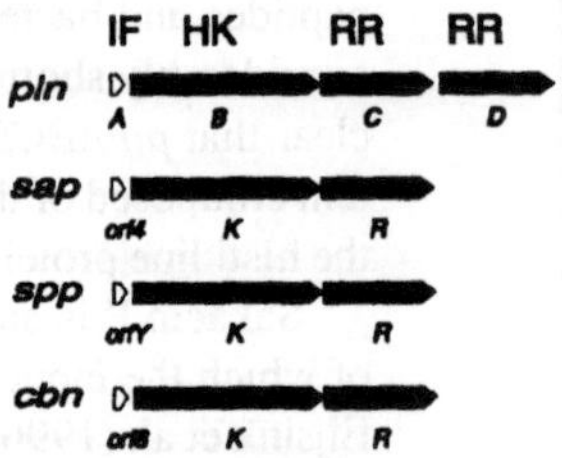

B:

System	Peptide sequence	
Plantaricin:	KSSAYSLQMGATAIKQVKKLFKKWGW	(PlnA)
Sakacin P:	MAGNSSNFIHKIKQIFTHR	(OrfY)
Sakacin A:	TNRNYGKPNKDIGTCIWSGFRHC	(Orf4)
Carnobacteriocin A:	SKNSQIGKSTSSISKCVFSFFKKC	(Orf6)

Figure 3. The three-component regulatory system of class II bacteriocins.
A: The genetic organization of the three genes involved in induction. IF is the induction factor, HK is the histidine protein kinase and RR is the respond regulator.
pln: plantaricin system; *spp*: sakacin A system; *sap*: sakacin P system; *cbn*: carnobacteriocin system.
B: The amino acid sequences of the putative induction factors.

seems to use bacteriocin-like molecules different from the bacteriocins, nisin is autoregulating its own production which is an intriguing difference raising new questions about the evolutionary aspect and biological role of bacteriocins.

Stiles and coworkers (Saucier et al., 1995) have reported that the production of bacteriocins by *Carnobacterium piscicola* LV17 is regulated by the bacteriocins themselves and they conclude that the induction might account for the difference in bacteriocin production observed in liquid and on solid growth media. However, the operons involved in production of these carnobacteriocins could encode several bacteriocin-like peptides whose roles have not been fully elucidated (Quadri et al., 1995a). It remains therefore unclear which peptide molecules are truly responsible for the regulation of the carnobacteriocin activity. It has not been clarified whether one peptide participates solely in the regulation or, alternatively, whether one peptide is serving the dual function of being a bacteriocin and an induction factor as seen with nisin (Kuipers et al., 1995). The organization of the bacteriocin gene cluster in *Carnobacterium piscicola* LV17 suggests that *orf6* encodes the actual induction factor (*cbn* system, Figure 3A). This assumption is based on the following observations *i)* *orf6* encodes apparently a bacteriocin-like peptide with a double-glycine leader, *ii)* its mature form (24 amino acids, Figure 3B) is shorter than a regular peptide- bacteriocin and, *iii)* *orf6* is located directly upstream of the histidine kinase gene of a two component regulatory system (Figure 3A) which are consistent with the notion that a three-component regulatory system regulates the trancription of carnobacteriocin B2 in *C. piscicola* LV17 analogous to the above-mentioned class II bacteriocins.

The induction factor is required not only for the induction of bacteriocin synthesis but is also needed to maintain bacteriocin production. By removing IF from the growth medium, the bacteriocin production stops immediately. Bacteriocin production is reestablished if sufficient amount of IF is allowed to be synthesized or supplied the growth medium (Figure 2). The induction cycle leading to bacteriocin production is shown in Figure 4 Among the regulated class II bacteriocins an autoregulatory circuit consisting of an induction factor, a sensor histidine kinase (HK), and a response regu-

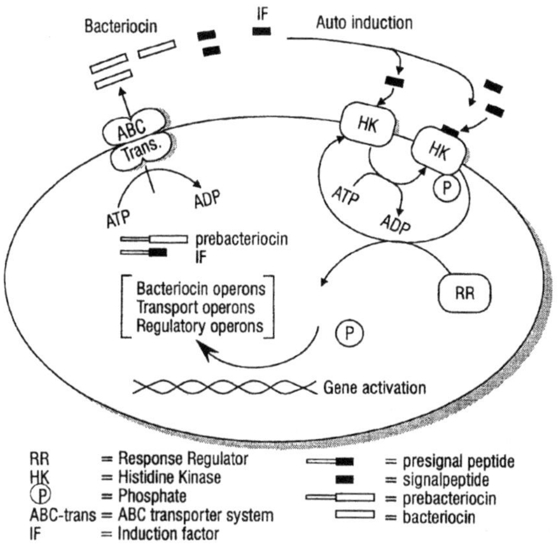

Figure 4. A schematic overview of the regulation/production of class II bacteriocins in a cell.

lator (RR) has been shown to regulate the bacteriocin production.

Two-component signal transduction has been studied thoroughly in many systems (Hock & Silhavy, 1995). The N-terminal part of most HK proteins accommodates an input domain with sensory function, which is often located at the periplasmic side of the inner membrane, while a C-terminal transmitter, located on the cytoplasmic side, contains an autokinase activity and a conserved histidine residue as a site for phosphorylation. The input domain and transmitter are linked by membrane-spanning segments. Computer-assisted analysis the HK associated with the class II bacteriocins identified 6 putative transmembrane helices which are different from other histidine protein kinases. The RR proteins are also build up of two different modules: an N-terminal receiver containing a conserved aspartic residue as the site for phosphorylation, and a C-terminal output domain, directly mediating an adaptive response. Communication between the HKs and their RRs involves phosphorylation and dephosphorylation. When a specific signal is received by the input domain, the signal is processed and transferred to the connected transmitter thereby activating its autokinase activity and leading to phosphorylation of the conserved histidine residue. The phosphorylated histidine residue in turn serves as a high-energy intermediate for subsequent phosphorylation (catalyzed by HK kinase activity) of the conserved aspartic residue on the RR receiver domain. Phosphorylation of the receiver domain is assumed to cause intramolecular changes in the output domain, which then becomes activated to trigger an output signal. Because the domains involved in phosphorylation of HK and RR is well conserved in the 'bacteriocin' regulatory proteins, it is very likely that the same residues are taking part in the phosphotransfer reaction during the bacteriocin induction response (Figure 4).

This three-component regulatory system also regulates the transcription of the secretion/processing machinery operon needed for of the bacteriocin synthesis (Diep et al., 1996).

The factors triggering bacteriocin production have already been given several names such as induction factor, regulatory peptide/factor, signal molecule, and bacteriocin-like peptide. From a formal point of view the proper designation of these peptides should be pheromones, as has been done for the competence factor of *Streptococcus pneumonia* recently (Håvarstein et al., 1995). According to the accepted definition, a pheromone is: 'a substance which is secreted to the outside by an individual and received by a second individual of the same species, in which they release a specific action , for example, a definite behavior or a development process. The principle of minute amounts being effective holds...' (Karlson & Lüscher, 1959). However, before this terminology should be adopted there should be consensus amongst the scientist in the field.

The unsolved problem is how bacteriocin production is activated or, more correctly, how the autoinduction system which induces bacteriocin production is activated. Two models may explain how bacteriocin synthesis is induced: A) If one assumes that induction factors are constitutively produced in low amounts, a gradual accumulation of IF will take place during growth. When the concentration of IF exceeds the threshold for autoinduction of IF the subsequent result will be a burst of bacteriocin production. This model is consistent with a cell density control mechanism which has been implied both in induction of competence in *B. subtilis* (Magnusson et al., 1994) and *S. pneumonia* (Håvarstein et al.) and regulation of virulence of *S. aureus* (Balaban & Novick, 1995; Guangyong et al. 1995). The alternative model is B): a constitutive production of induction factor is balancing at a concentration just below the activation threshold required for induction. Changes in environmental conditions (such as changes in the nutritional or physical/chemical growth condition) may cause a short and temporary increase in production of IF which leads to autoinduction of IF and subsequently bacteriocin production in the culture. This model excludes that a cell density monitoring system is the biological mechanism causing the induction but, instead, it suggests that external environmental factors independent of cell growth are actually triggering IF and bacteriocin production. This may be the cell's response to harsh environmental changes. Presently, however, no experimental data support either model.

Genetic organization of the regulated bacteriocins

Among the non-lantibiotics, the bacteriocin production by *L. plantarum* C11 is one of the most thoroughly studied with regard to the genetics and regulation. An approximately 16.5-kb chromosomally located DNA fragment most likely conferring bacteriocin production has been cloned, sequenced and analyzed with respect to transcription and transcription regulation (Diep et al., 1996). This fragment contains altogether 22 open reading frames (ORFs), of which 21 ORFs are distributed in five separate operons (*plnJKLR*, *plnMNOP*, *plnABCD*, *plnEFI* and *plnGHSTUV*) as seen in Figure 5.

lnABCD is the autoregulatory operon in the *pln* system (Diep et al., 1994, 1995). *plnA* encodes a prepeptide consisting of a double-glycine leader and a 26-amino acid C-terminal mature peptide, termed plantaricin A. This bacteriocin-like peptide serves as an induction factor triggering bacteriocin production and its own synthesis by activating transcription of all the five *pln* operons. The three downstream genes *plnBCD* encode a histidine protein kinase (HK) and two response regulators (RRs), respectively, making up a complete signal transducing pathway (Figure 3A). The three regulatory proteins are believed to play a crucial role in monitoring the concentration of plantaricin A and mediating the subsequent process that ultimately leads to induction of bacteriocin production.

plnEFI, *plnJKLR* and *plnMNOP* are all putative bacteriocin operons. The gene pairs, *plnEF* and *plnJK*, encode bacteriocin-like peptides with double-glycine leaders, and are followed by genes *plnI* and *plnL*, respectively encoding cationic, hydrophobic proteins which most probably confer immunity. The third putative bacteriocin operon starting with one such bacteriocin-like gene (*plnN*) is found together with two ORFs (*plnM* and *plnP*) encoding cationic hydrophobic proteins which either of them could confer immunity. The remaining ORFs, *plnR* and *plnO*, on these operons, encode proteins of unknown function. Based on the genetic information obtained from these genes, the operons *plnEFI*, *plnJKLR* and *plnMNOP* are suggested to encode two bacteriocins of two-peptide type (tentatively named PlnEF and PlnJK) and a bacteriocin of one-peptide type (tentatively named PlnN), respectively. The last operon (*plnGHSTUV*) which is only partially sequenced, starts with two ORFs (*plnGH*) encoding proteins homologous to those constituting the export machinery for the lactococcins which possess precursor double-glycine leaders. It is therefore assumed that the six bacteriocin-like peptides including the induction factor (PlnA, -E, -F, -J, -K and -N) are processed and externalized by this PlnGH-constituting export system. The four downstream genes *plnSTUV* and *orfl* which is located between the operons *plnMNOP* and *plnABCD*, encode putative hydrophobic proteins from which no functions hav yet been suggested. (Diep et al., 1996).

It is noteworthy that the genetic organization as found within the plantaricin A operon (*plnABCD*) (i.e., the HK and RR genes being organized in tandem with a small gene located just upstream of the HK gene), also occurs in some class II bacteriocin regulons (*spp*, *sap* and *cbn*, which will be discussed below), as well as in the functionally related regulatory systems, *agr* and *com*, the latter two controlling stationary expression of virulence and a number of exoproteins in *Staphylococcus aureus* (Morfeldt et al., 1988; Balaban et al., 1995; Guanggyong et al., 1995), and the development of competence for transformation in *Streptococcus pneumoniae* (Håvarstein et al., 1995; Pestova et al., 1996), respectively. In those systems (*agr* and *com*) where detailed studies have performed, the preceding small gene encodes a peptide serving as an induction factor triggering its own transcription as well as transcription of their target genes in a similar manner as found with plantaricin A. Moreover, all these regulons control pathways which are generally regarded as being accessory, i.e., not essential for bacterial growth. Thus, the striking resemblance in the genetic organization of the three regulatory components, induction factor (IF), HK and RR, strongly suggests that these regulatory cassettes confer a specific gene regulatory system.

Sakacin P production by *L. sake* strain (LTH673) is the second bacteriocin shown to be tancriptionally regulated (Eijsink et al., 1996). The clustered operons involved in the synthesis of sakacin P have also been sequenced in two different producer strains

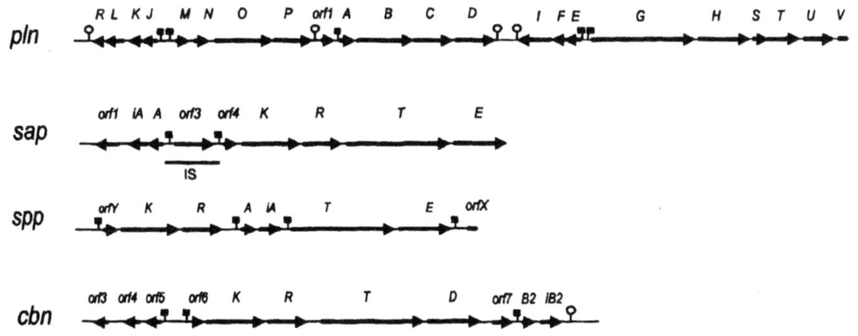

Figure 5. Genetic organization of the gene clusters of the *pln, sap, spp* and *cbn* bacteriocin systems,. (See text for details). Arrows indicate open reading frames (ORFs) *plnV* of *pln-* system and *orfX* of *spp*-system are incomplete ORFs. Filled rectangles indicate promoter or promoter-like structures with regulatory- like boxes containing direct repeats, and lollipops indicate potential transcriptional terminators. The data are obtained form the references of Diep et al. 1996 (*pln*), Axelsson & Holck 1995 (*sap*), Huehne et al. 1996, Brurberg et al. 1996 (*spp*) and Quadri et al. 1995a.

(Huehne et al., 1996, Brurberg et al., 1996) (Figure 5). As observed for other systems, the bacteriocin structural gene (*sppA*) and cognate immunity gene (*spiA*) are organized in tandem in an operon-like structure. Upstream of the sakacin P operon are three genes (*orfY*, *sppK, -R*) corresponding to an IKR autoregulatory cassette (Figures. 3A, 5). *orfY* encodes a bacteriocin-like peptide with double-glycine leader, whereas *sppK* and *sppR* encode an HK and a RR protein, respectively. The biological role of this three-component regulatory system has recently been examined by Eijsink et al. (1996) who showed that the mature product of *orfY*, as found with plantaricin A, serves as an induction factor in bacteriocin production. Based on the analogy in the organization of the regulatory operons, one must assume that *orf4* also has a similar regulatory role in the *sap* system as *plnA* has in the plantaricin system.

L. sake Lb706 produces a bacteriocin known as sakacin A (Axelsson & Holck, 1995). This bacteriocin belongs to the pediocin-like family as judged from the amino acid sequence and its antilisterial activity. An approximately 8.7-kb plasmid-derived fragment containing all the genetic information required for bacteriocin production and immunity has been subcloned and sequenced. The genes involved are organized into two operons separated by an IS-element (Figure 5) (Axelsson et al., 1994). The first operon encompasses the sakacin A structural gene (*sapA*) followed by the downstream immunity gene (*saiA*). The second operon is apparently composed of five genes, of which the first three genes (*orf4, sapK* and *sapR*) can constitute a three-component regulatory system as they encode a small bacteriocin-like peptide with a double-glycine leader, an HK and an RR, respectively (Figures 3, 5). The last two ORFs (*sapT* and *sapE*) of the operon encode proteins constituting the dedicated ABC-transport system. Mutations in *sapT* and *sapE* abolished bacteriocin production only, while mutations in either of *sapK* or *sapR* resulted in loss of both bacteriocin production and immunity (Axelsson & Holck, 1995), indicating that the signal transducing system is essential for expression of both operons. Based on these observations, is was suggested the RR component (SapR) to be a transcriptional regulator activating transcription of both operons. Preliminary data suggests that an induction factor is secreted into the growth medium because growth medium from a *L. sake* culture is able to induce bacteriocin production (Diep, unpublished data). However, it still has to be shown that the factor is indeed encoded by *orf4*.

The fourth class II regulated bacteriocin-producing system is found in *Carnobacterium piscicola* LV17B which produces two bacteriocins termed carnobacteriocin BM1 and B2, both of which belong to the pediocin-family (Quadri et al., 1994, 1995a, 1995b). The BM1 operon consists of the structural gene (*cbnBM1*) and its cognate immunity gene (*cbiBM1*), and is located on the chromosome, while the B2 operon consisting of a similar gene pair (*cbnB2* and *cbiB2*) is located on a 61-kb resident plasmid (pCP40). Upstream of bacteriocin B2 operon, a second operon is located constituting a putative three- component autoregulatory cassette (*orf6, cbnK* and *cbnR*) encoding a bacteriocin-like peptide with double-glycine leader, an HK and an RR, respectively, and a gene pair (*cbnT* and *cbnD*) encoding the ABC-export machinery (Figure 5). The components of the putative three-component autoregulatory system is believed to serve a regulatory role as found for their counterparts in *pln* and *spp*. Interestingly, two more ORFs (*orf4* and *orf5*) encoding bacteriocin-like peptides with double-glycine leaders are located further upstream. However, what role, if any, these two ORFs have is yet unknown.

The histidine protein kinase (HK) and the response regulator (RR) are the two components found in numerous bacterial signal transduction pathways. The HK constitutes the environmental sensor which are most likely the induction factor (IF) in these bacteriocin systems. Through a phosphorylation reaction HK communicates with RR which subsequently interact with the target genes. Many RR proteins contain a potential DNA binding motif (helix-turn-helix) (Bourret et al., 1991) and therefore the RR proteins have been proposed to be transcriptional regulators. This has not yet been shown for the RR found in the regulatory systems of the bacteriocins.

It is generally recognized that most transcription regulators control their target operon by direct binding to sites close to or within the promoter region, thereby stimulating or inhibiting binding of RNA polymerase to the promoter. Features such as inverted repeats or direct repeats in the promoter regions are frequently observed in binding sites for transcription regulators. Sequence alignment of the promoter regions of operons associated with the four bacteriocin systems (*pln*, *sap*, *spp*, and *cbn*) identified the –10 sites and the 35 sites (Figure 6). Just 2–9 basepairs upstream of the –35 site two conserved repeats (R (right) and L (left)), each consisting of 9 basepairs, were found the two repeats were separated by 12–13 nucleotides (Diep et al., 1996). Transcription regulators are often bound as dimers to target DNA repeats. The direct repeats found within the conserved regulatory-like boxes in the promoters are separated by two helical turns (Figure 6). Based on this analysis, we strongly believe that the regulatory-like boxes found in the promoters serve as a binding site for a transcription regulator which probably is the response regulator.

Perspectives

Production of non-lantibiotic bacteriocins seems to be widespread amongst lactic acid bacteria, and most probably these bacteriocins are also produced by other Gram-positive bacteria. In Gram-negative bacteria ColV and microcin 24 of *E. coli* can, on the basis of their properties, be classified as class II bacteriocins. Bacteriocins are used in industrial food fermentation processes and in certain food products. Due to numerous factors such as genetic instability, regulation, low production, inactivation etc., bacteriocin-producing starter cultures have not been fully successful in food and feed fermentation processes. Some of the problems encountered could be due to intrinsic regulation of bacteriocin production which could be an obstacle for effective bacteriocin synthesis in food fermentation processes. It is now possible to express bacteriocins constitutively and several reports describe how to clone and express bacteriocins in new hosts (Chikindas et al., 1995; Allison et al., 1995; Huehne et al., 1996). It is also possible to secrete double-glycine-leader-dependent bacteriocins with the *sec*- dependent system by replacing the leaders which may be a promising approach to produce multiple bacteriocins by the *sec*-dependent system in one LAB strain (Stiles, personal communication). Protein engineering of class II bacteriocins should be a feasible way to achieve more efficient bacteriocin production and to improve the stability and activity of class II bacteriocins. Hybrid bacteriocins of the pediocin-like bacteriocins have already been chemically synthesized. The biological activities of such hybrids were altered compared to the native bacteriocins (Fimland et al., 1996). With the large number and the heterogeneity of class II bacteriocins one should expect that new bacteriocins and new production systems of these compounds will appear in the near future.

While class II bacteriocins have been studied intensively during the last 5 to 6 years, we still feel it is pertinent to raise the question 'What is the physiological/biological role of these compounds?'. Based on the information available to date, the biological function of bacteriocins is an antimicrobial one, enabling their producer to compete with other bacteria for survival. However, it has been shown that bacteriocin-like substances do not always possess antimicrobial activity, but that some compounds may also function as induction signals for gene expression. In the case of nisin, the bacteriocin carries both functions, being antimicrobial agent as well as induction factor (Kuipers et al., 1995). The two-peptide lantibiotic cytolysin produced by *Enterococcus faecalis* exerts both antimicrobial and haemolytic activity (Gilmore et al., 1994) which also extends our conceptual knowledge of the functional diversity of bacteriocin-like compounds.

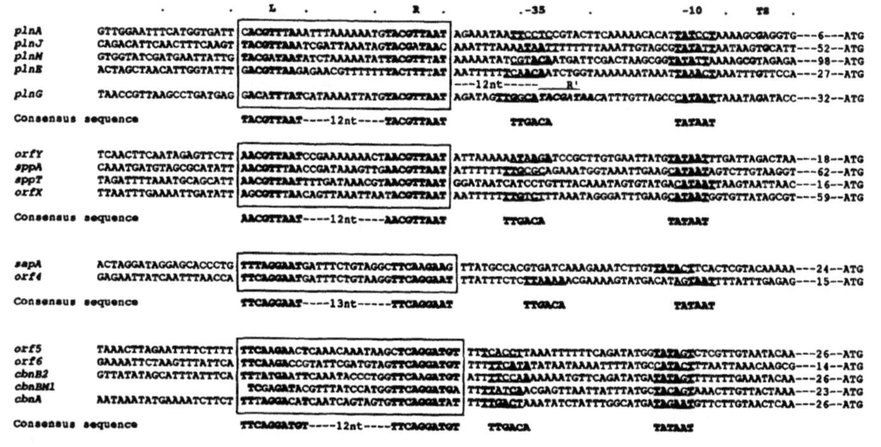

Figure 6. Alignment of the promoter sequences from the regulated bacteriocin operons. The *L. plantarum* C11 *pln* promoters (Dzung et al. 1996): *plnA, -J, -M, -E* and *-G*; the *L. sake spp* promoters (Huehne et al. 1996, Brurberg et al. 1996): *orfY, sppA, -T* and *orfX*; the *L. sake sap* promoters (Axelsson & Holck, 1995): *sapA* and *orf4*; and the *Carnobacterium piscicola cbn* promoters (Quadri et al. 1995): *orf5, orf6, cbnB2, cbnBM1* and *cbnA*. *L. sake orfY* and *orfX* were identified with coordinates 443– 551, and 7489–7597, respectively, in the DNA sequence obtained. The –35 and –10 sites are underlined whereas the conserved bases within these sites are indicated by bold letters. Similarly, conserved regulatory- like boxes (CRLBs) containing direct (L and R) repeats are boxed and the invariant bases within the repeats are indicated by bold letters. Consensus sequence of the direct repeats is indicated at the bottom line in each promoter group. The start of the transcription of the *pln* genes (TSs; indicated by bold letters) has been experimentally determined (Dzung et al. 1996,). The first codons (ATG/TTG) are indicated at the end of each sequence, whereas the preceding figures refer to the number of basepairs in the region in between. Note that the –35 sites are, almost in all cases, less conserved than that compared with the –10 sites, and that the *plnG* promoter contains an extra direct repeat (R'; indicated by italic letters) located just downstream of its –35 site.

Studies relevant to the functionality of bacteriocin-like peptides have recently been published.

Streptococcus pneumonia is known to induce its own competence for the uptake of DNA by a secreted competence factor (CF). The competence factor is actually a bacteriocin-like peptide secreted and processed like a class II bacteriocin (Håvarstein et al., 1995). The recent results concerning the regulation of bacteriocin production by bacteriocin-like peptides and the analogies with regulatory systems known from other bacteria, raise intriguing questions about the evolutionary history and the biological function of bacteriocins and bacteriocin-like peptides and it is conceivable that these types of molecules serve a much more diverse biological function than previously anticipated.

Acknowledgments

The work on bacteriocin in our laboratory has been funded by the Norwegian Research Council, the Nordic Industrial Fund and by the Biotech Program of the Commission of the European Community.

References

Abee, T 1995. Pore-forming bacteriocins of Gram-positive bacteria and self-protection mechanisms of producer organisms. FEMS Microbiol. Lett. 129: 1–10

Allison GE, C Fremaux and TR Klaenhammer 1994. Expansion of bacteriocin activity and host range upon complementation of two peptides encoded within the Lactacin F operon. J. Bacteriol. 176: 2235–2241

Allison GE, RW Worobo, ME Stiles and TR Klaenhammer 1995. Heterologous expresion of the lactacin F peptides by *Carnobacterium piscicola* LV17. Appl. Environ. Microbiol. 61: 1371–1377

Aymerich T, H Holo, LS Håvarstein, M Hugas, M Garriga, IF Nes 1996. Biochemical and genetic characterization of enterocin A from *Enterococcus faecium*, a new antilisterial bacteriocin in the pediocin family of bacteriocins. Appl. Environ. Microbiol. 62: 1676–1682

Axelsson L, A Holck, SE Birkeland, T Aukrust and H Blom 1993. Cloning and nucleotide sequencing of a gene from *Lactobacillus sake* LB706 necessary for sakacin A production and immunity. Appl. Environ. Microbiol. 59: 2868–2875

Axelsson L and A Holck 1995. The genes involved in production of and immunity to sakacin A, a bacteriocin from *Lactobacillus sake* Lb706. J. Bacteriol. 177: 2125–2137

Balaban N and RP Novick 1995. Autocrine regulation of toxin synthesis by *Staphylococcus aureus*. Proc. Natl. Acad. Sci. USA 92: 1619–1623

Bourret RB, KA Borkovich and MI Simon 1991. Signal transduction pathways involving protein phosphorylation in prokaryotes. An. Rev. Biochem. 60: 401–441

Brurberg MB, IF Nes and V Eijsink 1996. Unpublished results

Chikindas MJ, MJ Garcia Garcera, AM Driessen, AM Lederboer, J Nissen-Meyer, IF Nes, T Abee, WN Konings and G Venema 1993. Pediocin PA-1, a bacteriocin from *Pediococcus acidilactici* PAC1.0, forms hydrophillic pores in the cytoplasmatic membrane of target cells. Appl. Environ. Microbiol. 59: 3577–3584

Chikindas ML, K Venema, AM Ledeboer, G Venema and J Kok 1995. Expression of lactococcin A and pediocin PA–1 in heterologous hosts. Letters in Applied Microbiol. 21: 183–189

De Vos WM, OP Kuipers, JR van der Meer and RJ Siezen 1995. Maturation pathway of nisin and other lantibiotics: post- translationally modified antimicrobial peptides exported by gram- positive bacteria. Mol. Microbiol. 17: 427–337

Diep DB, LS Håvarstein, J Nissen-Meyer and IF Nes 1994. The gene encoding plantaricin, a bacteriocin from *Lactobacillus plantarum* C11, is located on the same transcription unit as an *agr*-like regulatory system. Appl. Environ. Microbiol. 60: 160–166

Diep DB, LS Håvarstein and IF Nes 1995. A bacteriocin-like peptide induces bacteriocin synthesis in *L. plantarum* C11. Mol. Microbiol. 18: 631–639

— 1996. Characterization of the locus responsible for the bacteriocin production in *Lactobacillus plantarum* C11. J. Bacteriol. 178: 4472–4483

Eisjink VGH, MB Brurberg, PH Middelhoven and IF Nes 1996. Induction of bacteriocin production in *Lactobacillus sake* by a secreted peptide. J. Bacteriol. 178: 2232–2237

Engelke G, Z Gutowski-Eckel, P Kiesau, K Siegers, M Hammelmann and KD Entien 1994. Regulation of nisin biosynthesis and immunity of *Lactococcus lactic* 6F3. Appl. Environ. Microbiol. 60: 814–825

Fath MJ, LH Zhang, J Rush and R Kolter 1994. Purification and characterization of colicin V from *Eschericia coli* culture supernatants. Biochemistry 33: 6911–6917

Fimland G, OR Blingso, K Sletten, G Jung, IF Nes, J Nissen-Meyer 1996. New biological activity hybrid bacteriocins constructed by combining regions from various pediocin-like bacteriocins. Appl. Environ. Microbiol. 62: 000-000

Franke CM, KJ Leenhout, AJ Haandrikman, J Kok, G Venema and K Venema 1996. Topology of LcnD, a protein implicated in the transport of bacteriocins from *Lactococcus lactic*. J. Bacteriol. 176: 1766–1769

Fremaux C, C Ahn and TR Klaenhammer 1993. Molecular analysis of the lactocin F operon. Appl. Environ. Microbiol. 59: 3906–3915

Gilmore MS, RA Segarra, MC Booth, CP Bogie, LR Hall and DB Clewell 1994. Genetic structure of the *Enterococcus faecalis* plasmid pAD1-encoded cytolytic system and its relationship to lantibiotic determinants. J. Bacteriol. 176: 7335–7344

Gilson L, HK Mahanty, R Kolter (1990). Genetic analysis of an MDR-like export system: the secretion of colicin V. EMBO J. 9: 3875–3884

Guangyong J, RC Blavis and RP Novick (1995). Cell density of staphylococcal virulence mediated by an octapeptide pheromone. Proc. Natl. Acad. Sci. USA 92: 12055–12059

Hastings JW, M Sailer, K Johnson, KL Roy, JC Vederas and ME Stiles 1991. Characterization of leucocin A-UAL 187 and cloning of the bacteriocin gene from *Leuconostoc gelidum*. J. Bacteriol. 173: 7491–7500

Hechard Y, DB Derijard, F Letellier and Y Cenatiempo 1992. Characterization and purification of mesentericin Y105, an anti- *Listeria* bacteriocin from *Leuconostoc mesenteroides*. J. Gen. Microbiol. 138: 2725–2731

Henderson JT, AL Chopko and PD van Wasserman 1992. Purification and primary structure of pediocin PA–1 produced by *Pediococcus acidilactici* PAC1.0. Arch. Biochem. Biophys. 295: 5–12

Hoch JA and TJ Silhavy (Eds.), 1995. Two -Component Signal Transduction. ASM Press, Washington, D.C. USA

Holck A, L Axelsson, SE Birkeland, T Aukrust and H Blom 1992. Purification and amino acid sequence of sakacin A, a bacteriocin from *Lactobacillus sake* Lb 706. J. Gen. Microbiol. 138: 2715–2720

Holo H. (1996) Resistance and immunity to bacteriocins of lactic acid bacteria. Manuscript submitted to Can. J. Microbiol.

Holo H, Ø Nilssen and IF Nes 1991. Lactococcin A, a new bacteriocin from *Lactococcus lactis* subsp. *cremoris*: isolation and characterization of the protein and its gene. J. Bacteriol. 173: 3879–3887

Huehne K, A Holck, L Axelson and L Kroeckel 1996. Analysis of sakacin P gene cluster from *Lactobacillus sake* LB674 and its expression in sakacin P negative *L. sake* strains. Microbiology. 142: 1437–1448

Håvarstein LS, H Holo and IF Nes 1994. The leader peptide of colicin V shares consensus sequences with leader peptides that are common amongst peptide bacteriocins produced by Gram-positive bacteria. Microbiol. 140: 2383–2389

Håvarstein LS, G Coomaraswamy and DA Morrison 1995. An unidentified heptadecapeptide pheromone induces competence for genetic transformation in *Streptococcus pneumoniae*. Proc. Natl. Acad. Sci. USA 92: 11140–11144

Håvarstein LS, BD Diep and IF Nes 1995. A Family of bacteriocin ABC transporters carry out proteolytic processing of their substrates concomitant with export. Mol. Microbiol. 16: 229-240

Jack RW, JR Tagg and B Ray (1995). Bacteriocins of Gram- Positive Bacteria. Microbiol. Rev. 59: 171–200

Jiménez-Diaz R, JL Ruiz-Barba, DP Cathcart, H Holo, IF Nes, KH Sletten and PJ Warner 1995. Purification and partial amino acid sequence of plantaricin S , a bacteriocin produced by *Lactobacillus plantarum* LPCO10, the activity of which depends on the complementary action of two peptides. Appl. Environ. Microbiol. 61: 4459–4463

Jiménez-Diaz R, RM Ríos-Sánchez, M Desmazeaud, JL Ruiz-Barba and JC Piard 1993. PlantaricinS and T, two new bactriocins produced by *Lactobacillus plantarum* LPCO10 isolated from a green olive fermentation. Appl. Environ. Microbiol. 59: 1416–1424

Kanatani K, M Oshimura and K Sano 1995. Isolation and charcterization of acidocin A and cloning of the bacteriocin gene from *Lactobacillus acidophilus*. Appl. Environ. Microbiol. 61: 1061–1067

Karlson P & M. Lüscher 1959. Pheromones a new term of class of biologically active substances? 183: 55-56

Klaenhammer TR 1993. Genetics of bacteriocins produced by lactic acid bacteria. FEMS Microbiol. Rev. 12: 39–86

Kuipers OP, HS Rollema, WMG Yap, HJ Boot, RJ Siezen and WM deVos 1992. Engineering dehydrated amino acid residues in the antimicrobial peptide nisin. J. Biol. Chem. 267: 24340–24346

Kuipers OP, MM Beerthuyzen, PGGA deRuyter, EJ Luesink and WM deVos 1995. Autoregulation of nisin biosynthesis in *Lactococcus lactic* by signal transduction. J. Biol. Chem. 270: 281–291

Konings RNH and CW Hilbers (Eds.), 1996. Lantibiotics: A Unique Group of Antibiotic Peptides. Antonie van Leeuwenhoek Int. J. Gen. Molec. Microbiol. 69: Issue 2

Larsen AG and B Nørrung 1993. Inhibition of *Listeria monocytogenes* by bavaricin A, a bacteriocin produced by *Lactobacillus bavaricus* MI401. Lett. Appl. Microbiol. 17: 132–134

Leer RL, JMBM van der Vossen, M van Giezen, JM van Noort and PH Pouwels (1995): Genetic analysis of acidocin B, a novel bacteriocin produced by *Lactobacillus acidophilus*. Microbiol. 141: 1629–1635

Magnusson R, J Solomon and AD Grossman 1994. Biochemical and genetic characterization of a competence pheromone from B. subtilis. Cell 77: 207–216

Marugg JD, CF Gonzales, BS Kunka, AM Ledeboer, MJ Pucci, MY Toonen, SA Walker, LCM Zoetmulder and PA Vandenbergh 1992. Cloning, expression, and nucleotide sequence of genes involved in production of pediocin PA–1, a bacteriocin from *Pediococcus acidilactici* PAC1.0. Appl. Environ. Microbiol. 58: 2360–2367

Morfeldt E, L Janzon, S Arvidson and S Löfdalh 1988. Cloning of a chromosomal locus (*exp*) which regulates the expression of several exoprotein genes in *Staphylococcus aureus*. Mol. Gen. Genet. 211: 435–440

Motlagh AM, AK Buhunia, F Szostek, TR Hansen, MC Johnson, B Ray 1992. Nucleotide and amino acid sequence of *pap* gene pediocin AcH) produced in *Pediococcus acididlactici* H. Lett. Appl. Microbiol. 15: 45–48

Muriana PM and TR Klaenhammer 1991. Cloning, phenotypic expression, and DNA sequence of the gene for lactacin F, an antimicrobial peptide produced by *Lactobacillus* spp. J. Bacteriol. 173: 1779–1788

Nes IF, LS Håvarstein and H Holo 1995. Genetics of non-lantibiotics bacteriocins. *In* Genetics of Streptococci, Enterococci and Lactococci. (J. J. Ferretti, M. S. Gilmore, T. R. Klaenhammer, and F. Brown, Eds.). Karger, New York, pp. 645–651

Nieto Lozano JC, J Nissen-Meyer, K Sletten, C Peláz and IF Nes 1992. Purification and amino acid sequence of a bacteriocin produced by *Pediococcus acidilactici*. J. Gen. Microbiol. 138: 1985–1990

Nissen-Meyer J, H Holo, LS Håvarstein, K Sletten and IF Nes 1992. A novel lactococcal bacteriocin whose activity depends on the complementary action of two peptides. J. Bacteriol. 174: 5686–5692

Nissen-Meyer J, LS Håvarstein, H Holo, K Sletten and IF Nes 1993a. Association of the lactococcin A immunity factor with the cell membrane: purification and characterization of the immunity factor. J. Gen. Microbiol. 139: 1503–1509

Nissen-Meyer J, AG Larsen, K Sletten, M Daeschel and IF Nes 1993b. Purifiaction and characterization of plantaricin A, a *Lactobacillus plantarum* bacteriocin whose activity depends on the action of two peptides. J. Gen. Microbiol. 139: 1973–1978

O'Brian GJ and HK Mahanty 1996. Complete nucleotide sequence of microcin 24 genetic region and analysis of a new ABC transporter. Accession number ECU47048

Parkinson JS and EC Kofoid 1992. Communication modules in bacterial signaling proteins. Annu. Rev. Genet. 26: 71–112

Pestova EV, LS Håvarstein and DA Morrison 1996. Regulation of transformability by an auto-induced peptide pheromone and a two-component regulatory system. Molec. Microbiol. 21: 855-864

Piard JC, OP Kuipers, HS Rolema, MJ Deamzeaud and WM deVos 1993. Structure, organization and expression of the lct gene for lacticin 481, a novel lantibiotic produced by *Lactococcus lactic*. J. Biol. Chem. 268: 16361–16368

Quadri LEN, KL Roy, JC Vederas and ME Stiles 1995a. Characterization of four genes involved in the production of antimicrobial peptides by *Carnobacterium piscicola* LV17B. Published DNA sequence with accession nr. 147121.em-bl.new

Quadri LEN, M Sailer, MR Terebiznik, KL Roy, JC Vederas and ME Stiles 1995b. Characterization of the protein conferring immunity to the antimicrobial peptide carnobacteriocin B2 and expression of the carnobacteriocins B2 and BM1. J. Bacteriol. 177: 1144–1151

Quadri LEN, M Sailers, KL Roy, JC Vederas and ME Stiles 1994. Chemical and genetic characterization of bacteriocins produced by *Carnobacterium piscicola* LV17B. J. Biol. Chem. 269: 12204–12211

Ray B 1992. Pediocin(s) of *Pediococcus acidilactici* as a Food Biopreservative. Page 265 - 322. In: Food Biopreservatives of Microbial Origin. Eds. B.Ray and M.Daeschel. CRC Press Inc., Bocan Raton, Forida

Ross KF, CW Ronson and JR Tagg 1993. Isolation and characterization of the lantibiotic salavaricin A and its structural gene *salA* from *Streptococccus salivarius* 20P3. Appl. Environ. Microbiol. 60: 1652–1657

Sahl HG, RW Jack and G Bierbaum (1995): Biosynthesis and biological activities of lantibiotics with unique post-translational modifications. Eur. J. Biochem. 230: 827–853

Saucier L, A Poon and ME Stiles (1995). Induction of bacteriocin in *Carnobacterium piscicola* LV17. J. Appl. Bacteriol. 78: 684–690

Stock JB, AJ Ninfa and AM Stock 1989. Protein phosphorylation and regulation of adaptive responses in bacteria. Microbiol. Rev. 53: 450–490

Stoddard GW, JP Petzel, MJ Van Belkum, J Kok and LL McKay 1992. Molecular analyses of the lactococcin A gene cluster from *Lactococcus lactis* subsp. *lactis* biovar *diacetylactis* WM4. Appl. Environ. Microbiol. 58: 1952–1961

Tichaczek PS, J Nissen-Meyer, IF Nes, RF Vogel and WP Hammes 1992. Characterization of the bacteriocins curvacin A from *Lactobacillus curvatus* LTH1174 and sakacin from *L. sake* LTH673. Syst. Appl. Microbiol. 15: 460–468

Tichaczek PS, RF Vogel and WP Hammes 1993. Cloning and sequencing of *cur*A encoding curvacin A, the bacteriocin produced by *Lactobacillus curvatus* LTH1174. Arch. Microbiol. 160: 279–283

— 1994. Cloning and sequencing of *sakP* encoding sakacin P, the bacteriocin produced by *Lactobacillus sake* LTH 673. Microbiology 140: 361–367

Van Belkum MJ, Hayema BJ, Jeeninga RE, Kok J and G Venema (1991). Organization and nucleotide sequences of two lactococcal bacteriocin operons. Applied and Environmental Microbiology 57: 492–498

Van Belkum MJ, J Kok and G Venema 1992. Cloning, sequencing, and expression in *Escherichia coli* of *lcn*B, a third bacteriocin determinant from the lactococcal bacteriocin plasmid p9B4–6. Appl. Environ. Microbiol. 58: 572–577

Van Belkum MJ and ME Stiles (1995) Molecular characterization of genes involved in the production of the bacteriocin leucocin A from *Leuconostoc gelidum*. Appl. Environ. Microbiol. 61: 3573–3579

Van der Mer JR, J Polman, MM Beerthuyzen, RJ Siezen, O Kuipers and WM de Vos 1993. Characterization of the *Lactococcus lactis* nisin A operon genes *nisP*, encoding a subtilisin-like seine protease involved in precursor processing, and *nisR* , encoding a regulatory protein involved in nisin biosynthesis. J. Bacteriol. 175: 2578–2588

Venema K, T Abee, AJ Haandrikman, KJ Leenhouts, J Kok, WN Konings and G Venema 1993. Mode of action of lactococcin B, a thiol-activated bacteriocin from *Lactococcus lactis*. Appl. Environ. Microbiol. 59: 1041–1048

Venema K, RE Haverkort, T Abee, AJ Haandrikman, KJ Leenhouts, LD Leij, G Venema and J Kok 1994. Mode of action of LciA, the lactococcin A immunity protein. Mol. Microbiol. 16: 521–532

Venema K, J Kok, JD Marugg, MY Toonen, AM Ledeboer, G Venema and ML Chikindas 1995a. Functional analysis of the pediocin operon of *Pediococcus acidilactici* PAC1.0:PedB is the immunity protein and PedD is the precursor processing enzyme. Mol. Microbiol. 17: 515–522

Venema K, MHA Dost, G Venema and J Kok 1996. Mutational analysis and chemical modification of lactococcin B, a bacteriocin produced by *Lactococcus lactis*. Manuscript

Venema K, MHA Dost, PAH Beun, AJ Haandrikman, G Venema and J Kok 1996. The genes for secretion of lactococcins are located on the chromosom of *Lactococcus lactis IL1403*. Appl. Environ. Microbiol. 62: 1689–1692

Worobo RW, T Henkel, M Sailer, KL Roy, JC Vederas and ME Stiles 1994. Characteristics and genetic determinant of a hydrophobic peptide bacteriocin, carnobacteriocin A, produced by *Carnobacterium piscicola* LV17A. Microbiology 140: 517–526

Antonie van Leeuwenhoek **70:** 129–145, 1996.

Inducible gene expression and environmentally regulated genes in lactic acid bacteria

Jan Kok
Department of Genetics, Groningen Biomolecular Sciences and Biotechnology Institute, University of Groningen, Kerklaan 30, 9751 NN Haren, the Netherlands

Key words: gene expression, gene regulation, *Lactococcus*, *Lactobacillus*, *Lactus* acid bacteria

Summary

Relatively recently, a number of genes and operons have been identified in lactic acid bacteria that are inducible and respond to environmental factors. Some of these genes/operons had been isolated and analysed because of their importance in the fermentation industry and, consequently, their transcription was studied and found to be regulatable. Examples are the lactose operon, the operon for nisin production, and genes in the proteolytic pathway of *Lactococcus lactis*, as well as xylose metabolism in *Lactobacillus pentosus*. Some other operons were specifically targetted with the aim to compare their mode of regulation with known regulatory mechanisms in other well-studied bacteria. These studies, dealing with the biosynthesis of histidine, tryptophan, and of the branched chain amino acids in *L. lactis*, have given new insights in gene regulation and in the occurrence of auxotrophy in these bacteria. Also, nucleotide sequence analyses of a number of lactococcal bacteriophages was recently initiated to, among other things, specifically learn more about regulation of the phage life cycle. Yet another approach in the analysis of regulated genes is the 'random' selection of genetic elements that respond to environmental stimuli and the first of such sequences from lactic acid bacteria have been identified and characterized. The potential of these regulatory elements in fundamental research and practical (industrial) applications will be discussed.

Introduction

Bacteria continuously monitor the availability of essential nutrients (e.g. sources of carbon, nitrogen, phosphorus and sulphur, trace metals and certain ions) by measuring extracellular concentrations and (fluxes in) intracellular pools. They have evolved complex regulatory networks enabling quick responses to changes in their environment. Generally, the response involves modulation of gene expression, allowing the bacteria to adapt to a wide variety of stimuli.

Gene expression in lactic acid bacteria (LAB) has received much attention during the last decade. These studies have addressed transcription initiation and termination as well as translation initiation and codon usage. Details of this work can be found in a number of excellent reviews covering this field, and the literature references therein (de Vos & Simons, 1994; van de Guchte, 1992; Chopin, 1993; Mercenier et al., 1994). For a number of LAB species, notably *L. lactis*, this has led to the definition of the canonical vegetative promoter. The lactococcal promoter consensus conforms to that of the *Escherichia coli* and *Bacillus subtilis* vegetative promoters: TTGACA–17/18-TATAAT. A number of the *L. lactis* promoters have been used to construct gene expression vectors for LAB with which many homologous and heterologous proteins in LAB have been (over)expressed (de Vos & Simons, 1994).

Only relatively recently studies on the control of gene expression in LAB were initiated. This is a reflection of the fact that, with the expanding genetic 'tool box' for LAB, increasingly more genes and operons from LAB are being cloned and sequenced. As a next step, the in depth analysis of many of these genes and their products is now undertaken. Methods are available for the targetted (knock-out) mutation of genes on the chromosome or the introduction of transcriptional or translational fusions of a gene of interest with reporter genes such as *E. coli lacZ* in one copy in the chromosome (Leenhouts & Venema, 1993). Both techniques are very useful for answering specific questions relating to regulation of gene expression. In a number of cases gene regulation has been directly studied at the mRNA level. In this way, several important industrial traits have been examined and the results of these studies will be detailed here.

Approaches which specifically address regulation of gene expression in LAB include the cloning from these organisms of genes that are known to be controlled in other organisms (e.g. amino acid biosynthesis operons). Transposons and integration plasmids carrying a promoterless *lacZ* reporter gene have been used to randomly target environmentally regulated or stress-induced chromosomal loci. Also, the effect of a variety of stress conditions on protein production in LAB is presently being investigated by one- and two-dimensional gelelectrophoresis (Kunji et al., 1993; Ayffray et al., 1992; Hartke et al., 1994, 1995). Several genes involved in one of the stress conditions, heat shock, have been cloned and sequenced (van Asseldonk et al.; 1993, Eaton et al., 1993; Kim & Batt, 1993).

The field of control of gene regulation in LAB is rapidly expanding as will be clear from the present evaluation. Only those cases which have been examined into appreciable detail at the DNA and mRNA levels will be discussed here, as well as some of the strategies employed to isolate regulated genes.

Random selection of regulated promoters

Several transposons with promoterless reporter genes have been developed for the analysis of expression of single-copy (chromosomal) genes in bacteria. For instance, derivatives of the *Enterococcus faecalis* transposon Tn*917* carrying the *E. coli lacZ* gene devoid of its own promoter have been used in the analysis of control of gene expression in *B. subtilis* (Youngman, 1987). Variants of Tn*917* were shown to also randomly transpose in *L. lactis* (Israelsen & Hansen, 1993). A collection of *L. lactis* strains with random fusions of *lacZ* to chromosomal loci was screened for β-galactosidase (LacZ) activity under standard growth conditions (Israelsen et al., 1995). A number of clones showed altered LacZ activities when growth conditions, such as temperature, medium pH and/or arginine content, were changed. One of the integrants appeared to produce β-Gal at pH 5.2 but not at pH 7.0, produced more of the enzyme at 15 °C than at 30 °C, while the highest activities were observed in the stationary phase. Chromosomal DNA (9.7 kb) upstream of the integrated *lacZ* gene was rescued from the integrant and introduced in a promoter-probe vector. On this plasmid the fragment displayed the same pattern of gene regulation as in the single-copy situation on the chromosome of the integrant. At present, no data are available as to the identity and the molecular details of regulation of this interesting promoter.

From the lactococcal plasmid vector pWV01 a series of non-replicative derivatives have been constructed that can be used for chromosomal integration in, among others, LAB (Leenhouts & Venema, 1993). One derivative, pORI13, carries a promoterless *lacZ* gene preceded by lactococcal translation signals and a multiple cloning site. Stop codons were introduced upstream of *lacZ* to allow for transcriptional fusions only (J.W. Sanders, pers. comm.). A library of random partial restriction fragments of *L. lactis* chromosomal DNA in this vector was made in *E. coli* and subsequently used to make a bank of integrants of *L. lactis*. Screening for LacZ activity in the presence or absence of 0.3 M NaCl gave colonies with all four possible phenotypes. One colony displaying LacZ activity only on plates with NaCl was taken for further analysis. The original insert in pORI13 in this clone was rescued and shown to comprize 10 kb of chromosomal DNA. The salt-inducible promoter was identified by deletion mapping, nucleotide sequence analysis and primer extension experiments. *lacZ* had been fused to the first 360 bp of an open reading frame X (ORFX) of which the deduced product shows homology with membrane proteins. The minimal region needed for salt-induced LacZ activity contained a gene for a regulator, *rggL*, followed by a 21-bp inverted repeat structure at position –35 of the actual promoter. No canonical –35 hexanucleotide could be identified, but a good –10 sequence was present, 7 bp downstream of which transcription started in the presence of salt. No primer extension product was observed when cells were grown in the absence of NaCl and only a very weak signal was obtained in Northern blots. In an *rggL* insertion mutant very little transcript was detected in Northern blots of mRNA from cells grown both in the presence or absence of NaCl, indicating that RggL acts as an activator in the NaCl-dependent expression of ORFX. The transcript was approximately 3 kb in size (J.W. Sanders, pers. comm.).

Screening for two-component regulatory systems

Systems used by bacterial cells to adapt to changes in their environment often involve two families of signal transduction proteins, namely sensory kinases and response regulators. The first component monitors an environmental parameter while the latter generally directs changes in gene expression (for review, see Stock et al., 1995). A complementation strategy was employed to randomly clone sensory kinases from *L. lactis* (M. O'Connell Motherway, G.F. Fitzgerald and D. van Sinderen, pers. comm.). An *E. coli phoR creC* double mutant lacks the two sensory kinases needed for expression of the phosphate regulon. Consequently, the strain is alkaline phosphatase negative ($PhoA^-$), and grows to white colonies on LB plates containing the chromogenic PhoA substrate 5-bromo 4-chloro 3-indolyl phosphate. Chromosomal DNA fragments of *L. lactis* cloned in an *E. coli* replicon which were able to complement the PhoA- deficiency were identified in several blue *E. coli(phoR creC)* colonies. In this way, the (partial) genes of five different kinases were isolated and sequenced. In two of the clones the corresponding response regulators were present in the upstream DNA sequences. The next important step will be to determine, on the basis of protein homologies and by making mutants and analysing their phenotypes, in which signalling pathway the genes are involved (M. O'Connell Motherway, G.F. Fitzgerald & D. van Sinderen, pers. comm.).

T7 RNA polymerase overexpression system

Pending the identification of homologous highly inducible gene expression systems for LAB, the T7 RNA polymerase overexpression system has been developed for use in *L. lactis* (Studier et al., 1990; Wells et al., 1993). Use is made of the lactose-inducible *lac* promoter and *lacR* repressor gene (see below) for the controled expression of the *E. coli* bacteriophage T7 RNA polymerase in *L. lactis*. This is accomplished by shifting cells from a glucose- to a lactose-containing medium. Target genes cloned behind the phage T7 promoter of gene *10* are overexpressed to up to 22% of total soluble protein (Wells et al., 1993). In a further advancement of the system two inducible expression-secretion vectors have been constructed, one of which was used to secrete the model protein tetanus toxin fragment C (Wells et al., 1993). Presently, the system employs three plasmids to provide all the necessary functions.

Lactose metabolism

Lactose metabolism in LAB is either initiated by a lactose permease system or a phosphoenolpyruvate dependent lactose phosphotransferase system (lactose PEP-PTS) (McKay et al., 1970). Lactococci used in dairy industry transport lactose exclusively via the lactose PEP-PTS. In this system, the lactose-specific components are the membrane-associated enzyme EII^{lac} (LacE) and the soluble protein $EIII^{lac}$ (LacF). Lactose is phosphorylated during uptake and the resulting lactose–6-phosphate is converted by phospho-β-galactosidase (LacG) to glucose, which is further taken down by the Embden-Meyerhof-Parnass route, and galactose–6-phosphate (McKay et al., 1970). Galactose–6-phosphate is metabolized via the tagatose–6-phosphate pathway by the consecutive action of galactose–6-phosphate isomerase (LacAB), tagatose–6-phosphate kinase (LacC), and tagatose 1,6-diphosphate aldolase (LacD) (Bissett & Anderson, 1974).

Although it was already known for quite some time that key enzymes in the breakdown of lactose by *L. lactis* could be induced by growth on lactose or galactose (Bisset & Anderson, 1974; Leblanc et al., 1979; McKay; Maeda & Gasson, 1989), it is only of recent date that the underlying mechanism is understood in considerable detail. The unraveling of the control of lactose utilization had to await the identification and isolation of the genes involved. In *L. lactis* ssp. *cremoris* strain NCDO712, the structural genes for the lactose-specific PTS enzymes (LacEF) as well as those for phospho-β-galactosidase and the enzymes of the tagatose–6- phosphate pathway (LacABCD) are organized in a 7.8-kb *lac* operon: *lacABCDFEGX* (de Vos & Gasson, 1989; de Vos et al., 1990; van Rooijen et al., 1991; see Figure 1). An oppositely oriented gene immediately upstream of *lacA*, *lacR*, specifies a repressor protein (van Rooijen & de Vos, 1990). Transcriptional analyses revealed that two lactose-inducible transcripts of 6 and 8.5 kb are formed, comprising the *lacABCDFE* and *lacABCDFEGX* genes, respectively (de Vos et al., 1990; van Rooijen & de Vos, 1990, van Rooijen et al., 1991). Both transcripts are initiated at a single lactose-inducible promoter upstream of *lacA* (van Rooijen et al., 1992). An inverted repeat between *lacE* and *lacG* could act as an intracistronic terminator for the 6-kb

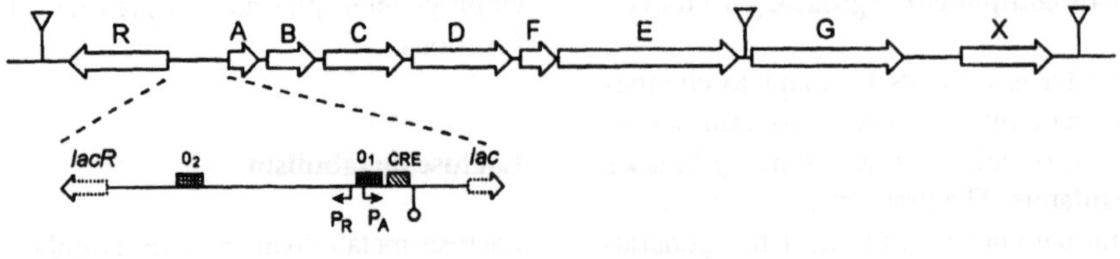

Figure 1. The lactose regulon of *L. lactis*. The function of the various gene products is given in the text. The *lacRA* intergenic region is enlarged to show the relative positions of the two operators *O1* and *O2*, *CRE*, the *lacR* promoter (P_R), the *lacABCDFEGX* promoter (P_A), and a possible stem-loop structure (lollypop). Stilt-triangles represent (putative) transcription terminators. Adapted from van Rooyen (1993).

transcript with partial readthrough explaining the presence of the 8.5-kb mRNA (de Vos & Gasson, 1989; de Vos et al., 1990). *lacR* is transcribed as a monocistronic mRNA of 1.2 kb in the presence of glucose, while synthesis of this mRNA is repressed 5-fold during growth on lactose. The intercistronic non-coding region between *lac* and *lacR* contains the promoters for both the *lac* operon and *lacR*. *lac* operon mRNA initiates at a G residue 94 bp upstream of the AUG start codon of *lacA*, both in the presence of glucose or lactose, although 5 to 10 times more of the transcripts was present in lactose-grown cells. The relevance of low levels of constitutive transcription remains to be elucidated (van Rooijen et al., 1992). The transcription start site (TSS) of *lacR* was located at an A residue 305 bases upstream of the AUG start codon both in the presence of lactose or glucose. Both promoters conform to the lactococcal promoter consensus. The activity of the *lac* promoter was enhanced 5- to 38-fold by immediate down- and upstream sequences. A possible stem-loop structure in the 5' non-coding region of the *lac* transcript was postulated to be involved in mRNA stability. The stimulating region upstream of *lac* TSS may either be involved in DNA bending or could be the target for a transcription activating protein (van Rooijen et al., 1992).

Transcriptional control of the *lac* operon is provided by LacR (van Rooijen & de Vos, 1990; van Rooijen et al., 1992). The protein specifically represses the lactose catabolic genes of *L. lactis*: when multiple copies of *lacR* were present in cells grown on lactose reduced growth rates were observed, whereas no effect was seen during growth on glucose. When the *lac* promoter was fused to the reporter gene *cat*–86 for chloramphenicol acetyl transferase (CAT), reduced CAT activities were measured when *lacR* was present both *in cis* and *in trans*. In a strain carrying a chromosomal copy of *lacR* and multiple copies of the *lac* promoter on a plasmid, derepression of the chromosomal *lac* operon was observed as a result of titration of LacR by the *lac* promoter. Conversely, super-repression and reduced LacG activities were seen during growth on glucose and lactose, respectively, in a strain carrying a single copy of the *lac* promoter on the chromosome and multiple copies of *lacR*. DNAse I footprinting and DNA homology searches identified two operators in the *lac* promoter region (van Rooijen & de Vos, 1990; see Figure 1). LacR protected the sequences from −31 to +6 (*lacO1*), and −313 to −279 (*lacO2*) relative to the *lac* operon TSS. A TGTTT motif present in both operators was postulated to constitute the LacR recognition sequence. In *lacO1*, but not in *lacO2*, the motif is part of an inverted repeat that shows homology with the *deoO1* operator (van Rooijen & de Vos, 1990, 1992). From gel mobility shift assays it was concluded that *lacO1* has a higher affinity for LacR than *lacO2*. Both operators are needed to repress transcription initiation of *lac* during growth on glucose, indicating that LacR binds to both *lacO1* and *lacO2*. The latter might be involved in the formation of a repression loop, such as has been proposed for the control of several *E. coli* operons (van Rooijen & de Vos, 1990; Matthews, 1992). The gel mobility shift assay was also used to identify the inducer of *lac* operon expression: of the phosphorylated intermediates formed during growth on lactose and galactose (both derepress *lac* operon expression [LeBlanc et al., 1979; Park & McKay, 1982]) only tagatose–6-phosphate inhibited formation of the LacR-operator complex.

A sequence 5 bp downstream of the *lacO1* operator shows strong homology to CRE, a catabolite-responsive element postulated to be involved in catabolite repression in *Bacillus* species (van Rooyen, 1993; see below for a more detailed explanation). Whether this sequence is the binding site for a lactococcal equivalent of the catabolite control protein CcpA and whether *lac* expression is under catabolite control is currently under investigation. A gene for a CcpA-like protein has recently been cloned using antibodies against *B. megaterium* CcpA (E.J. Luesink & O.P. Kuipers, pers. comm.).

LacR is a 29-kDa protein with a high overall homology to repressors of the DeoR family (van Rooijen & de Vos, 1990). The homology was most pronounced with the helix-turn-helix (HTH) region in DeoR, the repressor of the *E. coli* deoxyribose operon. This region in DeoR is thought to bind to the operator of the *deo* operon (Lehming et al., 1988). The HTH motif was mutagenized to identify amino acid residues in LacR involved in operator binding. The methionine and arginine residues at positions 34 and 38, respectively, were shown to be directly involved in DNA binding. These studies also confirmed what had been shown before *in vitro*, namely that LacR is most probably active as a dimer or multimer *in vivo* (van Rooijen & de Vos, 1993; van Rooijen, 1993). Another region of high homology is present between amino acid residues 212 and 222 of LacR. This region also showed homology with enzymes of pro- and eukayrotic organisms involved in sugar or nucleoside metabolism. The postulate that the latter region in LacR might, thus, be involved in the response to tagatose–6-phosphate was substantiated by site-directed mutagenesis experiments (van Rooijen & de Vos, 1990; van Rooijen et al., 1993). Replacement of the lysine residues at positions 72, 80, and 213, as well as the aspartic acid at position 210, by alanine led to repressed LacG activities during growth on lactose. The mutated LacR proteins did bind normally to the *lac* operators but they did not dissociate from the operators in the presence of tagatose–6- phosphate. Whether this behaviour of LacR has to be attributed to a decreased affinity for the inducer or to the inability to undergo a possibly essential conformational change upon inducer binding still has to be sorted out.

All of the above results were unified in a three-stage model for regulation of *lac* operon expression in *L. lactis* (van Rooijen, 1993):

First, binding of LacR repressor to operator *lacO1* during growth on glucose results in autoactivation of *lacR* expression. Although not proven directly, it is very likely that LacR and *lacO1* are both involved in the activation of transcription of *lacR*. Repression of *lac* operon expression probably does not occur at this stage in the process. Second, at increasing concentrations of LacR during growth on glucose, *lacO2* will be bound by repressor. Binding of LacR to both operators results in repression of transcription of the *lac* operon, which might involve the formation of a DNA loop between *lacO1* and *lacO2*. Activation of *lacR* expression by *lacO1* and repression of *lacR* expression by binding of LacR to *lacO2* leads to a certain steady state concentration of the repressor. The third step in the process is taken upon binding of the inducer tagatose–6-phosphate to LacR. When this happens, the LacR- *lacO1lacO2* complex dissociates resulting in the initiation of transcription of the *lac* operon.

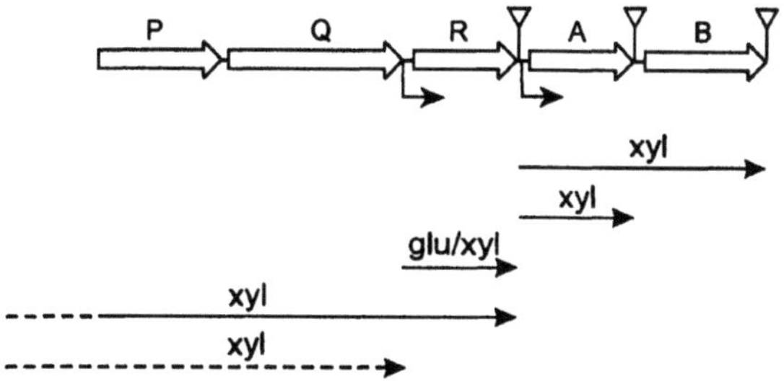

Figure 2. The xylose operon of *Lb. pentosus*. Below the operon the sizes and positions of the mRNAs are given. Xyl: produced in the presence of xylose; glu/xyl: mRNA present with glucose or xylose in the growth medium. Bent arrows show the position and direction of promoters of which the TSSs have been determined. Stilt-triangles: (putative) terminators. Adapted from Lokman et al. (1991, 1994).

Xylose metabolism

Pentoses are usually fermented by heterofermentative LAB. These sugars are internalized by specific permeases and converted to the central intermediate D-xylulose–5-phosphate which is subsequently fermented to lactate and acetate (Kandler, 1983). *Lactobacillus pentosus* is one of the very few *Lactobacillus* species able to use the pentose D-xylose as an energy source. Once intracellular, D-xylose is substrate for D-xylose isomerase and D-xylose kinase. It has been shown already a long time ago that D-xylose isomerase of *Lb. pentosus* is induced during growth in xylose-containing media (Mitsuhashi & Lampen, 1953). Lokman & coworkers (1991, 1994) have recently cloned and analyzed the genetic elements involved in xylose metabolism in *Lb. pentosus* MD353. Five genes in the order *xylPQRAB* were identified (see Figure 2). The putative function of the *xylP* and *xylQ* genes will be discussed below. *xylR*, *xylA* and *xylB* encode a repressor protein, D-xylose isomerase, and D-xylose kinase, respectively.

Northern blot analyses revealed that two xylose-inducible transcripts are produced from *xylA* (Lokman et al., 1994; see Figure 2). The major transcript of 1.5 kb only encompasses *xylA*, while a minor mRNA of 3 kb is produced from *xylA* and *xylB*. Neither transcript is found when cells are grown on glucose. Ribose and arabinose also repress transcription of *xylAB*, which is indicative of the operation of a general catabolite repression mechanism. *xylR* is transcribed as a 1.2-kb mRNA both in the presence of glucose and xylose. In addition, a transcript of more than 7 kb encompassing *xylR*, *xylQ* and *xylP* is synthesized during growth with xylose, but not with glucose. At least ten-fold more *xylR* mRNA, nearly all part of this polycistronic messenger, is present under inducing conditions. The long mRNA is initiated at a xylose-inducible promoter 140 bp upstream of *xylP* (S. Chaillou & P. Pouwels, pers. comm.). Xylose-specific transcripts of 5 and 2.4 kb also reacted with *xylP*- and *xylQ*-specific probes.

The promoters and TSSs of *xylR* and *xylAB* have been determined using mRNA isolated from xylose-grown cells (Lokman et al., 1994). The *xylAB* TSS is located 42 bp upstream of the AUG start codon of *xylA*, while that for *xylR* is 83 bp away from the GUG start codon. Fusions of both promoters with the *cat*–86 reporter gene were made and CAT activities were measured during growth of cells with xylose or glucose. The *xylAB* promoter was induced 60- to 80-fold in the presence of xylose. It was repressed 15 to 25 times when glucose was added to the xylose- containing medium. *cat*–86 gene expression was independent of the sugar source (glucose and/or xylose) when it was controlled by the *xylR* promoter. Expression from the *xylR* promoter was 10- fold less efficient than from the *xylAB* promoter. No promoter activity was detected with a DNA fragment encompassing the *xylA*- *xylB* intergenic region, which is in agreement with the fact that no mRNA was found to be initiated in this region.

The 43-kDa XylR repressor of *Lb. pentosus* is similar to *B. subtilis* XylR, a repressor involved in regulation of D-xylose gene expression (Lokman et al., 1991). In the N-terminus of *Lb. pentosus* XylR, from amino acid residues 30 to 50, an HTH motif was identified. Deletion of *xylR* results in constitutive expression of *xylAB*, indicating that XylR is a repressor (C. Lokman & P. Pouwels, pers. comm.). A region immediately upstream of the *xylA* Shine-Dalgarno sequence shows (limited) homology with the *B. subtilis xyl* operator, and may bind XylR. The *xylR* mutant showed an increased lag-phase in xylose-containing medium, suggesting that under these conditions XylR is an activator.

A second region in the *xylRA* intergenic region, located 68 bp upstream of the *xylA* start codon and overlapping the –35 sequence of the *xylAB* promoter (element II, Lokman et al., 1994), shows homology with CRE, the consensus sequence involved in catabolite (glucose) repression (Weickert & Chambliss, 1990; for reviews, see Fisher & Sohnenshein, 1991; Hueck & Hillen, 1995). When the *xylRA* intergenic region was introduced in *Lb. pentosus* on a plasmid, 5 times more *xylA* transcript was found in the presence of xylose as compared to the wild-type strain. This is suggestive of the titration of a *trans*-acting negative factor. It also suggests that in the wild-type strain, even under inducing conditions, a repression factor is bound to specific sequences in the *xylRA* intergenic region. Catabolite repression in various species of Gram-positive bacteria is affected by the *trans*-acting factors CcpA (the catabolite control protein) and Hpr (the heat-stable protein involved in PEP-PTS phosphate transfer). CcpA is a member of the GalR/LacI family of transcriptional regulators and, probably, specifically interacts with CRE (Hueck & Hillen, 1995). Using a PCR strategy, the gene specifying *Lb. pentosus* CcpA was isolated and sequenced. In an *Lb. pentosus ccpA* disruption mutant, glucose repression was almost completely eliminated (C. Lokman & P. Pouwels, pers. comm.), supporting the idea of a globally active mechanism of catabolite repression in Gram-positive bacteria (Hueck & Hillen, 1995).

As to the function of XylP and XylQ, the following has recently been observed (S. Chaillou & P. Pouwels, pers. comm.). Mutants with disrupted *xylQ* or *xylPQ* genes grow, despite a longer lag- phase, better than a wild-type strain in the presence of xylose. Whereas *xylAB* and *xylPQ* mRNA levels are low during lag-phase in the wild-type strain, and strongly increased once exponential growth commences, both disruption mutants showed higher expression levels during the lag-phase and sub-maximal transcriptional activity during exponential growth. Also, D-xylose uptake was shown to be increased in the *xylPQ* mutant. XylQ is thought to be required for high-level expression of the *xyl* genes. Although the protein shows homology with membrane transport proteins, XylP is not considered to be the actual xylose permease but, together with XylQ, is thought to be involved in regulation of xylose transport. In support of this hypothesis is the fact that XylQ shows similarities to domains conserved in two-component regulator proteins and, more specifically, with class III transcriptional activators (S. Chaillou & P. Pouwels, pers. comm.).

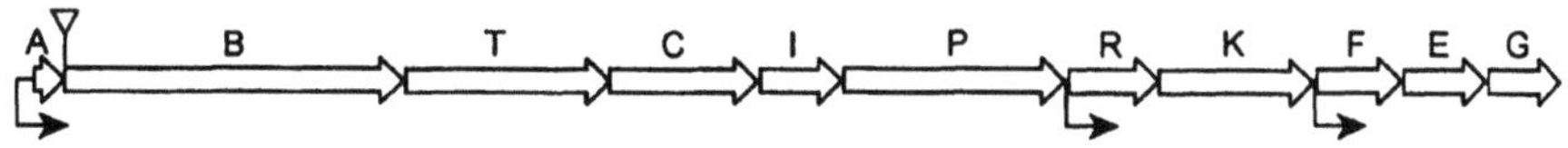

Figure 3. The nisin A regulon of *L. lactis*. The function of the gene products is given in the text. For explanation of symbols, see Figure 2. Compiled from literature mentioned in the text and personal communications of O.P. Kuipers & P. Saris.

Nisin production

Nisin is a ribosomally synthesized antimicrobial peptide which undergoes extensive posttranslational modification leading, ultimately, to the mature, extracellular, membrane-active bacteriocin used so widely in the food industry. A cluster of eleven genes in *L. lactis*, *nisABTCIPRKFEG*, is involved in all aspects of biosynthesis of this lantibiotic (see Figure 3). *nisA* is the structural gene, *nisBC* are postulated to be involved in the modification reactions, *nisT* and *nisP* translocate and cleave precursor nisin, respectively, while *nisI* and *nisFEG* have been implicated, by different mechanisms, in immunity towards nisin. Regulation of nisin biosynthesis is provided by the proteins encoded by *nisR* and *nisK*. These proteins belong to the class of two-component regulatory systems in which, in this case, NisR is the response regulator and NisK the signal sensor or sensor histidine kinase (for excellent reviews, see Schnell et al., 1988; Jung 1991; Sahl et al., 1995; de Vos et al., 1995).

Three promoters, upstream of *nisA*, *nisR* and *nisF*, have been mapped by primer extension (Kuipers et al., 1993; Kuipers et al., 1995; de Ruyter et al., 1996; P. Saris, pers. comm.). From the *nisA* gene a 260-bp transcript is produced which is initiated at a G residue 42 bp upstream of the AUG start codon (Kuipers et al., 1993). At a proper distance from this TSS a canonical *L. lactis* promoter sequence was identified. Interestingly, *nisA* transcription was completely abolished when a 4-bp internal deletion was made in the chromosomally located *nisA* gene. When a plasmid carrying the intact *nisA* gene was introduced in the Δ*nisA* strain, a transcript of Δ*nisA* was observed. Apparently, nisin or one of its precursors is required for transcription of its own gene. In a follow-up study (Kuipers et al., 1995), this phenomenon was further analyzed. Northern blotting using mRNA from the Δ*nisA* strain revealed that, indeed, Δ*nisA* transcription was restored upon addition of nisin to the culture medium. The amount of transcript was proportional to the amount of nisin added. Nisin Z, a natural His27Asn variant of nisin A (Mulders et al., 1991), and several of its mutants were also able to induce transcription. Transcriptional activation varied several 100-fold depending on the actual mutation, with the Dhb2Dha and Met17Trp mutants of nisin Z being more potent inducers than nisin Z itself. Related peptides like the lantibiotics subtilin, lacticin 481, and Pep5, as well as the unmodified synthetic precursor of nisin A did not induce transcription. By fusing a *nisA* promoter fragment to the promoterless *E. coli* reporter gene *gusA* for β-glucuronidase, induction capacities could be quantitated and it was established that less than 5 molecules of the best inducer (nisin Z mutant Dhb2Dha) are sufficient to activate Δ*nisA* transcription. Induction capacity and antimicrobial potency are two different, independent characteristics of the nisin molecule. Synthetic nisin A fragments were used to show that induction capacity resided in the first 11 residues, comprising the first two ring structures, of nisin A. Moreover, a hampered biosynthesis (modification) of nisin interferes with *nisA* transcription: an in-frame deletion of *nisB*, one of the putative modification genes, completely abolished *nisA* transcription, which could be restored by the extracellular addition of nisin (Kuipers et al., 1996).

Inactivation of the chromosomal copy of *nisK* led to the inability of any of the nisin variants added externally to induce transcription of Δ*nisA* (Kuipers et al., 1995). Also, introduction in the *nisK* deletion strain of a plasmid with an intact nisin structural gene did not lead to nisin production, whereas such a plasmid did restore bacteriocin production in the Δ*nisA* strain (which carries an intact copy of *nisK*). These results indicate that NisK is an essential component in the signal transduction pathway and, most probably, directly interacts with the nisin molecule. Deletion studies have shown that *nisR* is also essential for the production of nisin (van der Meer et al., 1993). One of the strains used in this study carried the *nisABTCIR* genes on a multicopy plasmid and was shown to secrete fully modified precursor nisin. This result indicates, among other things, that overexpression of *nisR* alone is enough for activation of transcription of *nisA* and the downstream genes, the latter through partial read-through of an inverted repeat in the *nisAB* intergenic region (see Figure 3). When a DNA fragment carrying the DNA region from the *nisA* promoter down to the 5'-part of *nisB* (including Δ*nisA*) was fused to *gusA*, β-glucuronidase activity was only detected after induction with nisin. The level of activity, however, was 50-fold lower than that observed when *gusA* was fused directly to the *nisA* promoter. Deletion of the *nisA* promoter completely abolished GusA activity, both in the presence and absence of nisin, showing that *nisB* and probably also *nisTCIP* expression depend on read- through from the *nisA* promoter and are under nisin control. When the *nisA* promoter fragment was present in multicopies next to a single chromosomal copy of the *nis* operon, 50-fold less nisin was produced and immunity was severely reduced. Probably, NisR is titrated by a putative NisR binding site in the *nisA* promoter region (Kuipers et al., 1995). The *nisRK* genes have their own promoter. The TSS was localized to an A residue 26 nucleotides upstream of the start codon of *nisR*. *nisRK* are sufficient for signal transduction: integration of only these two genes into the chromosome of *L. lactis* results in a nisin-inducible strain.

Recently, it has been shown that a promoter upstream of *nisF* shows homology to the *nisA* promoter sequence. This suggests that both promoters are under the same mechanism of control. Indeed, externally added nisin increases nisin immunity (O. Kuipers et al., 1995; de Ruyter et al., 1996; P. Saris, pers. comm.).

Proteolysis

Proteolysis by lactic acid bacteria has received much attention through the years. After the identification and characterization of some 20 genes involved in this trait, the next challenge in this field is the elucidation of the interrelationship, if any, between all these genes. In a number of cases, genes for putative regulator proteins have been identified in the vicinity of genes involved in casein utilization. Immediately downstream of the gene for the general aminopeptidase PepC of *L. lactis* MG1363 two genes are present, the products of which show similarity to LysR- and MerR-type regulators (I. Mierau, pers. comm.). Upstream of the *opp* operon encoding the oligopeptide transport system of the same strain a possible regulator gene was identified encoding a protein of the FNR/CRP family of regulators (A. de Jong, S. Tynkkynen, & J. Kok, unpublished). In the immediate vicinity of the *Lb. delbrueckii* ssp. *lactis* peptidase genes *pepI* and *pepQ* two genes for putative regulators were identified (J.R. Klein and B. Henrich, pers. comm.). In all of these cases, actual involvement of the genes in regulation has still to be proven.

Some strains of *L. lactis* produce more proteinase (PrtP) in a milk-based medium than in rich broth, while strain-specific differences in proteinase production have also been observed (Hugenholtz et al., 1984; Exterkate, 1985; Bruinenberg et al., 1992). In these studies, regulation was examined at the enzyme level with the pitfall that active enzyme production could be influenced by unrelated factors in proteinase secretion, processing and/or stability. Indeed, growth rate-dependent autoproteolysis of PrtP has recently been observed, with a decrease in specific growth rate leading to a sharp increase in PrtP breakdown (Meijer et al., 1996). Marugg et al. (1995) have made transcriptional fusions of the *prtP* and *prtM* promoters with the *gusA* gene in order to study transcriptional control of proteinase gene expression. The *prtP* and *prtM* genes are divergently orientated and their promoters are partially overlapping in a face-to-face fashion (see Figure 4). Both promoters were shown to be regulated by the peptide content of the medium. GusA activities under control of either promoter decreased approximately 10- fold with increasing peptide concentrations. Minimal expression was seen in a rich broth medium. These results were matched by the outcome of quantitative primer extension analyses: highest *prtP*- and *prtM*-specific mRNA levels were observed in media containing low amounts of peptides, with an approximately 8-fold decrease upon growth in a medium with a high peptide concentration (J. Marugg, pers. comm.). These data show that medium-dependent expression of the *prtP* and *prtM* promoters is controlled at the level of transcription initiation.

Peptide-dependent regulation of *prt* was further examined by adding specific (di- and tri)peptides to the growth medium of a strain carrying the *prtP-gusA* fusion (Marugg et al., 1995). Of 11 peptides tested only leucylproline and prolylleucine negatively affected GusA activity. Repression was a very rapid and transient process: addition of prolylleucine to a steady state chemostat culture led to an immediate halt in GusA production which resumed after a number of hours. Like a wild-type strain, a strain of *L. lactis* in which the di- /tripeptide transporter DtpT was mutated produced high levels of GusA from the *prtP-gusA* fusion under low-peptide conditions. However, GusA levels were not as low as in the wild-type under high-peptide conditions. This effect was even more extreme in a *dtpT alaT* double mutant in which GusA levels

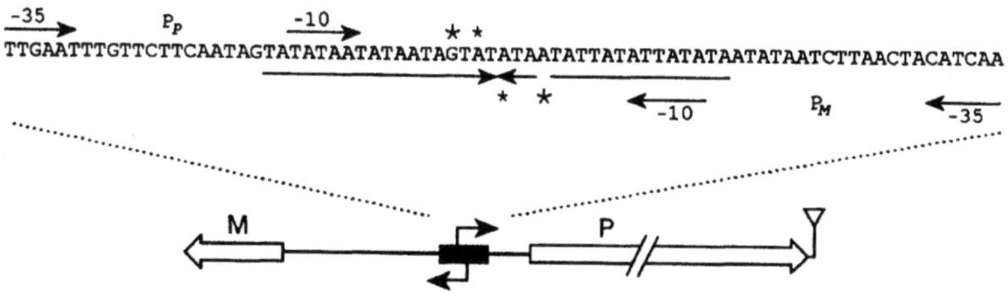

Figure 4. Organization of the *prtP* and *prtM* genes of *L. lactis*. Right and left turn arrows represent the *prtP* and *prtM* promoters, P_P and P_M, respectively. The −10 and −35 sequences of both promoters, as well as their TSSs (asterisks), are indicated in the nucleotide sequence at the top. The major TSSs are shown by the large asterisks. A region of dyad symmetry overlapping the transcription initiation sites of both promoters is also indicated.

were approximately four- to five- fold derepressed in a rich broth medium. A mutant in Opp, the oligopeptide transport system, behaved like the wild-type under all growth conditions tested. Apparently, uptake of small (di/tri)peptides, and possibly amino acids, plays an important role in the control of *prtP* promoter activity.

Deletion and mutation analyses of the *prt* promoter region revealed that a sequence of 90 bp containing both *prt* promoters is sufficient for their full expression and regulation (J. Marugg, pers. comm.). A region of dyad symmetry positioned around the transcription initiation sites of the two promoters is present in all *prtPM* determinants sequenced so far and was shown to be involved in *prt* regulation. Removal of half of the dyad repeat resulted in a nearly constitutive expression from the *prtP* promoter. Insertion of small linkers disrupting the palindrome led to increased mRNA levels and derepressed GusA activities at high peptide concentrations, while expression of both promoters at low peptide concentrations was hardly affected.

The following model was proposed to accommodate all the above results (Marugg et al., 1995; Meijer et al., 1996). A repressor protein is postulated to bind to the *cis* sequence of dyad symmetry in the *prt* promoter region (operator). Interaction of effector molecule(s) with this putative repressor would increase its affinity for the operator, leading to repression of transcription (J. Marugg, pers. comm.). Specific dipeptides (such as prolylleucine) or their derivatives may constitute the effector molecules. Because of the high intracellular peptidase levels in *L. lactis*, the possibility that the cell senses a temporary increase in intracellular peptide concentration is favoured.

Meijer et al. (1996) studied the activities of PrtP and two intracellular peptidases, the X-prolyl-dipeptidyl aminopeptidase PepXP and the general aminopeptidase PepN, under various growth conditions and in two different strains of *L. lactis*. In both strains, specific PrtP activities were highest at the end of the exponential phase of cells growing in milk. In a milk-based medium, specific PrtP activities decreased when the peptide content was increased. In contrast, a difference in the activity levels of PepXP and PepN was observed between the two strains. While PepXP and PepN activity levels showed a medium dependency similar to taht of PrtP in strain MG1363, PepN seemed to be hardly affected and PepXP only slightly so in strain SK1128. PepN and PepXP activity levels in MG1363 were repressed by the addition of 0.5 mM of the dipeptide prolylleucine. In glucose- limited continuous culture experiments, addition of peptides resulted in an immediate decline in the level of PepN activity in MG1363 and, to a much lesser extent, of that of PepXP. Prolylleucine had a similar repressive effect on PepN, whereas PepXP seemed to be insensitive to the addition of the dipeptide.

Amino acid biosynthesis

Amino acid biosynthesis has been studied in many organisms and these studies have revealed a wealth of information regarding gene organization and gene regulation. Recently, the operons for three amino acid biosynthetic pathways from *L. lactis* have been cloned, sequenced, and analyzed (Bardowski et al., 1992; Delorme et al., 1992, 1993; Godon et al., 1992, 1993). Tryptophan biosynthesis has been studied in a dairy strain, while the histidine and branched chain amino acid pathways were elucidated in a prototrophic non-dairy strain of *L. lactis*. These studies have given insight in why dairy lactococcal strains are more fastidious than *L. lactis* strains isolated from plant material (Godon et al., 1993; Delorme et al., 1993), and have

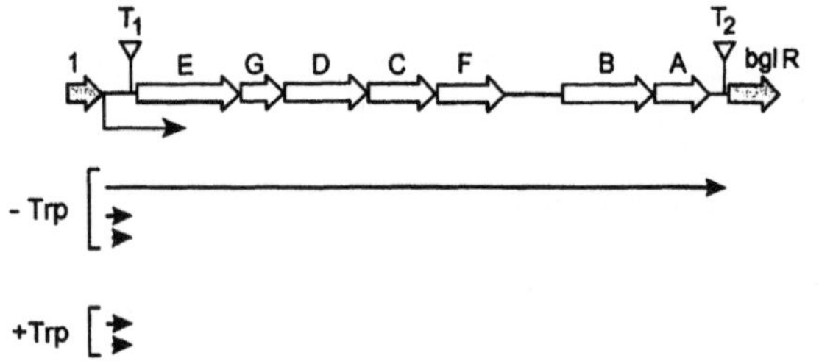

Figure 5. The tryptophanes biosynthesis operon of *L. lactis*. Transcription terminators T1 and T2 are indicated as well as the mRNAs produced in media with or without Trp. An ORF of unknown function (1) precedes the operon while it is followed by *bglR*, encoding a regulator involved in β-glucoside utilization (see text). Adapted from Chopin (1993).

also provided data on the regulation of these operons, as will be discussed now.

The tryptopan biosynthesis operon of *L. lactis* IL1403 contains seven structural genes in the order *trpEGDCFBA* (Bardowski et al., 1992; see Figure 5). All genes are preceded by proper ribosome binding sites, while a consensus vegetative promoter sequence is located upstream of *trpE*. The *trp* mRNA contains a 439-bp non- coding leader region. Three transcripts of 8000, 290, and 160 bases are initiated from the *trp* promoter. The 8-kb transcript encompasses the entire *trp* operon. The transcripts are produced by termination at a specific terminator (T2 for the long mRNA, T1 for the two short products, see Figure 5) and by site-specific endoribonucleolytic activity for the 160-b transcript. Several factors affect production of these transcripts: (1) the 8-kb transcript is only produced in the absence of tryptophan; (2) energy limitation of the cells prevents the production of the 8-kb and the 290-b mRNAs, even in the absence of Trp; (3) transfer of cells to fresh medium strongly stimulates transcription initiation at the *trp* promoter (A. Chopin & R.R. Raya, pers. comm.).

The lactococcal *trp* leader sequence displays sequence and predicted secondary structure conservation with leader regions of aminoacyl-tRNA synthetase genes and some amino acid biosynthesis operons in a number of Gram-positive bacteria. Upon starvation of bacterial cells for a certain amino acid, the cognate tRNA synthetase is generally induced. This response is specific: induction is only seen after limitation of the corresponding amino acid and not after general amino acid starvation (Nass & Neidhardt, 1967). A model has recently been proposed for the regulation of tRNA synthetase genes in *B. subtilis* (Grundy & Henkin, 1993, 1994). In this model, an antiterminator structure in the tRNA synthetase leader sequence involving a conserved 14-bp 'T-box' sequence is thought to be stabilized by direct interaction with the corresponding non-charged tRNA, resulting in transcription of the synthetase gene. Interaction is through the so-called 'specifier codon', a sequence present in a conserved part of the leader which determines the specificity of the response to amino acid limitation, and through a sequence that is complementary to sequences in the acceptor arm of tRNA. Several aspects of this model can be applied to lactococcal *trp* operon control but it does not fully explain all intricacies of *trp* operon expression (M. van de Guchte & A. Chopin, pers. comm.). A transcription terminator and possible antiterminator can be identified in the *trp* leader. Limited homology to TRAP (formerly MtrB protein) binding sequences of *B. subtilis* is also present, but it is not located in the antiterminator structure. The 8-kb *trp* mRNA is induced 50-fold under Trp limitation. This was shown to be due to transcription attenuation. Spontaneous *cis*-mutations resulting in constitutive *trp* operon expression were almost exclusively located in the terminator structure upstream of the structural genes (H. Frenkiel, pers. comm.). $tRNA^{trp}$ appears to play a key role in the sensing of Trp levels in the cell and, under Trp limitation, the uncharged form of $tRNA^{trp}$ is thought to bind to the non-translated leader transcript. This binding would stabilize the antiterminator structure, allowing transcription to proceed past T1 and over the entire length of the *trp* operon. Two elements are important in the mRNA-tRNA interaction, namely the specifier codon and a sequence in the antiterminator that shows complementarity with the acceptor arm of $tRNA^{trp}$. The specifier codon largely determines the efficiency and specificity of the response. Whereas the codon-anticodon interaction is not strictly indispensible, that involving the tRNA acceptor arm seems to be essential. In addition to these RNA interactions, results have been obtained that suggest the involvement of additional factor(s) (M. van de Guchte, pers. comm.).

The genes for the biosynthesis of the branched chain amino acids leucine, isoleucine, and valine (BCAA) of *L. lactis* NCDO2118 form a large cluster of two units (*leu* and *ilv* separated by 121 bp) with the gene order *leuABCD-ilvDBNCA* (Godon et al., 1992; see Figure 6). The entire cluster is needed for the biosynthesis of leucine while the *ilv* genes are required for the synthesis of isoleucine and valine. An *ilvE* homologue is absent from the operon indicating

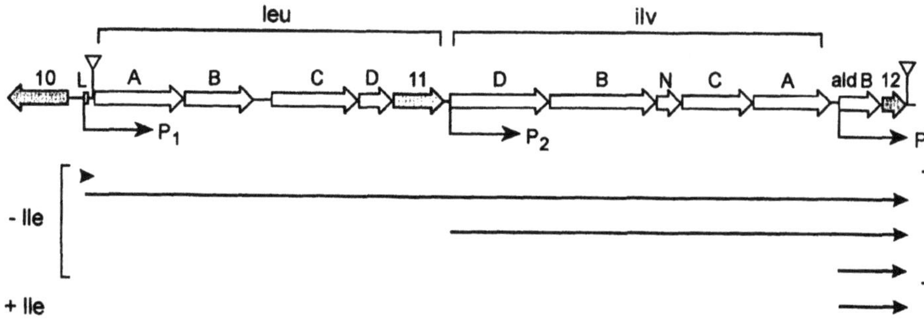

Figure 6. Organization of the two gene clusters of *L. lactis* involved in biosynthesis of branched chain amino acids. Shaded and numbered arrows depict ORFs with unknown function. Note that the *leu-ilv* genes and the *his* gene cluster are adjacent and separated by ORF10 (see Figure 7). The three promoters P1, P2, and P3 are indicated as are two functional terminators (stilt-triangles). *aldB* encodes α-acetolactate decarboxylase. Messenger RNAs detectable under different growth conditions are shown below the gene cluster. Adapted from Chopin (1993).

that the last step in BCAA synthesis (a transamination) is performed by a nonspecific transaminase or is specified elsewhere on the chromosome. Between *leuD* and *ilvD* an ORF is present the product of which shows homology to ATP-binding-cassette (ABC) proteins. The role, if any, of this putative protein in BCAA biosynthesis is unknown. The last gene of the cluster is followed by *aldB*, encoding α-acetolactate decarboxylase and an ORF of unknown function (ORF12).

Transcription analyses and *luxAB* fusion studies have revealed the presence of three functional promoters and two transcription terminators, one between P1 and *leuA* and one downstream of ORF12 (Godon et al., 1993; Renault et al., 1995; P. Renault, pers. comm.; see Figure 6). Promoters P1 (upstream of the entire cluster) and P2 (upstream of *ilvD*) initiate transcription only in the absence of isoleucine, resulting in 14.5- and 7.7-kb transcripts, respectively. Both transcripts terminate downstream of ORF12. Isoleucine represses both promoters approximately 10- to 20-fold and constitutive mutants are presently analysed to identify the putative repressor protein (P. Renault, pers. comm.). Leucine negatively influences transcription from P1. A transcript initiated at P1 could fold in either of two ways. One of these leads to the formation of a rho-independent transcription terminator upstream of *leuA* whereas the other does not. This configuration strongly suggest that the cluster is controlled by transcriptional attenuation (Kolter & Yanofski, 1982). Indeed, a small ORF, *leuL*, present in the leader of the *leu-ilv* transcript specifies a leader peptide of 16 amino acids with three consecutive leucines followed by an isoleucine (Godon et al., 1992). The model for regulation of biosynthesis of BCAA in *L. lactis* is the classical one (Yanofski, 1987): limited BCAA results in limited availability of the cognate charged tRNA. The consequent reduction in translation of the corresponding codons in the coding region for the LeuL leader peptide will cause ribosome stalling on the mRNA. Due to the position of the LeuL coding region, ribosome stalling will prevent the formation of the transcription terminator and the *leu- ilv* operon will be transcribed. In the presence of excess BCAA rapid translation of *leuL* mRNA will allow the terminator structure to be formed and to stop transcription (N. Goupil & P. Renault, pers. comm.).

The position of *aldB* immediately downstream of the *leu-ilv* cluster and the fact that it is cotranscribed with these genes (Godon et al., 1993) suggests a role of α-acetolactate decarboxylase (AldB) in BCAA biosynthesis. Indeed, the enzyme converts the leucine- and valine precursor acetolactate to acetoin and, thus, influences the flux of acetolactate. AldB enzyme activity is subject to allosteric activation by leucine (Phalip et al., 1994). This observation could explain why *L. lactis* can not grow in a leucine-rich medium without valine: acetolactate degradation by AldB leads to valine starvation. *aldB* is transcribed from P3 in the presence of BCAA but transcription is shut off when all three BCAA are lacking. The isoleucine-repressible transcripts initiated at P1 and P2 also cover *aldB*. *aldB* translation is negatively influenced by a strong stem-and-loop structure encompassing the ribosome binding site of *aldB*, sequences immediately upstream of this structure, and translation of the upstream *ilvA* gene. A 500-fold increase in *aldB* mRNA translation (as measured with a translational fusion to *luxAB*) was observed when the upstream region was deleted

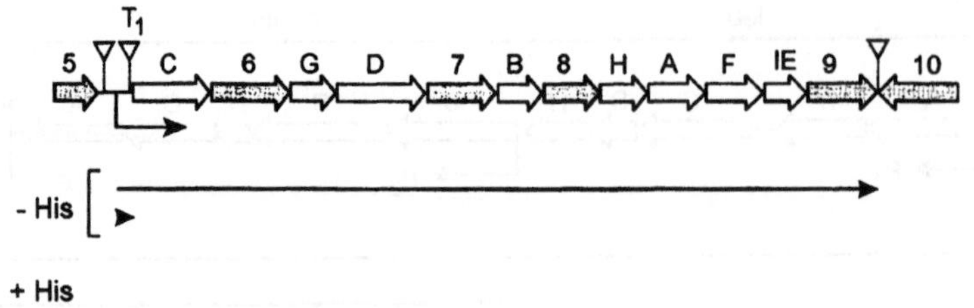

Figure 7. The *L. lactis* histidine biosynthetic operon. ORFs with unknown or putative functions are shaded. Three terminators, one of which (T1) is discussed in the text are depicted. The 10-kb and 250-b mRNAs produced in the absence of His are indicated. Adapted from Chopin (1993).

(Renault et al., 1995; N. Goupil & P. Renault, pers. comm.).

The histidine operon of *L. lactis* NCDO2118 has the gene order *hisCGDBHAFIE*. Interestingly, four additional ORFs are present in what seems to be one transcriptional unit of 10 kb (C. Delorme et al., 1992, 1993; see Figure 7). The function of ORFs 7 and 9 is unknown. The ORF8 product is possibly involved in dephosphorylation of histidinol phosphate, a function that seems to be missing from the lactococcal *hisB* product. The role of ORF6 will be detailed now.

The first gene in the operon, *hisC*, is preceded by a canonical lactococcal promoter. In the absence of His, two transcripts of 10 kb and 250 b are initiated from this promoter (Delorme et al., 1993; Renault et al., 1995). The long mRNA covers the entire *his* operon, while the short one stops at the rho-independent terminator T1 immediately upstream of *hisC* (see Figure 7). His virtually completely abolished synthesis of the 10-kb transcript. Like the *trp* operon leader sequence, the *his* leader does not specify a leader peptide but contains a consensus T-box immediately upstream of T1. A histidine codon at the proper position and all structural features proposed to be involved in T-box-mediated regulation are present (Renault et al., 1995). The role of the putative ORF6 product in regulation of *his* operon expression has recently been examined (Renault et al., 1995). This work was initiated following the observation that the ORF6 protein shows homology with *E. coli* histidinyl-tRNA synthetase, which catalyzes aminoacylation of $tRNA^{his}$ (Delorme et al., 1992). The ORF6 product is about 150 amino acids shorter, lacking one of the conserved motifs required for the activity of class II type tRNA synthetases and will probably not charge tRNA. ORF6 can be disrupted without affecting cell viability. Cells in which the ORF6 protein is overexpressed grow normally in rich broth and in a chemically defined medium with His. They grow slowly in the absence of His and the activity of the *hisD* product, histidinol dehydrogenase, is reduced 4- to 5-fold under these conditions. Overexpression of the ORF6 product decreases the amount of the 10-kb transcript to a level that is below detection even in the absence of His. All these results suggest that the ORF6 product inhibits *his* operon expression by increasing termination at T1. Alternatively, it may inhibit the $tRNA^{his}$-dependent antitermination at this terminator (Renault et al., 1995). As described above for *trp* regulation, T-box regulation involves the stabilization of an antiterminator structure by uncharged tRNA (increased due to the low level of the cognate synthetase or amino acids). The simplest model entertained at the moment for *L. lactis his* operon regulation assumes binding of the ORF6 protein to uncharged $tRNA^{his}$ (increased due to a low His level). This interaction would prevent association of the tRNA with the T-box structure, allowing ternubatir ti be firned and to prevent *his* expression. The level of attenuation control is approximately 10-fold. An additional level of control (approximately 20-fold) is exerted by a repressor of the *his* promoter (C. Delorme & P. Renault, pers. comm.).

β-Glucoside utilization

Next to 'classical' attenuation of the *leu-ilv* operon by ribosome stalling during translation of a leader peptide, and antitermination promoted by uncharged tRNAs (*trp* and *his* operons), a third mechanism controlling transcript elongation exists in *L. lactis*. A gene immediately downstream of the *trp* operon, *bglR* (see Figure 5) encodes a functional regulator of the BglG family (Bardowski et al., 1994). Proteins in this family positively control utilization of different sugars. Transcription antitermination of *E. coli* BglG synthesis is exerted by binding of the protein to a conserved RNA sequence partially overlapping the transcription terminator (Houman et al., 1990). This configuration, a transcription terminator and a 5' overlapping sequence with high similarity to the RNA binding site of these systems was present upstream of *bglR* (Bardowski et al., 1994). *bglR* is positively autoregulated: in a strain carrying a *bglR::lacZ* fusion on the chromosome, constitutive overexpression of *bglR* from a plasmid resulted in 2- to 3-fold increased LacZ levels in the presence of β-glucoside sugars. *L. lactis* BglR can functionally replace *E. coli* BglG. Expression of a *bglR::lacZ* fusion was increased by β- glucosides, while *bglR* disruption mutants are impaired in growth on some β-glucosides. All of these data indicate that BglR is an RNA-binding antiterminator protein (Bardowski et al., 1994).

Bacteriophages

Bacteriophages attacking lactic acid bacteria represent a potential pool of regulatory mechanisms. Recently, the molecular analysis of a number of these phages has been undertaken and nucleotide sequencing has revealed several genes specifying putative regulatory proteins. One of these, *bpi*, was identified in the genome of the temperate lactococcal bacteriophage BK5-T. The product of this gene has been shown to inhibit the activity of a number of BK5-T promoters (Lakshmidevi et al., 1990). The mechanism by which Bpi works is unknown. Another putative repressor encoded by BK5-T is the product of ORF297 (Boyce et al., 1995). A putative repressor has also been described for phage Tuc2009 (van de Guchte et al., 1994). The only phage-encoded protein that has been shown to function as a repressor of gene expression is Rro (*r*epressor *o*f *r*1t) of the *L. lactis* temperate bacteriophage r1t (Nauta et al., 1996).

The regulatory region of bacteriophage r1t consists of two divergently oriented genes, *rro* and *tec* (van Sinderen et al., 1996; Nauta et al., 1996; see Figure 8). The *rro* gene is driven by promoter P_1, while P_2 drives *tec* and downstream genes. The TSSs of both consensus lactococal promoters have been determined. The intergenic region between *rro* and *tec* contains two almost perfectly matching 21-bp direct repeats with internal dyad symmetry. These repeats, O_2 and O_3, overlap with the –35 sequences of P_2 and P_1, respectively, and may function as binding sites (operators) for Rro. A third putative operator (O_1) is located in the coding region of *tec*. Alignment of the operator sequences led

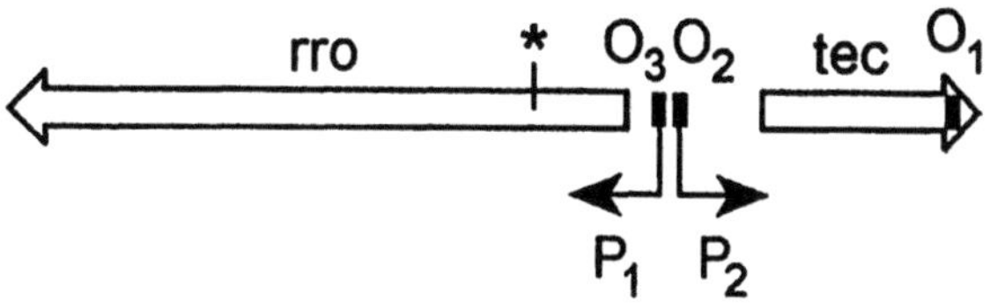

Figure 8. The regulatory region of the temperate bacteriophage r1t of *L. lactis*. The divergent genes *rro* and *tec* are transcribed from the promoters P_1 and P_2, respectively. Operators O_1, O_2, and O_3 are indicated by black boxes. The asterisk shows the position of the frame shift mutation inactivating Rro. Adapted from Nauta et al. (1996).

to the identification of an 11-bp consensus half-site with 7 invariable nucleotides. Rro was overproduced in *E. coli* and shown to specifically bind to a 21-bp synthetic double-stranded O_1 site in gel mobility shift assays (Nauta et al., 1996). It was unable to bind to a mutated copy of O_1 in which one of the invariable residues had been changed. Rro binding to synthetic 21-bp O_2 and O_3 fragments is less efficient (A. Nauta, pers. comm.).

Rro displays similarity with the CI and C2 repressors of the *E. coli* bacteriophages 434 and P22, especially with the regions involved in RecA-mediated autocleavage (Sauer et al., 1982; Little, 1993). Significant homology was also observed with regulator proteins involved in SOS induction. An HTH motif was identified in the extreme N-terminus of Rro by computer analysis. The C-terminal part of Rro is almost identical to the C-termini of the putative repressors CI and the product of ORF297 of the *L. lactis* bacteriophages Tuc2009 and BK5-T (van de Guchte et al., 1994; Boyce et al., 1995). Homology is restricted to the C-terminal halves of the proteins and does not include the HTH region, which is thought to be involved in target DNA recognition. The putative *tec* gene product contains 80 amino acid residues, may contain a HTH region, and is speculated to be the analogue of the *E. coli* Cro protein, hence its name *t*opological *e*quivalent of *C*ro (Nauta et al., 1996). The possible Cro-like proteins specified by ORF63 and ORF5 of BK5-T and Tuc2009 are completely different from Tec (Boyce et al., 1995; van de Guchte et al., 1994; Nauta et al., 1996).

By making a transcriptional fusion of *E. coli lacZ* with the ORF immediately downstream of *tec* it was shown that conditions that normally cause bacteriophage r1t to enter the lytic cycle (e.g. DNA damage caused by mitomycin C) induce expression of *lacZ*. Within three hours after mitomycin C addition to the culture carrying the plasmid in question, LacZ activity had increased 70-fold. When *rro* was inactivated by a frameshift mutation, *lacZ* was expressed constitutively at a high level, suggesting that Rro represses promoter P_2 (Nauta et al., 1996).

In conclusion, it seems that the strategy of developmental control by phage r1t is similar to that used by the lambdoid phages of *E. coli*. Rro most probably binds to the operators and shuts off expression of the lytic genes that are all located downstream of and in the same orientation as *tec*. In fact, this involves 47 of the 50 ORFs recognized in the r1t genome sequence. The genes likely to be required for the establishment and maintenance of lysogeny, namely *rro* and the integrase gene, and an ORF of unknown function, are located together on the opposite DNA strand (van Sinderen et al., 1996). Upon induction, *tec* is probably the first gene to be transcribed. Although DNA binding studies have not yet been performed with Tec, it is anticipated that the protein binds to the operators. Tec would thus prevent transcription of the lysogenic genes and direct the phage into the lytic cycle (Nauta et al., 1996).

Conclusions

Regulation of gene expression in lactic acid bacteria has received much attention over the last few years. As described here, this has resulted in a wealth of new information regarding several important traits of LAB. Both from a fundamental point of view as well as from a more applied angle, studies regarding gene expression and the control mechanisms involved are highly important. From the fundamental viewpoint, in addition to all that has been described already for the two prototype bacteria *E. coli* and *B. subtilis*, we can still learn and discover new principles in gene regulation. Studies on the genetics of amino acid biosynthesis, for instance, have uncovered that gene regulation mechanisms in *L. lactis* can markedly differ from those present in the other bacteria. The *E. coli*, *B. subtilis* and *L. lactis* *trp* operons appear to be regulated by three different mechanisms of termination and antitermination, namely translation of an RNA leader sequence, binding of a regulator protein to a non- translated leader, and binding of an uncharged tRNA to a non-translated leader, respectively. Also, the autoregulation of nisin is the first report of a peptide that induces transcription of its own structural gene via signal transduction. As for the industrial aspects of this work, a better understanding of the regulatory constraints of metabolic pathways would ultimately allow to willingly steer the routes and, thus, the actual fermentation processes in which the organism under study is employed. One example of such intervention is the redirection of α- acetolactate towards diacetyl in *aldB* mutants. These mutants could be positively selected for due to the knowledge of regulation of BCAA biosynthesis. A better knowledge of bacteriophage gene regulation holds the exiting possibility of (near) future targetting of specific stages in the phage life cycle with the aim to confer bacteriophage resistance to the bacterial host.

Whereas until recently genes in LAB could only be expressed via constitutive promoters, a number of the above described systems have been put to use for the expression of genes in a controllable way. The *nis* promoter can be used in a strain carrying the *nisRK* genes to overproduce proteins by the addition to the medium of small amounts of nisin or spent medium of a nisin-producer. As the promoter is tight, it was possible to even produce very toxic proteins that could not be expressed under a constitutive promoter (O. Kuipers, pers. comm.). The *rro*/*tec* system of phage r1t is currently further developed by the isolation of temperature-sensitive derivatives of the repressor. Their potential in temperature-induced gene expression in LAB is presently being examined (A. Nauta, pers. comm.). The salt- inducible promoter isolated from the chromosome of *L. lactis* has been used to induce expression of lytic genes in this organism by salt, with the consequent lysis of the cells and release of intracellular proteins and enzymes (J.W. Sanders, pers. comm). The vector systems are presently being used both for the elucidation of fundamental questions and for more applied purposes. They represent only the first generation of regulatable gene expression systems and refinements of these structures as well as new systems, controllable by other (food-compatible) effectors, are certainly foreseen in the near future.

Acknowledgements

I am endebted to my colleagues for communicating their results prior to publication. I greatly appreciated the valuable discussions with Pierre Renault, Maarten van de Guchte, Peter Pouwels, Oscar Kuipers, Arjen Nauta & Jan Willem Sanders. I thank Henk Mulder for preparing the figures. Part of the work was supported by the Royal Netherlands Academy of Arts and Sciences (KNAW).

References

Auffray Y, X Gansel, B Thammavongs & P Boutibonnes (1992) Heat shock-induced protein synthesis in *Lactococcus lactis* subsp. *lactis*. Curr. Microbiol 24: 281–284

Bardowski J, SD Ehrlich & A Chopin (1994) BglR protein, which belongs to the BglG family of transcriptional antiterminators, is involved in β-glucoside utilization in *Lactococcus lactis*. J. of Bacteriol. 18: 5681–5685

— (1992) Tryptophan biosynthesis genes in *Lactococcus lactis* subsp. *lactis*. J. Bacteriol. 174: 6563–6570

Bissett DL & RL Anderson (1974) Lactose and D-galactose metabolism in group N streptococi: presence of enzymes for both the D-galactose 1-phosphate and D-tagatose 6-phosphate pathways. J. Bacteriol. 117: 318–320

Boyce JD, BE Davidson & AJ Hillier (1995) Identification of prophage genes expressed in lysogens of the *Lactococcus lactis* bacteriophage BK5-T. Appl. Environ Microbiol 61: 4099–4104

Bruinenberg PG, P Vos & WM de Vos (1992) Proteinase overproduction in *Lactococcus lactis* strains: regulation and effect on growth and acidification in milk. Appl. Environ. Microbiol. 58: 78–84

Chopin A (1993) Organization & regulation of genes for amino acid biosynthesis in lactic acid bacteria. FEMS Microbiol. Rev. 12: 21–39

de Ruyter PGGA, OP Kuipers, MM Beerthuyzen, I van Alen-Boerrigter & WM de Vos (1996) Functional analysis of promoters in the nisin gene cluster of *Lactococcus lactis*. J. Bacteriol. 178: 3434–3439

de Vos WM, OP Kuipers, JR van der Meer & RJ Siezen (1995) Maturation pathway of nisin and other lantibiotics: post-translationally modified antimicrobial peptides exported by Gram-positive bacteria. Mol. Microbiol. 17: 427–437

de Vos WM & GFM Simons (1994) Gene cloning and expression systems in Lactococci. In: *Genetics and Biotechnology of Lactic Acid Bacteria*; (M.J. Gasson & W.M. de Vos, Eds.) pp. 52–105 Blackie Academic & Professional, Glasgow, UK

de Vos WM & MJ Gasson (1989) Structure and expression of the *Lactococcus lactis* gene for P-β-gal *(lacG)* in *Escherichia coli* and *L. lactis*. J. Gen. Microbiol. 135: 1833–1846

de Vos WM, I Boerrigter, RJ van Rooijen, B Reiche & W Hengstenberg (1990) Characterization of the lactose-specific enzymes of the phosphotransferase system in *Lactococcus lactis*. J. Biol. Chem. 265: 22554–22560

Delorme C, SD Ehrlich, JJ Godon & P Renault (1993) Gene inactivation in *Lactococcus lactis*. II. Histidine biosynthesis. J. Bacteriol. 175: 4391–4399

Delorme C, SD Ehrlich & P Renault (1992) Histidine biosynthesis genes in *Lactococcus lactis* subsp. *lactis*. J. Bacteriol 174: 6571–6579

Eaton T, C Shearman & M Gasson (1993) Cloning and sequence analysis of the *dnaK* gene region of *Lactococcus lactis* subsp. *lactis*. J. of Gen. Microbiol. 139: 3253–3264

Exterkate FA (1985) A dual-directed control of cell wall proteinase production in *Streptococcus cremoris* AM1: a possible mechanism of regulation during growth in milk. J. Dairy Sci. 68: 562–571

Fisher SH & AL Sohnenshein (1991) Control of carbon and nitrogen metabolism in *Bacillus subtilis*. Annu. Rev. Microbiol. 45: 107–135

Godon JJ, MC Chopin & SD Ehrlich (1992) Branched-chain amino acid biosynthesis genes in *Lactococcus lactis* subsp. *lactis*. J. Bacteriol. 174: 6580–6589

Godon JJ, C Delorme, J Bardowski, MC Chopin, SD Ehrlich & P Renault. (1993) Gene inactivation in *Lactococcus lactis*. I. Branched chain amino acid biosynthesis. J. Bacteriol. 175: 4383–4390

Grundy FJ & TM Henkin (1994) Conservation of a transcription antitermination mechanism in aminoacyl-tRNA synthetase and amino acid biosyntheis genes in Gram-positive bacteria. J. Mol. Biol. 235: 798–804

— (1993) tRNA as a positive regulator of transcription antitermination in *B. subtilis*. Cell 74: 475–482

Hartke A, S Bouche, X Gansel, P Boutibonnes & Y Auffray (1994) Starvation-induced stress resistance in *Lactococcus lactis* subsp. *lactis* IL1403. Appl. Environ. Microbiol. 60: 3474–3478

Hartke A, S Bouche, JM Laplace, A Benachour, P Boutibonnes and Y Auffray (1995) UV-inducible proteins and UV-induced cross- protection against acid, ethanol, H_2O_2 or heat treatments in *Lactococcus lactis* subsp. *lactis*. Arch. Microbiol. 163: 329–336

Houman F, MR Diaz-Torres & A Wright (1990) Transcriptional antitermination in the *bgl* operon of *Escherichia coli* is modulated by a specific RNA-binding protein. Cell 62: 1153–1163

Hueck CJ & W Hillen (1995) Catabolite repression in *Bacillus subtilis*: a global regulatory mechanism for the Gram-positive bacteria? Mol. Microbiol. 15: 395–401

Hugenholtz J, FA Exterkate & WN Konings (1984) The proteolytic systems of *Streptococcus cremoris*: an immunological analysis. Appl. Environ. Microbiol. 48: 1105–1110

Israelsen H & Hansen EB (1993) Insertion of transposon Tn*917* derivatives into the *Lactococcus lactis* subsp. *lactis* chromosome. Appl. Environ. Microbiol. 59: 21–26

Israelsen H, SM Madsen, A Vrang, EB Hansen & E Johansen (1995) Cloning and partial characterization of regulated promoters from *Lactococcus lactis* Tn*917-lacZ* integrants with the new promoter probe vector, pAK80. Appl. Environ. Microbiol. 61: 2540–2547

Jung G (1991) In: *Nisin and Novel Lantibiotics*: *Proceedings of the First International Workshop on Lantibiotics*. (Jung G & Sahl HG Eds). Leiden: Escom Publishers, pp. 1–25

Jung G & HG Sahl (1991) In: *Nisin and Novel Lantibiotics: Proceedings of the First International Workshop on Lantibiotics*, (Jung G & HG Sahl Eds.), Leiden: Escom Publishers, pp. 1–34

Kandler O (1983) Carbohydrate metabolism in lactic acid bacteria. Antonie van Leeuwenhoek 49: 209–224

Kim SG & CA Batt (1993) Cloning and sequencing of the *Lactococcus lactis* subsp. *lactis groESL* operon. Gene 127: 121–126

Kolter R & Yanofski C (1982) Attenuation in amino acid biosynthetic operons. Annu. Rev. Genet. 16: 113–134

Kuipers OP, Beerthuyzen MM, de Ruyter PGGA, Luesink EJ, & de Vos WM (1995) Autoregulation of nisin biosynthesis in *Lactococcus lactis* by signal transduction. J. Biol. Chem. 270: 27299–27304

Kuipers OP, MM Beerthuyzen, RJ Siezen & WM de Vos (1993) Characterization of the nisin gene cluster *nisABTCIPR* of *Lactococcus lactis*. Requirement of expression of the *nisA* and *nisI* genes for development of immunity. Eur. J. Biochem. 216: 281–291

Kunji ERS, T Ubbink, A Matin, B Poolman & WN Konings (1993) Physiological responses of *Lactococcus lactis* ML3 to alternating

conditions of growth and starvation. Arch. Microbiol. 159: 372–379
LeBlanc DJ, Crow VL, Lee LN & Garon CF (1979) Influence of the lactose plasmid on the metabolism of galactose by *Streptococcus lactis*. J. Bacteriol. 137: 878–884
Leenhouts KJ & G Venema (1993) Lactococcal plasmid vectors. In: *Plasmids. A Practical Approach* (K.G. Hardy, Ed.), pp. 65–92 Oxford University Press, New York
Lehming N, J Sartorius, S Oehler, B von Wilken-Bergmann & B Müller-Hill (1988) Recognition helices of *lac* and lambda repressor are oriented in opposite diretion and recognize similar DNA sequences. Proc. Natl. Acad. Sci. USA 85: 7947–7951
Little JW (1993) LexA cleavage and other self-processing reactions. J. Bacteriol. 175: 4943–4950
Lokman BC, P van Santen, JC Verdoes, J Krüse, RJ Leer, M Posno and PH Pouwels (1991) Organization and characterization of three genes involved in D-xylose catabolism in *Lactobacillus pentoses*. Mol. Gen. Genet. 230: 161–169
Lokman BC, RJ Leer, R van Sorge & PH Pouwels (1994) Promotor analysis and transcriptional regulation of *Lactobacillus pentosus* genes involved in xylose catabolism. Mol. Gen. Genet. 245: 117–125
Maeda S & MJ Gasson (1986) Cloning, expression and location of the *Streptococcus lactis* gene for phospho-β-D-galactosidase. J. Gen. Microbiol. 132: 331–340
Marugg JD, W Meijer, R van Kranenburg, P Laverman, PG Bruinenberg and WM de Vos (1995) Medium-dependent regulation of proteinase gene expression in *Lactococcus lactis*: control of transcription initiation by specific dipeptides. J. Bacteriol. 177: 2982–2989
Marugg JD, R van Kranenburg, P Laverman, GAM Rutten, & WM de Vos (1996) Identical transcriptional control of the divergently transcribed *prtP* and *prtM* genes that are required for proteinase production in *Lactococcus lactis* SK11. J. Bacteriol. 178: 1525–1531
Matthews KS (1992) DNA looping. Microbiol. Rev. 56: 123–136
McKay L, A Miller III, WE Sandine & PR Elliker (1970) Mechanisms of latose utilization by lactic streptococci: enzymatic and genetic analyses. J. Bacteriol. 102: 804–809
Meijer W, JD Marugg & J Hugenholtz (1996) Regulation of proteolytic enzyme activity in *Lactococcus lactis*. Appl. Environ. Microbiol. 62: 156–161
Mercenier A, PH Pouwels & BM Chassy (1994) Genetic engineering of lactobacilli, leuconostocs and *Streptococcus thermophilus*. In: *Genetics and Biotechnology of Lactic Acid Bacteria*, (M.J. Gasson & W.M. de Vos, Eds.), pp.254–293. Blackie Academic Professional, Glasgow, UK
Mitsuhashi S & JO Lampen (1953) Conversion of D-xylose to D-xylulose in extracts of *Lactobacillus pentosus*. J. Biol. Chem. 204: 1011–1018
Monnet C, V Phalip, P Schmitt & C Diviès (1994) Comparison of α-acetolactate synthase and α-acetolactate decarboxylase in *Lactococcus* spp. and *Leuconostoc* spp. Biotechnol. Lett. 16: 257–262
Mulders JWM, IJ Boerrigter, HS Rollema, RJ Siezen & WM de Vos (1991) Identification and characterization of the lantibiotic nisin Z, a natural nisin variant. Eur. J. Biochem. 201: 581–584
Nass G & FC Neidhardt (1967) Regulation of formation of aminoacyl-ribonucleic acid synthetases in *Escherichia coli*. Biochim. Biophys. Acta 134: 347–359
Nauta A, D van Sinderen, H Karsens, E Smit, G Venema & J Kok (1996) Inducible gene expression mediated by a repressor-operator system isolated from *Lactococcus lactis* bacteriophage rlt. Mol. Microbiol., in press
Park YH & LL McKay (1982) Distinct galactose phosphoenolpyruvate-dependent phosphotransferase system in *Streptococcus lactis*. J. Bacteriol. 149: 420–425
Phalip V, C Monnet, P Schmitt, P Renault, JJ Godon & C Diviès (1994) Purification and properties of the α- acetolactate decarboxylase from *Lactococcus lactis* subsp. *lactis* NCDO 2118. FEBS Lett. 351: 95–99
Renault P, JJ Godon, N Goupil, C Delorme, G Corthier & SD Ehrlich (1995) Metabolic operons in *Lactococci*. In: *Genetics of Streptococci, Enterococci and Lactococci*, (JJ Ferretti, MS Gilmore, TR Klaenhammer & F Brown, Eds.). Dev Biol Stand. Basel, Karger, 85: 431–440
Sahl HG, RW Jack & G Bierbaum (1995) Eur J Biochem 230: 827–853
Sauer RT, Pan J, Hopper P, Hehir K, Brown J & Poteete AR (1981) Primary structure of the phage P22 repressor and the gene *c2*. Biochemistry 20: 3591–3598
Schnell N, KD Entian, U Schneider, F Götz, H Zähner, R Kellner & G Jung (1988) Nature 333: 276–278
Siezen RJ, OP Kuipers & WM de Vos (1994) A lantibiotics database: Compilation and comparison of latibiotic gene clusters and encoded proteins. Minireview for 'Antonie van Leeuwenhoek' Proceedings of the 2e Workshop on Lantibiotics, Papendal, Arnhem, pp. 1–27
Stock JB, MG Surette, M Levit & P Park (1995) Two-component signal transduction systems: structure-function relationships and mechanisms of catalysis. In: *Two-component signal transduction* (J.A. Hoch & T.J. Silhavy, Eds.) pp. 25–51. Am. Soc. Microbiol. Washington, D.C
Studier FW, AH Rosenberg, JJ Dunn & JW Dubendorf (1990) Use of T7 RNA polymerase to direct expression of cloned genes. Meth. Enzymol. 185: 60–89
van Rooijen RJ, KJ Dechering, C Niek, J Wilmink & WM de Vos (1993) Lysines 72, 80 and 213 and aspartic acid 210 of the *Lactococcus lactis* LacR repressor are involved in the response to the inducer tagatose–6-phosphate leading to induction of *lac* operon expression. Prot. Engin. 6: 201–206
van Sinderen D, HA Karsens, J Kok, P Terpstra, MHJ Ruiters, G Venema & A Nauta (1996) Sequence analysis and molecular characterization of the temperate lactococcal bacteriophage rlt. Mol. Microbiol., in press
van Asseldonk M, A Simons, H Visser, WM de Vos & G Simons (1993) Cloning, nucleotide sequence, and regulatory analysis of the *Lactococcus lactis dnaJ* gene. J. of Bacteriol. 175: 1637–1644
van Rooyen RJ (1993) Characterization of the *Lactococcus lactis* lactose genes and regulation of their expression. Thesis. University of Wageningen, Wageningen, The Netherlands
van Rooijen RJ, S van Schalkwijk & W.M. de Vos (1991) Molecular cloning, characterization, and nucleotide sequence of the tagatose 6-phosphate pathway gene cluster of the lactose operon of *Lactococcus lactis*. J. Biol. Chem. 266: 7176–7181
van Rooijen RJ & WM de Vos (1992) Characterization of the *Lactococcus lactis* lactose operon promoter: contribution of flanking sequences and LacR repressor to promoter activity. J. Bacteriol. 174: 2273–2280
— (1990) Molecular cloning, transcriptional analysis and nucleotide sequence of *lacR*, a gene encoding the repressor of the lactose phosphotransferase system of *Lactococcus lactis*. J. Biol. Chem. 265: 18499–18503
van de Guchte M, J Kok & G Venema (1992) Gene expression in *Lactococcus lactis*. FEMS Microbiol. Rev. 88: 73–92
van der Meer RJ, J Polman, MM Beerthuyzen, RJ Siezen, OP Kuipers & WM de Vos (1993) Characterization of the *Lactococcus lactis* nisin A operon genes *nisP*, encoding a subtilisin-

like serine protease involved in precursor processing, and *nisR*, encoding a regulatory protein involved in nisin biosynthesis. J. Bact. 175: 2578–2588

Weickert MJ & Chambliss GH (1990) Site-directed mutagenesis of a catabolite repression operator sequence in *Bacillus subtilis*. Proc. Natl. Acad. Sci. USA 87: 6238–6242

Wells JM, PW Wilson, PM Norton, MJ Gasson & RWF Le Page (1993) *Lactococcus lactis*: high-level expression of tetanus toxin fragment C and protection against lethal challenge. Mol. Microbiol. 8: 1155–1162

Wells JM, PW Wilson, PM Norton & RWF Le Page (1993) A model system for the investigation of heterologous protein secretion pathways in *Lactococcus lactis*. Appl. Environ. Microbiol. 59: 3954–3959

Yanofski C (1987) Operon-specific control by transcription attenuation. Trends Genet. 3: 356–360

Youngman P (1987) Plasmid vectors for recovering and exploiting Tn*917* transpositions in *Bacillus subtilis* and other Gram- positive bacteria, p. 79–103. In: K. Hardy (ed.), Plasmids: a practical approach. IRL Press, Oxford, UK

Antonie van Leeuwenhoek **70:** 147–159, 1996.

Lytic systems in lactic acid bacteria and their bacteriophages

Michael J. Gasson
Department of Genetics and Microbiology, Institute of Food Research, Norwich Research Park, Colney, Norwich NR4 7UA, UK

Summary

Lytic systems of lactic acid bacteria and their bacteriophages are reviewed with an emphasis on molecular characterization. Details of enzyme biochemisrtry and the cloning and analysis of lytic genes are presented, with coverage of lactococcal prolate headed bacteriophages, lactococcal isometric bacteriophages, *Lactobacillus* bacteriophages and lactococcal autolysins. Some comments on the importance of autolysis in cheese ripening are included and the biotechological exploitation of cloned and characterized lytic genes is presented.

Introduction

In view of the general importance of bacteriophages as an industrial problem in the dairy industry and the likely significance of autolysis in intracellular enzyme release and flavour development, it is surprising that the lytic systems of lactic acid bacteria (LAB) and their bacteriophages have received relatively little in depth analysis. In general for bacteriophages, it is also true that their lysis mechanisms are not fully understood and this situation is especially extreme in the case of Gram-positive bacteria. Outside of the LAB significant work has been undertaken with the lytic systems of *Streptococcus pneumoniae* and *Bacillus subtilis* (Young, 1992) and this provides some background for comparison. Analysis of LAB bacteriophage lysins was pioneered by Reiter in the 1960's (Oram & Reiter, 1965; Reiter & Oram, 1963) but it is only in recent years that significant advances have been made in their molecular characterization, especially with the cloning and analysis of lytic genes. Earlier reviews of the subject area are limited, with Sable & Lortal (1995) dealing specifically with the LAB and Young (1992) providing a comprehensive coverage of bacteriophage lysis in all bacterial species. This review focuses on molecular analysis of the lysin and holin genes from LAB bacteriophages, covers the analysis of autolytic systems and includes the exploitation of cloned lysis genes in LAB.

Lytic systems of prolate headed bacteriophages of *Lactococcus lactis*

Lysins

The first analysis of the lytic capacity of lactococcal bacteriophages involved the partial purification of a lysin from bacteriophage ml3 (ϕvML3 of Shearman et al., 1989)). Oram and Reiter (1965) reported that the ml3 lysin was a muramidase, which like hen egg white lysozyme, cleaved the β1–4 bond between *N*-acetylmuramic acid and *N*-acetylglucosamine in the cell wall peptidoglycan. The lysin was shown to have pH and temperature optima of 6.5–6.8 and 37°C, respectively, and to be activated by both monovalent and divalent cations. The lysin was active against a wide spectrum of lactococci with variable efficiency and although a weak activity against enterococci was noted, other heterologous species that were tested proved to be insensitive. It is particularly noteworthy that the lytic spectrum was broader than that of the intact bacteriophage and this goes some way to explain the enhanced lytic spectrum of bacteriophage in mixed strain starters (Naylor & Cuzak, 1956; Mullan & Crawford, 1985c).

Mullan & Crawford (1985a; 1985b) also reported the purification and biochemical characterization of the lysin from *L. lactis* bacteriophage c2(w) and found it to be very similar to the lysin of bacteriophage ml3, with pH and temperature optima of 6.5–6.9 and 37°C,

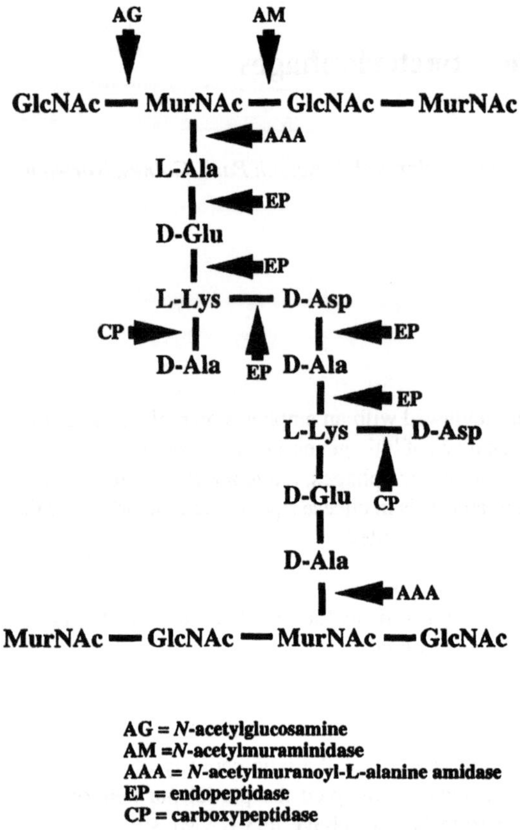

Figure 1. Impression of the lactococcal cell wall with target sites for lytic enzymes. Derived from Crow et al. (1995).

respectively and a broad spectrum of activity against lactococcal strains. Analysis of the cell wall linkage attacked by the lysin revealed the release of reducing sugars but no release of amino groups. The absence of evidence for glucosamine release led these authors to conclude that, like the ϕvML3 lysin, this lysin was an *N*-acetylmuramidase.

Molecular characterization of lactococcal bacteriophage lytic systems began with the cloning and DNA sequence analysis of the lysin gene from *L. lactis* bacteriophage ϕvML3 (Shearman et al., 1989). An *E. coli* bacteriophage lambda gene library of the lactococcal phage was constructed in the vector λgt10 and this was screened by overlaying the lambda plaques with a thick suspension of lysin-sensitive lactococcal cells. Bacteriophage lambda clones which expressed lysin activity were readily identified by their clear appearance, which was caused by their capacity to lyse the overlay of lactococcal cells as is shown in Figure 1. The DNA inserts in such positive clones were analyzed and a representative 1.2kb fragment of bacteriophage ϕvML3 DNA was subcloned onto a plasmid vector in *E. coli* and subject to deletion analysis and sequencing. A region of 832 bp of this fragment, bounded by *Dra*I and *Eco*RI restriction sites, was defined as the minimal DNA giving lysin expression and it was evident that the lysin gene was being expressed by transcription from an upstream bacteriophage promoter. *In vitro* transcription and translation and SDS PAGE was used to identify a 24 kDa protein product and both this and the lytic activity were abolished by a small internal deletion in the cloned DNA. The *E. coli* clone produced a very strong lytic activity that was related to the activity of the commercially available Gram-positive lytic enzyme mutanolysin. The expressed bacteriophage ϕvML3 lysin was calculated as 33,200 mutanolysin equivalent units per mg protein. Crude extracts from the lysin-expressing *E. coli* clone generated lactococcal protoplasts when susceptible cells were digested with osmotic buffering.

Analysis of the DNA sequence revealed the existence of one complete open reading frame that was proven to encode the lysin gene. The translation of the DNA sequence suggested the lysin to be a 187 amino acid protein that would have a predicted molecular weight of 21,090. However, the correction of a sequencing error (Shearman et al., 1994; Young, 1992) changed the carboxy-terminus of the lysin, adding 10 amino acids to the protein and a reinterpretation of the translational start extended the amino-terminal region, as is discussed in detail below. Thus, the corrected open reading frame encodes a protein of 226 amino acids with a predicted molecular weight of 25,352. Significant homology was found between the amino terminal region of the lactococcal lysin and gene 15 of the *B. subtilis* bacteriophages PZA and ϕ29. Originally, the latter gene was thought to play a role in phage morphogenesis rather than in lysis (Camacho et al., 1977) and this led to speculation that the bacteriophage ϕvML3 lysin protein had a dual function. However, it has been shown that gene 15 of these *Bacillus* bacteriophages is a lysin (Garvey et al., 1986; Young, 1992) and reexamination of the homology led to reassignment of the translational start for the bacteriophage ϕvML3 lysin. This was primarily based on the fact that the homology with the *Bacillus* gene was most significant if DNA upstream of the originally assigned translational start was also translated into amino acid sequence (Young, 1992).

```
          151
361   GAACTTTTAAAACAGACTGGACTTGTTCCGCTTGATTGTAAAGATGATAACATGGCTGAC
      E  L  L  K  Q  T  G  L  V  P  L  D  C  K  D  D  N  M  A  D
          DraI

421   GCTTATAACATTTTGACATATTGCGAACACTTGGGTTAGTTGTTCCCTTATAAAAAAACA
      A  Y  N  I  L  T  Y  C  E  H  L  G

                 213
481   ATAATAATAATTGGAGGTGGTAATATAAAAGTATCACAAAACGGTTTGAACTTGATTAAA
             PR 20           I  K  V  S  Q  N  G  L  N  L  I  K
                     PZA 15  M  Q  I  S  Q  A  G  I  N  L  I  K

            224
541   GAGTTTGAGGGTTGTAGGTTGACTGCTTATAAACCTGTACCGTGGGAACAAATGTACACT
      E  F  E  G  C  R  L  T  A  Y  K  P  V  P  W  E  Q  M  Y  T
      S  F  E  G  L  Q  L  K  A  Y  K  A  V  P  T  E  K  H  Y  T

601   ATCGGTTGGGGTCATTATGGAGTGACAGCAGGTACAACATGGACACAAGCTCAAGCAGAT
      I  G  W  G  H  Y  G  V  T  A  G  T  T  W  T  Q  A  Q  A  D
      I  G  Y  G  H  Y  GSDV S  P  R  Q  V  I  T  A  K  Q  A  E

                                          549 lysin-ßgal fusion
661   AGCCAGCTAGAGATTGACATCAATAATAAGTATGCACCTATGGTTGACGCTTACGTAAAA
      S  Q  L  E  I  D  I  N  N  K  Y  A  P  M  V  D  A  Y  V  K
      D  M  L  R  D  D  V  Q  A  F  .  .  .  .  V  D  V  D  G  V
                                                HincII

721   GGCAAAGCAAATCAAAATGAGTTTGACGCCTTAGTTTCATTGGCTTATAACTGTGGTAAT
      G  K  A  N  Q  N  E  F  D  A  L  V  S  L  A  Y  N  C  G  N

781   GTTTTCGTTGCTGACGGTTGGGCGCCTTTCTCACATGCTTATTGTGCTTCAATGATACCG
      V  F  V  A  D  G  W  A  P  F  S  H  A  Y  C  A  S  M  I  P

841   AAGTATCGTAATGCAGGCGGTCAAGTCTTACAAGGCTTAGTAAGACGCAGACAGGCAGAG
      K  Y  R  N  A  G  G  Q  V  L  Q  G  L  V  R  R  R  Q  A  E

901   CTTAACTTATTTAATAAACCAGTATCAAGTAATTCAAACCAAAACAATCAAACAGGAGGA
      L  N  L  F  N  K  P  V  S  S  N  S  N  Q  N  N  Q  T  G  G

961   ATGATAAAAATGTACCTTATTATAGGACTAGATAATTCAGGTAAAGCTAAACATTGGTAT
      M  I  K  M  Y  L  I  I  G  L  D  N  S  G  K  A  K  H  W  Y
      153      156

1021  GTTTCTGACGGTGTAAGTGTTCGTCATGTTCGTACAATTCGTATGTTGGAAAACTATCAA
      V  S  D  G  V  S  V  R  H  V  R  T  I  R  M  L  E  N  Y  Q

1081  AACAAATGGGCTAAACTTAACTTGCCAGTTGATACAATGTTTATTGCAGAAATCGAAGCA
      N  K  W  A  K  L  N  L  P  V  D  T  M  F  I  A  E  I  E  A

1141  GAGTTTGGACGTAAGATTGACATGGCTTCAGGAGAAGTGAAATAGGAGGAAGTGAATGAG
      E  F  G  R  K  I  D  M  A  S  G  E  V  K

1201  GGAATTC EcoRI  1207
```

Figure 2. Analysis of the prolate-headed bacteriophage lysin gene. Data is presented for the lysin gene of *L. lactis* bacteriophage ϕvML3. The homology with gene 15 of *Bacillus subtilis* bacteriophage PZA is indicated. Bold numbers above the sequence indicate the positions of deletion endpoints and a gene fusion. Plasmids pFI151 (151) and pFI213 (213) have a functional open reading frame that will express lysin activity, whereas pFI224 (224) is defective. Translational start codons are underlined and Shine Delgarno sequences are boxed (Shearman et al., 1994).

The revised structure for the bacteriophage ϕvML3 lysin gene was presented by Shearman et al. (1994), following analysis of the homology between the lactococcal and *Bacillus* lysins and a detailed molecular analysis of the function of the cloned DNA region. The start point for analysis was the 832 bp *Dra*I - *Eco*RI restriction fragment which expressed lytic activity. The upstream region of this fragment was subject to progressive BAL31 digestion leading to the construction of a series of deleted derivatives. It was found that all of these deletions resulted in loss of lytic activity, including cases where the originally assigned promoter region and translational start site were intact. In order to ensure that this inactivity was not simply due to the lack of an active promoter, all of the deleted fragments were placed downstream of an *E. coli* T7 promoter but, again, no activity was detected. Thus, it was concluded that the original definition of the ϕvML3 lysin gene was incorrect. The translated amino acid sequence was realigned with that of *B. subtilis* gene 15 as shown in Figure 2. A good ribosome binding site is present but there is no methionine start codon. The DNA sequence in this region was confirmed to be correct and it was concluded that the lactococcal lysin gene has an atypical start involving an isoleucine codon. The activity of this reassigned gene was proven by placing the open reading frame, including the native ribosome binding site, downstream of an *E. coli* T7 promoter. Inducible expression of lysin activity was demonstrated and it was significant that this depended on the integrity of the lactococcal ribosome binding site.

Data for the related prolate headed bacteriophage P001 is particularly relevant to the unusual translational start codon. In this bacteriophage an analogous sequence for the lysin gene is available. However, the putative start codon is AUC rather than AUA as in the ϕvML3 lysin gene, although both of these are isoleucine codons. In the case of bacteriophage P001, an amino terminal sequence for a putative lysin protein purified from a bacteriophage lysate has been obtained (Hertwig, 1990). There is an excellent match between this amino-terminal sequence (MKISQNGLNL) and that of the translated gene (IKISQNGLNL), strongly suggesting that lysin gene translation is initiated by the binding of formylmethionyl tRNA to an atypical AUA/AUC isoleucine codon. Cases of alternative start codons to AUG being used to initiate protein translation are well documented and include GUG, CUG and UUG, all of which can be explained by wobble on the first base of the formylmethionyl tRNA. In addition instances of AUA, ACG and AUU being used *in vivo* are documented (Shearman et al., 1994, Van Sinderen et al., 1996). Thus the use of an atypical start codon in the lactococcal lysin is a reasonable interpretation of the DNA sequence data. It is notable that in cases where an alternative start codon is used the efficiency of translation can be reduced and this may explain the relatively poor expression of the lactococcal lysin, even where a strong heterologous promoter is used (Shearman et al., 1994).

Several different prolate-headed bacteriophages have been subject to molecular characterization in varying degrees of detail. The presence of lysin genes with sufficient homology to facilitate Southern hybridization was demonstrated for bacteriophages ϕvML3, P001, P109, P029, P159. P167, P330, P177, P220 and c6A. In addition, the lysin genes were located on genome restriction maps for both ϕvML3 and P001 (Shearman et al., 1991). Complete genome sequences are available for the bacteriophages c2 (Jarvis et al., 1995; Lubbers et al., 1995; Ward et al., 1993) and bIL67 (Schouler et al., 1995) and for both bacteriophages the lysin gene has been located and oriented on a physical genome map. A 1.67kb fragment of bacteriophage c2 DNA has also been cloned on a plasmid vector in *E. coli* and this was shown to express a lytic activity against lactococci. When compared to the ϕvML3 lysin gene, the bacteriophage c2 lysin gene was identical, except for some silent substitutions. The reported slight difference in the carboxy-terminus of the c2 lysin protein (Ward et al., 1993) has been eliminated by correction of the original ϕvML3 lysin gene sequence (Shearman et al., 1994). From these analyses, and the separate characterization of bacteriophage P001 lysin (Hertwig, 1990) it is clear that the lysin gene sequences are highly conserved across the prolate-headed bacteriophages.

Beresford et al. (1993) undertook a temporal transcript mapping exercise for bacteriophage c2. For this, radioactively labelled cDNA, prepared from total RNA in bacteriophage c2 infected cultures, was hybridized with mapped restriction fragments of bacteriophage c2. This work revealed the sequence of transcripts that were produced during the bacteriophage infective cycle. The lysin gene was transcribed as part of a late polycistronic mRNA, which is consistent with the time that active lysin appeared in a bacteriophage infected culture, namely 15 minutes prior to lysis (Jarvis et al., 1995).

Putative holins

The lysin of prolate headed lactococcal bacteriophages is an intracellular protein that lacks an amino-terminal secretory leader. In common with other characterized bacteriophages, this implies that there must be a system for its passage through the cell membrane, so that it can reach its cell wall target. It is thus likely that the bacteriophage genome also includes the gene for a membrane disruptive holin and three different candidates for such a gene have been suggested.

The first of these involves the lysin gene itself which expresses a second small protein from within the same coding sequence. This was first observed in bacteriophage ϕvML3 when a series of deleted derivatives of the lysin gene were placed downstream of a T7 promoter and the expressed proteins were analyzed by SDS PAGE (Shearman et al., 1994). Two strong protein bands of 8–9kDa were detected upon induction of the T7 promoter and both proteins were separated on two dimensional gels, extracted and their amino-terminal sequences determined. Both sequences were identical (MYLIIGLDNS) and matched an internal sequence from within the lysin protein (Figure 2). The equivalent in-frame gene was preceded by a good ribosome binding site and would express a protein of 71 amino acids with a predicted molecular weight of 8,185. In the prolate headed bacteriophages c2 (Jarvis et al., 1995) and P001 (Hertwig, 1990) expression of a similar small protein has been observed. In the case of P001 this protein was purified from bacteriophage-induced lactococcal lysates and the amino terminal sequence was determined. This matched perfectly the sequence of the cloned gene product from bacteriophage ϕvML3 (Shearman et al., 1994) and expression of the protein during bacteriophage propagation suggests that it has a function.

In addition to the methionine residue that defines the start of the 71 amino acid protein, the amino acid sequence of the ϕvML3 lysin gene reveals another potential translational start methionine three residues upstream and the arrangement is reminiscent of a dual start motive, as has been observed in well-characterized holin genes, such as those of *E. coli* bacteriophages λ and P21, *Salmonella* bacteriophage P22 and *B. subtilis* bacteriophage ϕ29 (Young, 1992). This feature, and the small size of the protein, led Shearman et al. (1994) to speculate that the lysin gene expressed both a lytic enzyme and a holin from the same open reading frame. Whilst the putative holin is hydrophobic with a charged carboxy-terminus, it appears to lack the transmembrane domains that are a conserved feature of other holins.

In many other bacteriophage lytic systems the holin gene is linked to the lysin gene and this has prompted close examination of the genes adjacent to the ϕvML3 lysin gene. For bacteriophage c2, the open reading frame immediately upstream of the lysin gene has been suggested as the holin gene. Certainly the linkage is typical of other bacteriophage lytic systems and the 161 residue protein encoded by the gene has some features of a holin. It contains a sequence of 38 amino acids which is suggestive of two transmembrane domains with the potential for a turn and there is also a highly charged region close to the carboxy-terminus (Jarvis et al., 1995).

Complete sequence analysis of the bacteriophage bIL67 genome has led to the identification of a third potential holin gene (Schouler et al., 1994). This gene (ORF 37) is located at the end of the bacteriophage genome immediately adjacent to the *cos* site and it would be the last gene in the late transcript. The gene is the best candidate for encoding the holin of prolate-headed bacteriophages. Although it is not linked to the lysin gene, such an arrangement has previously been found in the case of *E. coli* bacteriophage T7 (Dunn & Studier 1983). The lactococcal gene encodes a protein with all the features of a holin. Two membrane spanning α-helical domains separated by a short loop containing a turn, a short hydrophilic amino-terminus and a highly charged carboxy-terminus are all present. In addition, a typical dual start motive can be identified. Most significantly, there is a region over 60 carboxy-terminal amino acids with good homology to the characterized holin of *B. subtilis* bacteriophage ϕ29. Lubbers et al. (1995) recently reported the presence of an analogous gene (*l17*) when they analyzed the open reading frames in the complete sequence of the bacteriophage c2 genome. In contrast to the other putative holins evidence for the function of the bacteriophage bIL69 ORF37 gene has recently been obtained by complementing the phenotype of a gene *S* defective mutant of bacteriophage lambda (M.-C. Chopin personal communication). The holin gene was cloned on an *E. coli* expression vector, downstream of an inducible promoter. The infection of this construct with an *S* mutant of bacteriophage lambda strain resulted in its propagation only when the promoter was induced. Thus, although genes for the other two proteins appear not to encode the major holin, it is conceivable that they play another role in the lytic system. This could be a regulatory function or provision of an additional lytic component.

With respect to the latter it is noteworthy that, in addition to a lysin and holin, a third component (R_Z) has been implicated in the lytic system of bacteriophage λ (Young, 1992).

Lytic systems in isometric bacteriophages of *Lactococcus lactis*

Virulent bacteriophages

Platteeuw & DeVos (1992) reported the cloning and sequence analysis of a lysin gene from the virulent isometric bacteriophage ϕUS3. A gene library of the lactococcal bacteriophage DNA was constructed in a plasmid expression vector carrying the lambda P_L promoter. This was induced by thermal inactivation of a *cI857*-encoded repressor that was present in the host strain. Production of lytic activity against *L. lactis* was detected using an overlay procedure, as described by Shearman et al. (1989). Four lysin- expressing clones were obtained and the cloned DNA fragments were located on a restriction map of the bacteriophage ϕUS3 genome. The region of DNA common to all four clones was contained within a 1.6kb restriction fragment which was subjected to DNA sequence analysis. The defined lysin gene, *lytA*, encoded a 258 amino acid protein with a calculated molecular weight of 28,977. The *lytA* gene was overexpressed in *E. coli* under control of a T7 promoter and SDS PAGE analysis of proteins revealed the LytA gene product to be a 29 kDa molecule. The lytic spectrum of the ϕUS3 lysin was found to include all except one of 35 tested strains of *L. lactis*, but no activity was found against *Leuconostoc* and *Lactobacillus* strains. Homology searches with the amino acid sequence of LytA revealed a match with the amino-terminal region of *S. pneumoniae* autolysin and this led to the conclusion that, unlike the other characterized lactococcal lysins, which are lysozyme-like muramidases, LytA is a *N*-acetylmuramoyl-L-alanine amidase. A second complete open reading frame was also present on the lysin-expressing ϕUS3 clone. This gene was located upstream of *lytA* and encodes a 66 amino acid hydrophobic protein with a calculated molecular weight of 7,691. The protein contains two putative membrane-spanning helices and is probably a holin.

The early literature (Oram & Reiter, 1965; Tourville & Johnstone, 1966; Tourville & Tokuba, 1967) reports the purification and biochemical characterization of a lysin from bacteriophage c10 but no molecular analysis of the gene has been undertaken.

Temperate bacteriophages

The lysin gene of the small isometric temperate bacteriophage Tuc2009 was identified by Arendt et al. (1994), following analysis of genome sequence data. An open reading frame was found that encodes a protein with homology to the lysins of *Streptococcus pneumoniae* bacteriophages Cp–1, Cp–7 and Cp–9 (26% identity over 331 residues) as well as *Lactobacillus bulgaricus* bacteriophage mv1 (26% identity over 159 residues). The gene encodes a 428 amino acid protein with a calculated molecular weight of 46,300. The function of the lysin gene was confirmed by its expression in *E. coli* under the control of a T7 promoter. SDS PAGE analysis revealed the expression of a 49 kDa protein and, as with other lysin genes, a lytic activity against lactococci was observed.

A putative holin gene was also identified immediately upstream of the lysin gene. The stop codon of this gene overlaps the start codon of the lysin gene and it encodes an 88 amino acid protein. The typical high hydrophobicity and two transmembrane domains are present, as is a highly charged carboxy-terminal domain. In addition, the protein has homology with the putative holin of *S. pneumoniae* bacteriophage EJ–1 (50% identity over 84 residues). A dual start motive, that is associated with regulation of the holin in several well characterized holin genes, was not found in the holin gene of bacteriophage Tuc2009.

The *L. lactis* temperate bacteriophage ϕLC3 was studied by Birkeland (1994) and linked genes for a lysin and holin were cloned and characterized. These genes are very similar to those of bacteriophage Tuc2009 (Arendt et al., 1994). The lysin gene, *lysA*, encoded a protein of 429 amino acids, which had homology to the lysins of *S. pneumoniae* bacteriophages Cp–1,Cp–7 and Cp–9 ((61% identity over 65 residues) as well as to the lysins of *Lb. delbrueckii* bacteriophage mv1. In common with these homologous lysins, the bacteriophage ϕLC3 lysin is thus likely to be a muramidase. The region of homology with these other lysins was in the amino region of LysA but the carboxy- region also had homology with *B. subtilis* bacteriophages PZA and ϕ29. This was associated with two repeat sequences of 42–43 amino acids which may play a role in cell wall binding. The holin gene, *lysB*, was located upstream of *lysA* and encoded an 88 amino acid protein with two transmembrane domains and highly charged amino- and carboxy- termini. Although holins are not usually conserved at the amino acid sequence level, the ϕLC3 holin was found to have homology to the holin gene of *S. pneumoniae* bacteriophage EJ-I.

Recently, Van Sinderen et al. (1996) reported the complete genome sequence of the temperate lactococcal bacteriophage r1t. Open reading frames for both a lysin and a holin gene were identified. The lysin gene encoded a protein of 227 amino acids with a molecular weight of 30,200. This protein had homology with the active site-containing amino-terminal region of lytic enzymes known to have *N*- acetylmuramoyl-L-alanine amidase activity. In addition, homology to the carboxy-terminal region of the characterized lysins of prolate- headed lactococcal bacteriophages was found. The holin gene was located immediately upstream of the lysin gene and encoded a protein of 76 amino acids with a molecular weight of 7,700. This protein exhibited the established conserved features of a membrane disruptive hoin, including two hydrophobic membrane-spanning domains separated by a short β-turn region, a hydrophobic amino-terminus and a highly charged carboxy-terminus.

Lytic systems in bacteriophages of *Lactobacillus*

Lb. casei bacteriophage PL1

The first characterization of a *Lactobacillus* lysin involved the purification of bacteriophage PL1 lysin from lysates of *Lb. casei.* (Watanabe et al., 1984; Hayashida et al., 1987). The enzyme was shown to be another *N*-acetylmuramidase with pH and temperature optima of 6.0–6.5 and 45°C, respectively. The molecular weight of the lysin was estimated to be 37,000 by SDS PAGE and amino acid analysis. The carboxy-terminal amino acid was shown to be tyrosine and the amino-terminal sequence was determined as AYPINKEFALGANEGXKQVANXLYIIL. The activity spectrum of this lysin was narrow being strongly effective against the host strain *Lb. casei* ATCC27092, but inactive or very weakly active against several other *Lb.casei* strains, *Lb. plantarum*, *S. pyogenes*, *S. mutans*, *M. luteus*, *B. megaterium*, *B. subtilis* and *E. coli*.

Lb. lactis bacteriophage LL-H

Trautwetter et al. (1986) analyzed the genome of the *Lb.lactis* virulent bacteriophage LL-H, producing a restriction map and expressing some genes as *E. coli* clones. In the course of these experiments one clone, pTHP2, was shown to generate a clear halo of lysis on the host strain *Lb. casei* LL23. Thus, the lysin gene of this bacteriophage was located on the genome restriction map. The cloned lysin was shown to be active on four other strains of *Lb. lactis* and *B. subtilis* but within other species of lactobacilli the lytic spectrum was limited with activity detected against only three of six *Lb. helveticus* strains and four of ten *Lb. bulgaricus* strains. All of the tested strains of *Lb. plantarum* and *Lb. casei* were resistant to the lysin.

Lb.bulgaricus bacteriophage mv1

Boizet et al. (1990) studied the lysin of *Lb. bulgaricus* temperate bacteriophage mv1, which is closely related to the virulent *Lb. lactis* bacteriophage LL-H, described above (Mata et al., 1986; Lahbib-Mansais et al., 1988). These bacteriophages belong to a conserved group that has been designated 'group a.'. The cloned fragment of DNA expressing the lysin of the LL-H bacteriophage was used as a hybridization probe and this demonstrated that a conserved lysin gene is present in other 'group a 'bacteriophages, including bacteriophage mv1 of *Lb. bulgaricus*. The gene from the latter strain was cloned onto the *E. coli* plasmid vector pBR322 and it was detected on the basis of the expression of lytic activity against *Lb. delbrueckii* subsp. *lactis* LKT. The lysin gene was located on the cloned fragment by deletion mapping and it appeared to be expressed from an adjacent bacteriophage promoter. The DNA sequence of the smallest subclone that expressed lytic activity was determined. The lysin gene *lysA* was defined and would encode a lysin protein of 195 amino acid residues with a predicted molecular weight of 21,120. Analysis of the gene product by transcription and translation and SDS PAGE revealed a protein product of 24 kDa. The cloned lysin gene was shown to complement an *R* gene mutant of *E. coli* bacteriophage lambda. The *R* gene is known to encode a lysin, thus confirming a similar role for *lysA* of bacteriophage mv1 in *Lactobacillus*. The amino acid sequence of the *lysA* gene product was compared to that of other characterized lytic proteins and homology was found with a muramidase of the fungus *Chalaropsis* (Flech et al., 1975) and the lysin of *S. pneumoniae* bacteriophage Cp–1 (Garcia et al., 1988).

Lb. gasseri bacteriophage ϕadh

Recently Henrich et al. (1995) cloned the lytic genes of *Lb. gasseri* temperate bacteriophage ϕadh by complementing an *E. coli* bacteriophage lambda mutant in the *S* gene. The latter gene encodes a holin and the complementation strategy involved construction of a bacteriophage ϕadh DNA library in a derivative of the cloning vector λgt11 that carried an amber mutation in the *S* gene. Plaque formation on a non-suppressing host strain of *E. coli* was used to identify clones that complemented the defect in gene *S*. Sequence analysis of one cloned fragment revealed the presence of two open reading frames encoding a lysin and holin. Together, these genes constitute the lytic system of bacteriophage ϕadh. The lysin gene would produce a protein of 317 amino acids with a molecular weight of 34,706 and the holin gene would generate a protein of 114 amino acids with a molecular weight of 12,886. The lysin exhibited significant homology to other characterized lytic enzymes, including the lysins of *Lb. bulgaricus* bacteriophage mv–1 (38% identity over 167 residues) and *S. pneumoniae* bacteriophages Cp–1 (39% identity over 66 residues), Cp–7 (38% identity over 59 residues) and Cp–9 (37% identity over 64 residues).

The bacteriophage ϕadh holin lacked sequence homology with previously characterized holins but it does exhibit the well-conserved structural features of a holin. These include high hydrophobicity and a pair of potential transmembrane domains separated by a β turn and a charged carboxy-terminus. These features, together with its capacity to complement a known holin mutation in bacteriophage lambda, establish its role in the *Lactobacillus* bacteriophage.

Lytic systems in bacteriophages of other LAB

The lytic systems of bacteriophages attacking other species of LAB appear not to have been characterized. The literature includes a single reference (Shin & Sato, 1980) that reports the partial purification of a lysin from a strain of *Leuconostoc dextranicum*. In common with other LAB lysins this exhibited a broader spectrum of lytic activity against *Leuconostoc* strains than the intact bacteriophage.

Autolytic systems in LAB

The normal growth and division of bacterial cells involves the activities of cell wall hydrolases that can break down the peptidoglycan. These enzymes probably play a role in cell separation, cell wall turnover, sporulation, the development of competence, flagella formation and the activity of some antibiotics (Doyle et al., 1988). In a variety of bacteria, the autolytic machinery has been shown to involve several different hydrolytic activities that include *N*-acetylmuramidase, *N*-acetylglucosamidase, *N*-acetylmuramyl-L-alanine amidase, endopeptidase and transglycosylase.

In the lactococci, some biochemical characterization of autolysis and autolytic activity has been undertaken. For *L. lactis* subsp *cremoris*, autolytic activity has been found to be maximal at neutral pH during exponential growth. The nature of the autolytic enzyme activity was shown to be an *N*-acetylmuramidase and it was inhibited by lipoteichoic acid and cardiolipin but activated by trypsin (Mou et al., 1976; Niskasaari, 1989). It has also been noted that the cell walls of exponentially growing lactococci autolyse at the equatorial ring (Mou et al., 1976) and filament formation has been associated with decreased autolytic activity (McDonald, 1971; Langsrud et al., 1987) suggesting that these enzymes play a role in lactococcal cell separation.

Ostlie et al. (1995) used SDS PAGE to separate lytic enzymes in lactococcal cells. A series of different strains, including both *lactis* and *cremoris* subspecies and the *diacetylactis* biovar, were screened for activity against *M. luteus* cells using an agar plate assay and renaturing SDS PAGE zymograms. In the latter, between two and five individual bands of variable intensity were detected, with molecular weights ranging from 32,000 to 53,000. A time course experiment with *L. lactis* subsp. *lactis* biovar *diacetylactis* INF-E2–6 showed temporal variation in the spectrum of lytic enzymes. One major 47 kDa band and two weaker 32 kDa and 39 kDa bands were present during exponential phase and these exhibited increased intensity in stationary phase when an additional 53 kDa band also appeared. When lactococcal cells were substituted as substrate, some changes to the pattern of bands were observed in a variety of different strains. Cell fractions were also tested for bacteriolytic activity. All tested strains had enzymes of 47 and 53 kDa in their cell extracts, SDS- treated cell extracts and cell wall fraction. In addition, the clear lytic zones detected around lactococcal colonies in the agar plate assay were eluted and subject to renaturing SDS PAGE analysis, which revealed the presence of only the 47 kDa enzyme. As has been described previously, these authors found autolytic activity to be maximal in exponential-phase cells with pH and temperature optima of 6.0–7.5 and 37°C, respectively. The specificity of the autolysins produced by three strains was also investigated by detection of newly exposed groups in the peptidoglycan, following hydrolysis of the cell wall. Two strains contained a glycosidase and all contained an *N*-acetylmuramyl-L-alanine amidase or endopeptidase.

The molecular genetic characterization of the autolytic system of lactococci has recently been initiated, with the cloning and DNA sequence analysis of the lactococcal gene for the major peptidoglycan hydrolase (Buist et al., 1995a). Preliminary biochemical analysis indicated that *L. lactis* MG1363 contained an *N*-acetylmuramidase and renaturing SDS PAGE analysis revealed the presence of prominent lytic enzymes of 46 kDa and 41 kDa in cell extracts, with the former activity also being found in the culture supernatant. A gene library of *L. lactis* MG1363 in an *E. coli* plasmid vector was screened in a plate assay for lytic activity against *M. lysodeikticus* cells. Thirteen positive clones were obtained and these were analyzed for lytic enzyme production using renaturing SDS PAGE. Enzymes with molecular weights of 41 kDa and 46 kDa were detected in the positive clones. Restriction enzyme analysis showed that all the positive clones shared a 4.1 kb fragment and a clone containing solely this fragment exhibited growth problems, with foaming and cell flocculation. A 2,428 bp region of this fragment including the lytic gene, *acmA*, was sequenced. The *acmA* gene would encode a 437 amino acid residue protein with a calculated molecular weight of 46,564. A 57 amino acid signal peptide was identified at the amino- terminus and this included a membrane spanning domain. Cleavage of the signal sequence would generate a protein with a molecular weight of 40,264. Thus these two proteins show a good match with the 46 kDa and 41 kDa enzymes identified in the biochemical analysis. Homology was found between the lactococcal AcmA and muramidase–2 of *E. hirae* and the autolysin of *E. faecalis* (56% identity over 156 residues in the amino terminal region). The lactococcal AcmA enzyme was thus concluded to be a lysozyme with *N*-acetylmuramidase activity. Three carboxy terminal repeats of 44 amino acids were present separated by intervening sequences that were rich in serine, threonine and asparagine. These repeats have homology with the same region of the cell wall hydrolases of *E. hirae* and *E. faecalis* as well as protein p60 of *L. monocytogenes*. In several other lactococcal strains, the presence of the 46 kDa and 41 kDa enzymes and the *acmA* gene was demonstrated by renaturing SDS PAGE and PCR amplification, respectively.

A deletion within the *acmA* gene was introduced in the chromosome of *L. lactis* MG1363 by replacement recombination, demonstrating that the gene is not essential. The deletion eliminated the production of all cell wall hydrolase activity when examined by renaturing SDS PAGE, suggesting that the multiple clearing bands all originated from AcmA. Unlike *L. lactis* MG1363, the *acmA* - deletion strain did not lyse during prolonged stationary phase growth. Examination by light microscopy showed mutant cells to be arranged in unusually long chains, suggesting that AcmA plays a role in cell separation.

Relatively little information is available on the autolytic systems of other LAB. Lortal et al. (1991) reported a study of cell wall structure in *Lb. helveticus* ATCC12046 and determined conditions leading to autolysis. The specificity of the autolysin involved in cell wall degradation was defined as an *N*-acetylmuramidase. Valence & Lortal (1995) used renaturing SDS PAGE to investigate the autolysins of *Lb. helveticus* ISLC5. An enzyme of 42 kDa and a group of similarly sized active proteins of 29 to 33 kDa were identified. Zymograms for 30 different strains of *Lb. helveticus* were also produced leading to the conclusion that the autolysins are highly conserved.

Relevance of starter lysis in flavour generation

The importance of starter cell lysis in the generation of mature cheese flavour is an area of longstanding debate that is presently receiving renewed interest. Sensitivity to cooking temperature (thermolysis) and subsequent death in the cheese matrix are established as key properties of lactococcal starter strains, such as AM1, AM2 and US3, that consistently generate cheese of superior quality (Martley et al., 1972). The availability of molecular genetic approaches to the construction of novel strains provides new opportunities to analyze the role of lysis, as well as the promise of useful engineered starters. Crow et al. (1995) provide a comprehensive review of current knowledge on the mechanisms and role of LAB autolysis in cheese ripening and the discussion here is limited to the selection or construction of strains.

The perceived value of an autolytic phenotype is the early release of intracellular enzymes into the cheese curd, providing the potential for enhanced or accelerated flavour development. With respect to naturally thermolytic strains, Wiederholt & Steele (1993) explored the role of prophage induction in the phenotype and concluded that more was involved. Several approaches have been taken to isolate new strains than exhibit early lysis. The application of a genetic engineering approach involving cloned genes for LAB lytic enzymes is discussed in the following section. Feirtag & McKay (1987) used the more conventional approach of chemical mutagenesis to isolate thermolytic strains. Following EMS mutagenesis of *L. lactis*, colonies were sought which grew at 30 °C but not at 40 °C. Amongst these, two mutants were obtained which, on the basis of optical density loss and the release of intracellular β- galactosidase, lysed at the higher temperature. However, lysis takes place in only a small percentage of cells in the culture. Crow et al. (1995) report the selection of strains that lyse upon reaching stationary phase when grown in a medium with limiting carbohydrate and for these strains a higher percentage of cells were disrupted. The autolysins involved in the lysis of one such strain are presently being characterized.

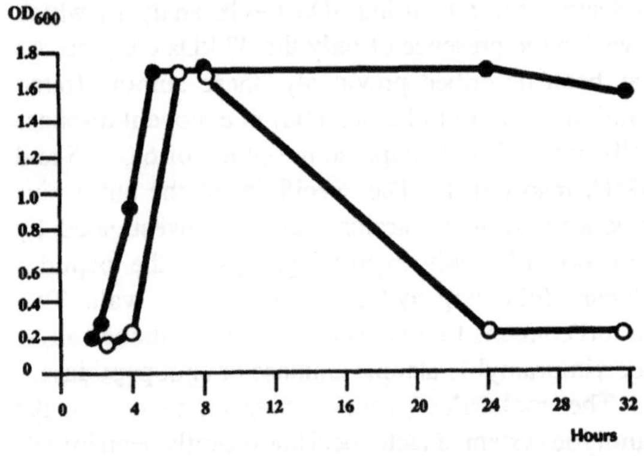

Figure 3. Autolytic *L. lactis* strain expressing a cloned bacteriophage lysin gene. Control strain is indicated by the black circles. A strain expressing lactococcal lysin grows normally through exponential phase but lyses prematurely in the stationary phase (Shearman et al., 1992).

Exploitation of cloned lytic genes

The availability of lactococcal lytic enzymes, holins and their genes, provides the opportunity to develop new technologies for the LAB. The superior lytic activity of lysins in comparison to hen egg white lysozyme has stimulated their use to generate protoplasts, as described above for lactococci using bacteriophage ϕML3 lysin. Watanabe et al. (1987, 1992) similarly used the lysin of bacteriophage PL1 to produce and regenerate protoplasts of *Lb. casei* and, furthermore, used these protoplasts to demonstrate the transfection of PL1 phage DNA.

One of the most significant biotechnological applications for bacteriophage lytic systems and autolysins is the development of controlled starter cell lysis, primarily to facilitate intracellular enzyme release. The first demonstration of this approach involved expression of the ϕML3 lysin in lactococci (Shearman et al., 1992). The gene was introduced on a plasmid vector, together with its bacteriophage-derived upstream sequences. It caused an autolytic phenotype, in which the strain grew normally through exponential phase but, unlike a control strain, it lysed spontaneously in the stationary phase as shown in Figure 3. Shearman et al. (1994) used lysin gene fusions to *E. coli* β-galactosidase to demonstrate constitutive expression of the ϕML3 lysin under control of its own promoter and this eliminated the possibility that the autolytic phenotype resulted from late expression of the lysin gene. The simplest explanation for the phenotype is the intracellular containment of expressed lysin during the exponential growth phase, resulting in stationary phase cells loaded with lytic enzyme. The spontaneous lysis of just one cell would trigger a cascade of lysis as more and more lysin was released. This explanation is consistent with the fact that lysins are not secreted, but rather rely on membrane disruptive holins to access their peptidoglycan substrate. Maintenance of autolytic lysin-expressing strains was readily achieved by the provision of osmotic buffering to prevent initiation of the lytic cascade and under these conditions the cultures could be frozen and revived without loss of the phenotype.

Other lytic systems have been exploited for controlled enzyme release but this has generally involved the fusion of lytic genes to a controllable promoter. The use of a lysin and holin gene pair provides a potentially lethal combination that would be expected to promote cell lysis upon intracellular expression. The lysin and holin lytic systems of both temperate bacteriophage r1t (Nauta et al., 1995) and virulent bacteriophage ϕUS3 (Platteeuw & De Vos, 1992; W. De Vos personal communication) have been exploited in this way. The most impressive result to date was obtained when the latter

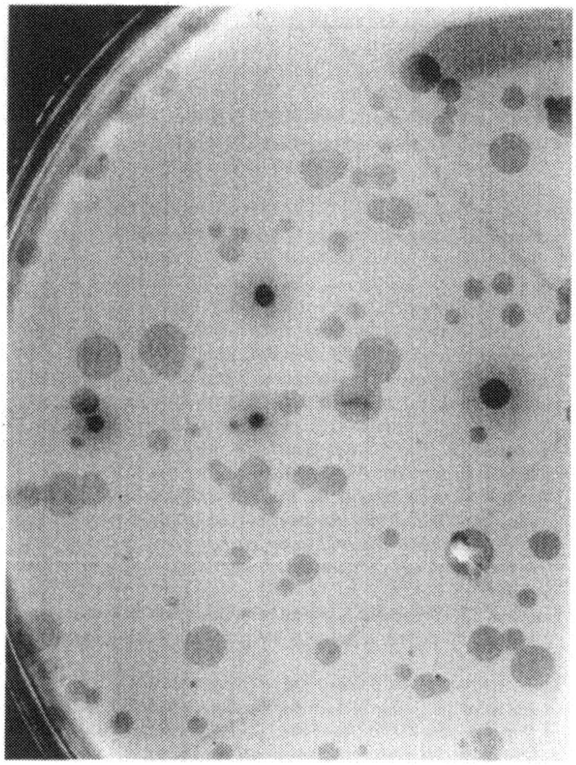

Figure 4. The identification of lambda bacteriophage clones producing lactococcal bacteriophage lysin. Bacteriophage lambda plaques are overlayed with a thick suspension of lactococcal cells. Plaques that express lysin create a window of lysis in the lawn making the plaque appear darker (Shearman et al., 1989).

genes were expressed under control of the P_{nisA} nisin operon promoter. This promoter is subject to very tight positive control that involves a two component regulatory system. In the absence of nisin, expression is completely repressed, but provision of nisin induces derepression facilitating expression of genes controlled by the nisin promoter. Using this system the lethal genes were successfully maintained in *L. lactis* in the absence of nisin but cultures were efficiently lysed when inducing levels of nisin were added.

Other inducible expression systems of interest include the recently described repressor and operator of temperate lactococcal bacteriophage r1t (Nauta et al., 1996) which is presently being developed into a thermoinducible system. Also a variety of environmentally regulated promoters, induced by salt, temperature or pH, show great promise for control of lytic genes during cheesemaking (Israelsen et al., 1992; Sanders et al., 1995).

In addition to bacteriophage-derived genes, the autolysin gene *acmA* (Buist et al., 1995b) is being used to develop potential enzyme release systems. The recently described inducible gene expression system of temperate lactococcal bacteriophage r1t (Nauta et al., 1996), has been used to control expression of the autolysin gene. In addition, the novel approach of exploiting bacteriocin genes has also been explored (Morgan et al., 1995). In this latter case a strain of *L. lactis* that produced the three lactococcins A,B and M was combined with sensitive strains so as to stimulate their premature lysis and intracellular enzyme release.

Heterologous lysin genes have also been expressed in lactococci including those for hen egg white lysozyme (Van de Guchte et al., 1989) and the *E. coli* bacteriophage T4 and lambda lysin (Van de Guchte, 1990). In addition, genes for lysins that are active against pathogenic and spoilage bacteria have been used to engineer novel antimicrobial phenotypes in LAB. The lysin gene from a *Listeria monocytogenes* bacteriophage was expressed in *L. lactis* under control of the lactococcal lactose operon promoter, both as a plasmid clone and following chromosomal integration. In both cases lactose regulated expression of an antimicrobial activity that lysed a wide range of *Listeria* strains was demonstrated (Payne et al., 1995). Release of the *Listeria* lysin has also been promoted by the expression of its gene in an autolytic derivative of *L. lactis* that also carries the bacteriophage ϕML3 lysin gene. In this case the anti-*Listeria* activity was spontaneously released into the supernatant of stationary phase cells, illustrating the potential of autolytic strains to deliver heterologous as well as homologous enzymes.

Concluding remarks

Until recently the lytic systems of LAB and their bacteriophages have been little studied, with literature reports being restricted to partial purification of lytic activities and limited biochemical analysis. However, over the last few years the genes involved in both autolysis and the cellular release of mature bacteriophages have been cloned and characterized. A combination of amino acid homology analyses and the overexpression and inactivation of individual genes is providing a better picture of what is involved. The basic characteristics of the lytic systems that have been characterized at a molecular level are summarized in Table 1 and Figure 1 provides an outline of the mucopeptide targets for the various distinct lytic enzymes. In both bacteriophage lytic systems and autolysins, the *N*-acetyl muramidase activity is by far the most commonly encountered lytic enzyme activity.

The availability of characterized lytic genes is providing the opportunity for biotechnological exploitation most notably in the development of starter strains with an engineered capacity to release their intracellular enzymes. It remains to be seen if this will have a beneficial effect on flavour development, but certainly it will provide novel tools with which to probe the significance of this property in dairy fermentations.

References

Arendt EK, Daly C, Fitzgerald GF & Van der Guchte (1994) Molecular characterization of lactococcal bacteriophage Tuc2009 and identification and analysis of genes encoding lysin, a putative holin, and two structural proteins. Appl. Environ. Microbiol. 60: 1875–1883

Beresford TPJ, Ward LJH & Jarvis AW (1993) Temporally regulated transcriptional expression of the genomes of lactococcal bacteriophages c2 and sk1. Appl. Environ. Microbiol. 59: 3708–3712

Birkeland NK (1994) Cloning, molecular characterization, and expression of the genes encoding the lytic functions of lactococcal bacteriophage ϕLC3 - a dual lysis system of modular design. Can. J. Microbiol. 40: 658–665

Boizet B, Lahbib -Mansais Y, Dupont L, Ritzenthaler P & Mata M (1990) Cloning, expression and sequence analysis of an endolysin- encoding gene of *Lactobacillus bulgaricus* bacteriophage mv1. Gene 94: 61–67

Buist G, Kok J, Leenhouts KJ, Dabrowska M, Venema G & Haandikman AJ (1995a) Molecular-cloning and nucleotide-sequence of the gene encoding the major peptidoglycan hydrolase of *Lactococcus lactis*. J. Bacteriol. 177: 1554–1563

Buist G, Karsens H, Nauta A, Van Sinderen D & Kok J (1995b) Autolysis of *L. lactis* caused by overproduction of the major autolysin AcmA. In: Abstracts of the International Lactic Acid Bacteria Conference, New Zealand, February 1995

Camacho A, Jimenez F, la Torre DE, Carrascosa JL, Mellado RP, Vasquez C, Vinuela E & Salas M (1977) Assembly of *Bacillus subtilis* phage ϕ29. 1. Mutants in cistrons coding for the structural proteins. Eur. J. Biochem. 73: 39–55

Crow VL, Coolbear T, Gopal PK, Martley FG, McKay LL & Riepe H (1995) The role of autolysis of lactic acid bacteria in the ripening of cheese. Int. Dairy J. 5: 855–875

Doyle RJ, Chaloupka J & Vinter V (1988) Turnover of cell walls in microorganisms. Microbiol. Rev. 52: 554–567

Dunn JJ & Studier FK (1983) Complete nucleotide sequence of bacteriophage T7 DNA and the locations of T7 genetic elements. J. Mol. Biol. 166: 477–535

Feirtag JM & McKay LL (1987) Isolation of *Streptococcus lactis* C2 mutants selected for temperature sensitivity and potential use in cheese manufacture. J. Dairy Sci. 70: 1773–1778

Flech JW, Inagami T & Hash JH (1975) The *N,O*- diacetylmuramidase of *Chalaropsis* species. J. Biol. Chem. 25: 3713–3720

Garcia E, Garcia JL, Garcia P, Arraras A, Sanchez-Puelles JM & Lopez R (1988) Molecular evolution of lytic enzymes of *Streptococcus pneumoniae* and its bacteriophages. Proc. Natl. Acad. Sci. USA 85: 914–918

Garvey KJ, Saedi MS & Ito I (1986) Nucleotide sequence of *Bacillus* phage ϕ29 genes 14v and 15: homology of gene 15 with other phage lysozymes. Nucl. Acids Res. 14: 10001–10008

Hayashida M, Watanabe K, Muramatsu T & Goto MA (1987) Further characterization of PL–1 phage-associated *N*-acetyl-muramidase of *Lactobacillus casei*. J. Gen. Microbiol. 133: 1343–1349

Henrich B, Binishofer B & Blasi U (1995) Primary structure and functional analysis of the lysis genes of *Lactobacillus gasseri* bacteriophage ϕadh. J. Bacteriol. 177: 723–732

Hertwig S (1990) PhD Thesis. University of Kiel, Germany

Israelsen H, Madsen S, & Hansen EB (1993) Cloning and characterization of regulated promoters from the *Lactococcus lactis* ssp. *lactis* chromosome.FEMS Microbiol. Rev. 12: P23

Jarvis AW, Lubbers MW, Beresford TPJ, Ward, LJH, Waterfield NR, Collins, LJ and Jarvis BDW (1995) Molecular biology of lactococcal bacteriophage c2. In Ferretti JJ, Gilmore MS, Klaenhammer TR, Brown F (Ed) Genetics of Streptococci, Enterococci and Lactococci. Dev Biol Stand. Basel, Karger (pp. 561–567)

Lahbib-Mansais Y, Boizet B, Dupont L, Mata M & Ritzenthaler P (1992) Characterization of a temperate bacteriophage of*Lactobacillus delbrueckii* subsp.*bulgaricus* and its interactions with the host cell chromosome. J. Gen. Microbiol. 138: 1139–1146

Langsrud T, Landaas A & Castberg HB (1987) Autolytic properties of different strains of group N streptococci. Milchwissenschaft 42: 556–560

Lortal S, Rousseau M, Boyaval P & van Heijenoort J (1991) Cell wall and autolytic system of *Lactobacillus helveticus* ATCC 12046. J. Gen. Microbiol. 137: 549–559

Lubbers MW, Waterfield NR, Beresford TPJ, LePage RW and Jarvis AW (1995) Sequencing and analysis of the prolate-headed lactococcal bacteriophage c2 genome and identification of the structural genes. Appl. Environ. Microbiol. 61: 4348–4356

Martley FG & Lawrence RC (1972) Cheddar cheese flavour. II. Characteristics of single strain starters associated with good or poor flavour development. N.Z. J. Dairy Sci. Technol. 7: 38–44

Mata M, Trautwetter A, Luthaud G & Ritzenthaler P (1986) Thirteen virulent and temperate bacteriophages of *Lactobacillus bulgaricus* and *Lactobacillus lactis* belong to a single DNA homology group. Appl. Environ. Microbiol. 52: 429–435

McDonald IJ (1971) Filamentous forms of *Streptococcus cremoris* and *Streptococcus lactis*: observations on structure and susceptibility to lysis. Can. J. Microbiol. 17: 897–902

Morgan S, Ross RP & Hill C (1995) Bacteriolytic activity caused by the presence of a novel lactococcal plasmid encoding lactococcin A, lactococcin B and lactococcin M. Appl. Environ. Microbiol. 61: 2995–3001

Mou L, Sullivan JJ & Jago GR (1976) Autolysis of *Streptococcus cremoris*. J. Dairy Res. 43:275–282

Mullan WMA & Crawford RJM (1985a) Partial purification and some properties of $\phi C_2(w)$ lysin, a lytic enzyme produced by phage-infected cells of *Streptococcus lactis* C_2. J. Dairy Res. 52: 123–138

Mullan WMA & Crawford RJM (1985b) Lysin production by $\phi C_2(w)$, a prolate phage for *Streptococcus lactis* C_2. J. Dairy Res. 52: 113–121

Mullan WMA & Crawford RJM (1985c) Factors affecting the lysis of group N streptococci by phage lysin. Milchwissenschaft 40: 342–345

Murga MLF, Holgado APD & Devaldez GF (1995) Influence of the incubation-temperature on the autolytic activity of *Lactobacillus acidophilus*. J. Appl. Bacteriol. 78: 426–429

Nauta A, Van Sinderen, D, Karsens H, Venema G & Kok J (1995) Repressor/operator region of the lactococcal temperate bacteriophage r1t and its use for the induced lysis of *Lactococcus lactis*. In: Abstracts of the International Dairy Lactic Acid Bacteria Conference, New Zealand, February 1995.

Nauta A, van Sinderen D, Karsens H, Smit E, Venema G & Kok J (1996) Inducible gene expression mediated by a repressor-operator system isolated from *Lactococcus lactis* bacteriophage r1t. Mol. Microbiol. 19: 1331–1341

Naylor J & Czulak J (1956) Host phage relationship of cheese starter organisms II. Effect of phage activity on heterologous strains of lactic streptococci. J. Dairy Res. 23: 126–130

Niskasaari K (1989) Characteristics of the autolysis of variants of *Lactococcus lactis* subsp. *cremoris*. J. Dairy Res. 56: 639–649

Oram JD & Reiter B (1965) Phage-associated lysins affecting group N and group D streptococci. J. Gen. Microbiol. 40: 57–70

Ostlie HM, Vegaraud G & Langsrud T (1995) Autolysis of lactococci: detection of lytic enzymes by polyacrylamide gel electrophoresis and characterization in buffer systems. Appl. Environ. Microbiol. 61: 3598–3603

Payne J, MacCormick CA, Griffin HG & Gasson MJ (1996) Exploitation of a chromosomally integrated lactose operon for controlled gene expression in *Lactococcus lactis*. FEMS Microbiol. Lett. 136: 19–24

Platteeuw C & DeVos WM (1992) Location, characterization and expression of lytic enzyme-encoding gene *lytA*, of *Lactococcus lactis* bacteriophage ϕUS3. Gene 118: 115–120

Reiter B & Oram JD (1963) Group N streptococcal phage lysin. J. Gen. Microbiol. 32: 29–32

Sable S and Lortal S (1995) The lysins of bacteriophages infecting lactic acid bacteria. Appl. Microbiol. Biot. 43: 1–6

Sanders JW, Venema G, Kok, J & Leenhouts KJ (1995) Development of a random integration-expression system for *Lactococcus lactis*: identification of a salt inducible promoter. In: Abstracts of the International Dairy Lactic Acid Bacteria Conference, New Zealand, February 1995

Schouler C, Ehrlich SD and Chopin M-C (1994) Sequence and organization of the lactococcal prolate headed bIL67 phage genome. Microbiol. 140: 3061–3069

Shearman CA, Underwood HM, Jury K & Gasson MJ (1989) Cloning and DNA sequence analysis of a *Lactococcus* bacteriophage lysin gene. Mol. Gen. Genet. 218: 214–221

Shearman CA, Hertwig S, Teuber M & Gasson MJ (1991) Characterization of the prolate-headed lactococcal bacteriophage ϕvML3: location of the lysin gene and its DNA homology with other prolate-headed phages. J. Gen. Microbiol. 137: 1285–1291

Shearman CA, Jury K & Gasson MJ (1992) Autolytic *Lactococcus lactis* expressing a lactococcal bacteriophage lysin gene. Biotechnol. 10:196–199

Shearman CA, Jury K & Gasson MJ (1994) Controlled expression and structural organization of a *Lactococcus lactis* bacteriophage lysin encoded by two overlapping genes. Appl. Environ. Microbiol. 60: 3063–3073

Shin C & Sato Y (1980) *Leuconostoc* phage-associated lysin acting on lactic acid bacteria. Jpn. J. Zootech Sci. 51: 443–446

Thomas TD, Jarvis BDW & Skipper NA (1974) Localization of proteinase(s) near the cell surface of *Streptococcus lactis*. J. Bacteriol. 118: 329–333

Tourville DR & Johnstone DB (1966) Lactic streptococcal phage-associated lysin. I Lysis of heterologous lactic streptococci by a phage- induced lysin. J. Dairy Sci. 49: 158–162

Tourville DR & Tokuba S (1967) Lactic streptococcal phage-associated lysin. II Purification and characterization. J. Dairy Sci. 50: 1019–1024

Trautwetter A, Ritzenthaler P, Alatassova T & Mata-Gilsinger M (1986) Physical and genetic characterization of the genome of *Lactobacillus lactis* bacteriophage LL-H. J. Virol. 59: 551–555

Valence F & Lortal S (1995) Zymogram and preliminary characterization of *Lactobacillus helveticus* autolysins. Appl. Environ. Microbiol. 61: 3391–3399

Van de Guchte M (1991) Heterologous gene expression in *Lactococcus lactis*, PhD thesis. University of Groningen, The Netherlands

Van de Guchte M, van der Vossen JMBM, Kok J & Venema G (1989) Construction of a lactococcal expression vector: expression of hen egg white lysozyme in *Lactococcus lactis* subsp. *lactis* Appl. Environ. Microbiol. 55: 224–228

Van Sinderen D, Karsens H, Kok J, Terpstra P, Ruiters MHJ, Venema G & Nauta A (1996) Sequence analysis and molecular characterization of the temperate lactococcal bacteriophage r1t. Mol. Microbiol. 19: 1343–1355.

Ward LJH, Beresford TPJ Lubbers MW, Jarvis BDW & Jarvis AW (1993) Sequence analysis of the lysin gene region of the prolate lactococcal bacteriophage c2. Can. J. Microbiol. 39: 767–774

Watanabe K, Hayashida M, Ishibashi K & Nakashima Y (1984) An *N*-acetylmuramidase induced by PL–1 phage infection of *Lactobacillus casei*. J. Gen. Microbiol. 130: 275–277

Watanabe K, Hayashida M, Nakashima Y & Hayashida S (1987) Preparation and regeneration of bacteriophage PL–1 enzyme-induced *Lactobacillus casei* protoplasts. Appl. Environ. Microbiol. 53: 2686–2688

Watanabe K, Kakita Y, Nakashima Y & Miake F (1992) Calcium requirement for protoplast transfection mediated by polyethylene glycol of *Lactobacillus casei* by PL–1 phage DNA. Biosci. Biotech. Bioch. 56: 1859–1862

Wiederholt KM & Steele JL (1993) Prophage curing and partial characterization of temperate bacteriophages from thermolytic strains of *Lactococcus lactis* spp *cremoris*. J. Dairy Sci. 76: 921–930

Young RY (1992) Bacteriophage lysis: mechanism and regulation. Microbiol. Rev. 56: 430–481

Antonie van Leeuwenhoek **70:** 161–183, 1996.

Genomic organization of lactic acid bacteria

Barrie E. Davidson*[,1], Nancy Kordias[1,3], Marian Dobos[2] & Alan J. Hillier[3]
[1]*Department of Biochemistry and Molecular Biology, The University of Melbourne, Parkville, Victoria 3052, Australia;* [2]*Department of Medical Laboratory Science, Royal Melbourne Institute of Technology, Melbourne, Victoria 3000, Australia; and* [3]*Commonwealth Scientific and Industrial Research Organization, Division of Food Science and Technology, Melbourne Laboratory, Highett, Victoria 3190, Australia (*author for correspondence)*

Key words: Lactococcus, Streptococcus thermophilus, Lactobacillus, Leuconostoc, Oenococcus, Pediococcus, Carnobacterium, genome, chromosome, plasmid, temperate phage, genetic map, IS elements, genome plasticity

Abstract

Current knowledge of the genomes of the lactic acid bacteria, *Lactococcus lactis* and *Streptococcus thermophilus*, and members of the genera *Lactobacillus, Leuconostoc, Pediococcus* and *Carnobacterium*, is reviewed. The genomes contain a chromosome within the size range of 1.8 to 3.4 Mbp. Plasmids are common in *Lactococcus lactis* (most strains carry 4-7 different plasmids), some of the lactobacilli and pediococci, but they are not frequently present in *S. thermophilus, Lactobacillus delbrueckii* subsp. *bulgaricus* or the intestinal lactobacilli. Five IS elements have been found in *L. lactis* and most strains carry multiple copies of at least two of them; some strains also carry a 68-kbp conjugative transposon. IS elements have been found in the genera *Lactobacillus* and *Leuconostoc*, but not in *S. thermophilus*. Prophages are also a normal component of the *L. lactis* genome and lysogeny is common in the lactobacilli, however it appears to be rare in *S. thermophilus*. Physical and genetic maps for two *L. lactis* subsp. *lactis* strains, two *L. lactis* subsp. *cremoris* strains and *S. thermophilus* A054 have been constructed and each reveals the presence of six *rrn* operons clustered in less than 40% of the chromosome. The *L. lactis* subsp. *cremoris* MG1363 map contains 115 genetic loci and the *S. thermophilus* map has 35. The maps indicate significant plasticity in the *L. lactis* subsp. *cremoris* chromosome in the form of a number of inversions and translocations. The cause(s) of these rearrangements is (are) not known. A number of potentially powerful genetic tools designed to analyse the *L. lactis* genome have been constructed in recent years. These tools enable gene inactivation, gene replacement and gene recovery experiments to be readily carried out with this organism, and potentially with other lactic acid bacteria and Gram-positive bacteria. Integration vectors based on temperate phage *attB* sites and the random insertion of IS elements have also been developed for *L. lactis* and the intestinal lactobacilli. In addition, a *L. lactis* sex factor that mobilizes the chromosome in a manner reminiscent to that seen with *Escherichia coli* Hfr strains has been discovered and characterized. With the availability of this new technology, research into the genome of the lactic acid bacteria is poised to undertake a period of extremely rapid information accrual.

Introduction

Organisms that produce lactic acid as a major metabolic by product have been used for millennia as fermenting agents for the preservation of food. The low pH generated by the acid and the action of other fermentation products inhibits the growth of many spoilage and pathogenic bacteria. Lactic acid fermentations are typically used to make cheese, yoghurt and a wide variety of fermented dairy products from milk, to produce sausages such as salami from meat, for pickling vegetables and olives, in the manufacture of soy sauce, sour dough breads, soda crackers and silage (McKay & Baldwin, 1990). The majority of organisms that are used are members of the genera *Lactococcus, Lactobacillus, Streptococcus, Pediococcus, Leuconostoc* and *Carnobacterium*. The analysis of the genomes of these organisms only commenced in the

1980s and progress since then has varied considerably from one genus and species to another. Thus, the genome of *Lactococcus lactis* is the most closely analyzed, while there is little published information describing genomes from the genera *Leuconostoc* and *Carnobacterium.*

The term genome is used in this article to cover the chromosome and plasmids of an organism and other genetic elements such as transposable elements and prophages which can be covalently part of these structures. The DNA of lytic (non-temperate) phages could also be considered to be a genomic component because of its ability to exchange by recombination with either chromosome or plasmids. However, this subject will not be discussed in this review because of size considerations, although the recent determination of the complete nucleotide sequences of the genomes of *L. lactis* phages c2 (prolate-headed) and sk1 (small-isometric-headed) makes this topic one of particular current interest (Jarvis et al., 1995; Chandry et al., 1996). The field of chromosomal mapping in the lactic acid bacteria, which is part of the present article, has been reviewed previously (Le Bourgeois et al., 1993) while other reviews on specific topics are cited in relevant sections of the text.

Bacterial chromosomes are the subject of significant research endeavours at present (Cole & Saint Girons, 1994), particularly with the recent milestone publications of the complete sequences of the *Haemophilus influenzae* Rd and *Mycoplasma genitalium* chromosomes (Fleischmann et al., 1995; Fraser et al., 1995). The huge amount of data that these achievements uncovered for scientists studying the genetics, biochemistry and physiology of these organisms are awesome. For the present, those interested in lactic acid bacteria must content themselves with considerably more modest data sets describing the chromosomes of these organisms.

The genus *Lactococcus*

The genus *Lactococcus*, which includes *L. lactis*, *L. garviae*, *L. plantarum*, *L. piscium* and *L. raffinolactis* (Schleifer et al., 1985; Williams et al., 1990), is characterized by genomes of low G+C content (36 - 38% (Kilpper-Bälz et al., 1982)). Most of the genetic analysis of the genus has been carried out on the two subspecies *L. lactis* subsp. *lactis* and *L. lactis* subsp. *cremoris* because of their widespread use as starter bacteria in dairy fermentations.

Genetic differentiation between L. lactis *subsp.* lactis *and* L. lactis *subsp.* cremoris

Traditionally, these two subspecies were differentiated on the basis of phenotypic characteristics: *L. lactis* subsp. *lactis* produces ammonia from arginine and grows at 40 °C, whereas *L. lactis* subsp. *cremoris* does neither. Jarvis & Jarvis (1981) used DNA-DNA hybridization studies to investigate homology between the genomes of 43 different *L. lactis* strains. They observed that the *L. lactis* subsp. *cremoris* strains fell into a single homology group while the *L. lactis* subsp. *lactis* strains formed another, except for three atypical strains (including *L. lactis* ML3 discussed below) which were more aligned with the *L. lactis* subsp. *cremoris* group. Because of the large number of strains examined, these initial studies provided important evidence that the two subspecies are distinct genetic entities. More recently, it has been demonstrated that the two subspecies can be distinguished by their 16S rRNA gene sequences and the extent of hybridization of their genomes with a number of gene-specific DNA probes from either one of the two subspecies (Salama et al., 1991; Godon et al., 1992b). Application of these methods to *L. lactis* C2 and related strains, which have a subspecies *lactis* phenotype, indicated that they are genetically members of the subspecies *cremoris*. Relatives of *L. lactis* C2 include *L. lactis* ML3 and the commonly used laboratory strains *L. lactis* MG1363 and *L. lactis* LM0230 (Davies et al., 1981; Gasson, 1983; Efstathiou & McKay, 1977). The 16S rRNA data for *L. lactis* ML3 are therefore in agreement with the hybridization data described above (Jarvis & Jarvis, 1981). An important conclusion from these analyses is that some *L. lactis* strains which have a subspecies *lactis* phenotype are genetically subspecies *cremoris*. Such strains will be referred to as *L. lactis* subsp. *cremoris* in this review, in line with current usage.

Measurements of the amount of hybridization between specific probes and DNA from each subspecies enabled Godon et al., (1992b) to estimate that the extent of divergence between the two subspecies is 20 - 30%. Comparison of the nucleotide sequences of 2.5 kbp of DNA from the *his* operon of *L. lactis* subsp. *lactis* NCDO 2118 and *L. lactis* subsp. *cremoris* NCDO 763 revealed an overall divergence of 27%, in agreement with this estimate (Delorme et al., 1994). A more detailed comparison of the two sequences revealed that the extent of divergence varied along the length of the sequenced DNA, and that conserved and variable regions were interspersed in a mosaic structure.

Components of the mosaic that were highly divergent between *L. lactis* subsp. *lactis* and *L. lactis* subsp. *cremoris* were well conserved within *L. lactis* subsp. *cremoris*. This was taken to indicate that these components entered the *L. lactis* subsp. *cremoris* genome by horizontal gene transfer from distantly related bacteria. The availability of more extensive nucleotide sequence data for the two subspecies will be invaluable in ascertaining the extent to which this mosaic structure is typical of the *L. lactis* subsp. *cremoris* chromosome.

In addition to strain typing, Jarvis & Jarvis (1981) used renaturation kinetics to obtain values of 2.8 - 3.1 Mbp for the size of the genome of *L. lactis* subsp. *cremoris* ML3 and two other strains (Table 1). Other estimates of *L. lactis* genome sizes, obtained by summing the sizes of fragments produced by digestion with infrequently cutting endonucleases such as *Sma*I and *Apa*I and separated by pulsed-field gel electrophoresis (PFGE), are in good agreement with the renaturation kinetic data (Table 1). It is worth noting that the PFGE procedure is subject to uncertainties when used for estimating genome sizes of organisms which carry one or more large circular plasmids that are not digested by the endonuclease. This is because circular and linear DNA molecules of the same size migrate with different mobilities when subjected to PFGE (Davidson et al., 1992; Ferdows & Barbour, 1989). Given the frequency with which large circular plasmids are present in *L. lactis*, it would be imprudent to conclude from the data in Table 1 that significant differences exist between the sizes of the *L. lactis* subsp. *lactis* and the *L. lactis* subsp. *cremoris* chromosomes or genomes.

Physical mapping of the L. lactis *chromosome*

In the absence of a suitable genetic transfer system, the only strategy initially available for analysing the *L. lactis* chromosome was to prepare a restriction map, using PFGE to analyze the products obtained by digesting high molecular weight chromosomal DNA with infrequently cutting restriction endonucleases (*Not*I, *Asc*I, I- *Ceu*I, *Sma*I, *Sgr*AI, *Apa*I and *Csp*I) (Chu et al., 1987; Hillier & Davidson, 1995). The ease with which a physical map can be determined by this approach depends to a large extent upon the number and distribution of appropriate digestion sites, properties that were found to be strain dependent in *L. lactis* (Le Bourgeois et al., 1989; Tanskanen et al., 1990). Indeed, the strain-specific nature of the restriction digest patterns has been used in the Australian dairy industry for the last seven years as the basis of a strain typing procedure. It was initially helpful to construct a physical map of the chromosome of *L. lactis* subsp. *lactis* DL11 because it has a favourable number and distribution of *Not*I and *Sma*I sites (Davidson et al., 1991; Tulloch et al., 1991). The resulting map established that *L. lactis* has one, circular chromosome of 2.58 Mbp (Tulloch et al., 1991). Subsequently, physical maps of the chromosomes of *L. lactis* subsp. *lactis* IL1403, *L. lactis* subsp. *cremoris* MG1363 and the industrial cheese starter strain *L. lactis* subsp. *cremoris* FG2 were determined (Le Bourgeois et al., 1992a; Le Bourgeois et al., 1995; Kordias et al., 1996). Some of this work was facilitated by using of the plasmid pRL1 which carries unique sites for *Apa*I, *Sma*I and *Not*I and integrates randomly via transposition into the *L. lactis* chromosome (Le Bourgeois et al., 1992b).

A comparison of the physical maps of the two subspecies *lactis* strains, DL11 and IL1403, revealed good correspondence in the positions of the *Sma*I restriction sites over approximately 60% of the chromosome, but only limited correspondence in the remaining 40% (Le Bourgeois et al., 1992a). The physical maps we have determined for two subspecies *cremoris* strains, MG1363 and FG2 (Kordias et al., 1996) are compared in Figure 1. An obvious feature of the comparison is that five of the six I-*Ceu*I sites have similar locations. Because I-*Ceu*I sites are found exclusively in rRNA (*rrn*) operons, similarities in their location reflect conservation in *rrn* operon distribution, a topic which is discussed in more detail below. *rrn* operons also contain *Sma*I sites, so five of the *Sma*I sites also have a similar location in the two maps (Figure 1). Apart from these similarities, there is no other striking correspondence between the locations of the restriction endonuclease sites in the chromosomes of the two organisms. The differences between the two maps could be the result of point mutations, insertions, deletions, inversions or transpositions, and we have not attempted to analyze the contributions of each of these mechanisms to the total pattern. However, genetic mapping (see below) indicates that several segments of the *L. lactis* subsp. *cremoris* FG2 and MG1363 chromosomes have been transposed with respect to one another. This process would be responsible for some of the differences between the two physical maps.

The chromosome sizes obtained from the physical mapping suggest a small amount of strain-dependent variability in the size of *L. lactis* chromosomes, within the range of 2.42-2.7 Mbp (Table 1). These values are bigger than those of the recently sequenced *Haemophilus influenzae* and *Mycoplasma genitalium*

Table 1. Genome and chromosome sizes of lactic acid bacteria[a].

Organism/strain	Size (Mbp)	Method of determination	Reference
L. lactis subsp. *lactis*			
DL11	2.58	Physical mapping	Tulloch et al., 1991
IL1403	2.42	Physical mapping	Le Bourgeois et al., 1992a
ML8	2.8	Renaturation kinetics	Jarvis & Jarvis, 1981
F166	2.5	PFGE	Le Bourgeois et al., 1989
C6	2.0	PFGE	Tanskanen et al., 1990
C10	2.3	PFGE	Tanskanen et al., 1990
P2	2.1	PFGE	Tanskanen et al., 1990
DRC2[b]	2.5	PFGE	Tanskanen et al., 1990
DRC3[b]	2.4	PFGE	Tanskanen et al., 1990
DRC4[b], 18–16[b]	2.3	PFGE	Tanskanen et al., 1990
L. lactis subsp. *cremoris*			
MG1363[c]	2.70, 2.56	Physical mapping	Davidson et al., 1995a; Le Bourgeois et al., 1995
FG2	2.69	Physical mapping	Davidson et al., 1995a
LM0230[c]	2.60	Physical mapping	Davidson et al., 1995a
ML3[c]	2.9	Renaturation kinetics	Jarvis & Jarvis, 1981
AM1	3.1	Renaturation kinetics	Jarvis & Jarvis, 1981
C2[c]	2.5, 2.7	PFGE	Le Bourgeois et al., 1989; Tanskanen et al., 1990
BK5	2.6, 2.6	PFGE	Le Bourgeois et al., 1989; Tanskanen et al., 1990
H2	2.6, 2.6	PFGE	Le Bourgeois et al., 1989; Tanskanen et al., 1990
C7, EB3	2.3	PFGE	Tanskanen et al., 1990
E8	2.2	PFGE	Tanskanen et al., 1990
WM1	2.3	PFGE	Tanskanen et al., 1990
134	2.5	PFGE	Tanskanen et al., 1990
480	2.4	PFGE	Tanskanen et al., 1990
S. thermophilus A054	1.82	Physical mapping	Roussel et al., 1994
Lb. acidophilus IP7613	1.85	PFGE	Roussel et al., 1993
Lb. gasseri IP102991	2.02	PFGE	Roussel et al., 1993
Lb. delbrueckii subsp. *bulgaricus*	2.3	PFGE	Leong-Morgenthaler et al., 1990
Lb. plantarum CM 1904	3.4	PFGE	Chevallier et al., 1994
O. oeni (11 strains)	1.8 - 2.1	PFGE	Lamoureux et al., 1993
O. oeni (30 strains)	1.86[d]	PFGE	Tenreiro et al., 1994
Leuc. mesenteroides L215	2.03	PFGE	Tenreiro et al., 1994
Leuc. mesenteroides L872	1.76	PFGE	Tenreiro et al., 1994
Leuc. mesenteroides L912	1.85	PFGE	Tenreiro et al., 1994
Leuc. citreum L4018	1.75	PFGE	Tenreiro et al., 1994
Leuc. citreum L4025	1.84	PFGE	Tenreiro et al., 1994
Leuc. gelidum L4026	2.17	PFGE	Tenreiro et al., 1994
Leuc. pseudomesenteroides L4027	2.05	PFGE	Tenreiro et al., 1994
P. acidilactici PAC1.0	1.86	PFGE	Luchansky et al., 1992
P. acidilactici LB42	2.13	PFGE	Luchansky et al., 1992
C. divergens	3.2	PFGE	Daniel, 1995

[a]Values determined by physical mapping are chromosome sizes. Values determined by PFGE analysis estimate chromosome or genome size (see text).
[b]*L. lactis* subsp. *lactis* biovar. diacetylactis strains.
[c]These strains are closely related, with similar, if not identical, chromosomes (Davies et al., 1981).
[d]Mean value for 30 strains with a range of 1.78–1.93 Mbp.

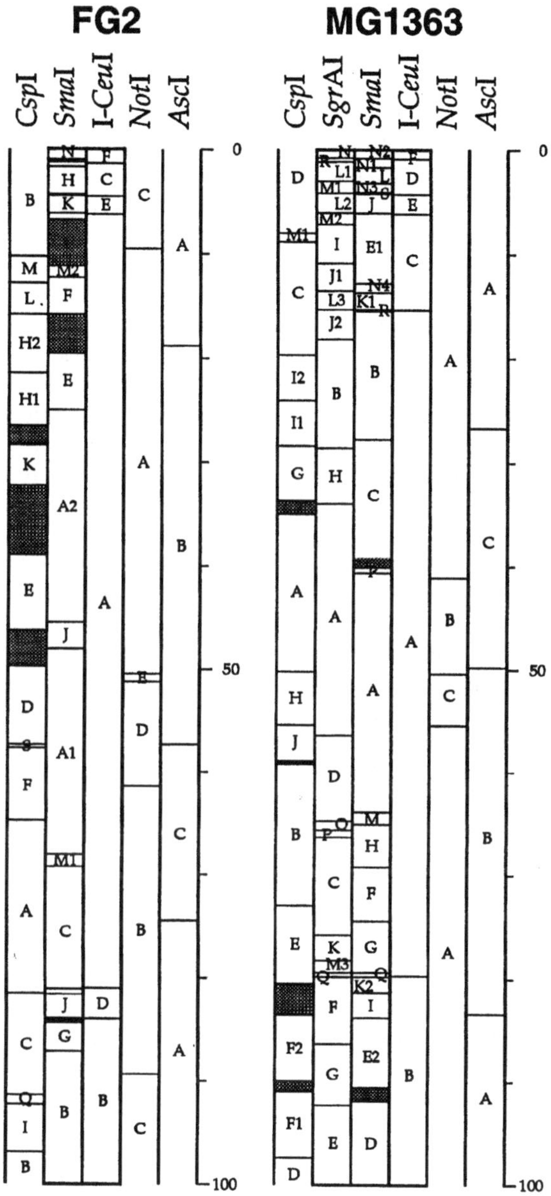

Figure 1. Comparison of the physical maps of the chromosome of *L. lactis* subsp. *cremoris* strains MG1363 and FG2 (Kordias et al., 1996). The circular map has been linearized at the I-*Ceu*I site in *rrnA*. Shaded areas indicate regions of uncertainty in the map.

chromosomes (Fleischmann et al., 1995; Fraser et al., 1995), but considerably smaller than those from *Escherichia coli* and *Bacillus subtilis* (Cole & Saint Girons, 1994).

Genetic mapping of the L. lactis *chromosome*

While physical maps are useful in establishing important features of the chromosome such as size and circularity (or linearity), their main use is as a foundation for the construction of genetic maps. The most accurate basis for the genetic map of a replicon is provided by its nucleotide sequence. In the absence of this information for the *L. lactis* chromosome, alternative approaches such as hybridization of Southern blots with specific gene probes were used. To avoid incorrect assignments, it is necessary to employ high-stringency hybridization conditions which in turn require that the gene probe be from *L. lactis* or a closely related bacterium. The resolution provided by this approach depends upon the resolution of the physical map. In order to map *L. lactis* genes precisely, Le Bourgeois et al., (1992b) developed a new vector, pRC1, containing unique sites for *Not*I, *Sma*I and *Apa*I. When a derivative of pRC1 carrying a cloned *L. lactis* copy of the gene is introduced into a *L. lactis* strain it undergoes site-specific recombination with the chromosomal copy of the gene. The introduced sites for the infrequent cutting restriction endonucleases can then be used to determine the precise map location of the gene and its orientation.

The first lactococcal genes mapped were those of the *rrn* operons in the *L. lactis* subsp. *lactis* DL11 chromosome (Tulloch et al., 1991). More detailed genetic maps have now been published for *L. lactis* subsp. *lactis* IL1403, *L. lactis* subsp. *lactis* DL11, *L. lactis* subsp. *cremoris* MG1363, and *L. lactis* subsp. *cremoris* FG2 (Le Bourgeois et al., 1992a; Davidson et al., 1995a; Le Bourgeois et al., 1995). The most detailed map is that of *L. lactis* subsp. *cremoris* MG1363, and Figure 2 shows our current version of this map which combines the previously reported information (Davidson et al., 1995a; Le Bourgeois et al., 1995) with our recent unpublished data. This map contains a total of 115 genetic loci, including the four in each *rrn* operon (see below). Forty seven of these loci have been mapped precisely and the direction of transcription of 43 of them has been determined. On the basis of the average gene sizes of 1042 and 1040 bp found for the *H. influenzae* and *M. genitalium* chromosomes (Fleischmann et al., 1995; Fraser et al., 1995), we can expect 2,400 genes in a 2.5-Mbp bacterial chromosome such as that of *L. lactis*. Thus, the present map for *L. lactis* subsp. *cremoris* MG1363 accounts for only 4-5% of the

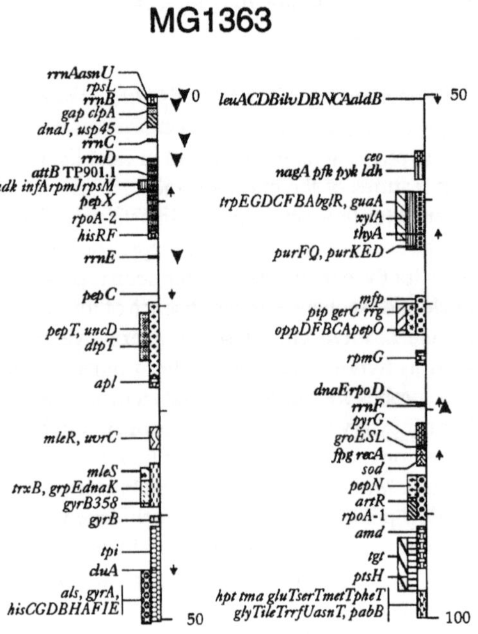

Figure 2. Genetic map of the *L. lactis* subsp. *cremoris* MG1363 chromosome. The circular map has been linearized at the I-*Ceu*I site in *rrnA*. The absence of a comma between two gene names indicates that the genes are cotranscribed in the same transcriptional unit. The names of genes whose location has been precisely established are shown in bold face. Arrowheads indicate the direction of transcription of *rrn* operons, and arrows indicate the direction of transcription of other transcriptional units. Gene locations are from Davidson et al., (1995a), Le Bourgeois et al., (1995) and Kordias et al., (1996).

expected genetic complement of this organism. Given that it is nearly nine years since the physical mapping studies on the *L. lactis* chromosome commenced, more rapid strategies for characterizing it and evaluating its genetic potential are clearly required.

One of the six *rrn* operons from *L. lactis* subsp. *lactis* IL1403 has been cloned and sequenced (Chiaruttini & Milet, 1993). It is a typical eubacterial *rrn* operon, containing one gene each for 23S rRNA (*rrl*), 16S rRNA (*rrs*), 5S rRNA (*rrf*), $tRNA^{Ala}$ (*alaT*), and $tRNA^{Asn}$ (*asnU*) in the order *rrs alaT rrl rrf asnU*. Further analyses with *L. lactis* subsp. *cremoris* MG1363 DNA indicated that the other five *rrn* operons also contain a $tRNA^{Ala}$ gene but no $tRNA^{Asn}$ gene, although a second $tRNA^{Asn}$ gene was detected elsewhere in the chromosome (Kordias et al., 1993; Davidson et al., 1995a). Thus, the *L. lactis* subsp. *cremoris* MG1363 chromosome has six $tRNA^{Ala}$ genes and at least two different types of *rrn* operon. The second $tRNA^{Asn}$ gene (*asnT*) was found by Nilsson & Johansen (1994) to be in an operon consisting of seven tRNA genes and a seventh 5S rRNA gene (*rrsU*) (Figure 2).

The location and orientation of the *rrn* operons in bacterial chromosomes are of interest because they can provide a clue to the location of *oriC*, the origin of chromosome replication (Tulloch et al., 1991). The four mapped *L. lactis* strains each contain six *rrn* operons clustered in 40% of the chromosome (Figure 2) in a manner analogous to that found in many other eubacterial chromosomes (Cole & Saint Girons, 1994). If replication and transcription of the *L. lactis rrn* operons were colinear, as it is in characterized eubacteria such as *E. coli* (Cole & Saint Girons, 1994), *oriC* would be located between *rrnA* and *rrnF*. This region of the chromosome is located on an I-*Ceu*I fragment of approximately 500 kbp which could be a good source of DNA for attempts at shotgun cloning of a functional *oriC*. One argument against *oriC* being in this region is that *gyrB*, which is often (but not always) close to *oriC* (Cole & Saint Girons, 1994), is located between *rrnE* and *rrnF* (Figure 2).

What does a comparison of the four available *L. lactis* genetic maps reveal? Although the genetic maps of the two subspecies *lactis* strains are indistinguishable at the current level of resolution (Kordias et al., 1996), indications of plasticity in the chromosome are seen when the subspecies *cremoris* is considered (Figure 3 and Davidson et al., (1995a)). For example, the inter subspecies comparison of the *L. lactis* subsp. *cremoris* MG1363 and *L. lactis* subsp. *lactis* DL11 maps reveals a large inversion (marked Img/dl in Figure 3) . A similar observation was noted when the genetic maps of *L. lactis* subsp. *cremoris* MG1363 and *L. lactis* subsp. *lactis* IL1403 were compared (Le Bourgeois et al., 1995). The limits of the inversion in *L. lactis* subsp. *cremoris* MG1363 are between *pepC* and *dtpT* on the left side and between either *pip* or the *opp* operon and *rpmG* on the right side. The length of the inversion is therefore between 45 and 55%, that is half, of the chromosome. At present there are no experimental data to indicate the mechanism by which this large scale chromosomal rearrangement was generated. There is no *rrn* operon at either of its limits, so the inversion was not generated by recombination between two *rrn* operons in a manner similar to the well characterized chromosomal inversion in the *E. coli* strain W3110 (Hill & Harnish, 1981; Hill et al., 1990). The possibility that it was mediated by recombination between two IS*1076* insertion elements has also been ruled out (Le Bourgeois et al., 1995), although one of the other IS elements found in the *L. lactis* chromosome (discussed below) may have been involved in a recombinative event. The phenomenon of recombinative events that rearrange the bacterial chromosome is considered in detail by Mahan et al., (1990).

There are also significant differences between *L. lactis* subsp. *cremoris* MG1363 and the two *L. lactis* subsp. *lactis* strains (DL11 and IL1403) in the length of the DNA that separates the *rrnB* and *rrnC* operons, and the *rrnE* and *rrnF* operons (Figure 3 and (Le Bourgeois et al., 1995)). On the basis of the current limited data, the size of the *rrnB*/*rrnC* separation is subspecies specific and its measurement could provide a useful means for subspecies identification.

A comparison of the maps of two subspecies *cremoris* strains provides additional evidence for genome plasticity (Kordias et al., 1996). Four discreet regions of the *L. lactis* subsp. *cremoris* FG2 chromosome (marked TIfg/mg1, TIfg/mg2, TIfg/mg3 and TIfg/mg4 in Figure 3) are inverted and occupy different locations with respect to the *L. lactis* subsp. *cremoris* MG1363 chromosome. The nature of the differences suggests that there have been four separate genetic events during the divergent evolution of these two strains of the same subspecies, with each event involving the transposition and inversion of between 30 and 350 kbp of DNA. The mechanism by which these events occurred is not known.

Other examples of plasticity in the *L. lactis* genome have also been noted (Davidson et al., 1995a; Kordias et al., 1996) and two of these are indicated in Figure 3. The first involves *L. lactis* subsp. *cremoris* LM0230, which was generated as a plasmid- and prophage-cured derivative of *L. lactis* subsp. *cremoris* C2 by nitrosoguanidine treatment and UV irradiation (Efstathiou & McKay, 1977). Its genealogy is similar to that of *L. lactis* subsp. *cremoris* MG1363 which was cured of plasmids and prophage by protoplast treatment of *L. lactis* subsp. *cremoris* 712, the progenitor of *L. lactis* subsp. *cremoris* C2 (Davies et al., 1981; Gasson, 1983). The restriction enzyme digestion patterns for strains 712 and C2 are identical, but those of strains LM0230 and MG1363 are different (Tanskanen et al., 1990; Davidson et al., 1995a). The differences result from three separate deletions of DNA from the *L. lactis* subsp. *cremoris* LM0230 chromosome, causing a total of 105 kbp to be absent (Figure 3). This observation emphasizes the susceptibility of the *L. lactis* subsp. *cremoris* chromosome to rearrangement.

The second example of plasticity (marked ∇FG in Figure 3) was observed when cultures of *L. lactis* subsp. *cremoris* FG2 were incubated in the presence of lytic phage to isolate phage resistant derivatives for use as industrial starter strains. Some of the derivatives which survived the challenge exhibited a restriction fragment length polymorphism of SmA2 (Figures 1 and 3), which involved the duplication of *gyrB* in one case, and of *uvrC* in another (Davidson et al., 1995a; Kordias et al., 1996). The rearrangements, which involved either an inversion or a translocation of the duplicated DNA, may have been part of a stress response to the presence or infection by phage in the same way that genome rearrangements in *Streptomyces ambofaciens* occur more frequently in stressed cells (Volff et al., 1993). Exposure of starter strains to infecting phage is normal during dairy fermentations and the resultant stress may be an important contributing factor to the rate and course of evolution of the *L. lactis* chromosome.

Plasmids in L. lactis

Plasmids are a normal component of the genome of industrial *L. lactis* strains. Indeed, certain plasmids are essential components because they carry the genetic determinants of industrially-relevant characteristics (McKay, 1983). It is possible, though conjectural, that these plasmids are a recent addition to the genome that enables *L. lactis* to use milk as a growth medium. The plasmids are diverse in size, function and distribution and many of them have received a great deal of research attention. It is not feasible in this article to provide more than a brief overview of this field which could be the subject of a separate review of its own. The following articles include excellent synopses of plasmid research areas and are recommended for their detailed listings of relevant articles on the subject: Lac^+ plasmids and their role in acid production, McKay (1983) and Davidson et al., (1995b); bacteriocin production and resistance, Klaenhammer (1993); Prt^+ plasmids, Tan et al., (1993) and Law et al., (1992); phage resistance, Hill (1993); citrate utilization, McKay (1983) and Hugenholtz (1993); ropiness, Neve et al., (1988); conjugative transfer, Mills et al., (1994).

L. lactis plasmids range in size from 3 to >130 kbp, and in number per cell from 2 to 11, although most strains contain between 4 and 7. As examples, *L. lactis* subsp. *cremoris* Wg2 has five plasmids of 2.2-24 kbp, *L. lactis* subsp. *lactis* K1 has five plasmids of about 2-50 kbp, and *L. lactis* subsp. *lactis* biovar. diacetylactis Bu2 has six plasmids of 4.4-51 kbp (Otto et al., 1982; Leenhouts et al., 1991b; Donkersloot & Thompson,

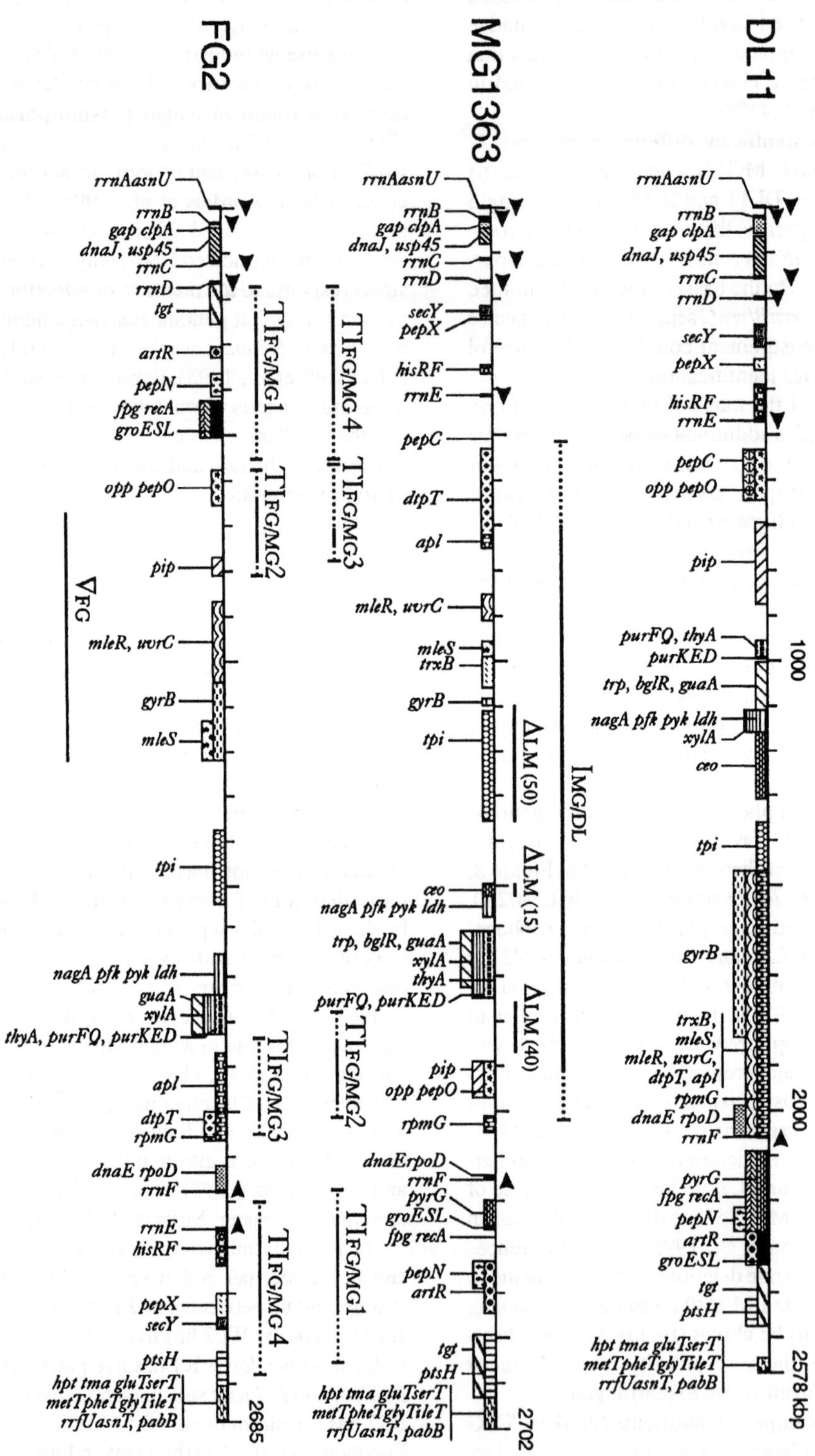
DL11
MG1363
FG2
rrnAasnU
rrnB
gap clpA
dnaJ, usp45
rrnC
rrnD
tgt
secY
pepX
artR
pepN
hisRF
rrnE
fpg recA
groESL
pepC
opp pepO
dtpT
apl
pip
mleR, uvrC
purFQ, thyA
purKED
1000
mleS
trxB
trp, bglR, guaA
gyrB
nagA pfk pyk ldh
xylA
tpi
ceo
thyA
purFQ, purKED
guaA
thyA, purFQ, purKED
trxB,
mleS,
mleR, uvrC,
dtpT, apl
rpmG
dnaE rpoD
dnaErpoD
2000
rrnF
pyrG
ptsH
hpt tma gluTserT
metTpheTglyTileT
rrfUasnT, pabB
2578 kbp
2702
2685
TIFG/MG1
TIFG/MG2
TIFG/MG3
TIFG/MG4
IMG/DL
ΔLM (50)
ΔLM (15)
ΔLM (40)
VFG

←

Figure 3. Comparison of the genetic maps of the chromosomes of *L. lactis* subsp. *lactis* DL11, *L. lactis* subsp. *cremoris* MG1363 and *L. lactis* subsp. *cremoris* FG2. The maps have been linearized and aligned at the I-*CeuI* site in *rrnA*. Genes are represented by the same fill pattern in each map. The two lines marked Img/dl indicate a region of the *L. lactis* subsp. *lactis* DL11 chromosome (top line) that is inverted with respect to the *L. lactis* subsp. *cremoris* MG1363 chromosome (lower line), with the maximal and minimal limits of the inversion indicated by dotted and solid lines, respectively. The lines marked δlm(50), δlm(15), and δlm(40) indicate regions of the *L. lactis* subsp. *cremoris* MG1363 chromosome that carry deletions of 50, 15 and 40 kbp, respectively, in the *L. lactis* subsp. *cremoris* LM0230 chromosome (Davidson et al., 1995a). The four lines immediately above the FG2 map, marked TIfg/mg1, TIfg/mg2, TIfg/mg3 and TIfg/mg4, indicate regions of the *L. lactis* subsp. *cremoris* FG2 chromosome that are transposed and inverted with respect to the *L. lactis* subsp. *cremoris* MG1363 chromosome, with the maximal and minimal limits of the inversion indicated by dotted and solid lines, respectively. The set of four identically marked lines above them indicate the corresponding regions of the *L. lactis* subsp. *cremoris* MG1363 chromosome. The line marked ∇fg indicates a region of the *L. lactis* subsp. *cremoris* FG2 chromosome in which insertions of >90 kbp of DNA were observed during the isolation of phage-resistant derivatives (Davidson et al., 1995a). Arrowheads indicate the direction of transcription of *rrn* operons. The MG1363 map is a simplified version of that shown in Figure 2, while gene locations for the DL11 and FG2 maps are from Tulloch et al., (1991), Davidson et al., (1995a), and Kordias et al., (1996).

1990; Jahns et al., 1991). McKay (1983) lists a number of Lac$^+$ plasmids within the size range of 45-70 kbp and Prt$^+$ plasmids of 13-24 kbp. A Cit$^+$ plasmid of 7.8 kbp has been characterized by partial nucleotide sequence determination (Jahns et al., 1991). Conjugative transfer by *L. lactis* plasmids is common, and mobilization of non-conjugative plasmids can occur through cointegrate formation (possibly involving IS elements (Polzin & Shimizu-Kadota, 1987)) with a conjugative replicon (Mills et al., 1994). The 75-kbp Lac$^+$ Prt$^+$ plasmid pCI301 from *L. lactis* subsp. *lactis* UC317 recombines readily with other plasmids and is also able to recombine with the *L. lactis* subsp. *cremoris* MG1363 chromosome, yielding a stable Lac$^+$ Prt$^+$ integrant (Hayes et al., 1990). Cryptic plasmids from *L. lactis* strains have been modified to generate a range of cloning vectors (De Vos, 1987; Kok 1991). The versatility of many of these vectors is enhanced because the lactococcal replicon can also function in *E. coli* and other lactic acid bacteria.

The presence of a number of different plasmids in the cells of any one *L. lactis* strain raises questions about their compatibility and the mechanism(s) that enable them to coexist. The answers to these questions have commercial as well as fundamental significance because replicon instability that results in the loss of a plasmid can result in unsatisfactory fermentations. It was shown by Seegers et al., (1994) that the majority of lactococcal plasmids exhibit nucleotide sequence homology with either one of two different types of replicon: the rolling circle type found in pWV01, or the theta type found in pWV02. It seems that only one plasmid species with a rolling circle replicon is able to exist in a cell, but different plasmids with a pWV02-type theta replicon are mutually compatible and several can stably coexist in the same cell. Thus, pWV01 is the only plasmid with a rolling circle replicon in *L. lactis* subsp. *cremoris* Wg2, whereas the other four plasmids found in this strain, pWV02, pWV04, pWV03 and pWV05 have homologous replicons of the theta type (presumed for pWV03). Future analyses will reveal whether the maintenance of these plasmids with related replicons is completely independent or whether there is some interaction which affects important properties such as copy number.

Transposable elements in L. lactis

The topic of transposable elements in the lactococci has been reviewed in detail by Romero & Klaenhammer (1993). Five distinct types of insertion sequence elements, IS*S1*, IS*904*, IS*981*, IS*905* and IS*982*, have been found in *L. lactis* (Table 2). They are a common feature of the *L. lactis* genome, being present in significant numbers in both plasmids and the chromosome (Schäfer et al., 1991). IS elements promote genetic exchange between organisms by providing sites for cointegrate formation and by acting as composite transposons. Cointegrate formation allows intercellular exchange of genes located on Tra$^-$ replicons if one of the fused replicons is Tra$^+$. The activity of a composite transposon has a similar end result when the transposon moves its enclosed genes from a Tra$^-$ to a Tra$^+$ replicon. The transposition of an *in vitro* constructed composite transposon carrying an Em^r determinant has been demonstrated directly in *L. lactis* (Romero & Klaenhammer, 1993). IS elements in *L. lactis* are commonly associated with genes that encode essential properties such as the Lac$^+$, Prt$^+$, nisin production, phage resistance, and Tra$^+$ phenotypes (Romero & Klaenhammer, 1993), suggesting that they have played an important role in the evolutionary adaptation of *L. lactis* to its current environment.

Given the prevalence of IS elements in the *L. lactis* genome, it is tempting to postulate that the inver-

Table 2. IS elements of lactic acid bacteria.

Element	Host range[a]	Size (bp)	Copies per genome[b]	ORF length (codons)	Reference
The genus *Lactococcus*					
IS*S1S*	*L. lactis* subsp. *cremoris* ML3, many *L. lactis* strains	808		226	Polzin & Shimizu-Kadota, 1987
IS*904*	*L. lactis* subsp. *cremoris* FI5876	1241	≥7	253	Dodd et al., 1990
IS*981*	*L. lactis* subsp. *cremoris* LM0230, all 23 *L. lactis* strains tested	1222	4 - 26	87, 280[c]	Polzin & McKay, 1991
IS*905*	*L. lactis* subsp. *cremoris* FI7304, many (but not all) *L. lactis* strains	1313	≥16	391	Dodd et al., 1994
IS*982*	*L. lactis* subsp. *cremoris* SK11, all 12 *L. lactis* strains tested	1003	1 - 12	296	Yu et al., 1995
The genus *Lactobacillus*					
ISL*1*	*Lb. casei*	1256		93, 274	Shimizu-Kadota et al., 1985
IS*1201*	*Lb. helveticus*	1387	3 - 16	369	Tailliez et al., 1994
ISL2	*Lb. helveticus* (not *Lb. delbrueckii, Lb. acidophilus* or *S. thermophilus*)	858	4 - 21[b]	268	Zwahlen & Mollet, 1994
IS*1163*	*Lb. sake* L45	1180	2	87, 302[c]	Skaugen & Nes, 1994
IS*1223*	*Lb. johnsonii* NCK61	1492		177, 313[c]	Walker & Klaenhammer, 1994
The genus *Leuconostoc*					
IS*1165*	*Leuc. mesenteroides* subsp. *cremoris* DB1165 *Leuc. lactis, O. oeni, Pediococcus* spp., *Lb. helveticus, Lb. casei* (not *L. lactis*)	1553	4 - 13	412	Johansen & Kibenich, 1992
IS*1070*	*Leuc. lactis* NZ6009	1027	≥15 (5)[d]	305	Vaughan & De Vos, 1995

[a]First entry is the strain from which the element was first isolated.
[b]Strain dependency normal.
[c]A putative transposase produced by a fusion of both ORFs (467 residues for IS*1223*; 356 residues for IS*1163*) may be produced by a programmed frameshift of −1 nucleotides during the translation of mRNA containing these two ORFs (Chandler & Fayet, 1993).
[d]Number in parentheses indicates plasmid-borne copies.

sions, insertions, deletions and translocations it has undergone (see above) have been IS element- mediated. This may be the case, however at this stage there is no direct experimental evidence to support such a suggestion. Detailed analyses of the chromosome at the break points of one or more of these rearrangements should provide insights into their causative mechanism.

Some strains of *L. lactis* carry a chromosomal copy of the 68-kbp conjugative transposon Tn*5276* (Dodd et al., 1990; Steen et al., 1991; Rauch & De Vos, 1992; Gireesh et al., 1992). In addition to the genetic determinants required for conjugal transfer, Tn*5276* carries genes encoding a number of phenotypic characters it expresses: nisin production, nisin resistance, ability to ferment sucrose, and reduced sensitivity to infection by small isometric phages such as sk1. Insertion of Tn*5276* in the recipient chromosome occurs preferentially at one site in an orientation-specific manner, although several secondary chromosomal sites and multiple copies of the transposon were observed (Rauch & De Vos, 1992; Gireesh et al., 1992). It can exist in strains of both *L. lactis* subspecies.

A related conjugative transposon (Tn*5206*) which carries *ceo* as well as the determinants for nisin production and resistance and sucrose fermentation has been described (Donkersloot & Thompson, 1990; Donkersloot & Thompson, 1995). The gene *ceo* encodes N^5-(carboxyethyl)ornithine synthase, an enzyme that has only been detected in some strains of *L. lactis* and no other organism. As Donkersloot & Thompson (1995) point out, the linkage between *ceo* and conjugative transposons of the Tn*5306*/ Tn*5276* type is not universal, because sucrose-fermenting strains exist that lack N^5-(carboxyethyl)ornithine synthase, and vice versa. Our *ceo* mapping data for strains that do not carry conjugative transposons of this type (Figure 3) are in agreement with this conclusion.

χ sites in the L. lactis *chromosome*

In *E. coli*, the octanucleotide sequence 5′-GCTGGTGG-3′ (known as a χ site) protects DNA from the exonucleolytic activity of the major *E. coli* exonuclease ExoV. Recent analyses have suggested that the sequence 5′-GCGCGTG-3′ performs a similar function in *L. lactis* (Biswas et al., 1995). This putative χ-like element occurs in the 42 kbp of sequenced *L. lactis* subsp. *lactis* IL1403 DNA at a frequency approximately 4 times that expected on a random basis. It was proposed that these sequences protect *L. lactis* DNA from degradation after DNA damage or after its uptake from other *L. lactis* cells. Foreign DNA which contains the sequence infrequently would be broken down. This mechanism would ensure that genetic exchange with related organisms would not be prevented.

L. lactis *prophages*

Many *L. lactis* strains are lysogenic, carrying one or more chromosomally-integrated prophages (Davidson et al., 1990; Klaenhammer & Fitzgerald, 1994). Temperate phages normally integrate into a specific site in the bacterial chromosome (*attB*) by a mechanism that involves phage-encoded proteins and IHF, a normal cellular component. Interest in the analysis of *L. lactis* temperate phages and their integration systems is partly driven by the possibility of using them as precision delivery systems to introduce genes into the *L. lactis* chromosome. This has been achieved by host-mediated integration into the prophage itself (Chopin et al., 1989), or by integration into *attB* with the help of the phage-encoded integration system (Christiansen et al., 1994; van de Guchte et al., 1994). Analysis of temperate phage control regions should also identify genetic switches that give regulated gene expression in the lactic acid bacteria. Molecular definition of the genomes of these phages by sequence analysis is now providing us with some of the longest contiguous sequences from the lactococcal genome (Boyce et al., 1995a; van Sinderen et al., 1996), comparable in size with the 30 kbp of DNA that contains the *his, leu, ile* and *ald* operons (Delorme et al., 1992; Godon et al., 1992a).

The most thoroughly investigated temperate *L. lactis* phages are BK5-T, ϕLC3, Tuc2009, r1t and TP901–1, all of which were originally isolated from *L. lactis* subsp. *cremoris* strains. They have genome sizes of 40, 33, 39, 33.4 and 38.4 kbp, respectively (Boyce et al., 1995a; Lillehaug et al., 1991; Arendt et al., 1994; van Sinderen et al., 1996; Christiansen et al., 1994). The *attB* core sequences for the first four phages are identical, TTCTTCATG, and a copy of this sequence is also present in the *L. lactis* subsp. *cremoris* MG1363 chromosome (Boyce et al., 1995b; Lillehaug & Birkeland, 1993; van de Guchte et al., 1994). Even though these phages cannot infect *L. lactis* subsp. *cremoris* MG1363, integration plasmids based on ϕLC3, Tuc2009 and TP901–1 are able to integrate into the MG1363 chromosome. This is a fortunate phenomenon, given that this strain is widely used in research laboratories and that its genetic map is the most detailed of any *L. lactis* map. For TP901–1, the location of the integration site is shown in the genetic map in Figure 2. Interestingly, *attB* in this site is near the 3′ end of an open reading frame (ORF) that encodes a protein of unknown function, but phage integration does not alter the length of the ORF although it changes the identity of the last five codons (Lillehaug & Birkeland, 1993). It is therefore possible that the lysogenic process does not inactivate the protein encoded by this ORF.

The nucleotide sequence of almost half of the BK5-T genome has been reported (Boyce et al., 1995a). The sequenced DNA contains 32 ORFs of more than 60 codons that are grouped into at least seven transcriptional units. The phage integrase gene (*int*) and a putative functional homologue of the phage λ regulatory protein cI were identified, along with a nearby region containing sequence features characteristic of temperate-phage immunity regions (Boyce et al., 1995b; Boyce et al., 1995c). The largest ORF (1904 codons) in the sequenced DNA is notable because it contains four perfect tandem repeats of 468 bp and a fifth incomplete tandem repeat. Because the full length repeat contains a precise number of codons, the DNA repeats are reflected in extensive repeated sequences in the amino acid sequence of the encoded protein. Embedded within these repeats are 64 copies of the sequence Gly-X-Y. It was proposed that this protein may be involved in cell lysis during the lytic cycle of BK5-T or cell wall hydrolysis to enable phage DNA injection.

Spontaneous deletion mutants of BK5-T were readily obtained when the phage was grown vegetatively and three were characterized at the molecular level. All three had deletions of two or more of the 468-bp perfect repeats in ORF1904, and two of them had a second deletion. One of these mutants which had a lytic phenotype due to the loss of *attP* was of interest in an industrial context because it demonstrated the ease with which a temperate lactococcal phage can yield a lytic derivative.

Recently, the entire nucleotide sequence of the r1t genome has been determined (van Sinderen et al., 1996). A total of 50 putative protein-encoding regions (ORFs), all oriented in the same direction except for three, were identified. One of the three divergently oriented ORFs (designated *rro*) was shown to encode a DNA-binding protein with the regulatory properties and susceptibility to mitomycin C treatment expected of a phage λ cI homologue (Nauta et al., 1996). Thus, Rro would be responsible for establishing and maintaining lysogeny by r1t. It is anticipated that temperature-sensitive mutants of *rro* will be useful in achieving regulated gene expression in *L. lactis* and other lactic acid bacteria. Several r1t ORFs were identified as encoding phage structural proteins, and sequence similarity considerations were used to ascribe functions to six others. These included ORFs specifying the enzyme dUTPase and proteins involved in DNA replication, phage release from the host (holin and lysin), and integration of the phage into the host chromosome (integrase). Of particular interest was the observation that one ORF appears to be located within a self-splicing group I intron; it was suggested that the protein encoded by this ORF is likely to be an endonuclease.

Tools for genetic analysis and chromosome mapping in L. lactis

Effective analyses of the genetics of an organism and its chromosome require efficient and reliable techniques for gene inactivation and gene replacement. It is common for these techniques to use transposon- mediated insertional mutagenesis and vectors that enable homologous recombination. A number of procedures that were useful in this regard for other bacteria, such as those based on the use of Tn*916*, have been applied to *L. lactis*, but found only limited success. The discovery and characterization of insertion sequences in *L. lactis* led to the development of suicide vectors carrying IS*S1*RS (pRL1) and the iso-IS*S1* element IS*946* (pTRK145) (Le Bourgeois et al., 1992b; Romero & Klaenhammer, 1992). The use of pRL1 in the physical mapping of the *L. lactis* subsp. *cremoris* MG1363 chromosome has been noted above. Random integration of pTRK145 into the *L. lactis* subsp. *cremoris* MG1363 chromosome was observed, demonstrating the potential of this vector for use in gene inactivation studies (Dinsmore et al., 1993).

Promising results have been obtained with systems based on *repA* deficient forms of the plasmid pWV01 (Biswas et al., 1993; Law et al., 1995). Plasmid pWV01, originally isolated from *L. lactis* subsp. *cremoris* Wg2 (Otto et al., 1982; Leenhouts et al., 1991b), has an absolute requirement for the action of the *repA* product for replication to proceed from the plus origin (ori^+). Derivatives of pWV01 that are $repA^-$ are unable to replicate and are not maintained in the cell, unless RepA is provided in *trans* by another replicon. Leenhouts et al., (1991a) constructed a δ*repA* derivative of pWV01 carrying a *L. lactis* chromosomal fragment that was stably maintained in *B. subtilis* 895::$repA^+$. When this plasmid (pINT1) was introduced by transformation into *L. lactis* subsp. *cremoris* MG1363, 20 integrants per μg of plasmid DNA were obtained. Integration had proceeded by a Campbell-like single cross over mechanism at the chromosomal site of the cloned fragment.

This vector system has now been further refined, yielding what promises to be a major breakthrough in available genetic tools for *L. lactis* and other lactic acid bacteria (Law et al., 1995). The new Ori^+ $RepA^-$ Em^r vector pORI19 that was constructed carries *lacZ*α and the multiple cloning site from pUC19. It was used to prepare a *L. lactis* library in an *E. coli*::$repA^+$ strain and the library DNA was transformed into *L. lactis* subsp. *cremoris* MG1363, yielding 10^3 Em^r colonies per μg of DNA. The yield was improved 10- fold when the transformation event was separated from the integration event by using *L. lactis* subsp. *cremoris* MG1363 (pVE6007) as the recipient for transformation. Plasmid pVE6007 carries a temperature-sensitive mutant of *repA*, isolated previously by *in vitro* mutagenesis (Maguin et al., 1992). The RepA activity provided by pVE6007 allowed the incoming recombinant pORI19 to be stabilized and selected for during a 180-min incubation at the permissive temperature in the presence of erythromycin. A subsequent temperature jump forced homologous recombination of the integration plasmid into the chromosome and the loss of pVE6007. The use of this two-step procedure should be particularly beneficial when the recipient organism has a low transformation frequency, as do many industrial *L. lactis* strains.

An additional useful feature of this system is that the integration plasmid can readily be recovered from the chromosome for subsequent characterization by resupplying active RepA. In a test of the system, a novel Mal^- *L. lactis* subsp. *cremoris* MG1363 mutant was isolated and the integration plasmid that caused the Mal^- phenotype was recovered and sequenced. The nucleotide sequence had significant homology with the sequences of genes involved in sugar uptake from organisms such as *Streptococcus mutans* and *E. coli*. The *Sma*I site in pORI19 could be used for precise mapping of the chromosomal insertion sites of integrants, if it were not lost during *L. lactis* library construction. Derivatives of pORI19 carrying sites for enzymes such as *Not*I, *Asc*I, I-*Ceu*I and *Sgr*AI would enhance the flexibility of this mapping procedure. The addition of a site for an enzyme which does not digest the *L. lactis* chromosome, for example I-*Sce*I, would yield integrants with a single site for that enzyme and provide a simple method for determining chromosome sizes of different *L. lactis* strains (Jumas-Bilak et al., 1995).

In their work with pWV01 $repA^{ts}$ derivatives, Biswas et al., (1993) also developed a vector (pG^+host5) that can be used for gene inactivation in *L. lactis* . An additional feature of this system was that they were able to obtain gene replacement by suitable treatment of integrants. Counting the integrant strain as the starting point, they obtained between 1 and 40% of cells that underwent replacement recombination, in the absence of any selection. The pG^+host vector has been exploited further as a delivery system for IS*S1* (Polzin & McKay, 1992). Highly efficient transposition of a number of pGh:IS*S1* derivatives into the chromosomes of *L. lactis* subsp. *lactis* IL1403 and *L. lactis* subsp. *cremoris* MG1363 was observed, and interrupted genes were subsequently isolated as plasmid clones in *E. coli*. The procedure gave satisfactory transposition in *S. thermophilus* and *Enterococcus faecalis*.

An impressive chromosomal walking strategy has been developed to exploit the chromosomal integration of genes cloned into Ori^+ $RepA^-$ Em^r plasmid of the pORI19 class (Godon et al., 1994). The plasmid, containing portion of *L. lactis cluA*, was transformed into *L. lactis* subsp. *cremoris* MG1363 (pVE6007) and integrants were isolated as described above. DNA from an integrant was then isolated, digested with restriction endonucleases, religated and transformed into *E. coli* to enable the recovery of a plasmid carrying the desired region of the chromosome adjacent to the starting point in *cluA*.

At this stage, the results with modified pWV01 derivatives such pORI19, pG^+host5 and the pGh:IS*S1* derivatives indicate that they should be particularly effective tools for obtaining gene inactivation and gene replacement in *L. lactis*. In addition, our current understanding of the ori^+ replicon and preliminary analyses with pGh5:IS*S1* suggest that they may also work well in other lactic acid bacteria and Gram-positive organisms.

An additional powerful tool for genetic mapping in *L. lactis* has resulted from the discovery by Gasson et al., (1995) of a cryptic 50- kbp sex factor in the *L. lactis* subsp. *cremoris* MG1363 chromosome. In a series of elegant experiments, intercellular transfer of different chromosomal markers was observed at frequencies which suggested that the sex factor mobilizes the chromosome in a unidirectional polar manner reminiscent of *E. coli* Hfr strains. The factor transfers readily between a variety of *L. lactis* strains, but not to other lactic acid bacteria. The discovery of this sex factor and the characterization of its insertion site opens the door for genetic mapping in *L. lactis* by a truly genetic approach.

Streptococcus thermophilus

S. thermophilus is used in dairy fermentations to manufacture yoghurt and some hard cheeses of the Swiss type. Phylogenetically, it is the most closely related of the lactic acid bacteria to *L. lactis*. Analyses of the genome and other genetic features of the organism (reviewed by Mercenier (1990)) have lagged behind the analyses of the lactococci. The genome contains both a chromosome (see below) and plasmids (Herman & McKay, 1985; Mercenier, 1990). Several analytical procedures, based on genomic properties, have been developed to assist in the identification of the various strains of the organism, especially those of commercial value. Colmin et al., (1991) developed a probe that, with the exception of *S. salivarius*, was specific for *S. thermophilus* and could be used to differentiate strains by detecting restriction fragment length polymorphisms. Ribotyping and PFGE have also been used as fingerprinting techniques for strain identification (Pèbay et al., 1992; Salzano et al., 1993).

Mapping of the S. thermophilus *chromosome*

PFGE was used to construct a physical map of the *S. thermophilus* A054 chromosome for the enzymes *Sfi*I, *Bss*HII and *Sma*I (Roussel et al., 1994). The analysis indicated that *S. thermophilus* A054 has one circular chromosome of 1.82 Mbp. This value was similar to that of 1.75 Mbp obtained earlier for *S. thermophilus* ST1 by summing the sizes of restriction fragments (Le Bourgeois et al., 1989) and confirms that the *S. thermophilus* chromosome is smaller than the *L. lactis* chromosome.

The physical map was used to prepare a genetic map by hybridizing Southern blots with *S. thermophilus* gene probes. Eighteen rRNA genes, 6 tRNA genes and 11 protein-encoding genes were mapped (Roussel et al., 1994). Six *rrn* operons were detected, clustered in a region that corresponds to 30% of the chromosome; five of them were transcribed in one direction and the sixth was transcribed divergently. The arrangement is similar to that in *L. lactis* subsp. *lactis* DL11, *L. lactis* subsp. *lactis* IL1403 and *L. lactis* subsp. *cremoris* MG1363 (Figure 3). The gene *ldh* occupies similar positions in the *S. thermophilus* A054 and *L. lactis* subsp. *cremoris* MG1363 maps, using the *rrn* operons as a reference, but the locations of *recA* and *pepC* are different. The other *S. thermophilus* genes that were localized have not been mapped in *L. lactis*.

Genomic instability has been observed in *S. thermophilus*. As an extension of the chromosome mapping studies described above, several clones of *S. thermophilus* A054 obtained by successive subculturing on solid medium were also mapped (Roussel et al., 1994). Both of them had lost 6 kbp of DNA containing the equivalent of an *rrn* operon, so that only five *rrn* operons were left, and one had also lost 5 kbp from another region of the chromosome. It was proposed that the first deletion occurred as a result of homologous recombination between two adjacent *rrn* operons; the genetic mechanism that caused the second deletion is unknown.

Plasmids and temperate phages in S. thermophilus

The plasmid content and biology of *S. thermophilus* has been reviewed by Mercenier (1990). In general, the plasmid-free state is the more common in this species and most observed plasmids are small and presumably cryptic. For example, of 23 *S. thermophilus* strains examined by Herman & McKay (1985) five contained 4 - 18 copies of a single small (2.2 - 3.5 kbp) plasmid and the remainder were plasmid- free. The largest *S. thermophilus* plasmid reported is of 25.5 kbp (Mercenier, 1990). Curing of this plasmid did not affect the ability of its host to utilize carbohydrates or its antibiotic resistance profile. However, a 6.9-kbp plasmid, pA33, which affects the morphology, milk coagulation rate and antibiotic and phage resistance profiles of its host has been discovered (Mercenier, 1990). The nucleotide sequence of pA33 was determined but the functions of its ORFs were not apparent. A small, cryptic plasmid from *S. thermophilus* ST108 has been used to develop a shuttle vector which displays good stability in both *S. thermophilus* and *E. coli* (Solaiman & Somkuti, 1993). It was anticipated that this plasmid will enable the development of food- grade vectors for *S. thermophilus*.

Initial efforts to induce infective temperate phages from *S. thermophilus* strains by mitomycin C treatment or UV irradiation were unsuccessful, suggesting that the induction conditions and/or indicator strains used were not appropriate, or that lysogeny is rare in the species (see reviews by Davidson et al., (1990) and Mercenier (1990)). Two cases of lysogeny have now been confirmed. In the first, two *S. thermophilus* strains that were identified as having DNA homology with lytic-phage DNA yielded defective (tailless) phage particles when treated with mitomycin C (Neve et al., (unpublished) quoted by Davidson et al., (1990) and Mercenier (1990)). The particles from each strain contained 45-kbp DNA molecules of identical restriction patterns and shared homology with *S. thermophilus* lytic phage P55 DNA. The second case was observed when mitomycin C treatment caused lysis of one of 45 *S. thermophilus* strains examined (Carminati & Giraffa, 1992). Two of the 45 strains were suitable indicator strains for the phage in liquid medium, but not on agar plates. The genome of this phage was mapped, its size determined (40.9 ± 0.5 kbp) and it was found not to have cohesive ends. The availability of DNA from two different temperate phages should allow the development of integration vectors and determination of the prevalence of lysogeny in *S. thermophilus* to be evaluated by hybridization studies.

Tools for genetic analysis and chromosome mapping in S. thermophilus

Genomic integration, along with gene replacement and integrative gene expression, has been achieved with *S. thermophilus* ST11 (Mollet et al., 1993). The recipient was transformed with a linearized version of an *E. coli* vector containing short stretches of a deleted form of the *S. thermophilus lac* region. Integration of multiple copies of the plasmid into the *lac* region of the chromosome occurred by homologous recombination and resolution of the integrated plasmids was then obtained by continual growth of the integrants without selection for the plasmid. The resolved integrants had undergone a gene replacement and carried the deleted form of the *lac* region. In other experiments, the *cat* gene was integrated into the lactose operon and measurements of chloramphenicol acetyltransferase activities of this group of integrants reflected the response of the operon to growth on different carbohydrates. These analyses establish that integrative gene replacement can be carried out with *S. thermophilus*. Use of RepA$^-$ vectors of the pG$^+$host5 type should improve the efficiency and widen the scope of this process, enabling significant advances in our knowledge of the *S. thermophilus* genome and its genetic potential.

The genus *Lactobacillus*

The lactobacilli are a diverse group of microorganisms of widespread importance (Chassy & Murphy, 1994). They are used for many food fermentations, and although their use in the dairy industry is of particular economic importance, they are also essential to the production of sausages, olives, pickled vegetables and silage. In addition, they are a rich source of bacteriocins (Klaenhammer, 1993; Jack et al., 1995), while their presence in the gut is claimed to have significant benefits for a number of aspects of human health such as the control of gastro-intestinal infections, the control of serum cholesterol levels and the production of anti-carcinogenic activities. Intestinal lactobacilli were originally known as *Lb. acidophilus*, but this former group of bacteria is now recognized as consisting of the six species *Lb. acidophilus, Lb. amylovorus, Lb. crispatus, Lb. gallinarum, Lb. gasseri* and *Lb. johnsonii* (Klaenhammer, 1995). A number of excellent reviews on different aspects of the lactobacilli have been published in recent years: an outline and evaluation of the methods for differentiating and typing the sixty described species of the genus *Lactobacillus* by Dykes & von Holy (1994), and two fine reviews which provide extremely thorough listings of the relevant literature on the genetics of the lactobacilli and their plasmids by Pouwels & Leer (1993) and Klaenhammer (1995).

Chromosomes and plasmids of the genus Lactobacillus

The diversity of the genus is exemplified by the range in the G+C contents of its component species (Sneath et al., 1989). *Lb. gasseri* is typical of species with a low G+C content (33-35%), while *Lb. delbrueckii* (49-51%) and *Lb. sharpeae* (53%) are at the other end of the range. *Lb. helveticus* has the intermediate base composition of 38-40% G+C. Genome sizes for various *Lactobacillus* spp. have been determined by summing the sizes of restriction fragments generated by infrequently cutting endonucleases (Table 1). Apart from *Lb. plantarum*, which has a chromosome larger than that of any other lactic acid bacterium, all the chromosomes are of approximately 2 Mbp. No chromosome maps have been published for any member of the genus, a gap in our knowledge which hopefully will be redressed in the near future. In excess of 65 genes have been cloned from fourteen *Lactobacillus* species and their sequences determined, but only for *Lb. casei* is the number in double figures (Pouwels & Leer, 1993; Klaenhammer, 1995). The *rrn* operon complement has been reported as ≥ 5 copies for *Lb. plantarum* and ≥ 4 copies for *Lb gasseri* (Chevallier et al., 1994; Roussel et al., 1993). On the basis of I-*Ceu*I digestion products we have concluded that *Lb. gasseri* contains six *rrn* operons and that *Lb. acidophilus* contains four (El-Osta et al., 1996). From the above synopsis it can be concluded that our knowledge of the lactobacillal chromosome is scant, and that there is currently little available detailed information on this important topic.

Since the first report of plasmids in *Lb. casei* (Chassy et al., 1976), a number of studies have been directed specifically at identifying and describing plasmids in members of the genus (Pouwels & Leer, 1993; Klaenhammer, 1995). Although some species carry one or a few small, cryptic plasmids, they are not commonly found in other species such as *Lb. delbrueckii* subsp. *bulgaricus* and the intestinal lactobacilli. Thus, the plasmid composition of the lactobacillal genome contrasts with that of the lactococcal genome (see above). These differences may reflect a longer period of adaptation by the lactobacilli to their current environmental niche than that experienced by the dairy lactococci . Of notable interest is the recent report of several linear plasmids (150 and 50 kbp in size) in two strains of *Lb. gasseri* (Roussel et al., 1993). We have also observed a 48-kbp linear plasmid in *Lb. gasseri* ATCC 33323 (El-Osta et al., 1996). Linear replicons are unusual in bacteria, although linear plasmids and a linear chromosome have been found in the spirochaetes of the genus *Borrelia* and some other eubacteria (Ferdows & Barbour, 1989; Davidson et al., 1992; Hinnebusch & Tilly, 1993; Cole & Saint Girons, 1994). They have not been reported previously for any lactic acid bacteria and this discovery raises interesting questions about the possible origin of these *Lb. gasseri* plasmids and the genes they carry.

IS elements in the lactobacilli

The first IS element to be characterized from a lactic acid bacterium was ISL*1* from *Lb. casei*, which was found to have the industrially important property of being able to convert a temperate phage into a lytic phage (Shimizu-Kadota et al., 1985). For some time ISL*1* was the only reported lactobacillal IS element, but there has been a flurry of research activity in this area in recent years and four new elements have been reported (Table 2). The restricted distribution range of IS*1201* has led to the suggestion that it could serve as a specific probe for *Lb. helveticus* (Tailliez et al., 1994). Although these IS elements are important for the insights that their analysis provides into the biology of transposition in bacteria, their value extends further because they can be used in integration vectors as genetic tools for the analysis of the *Lactobacillus* chromosome (see below).

An element that has some features typical of transposable elements but lacks inverted repeats at either end and exhibits no homology with the sequences of known transposases was found in the genome of the *Lb. delbrueckii* subsp. *lactis* phage LL-K (Forsman & Alatossava, 1994). The element, designated a KIS-element by its discoverers, is 1.5 kbp long and is not present in the almost identical phage LL-H. Whether it is a new type of transposable element is not yet clear.

Genetic instability

Apart from the genetic instability associated with the IS elements described above there is an interesting report on this subject by Mollet & Delley (1990), who found a relatively high frequency of spontaneous mutants in the β-galactosidase gene of *Lb. bulgaricus* N123 (identical to NCDO 1489). Most mutational events did not involve a DNA rearrangement, but 10 out of 107 carried deletions of various lengths. The deletions were located in a specific region of the gene that contains short direct repeats of 14 and 13 bp together with short inverted repeat sequences overlapping the direct repeats. The difference between *Lb. bulgaricus* N123 and other *Lb. bulgaricus* strains that do not exhibit this high mutation frequency could be due to differences in the DNA sequence of the β- galactosidase gene or differences in an undefined host factor that affects the deletion frequency.

Prophages of the genus Lactobacillus

There are many reported examples of lysogeny in the genus (see reviews by Séchaud et al., (1988) and Davidson et al., (1990)). The first temperate phage to be characterized at the molecular level was ϕadh (43 kbp) from *Lb. gasseri* ADH (Raya et al., 1992). The nucleotide sequences of *attB, attP, attL* and *attR* for ϕadh and its host were determined, and a 16-bp core region required for site specific integration of the phage in the host chromosome was identified. All of the phage functions needed for this integration (including the integrase gene *int*) were shown to be present in a 4.3-kbp DNA fragment which was used as the basis for the construction of the integration vector pTRK182 (see below). In subsequent analyses of ϕadh, the minimum amount of phage DNA required for integration was narrowed down to 1,969 bp (Fremaux et al. 1993). Recently, the 36-kbp temperate phage mv4 from *Lb. delbrueckii* subsp. *bulgaricus* LT4 has been subjected to a similar analysis (Dupont et al., 1995). Interestingly, there is no homology between the integration core region sequences for ϕadh and mv4, and the mv4 integrase is more closely related to *L. lactis* phage ϕLC3 than to ϕadh.

Tools for genetic analysis and chromosome mapping in the genus Lactobacillus

A number of useful tools that will facilitate the genetic analysis of the lactobacillal chromosome have been developed in recent years. Of particular interest for chromosome mapping is the IS*1223*-based vector pTRK327 described by Walker & Klaenhammer (1994). When this vector, which lacks a Gram-positive origin of replication, was introduced into several *Lb. gasseri* strains, the important phenomenon of random integration into the chromosome was observed. Random chromosomal integration by pTRK327 has also been reported for *Lb. acidophilus* (Klaenhammer, 1995), but does not occur with *Lb. johnsonii* which carries resident copies of IS*1223* in its genome (Walker & Klaenhammer, 1994). pTK327 and related vectors should significantly facilitate the construction of physical maps of the *Lb. gasseri* and *Lb. acidophilus* chromosomes.

Vectors based on the integrative properties of the temperate phages ϕadh and mv4 have also been developed (Raya et al., 1992; Dupont et al., 1995). These vectors enable site-specific integration of DNA at *attB* in the chromosomes of *Lb. gasseri, Lb. johnsonii* and *Lb. plantarum* (Raya et al., 1992; Fremaux et al., 1993; Klaenhammer, 1995; Dupont et al., 1995). *attB* for mv4 is located in a highly conserved sequence of a minor $tRNA^{Ser}$ gene, raising the possibility that mv4-based vectors could be used to integrate DNA in the chromosome of a broad range of lactic acid bacteria which may contain the same $tRNA^{Ser}$ gene (Dupont et al., 1995).

A third class of vector system, which takes advantage of homologous recombination to achieve gene replacement, has been used in a number of laboratories (Bhowmik & Steele, 1993; Bhowmik et al., 1993; Leer et al., 1993; Hols et al., 1994; Fitzsimons et al., 1994). In one example, the construction of strains of *Lb. plantarum* with improved silage-making properties has been explored by using pGK13 which is poorly maintained in this organism (Hols et al., 1994; Fitzsimons et al., 1994). When a pGK13 derivative containing the *Lb. amylovorus* α-amylase gene inserted within a copy of the *Lb. plantarum* gene *cbh* was introduced into *Lb. plantarum*, single-crossover integration of the plasmid into the host copy of *cbh* was observed. Subsequently, a double-crossover event that resulted in the loss of pGK13 DNA was generated during prolonged growth of the integrant in the absence of any selection for the pGK13 antibiotic resistance marker. The strains that were produced carried the α-amylase gene inserted in *cbh*. This example, and a similar manipulation in which an internally deleted version of *pepXP* replaced the chromosomal copy of the gene without permanently introducing any foreign DNA, clearly establish that the lactobacilli have the genetic and enzymatic machinery required to undergo recombination-mediated gene replacement. It is to be likely that future developments in this area will benefit significantly from the pORI19 and pG^{+}host class of vectors developed (as described above) for *L. lactis*.

The genera *Leuconostoc* and *Oenococcus*

Bacteria of the genus *Leuconostoc* are heterofermentative organisms, often used in food fermentations in conjunction with other lactic acid bacteria. In dairy fermentations, they contribute to flavour development, particularly in Dutch-style cheeses. Reviews on the role of leuconostocs in dairy applications, and their capacity to produce bacteriocins have been published (Vedamuthu, 1994; Stiles, 1994). Although *Leuconostoc oenus*, which is used for malolactic fermentations in wine, has recently been reclassified as *Oenococcus oeni* (Dicks et al., 1995), it is conveniently included in this section.

The analysis of the *Leuconostoc* and *Oenococcus* genomes, which have G+C contents of 40-42% (Dicks et al., 1990), is still in the early stages, and few genes have been characterized at the molecular level (David et al., 1992). Chromosome sizes have been estimated by summing the sizes of restriction fragments separated by PFGE, yielding the following values: *O. oeni*, 1.81-2.09 Mbp; *Leuc. mesenteroides*, 1.8-2.0 Mbp; *Leuc. citreum*, 1.7-1.8 Mbp; *Leuc. gelidum*, 2.2 Mbp; *Leuc. pseudomesenteroides*, 2.0 Mbp (Lamoureux et al., 1993; Tenreiro et al., 1994). Two unrelated insertion sequence elements have been discovered in the genus (Table 2), IS*1165* in *Leuc. mesenteroides* and IS*1070* in *Leuc. lactis*, and their nucleotide sequences reported (Johansen & Kibenich, 1992; Vaughan & De Vos, 1995). Plasmids are a common component of the *Leuconostoc* genome, and the smaller (2.9 kbp) of the two present in the type strain *Leuc. lactis* 533 has been completely sequenced (Coffey et al., 1994). It was found to be a cryptic, rolling circle plasmid capable of replication in a broad range of Gram- positive bacteria.

The genus *Pediococcus*

Members of the genus *Pediococcus* have attracted interest because of their importance in the production of fermented meats and silage, and because some *P. acidilactici* strains produce an anti-listerial bacteriocin of commercial application (Marugg et al., 1992; Motlagh et al., 1994). The genus (G+C content 37 - 44%) is not well studied genetically. For example, the GenBank database contains the sequences of only 19 *Pediococcus* spp. genes, including the genes for the *P. pentosaceus* raffinose operon and *P. acidilactici* l - and d - lactate dehydrogenases (Garmyn et al., 1995a; Garmyn et al., 1995b; Leenhouts et al., 1994). The size of the chromosome of several *P. acidilactici* strains was determined to be 1.9 - 2.1 Mbp by summing restriction fragment sizes (Luchansky et al., 1992). Plasmids are a common component of the pediococcal genome and the ability to produce bacteriocins and to utilize raffinose, melibiose and sucrose are plasmid-linked (Gonzalez & Kunka, 1986). The complete nucleotide sequence (8,777 bp) of the *P. acidilactici* AcH plasmid that carries the determinants for pediocin AcH production has been determined (Motlagh et al., 1994) and a plasmid of >30 kbp has been observed in a number of *P. acidilactici* strains (Luchansky et al., 1992). A cryptic plasmid from *Tetragenococcus halophila* (formerly *P. halophila*) was recently found to have a theta-type of replicon, novel for lactic acid bacteria, that can be maintained in *Enterococcus, Pediococcus, Lactobacillus* and *Leuconostoc*, but not *Lactococcus* (Benachour et al., 1995). This replicon has the potential to provide an additional family of cloning vectors for general use in the lactic acid bacteria except for the lactococci.

The genus *Carnobacterium*

Bacteria of the genus *Carnobacterium* are isolated frequently from chilled meats. They form a single phylogenetic group and resemble lactobacilli phenotypically, but have low levels of genetic relatedness with the type species of the genus, *Lb. delbrueckii*, and the other lactobacilli (Wallbanks et al., 1990). *Carnobacterium* spp. produce a diverse range of bacteriocins. For example, one lantibiotic (carnocin), and four cystibiotic bacteriocins (carnobacteriocins A, BM1 and B2, and piscicolin 126) have been isolated and characterized from isolates of *C. piscicola* (Stoffels et al., 1992; Worobo et al., 1994; Quadri et al., 1994; Jack et al., 1996), and the properties of another have been reported (Pilet et al., 1995). Genetically, the carnobacteria are amongst the least studied of the lactic acid bacteria. The only report on the structure of the chromosome is a size estimate of 3.2 Mbp for *C. divergens* obtained by summing restriction fragment sizes (Daniel, 1995).

Concluding comments and future directions

A clear picture of the genomes of a number of the lactic acid bacteria is now emerging, with the most detailed information being available for the *L. lactis* genome. Chromosome sizes for lactic acid bacteria are 1.8-3.4 Mbp, placing them in the lower half of the size range for bacterial chromosomes. *L. lactis* and *S. thermophilus* contain a single, circular chromosome, but chromosome number and geometry remain to be determined for the other lactic acid bacteria. Plasmids are common in *L. lactis* and some of the lactobacilli and pediococci, but are not frequently present in *S. thermophilus*, *Lactobacillus delbrueckii* subsp. *bulgaricus* or the intestinal lactobacilli. The observation of a linear plasmid in the *Lb. gasseri* genome warrants confirmation and further analysis in view of the comparative rarity of linear replicons in bacteria.

For *L. lactis*, physical and genetic maps of the chromosomes of two subspecies *lactis* strains and two subspecies *cremoris* strains have now been determined. An important consequence of these mapping studies was the revelation of several inversions and translocations of blocks of genes, indicative of a significant degree of plasticity in the *L. lactis* genome. Five different IS elements are present in *L. lactis*, and these are known to provide sites for cointegrate formation between plasmids. One or more of these elements could be responsible for the chromosome rearrangements. Further analysis of this phenomenon is required.

Four *L. lactis* temperate phages have been subjected to detailed analyses and the elucidation of some of their nucleotide sequences provides an important basis for analysing gene regulation in the species. Most significantly, work on these phages has led to the development of cloning vectors which allow site specific integration of DNA into the *L. lactis* chromosome. The controlled and stable introduction of genes by this approach has much relevance for improvements in the properties of industrial strains.

The development in recent years of new vectors that enable gene inactivation, gene replacement, and gene recovery in *L. lactis* is particularly noteworthy. Much of our present knowledge of lactic acid bacteria genomes would not have been gained without two technical advancements that came from scientists outside the field of lactic acid bacteria research. These were the development of PFGE which enabled separation and size determination for DNA fragments larger than 20 kbp, and the development of electroporation which converted the technique of transformation for the lactic acid bacteria from an irreproducible and mysterious procedure to a straightforward scientific protocol. The new $RepA^-$ ori^+ and $RepA^{ts}$ vectors promise to produce a similar acceleration in the rate of analysis of the *L. lactis* genome and its component genes. These tools have potential for use in the analysis of the genetics of other Gram-positive organisms, thereby providing a *quid pro quo* for the benefits received previously from workers outside the lactic acid bacteria field. The discovery and characterization of a sex factor in *L. lactis* subsp. *cremoris* MG1363 will also have a profound impact on *L. lactis* chromosome analysis through its use as a gene mapping tool.

Our knowledge of the genomes of other lactic acid bacteria continues to lag behind that of *L. lactis*. This situation will probably continue for many years, although investigations into the genetics of the intestinal lactobacilli could expand through interest in their commercial utilization. The physical and genetic mapping of the *S. thermophilus* chromosome represents a major recent achievement, and it is likely that maps for one or more of the lactobacilli will soon be available. Other encouraging advances have been made in the development of specific vectors for the analysis of the *S. thermophilus* and lactobacillal chromosomes. If the Rep^- ori^+ vector systems work efficiently in these other lactic acid bacteria, we can expect rapid improvements in our knowledge of their chromosomes and gene complements in the immediate future.

Knowledge of the sites of *oriC* and *terC* in the lactic acid bacteria chromosomes is required to optimize the positioning of novel genes in the chromosome. Studies with *E. coli* have indicated that gene dosage in actively growing cells is inversely proportional to the distance of the gene from *oriC*. The magnitude of this effect in lactic acid bacteria should be determined, because it is likely that the effectiveness of certain genes (for example, those encoding phage resistance) will be copy number dependent. Thus, chromosomal stability considerations suggest that genes required in elevated copy number would be better introduced as a single copy near *oriC* than in multiple copies at another location.

Unfortunately, despite the rapid initial advances in mapping the *L. lactis* chromosome (the map of the *L. lactis* subsp. *lactis* DL11 was one of the first physical maps to be determined for a bacterial chromosome) our knowledge of its chemical structure has advanced relatively slowly when compared with recent progress made in the analysis of the chromosomes of bacteria such as *H. influenzae, M. genitalium, E. coli* and *B. subtilis.* A genetic investigation of *L. lactis* or any of the other lactic acid bacteria would be achieved more effectively if the complete sequence of all genes of interest were available as the starting point for the analysis, rather than as the product of the first phase of the investigation. There is no doubt that organisms with known nucleotide sequences will be more attractive and tractable subjects for genetic, biochemical and physiological studies in the future than those where the chromosome is poorly-defined. It is to be hoped, therefore, that this Symposium on Lactic Acid Bacteria will be the forum in the not-too-distant future for a presentation of the complete nucleotide sequence of the chromosome of at least one of the lactic acid bacteria.

Acknowledgements

We are indebted to many scientists for making available sequence data or samples of lactococcal DNA for mapping studies, or for providing information prior to publication. Specifically we want to mention T. Araya-Kojima, C. Batt, M. Cancilla, A. Chopin, M.-C. Chopin, S. Condon, M. Denayrolles, W. de Vos, P. Duwat, T. Klaenhammer, T. Koivula, J. Kok, R. Llanos, D. Nilsson, P. Renault, P. Strøman, J. Thompson, G. Venema, K. Venema and L. Vining. BD also wishes to acknowledge I. Old and I. Saint Girons of the Pasteur Institute for many stimulating discussions about the structure of bacterial genomes. We are grateful to the Australian Dairy Research and Development Corporation and the Australian Research Council for financial support that enabled some of the work described above to be carried out, and to Clare Phillips for assistance with the manuscript preparation.

References

Arendt EK, Daly C, Fitzgerald GF & Guchte M van de (1994) Molecular characterization of lactococcal bacteriophage Tuc2009 and identification and analysis of genes encoding lysin, a putative holin, and two structural proteins. Appl. Environ. Microbiol. 60: 1875–1883

Benachour A, Frére J & Novel G (1995) pUCL287 plasmid from *Tetragenococcus halophila (Pediococcus halophila)* ATCC 33315 represents a new theta-type replicon family of lactic acid bacteria. FEMS Microbiol. Lett. 128: 167–176

Bhowmik T, Fernández L & Steele JL (1993) Gene replacement in *Lactobacillus helveticus*. J. Bacteriol. 175: 6341–6344

Bhowmik T & Steele JL (1993) Development of an electroporation procedure for gene disruption in *Lactobacillus helveticus* CNRZ 32. J. Gen. Microbiol. 139: 1433–1439

Biswas I, Gruss A, Ehrlich SD & Maguin E (1993) High- efficiency gene inactivation and replacement system for Gram-positive bacteria. J. Bacteriol. 175: 3628–3635

Biswas I, Maguin E, Ehrlich SD & Gruss A (1995) A 7-base pair sequence protects DNA from exonucleolytic degradation in *Lactococcus lactis*. Proc. Natl. Acad. Sci. USA 92: 2244–2248

Boyce JD, Davidson BE & Hillier AJ (1995a) Sequence analysis of the *Lactococcus lactis* temperate bacteriophage BK5-T and demonstration that the phage DNA has cohesive ends. Appl. Environ. Microbiol. 61: 4089–4098

—(1995b) Spontaneous deletion mutants of the *Lactococcus lactis* temperate bacteriophage BK5-T and localization of the BK5-T *attP* site. Appl. Environ. Microbiol. 61: 4105–4109

—(1995c) Identification of prophage genes expressed in lysogens of the *Lactococcus lactis* bacteriophage BK5-T. Appl. Environ. Microbiol. 61: 4099–4104

Carminati D & Giraffa G (1992) Evidence and characterization of temperate bacteriophage in *Streptococcus salivarius* subsp. *thermophilus* St18. J. Dairy Res. 59: 71–79

Chandler M & Fayet O (1993) Translational frameshifting in the control of transposition in bacteria. Mol. Microbiol. 7: 497–503

Chandry PS, Moore AS, Boyce JD, Davidson BE & Hillier AJ (1996) Nucleotide sequence analysis of the genome of sk1, an isometric-headed phage that infects *Lactococcus lactis*. (In preparation)

Chassy BM, Gibson E & Giuffrida A (1976) Evidence for extrachromosomal elements in *Lactobacillus*. J. Bacteriol. 127: 1576–1578

Chassy BM & Murphy CM (1994) *Lactococcus* and *Lactobacillus*. In: Sonenshein AL, Hoch JA & Losick R (Eds) *Bacillus subtilis* and other Gram-positive bacteria (pp 65–82) Amer. Soc. Microbiol., Washington, DC

Chevallier B, Hubert J-C & Kammerer B (1994) Determination of chromosome size and number of *rrn* loci in *Lactobacillus plantarum* by pulsed-field gel electrophoresis. FEMS Microbiol. Lett. 120: 51–56

Chiaruttini C & Milet M (1993) Gene organization, primary structure and RNA processing analysis of a ribosomal RNA operon in *Lactococcus lactis*. J. Mol. Biol. 230: 57–76

Chopin M-C, Chopin A, Rouault A & Galleron N (1989) Insertion and amplification of foreign genes in the *Lactococcus lactis* subsp. *lactis* chromosome. Appl. Environ. Microbiol. 55: 1769–1774

Christiansen B, Johnsen MG, Stenby E, Vogensen FK & Hammer K (1994) Characterization of the lactococcal temperate phage

TP901-1 and its site-specific integration. J. Bacteriol. 176: 1069–1076

Chu G, Vollrath D & Davis RW (1987) Separation of large DNA molecules by contour-clamped homogeneous electric fields. Science 234: 1582–1585

Coffey A, Harrington A, Kearney K, Daly C & Fitzgerald G (1994) Nucleotide sequence and structural organization of the small, broad-host-range plasmid pCI411 from *Leuconostoc lactis* 533. Microbiology 140: 2263–2269

Cole ST & Saint Girons I (1994) Bacterial genomics. FEMS Microbiol. Rev. 14: 139–160

Colmin C, Pébay M, Simonet JM & Decaris B (1991) A species-specific DNA probe obtained from *Streptococcus salivarius* subsp. *thermophilus* detects strain restriction polymorphism. FEMS Microbiol. Lett. 81: 123–128

Daniel P (1995) Sizing of the *Lactobacillus plantarum* genome and other lactic acid bacterial species by transverse alternating field electrophoresis. Curr. Microbiol. 30: 243–246

David S, Stevens H, Riel M van, Simons G & Vos WM de (1992) *Leuconostoc lactis* β-galactosidase is encoded by two overlapping genes. J. Bacteriol. 174: 4475–4481

Davidson BE, Powell IB & Hillier AJ (1990) Temperate bacteriophages and lysogeny in lactic acid bacteria. FEMS Microbiol. Rev. 87: 79–90

Davidson BE, Finch LR, Tulloch DL, Llanos R & Hillier AJ (1991) Physical and genetic mapping of the *Lactococcus lactis* chromosome. In: Dunny GM, Cleary PP & McKay LL (Eds) Genetics and molecular biology of streptococci, lactococci, and enterococci (pp 103–108) Amer. Soc. Microbiol., Washington, DC

Davidson BE, MacDougall J & Saint Girons I (1992) Physical map of the linear chromosome of the bacterium *Borrelia burgdorferi* 212, a causative agent of Lyme disease, and localization of rRNA genes. J. Bacteriol. 174: 3766–3774

Davidson BE, Kordias N, Baseggio N, Lim A, Dobos M & Hillier AJ (1995a) Genomic organization of lactococci. In: Ferretti JJ, Gilmore MS, Klaenhammer TR & Brown F (Eds) Genetics of streptococci, enterococci and lactococci. Dev. Biol. Stand. Vol. 85 (pp 411–422) Karger, Basel

Davidson BE, Llanos RM, Cancilla MR, Redman NC & Hillier AJ (1995b) Current research on the genetics of lactic acid production in lactic acid bacteria. Int. Dairy J. 5: 763–784

Davies FL, Underwood HM & Gasson MJ (1981) The value of plasmid profiles for strain identification in lactic streptococci and the relationship between *Streptococcus lactis* 712, ML3 and C2. J. Appl. Bacteriol. 51: 325–337

Vos WM de (1987) Gene cloning and expression in lactic streptococci. FEMS Microbiol. Rev. 46: 281–295

Delorme C, Ehrlich SD & Renault P (1992) Histidine biosynthesis genes in *Lactococcus lactis* subsp. *lactis*. J. Bacteriol. 174: 6571–6579

Delorme C, Godon JJ, Ehrlich SD & Renault P (1994) Mosaic structure of large regions of the *Lactococcus lactis* subsp. *cremoris* chromosome. Microbiology 140: 3053–3060

Dicks LM, Dellaglio F & Collins MD (1995) Proposal to reclassify *Leuconostoc oenus* as *Oenococcus oeni* [corrig.] gen. nov., comb. nov.. Int. J. Syst. Bacteriol. 45: 395–397

Dicks LMT, Vuuren HJJ van & Dellaglio F (1990) Taxonomy of *Leuconostoc* species particularly *Leuconostoc oenus* as revealed by numerical analysis of total soluble protein patterns, DNA base compositions and DNA - DNA hybridizations. Int. J. Syst. Bacteriol. 40: 83–91

Dinsmore PK, Romero DA & Klaenhammer TR (1993) Insertional mutagenesis in *Lactococcus lactis* subsp. *lactis* mediated by IS*946*. FEMS Microbiol. Lett. 107: 43–48

Dodd HM, Horn N & Gasson MJ (1990) Analysis of the genetic determinant for production of the peptide antibiotic nisin. J. Gen. Microbiol. 136: 555–566

—(1994) Characterization of IS*905*, a new multicopy insertion sequence identified in lactococci. J. Bacteriol. 176: 3393–3396

Donkersloot JA & Thompson J (1990) Simultaneous loss of N^5-(carboxyethyl)ornithine synthase, nisin production, and sucrose-fermenting ability by *Lactococcus lactis* K1. J. Bacteriol. 172: 4122–4126

—Cloning, expression, sequence analysis, and site-directed mutagenesis of the Tn*5306*-encoded N^5- (carboxyethyl)ornithine synthase from *Lactococcus lactis* K1. J. Biol. Chem. 270: 12226–12234

Dupont L, Boizet-Bonhoure B, Coddeville M, Auvray F & Ritzenthaler P (1995) Characterization of genetic elements required for site-specific integration of *Lactobacillus delbrueckii* subsp. *bulgaricus* bacteriophage mv4 and construction of an integration-proficient vector for *Lactobacillus plantarum*. J. Bacteriol. 177: 586–595

Dykes GA & von Holy A (1994) Strain typing in the genus *Lactobacillus*. Lett. Appl. Microbiol. 19: 63–66

Efstathiou JD & McKay LL (1977) Inorganic salts resistance associated with a lactose-fermenting plasmid in *Streptococcus lactis*. J. Bacteriol. 130: 257–265

El-Osta Y, Dobos M, Hillier AJ, Davidson BE (1996) Mapping studies with the *Lactobacillus acidophilus* ATCC 4356 and *L. gasseri* ATCC 33323 chromosomes. Proc. Fifth Symp. Lactic Acid Bact.

Ferdows MS & Barbour AG (1989) Megabase-sized linear DNA in the bacterium *Borrelia burgdorferi*, the Lyme disease agent. Proc. Natl. Acad. Sci. USA 86: 5969–5973

Fitzsimons A, Hols P, Jore J, Leer RJ, O'Connell M & Delcour J (1994) Development of an amylolytic *Lactobacillus plantarum* silage strain expressing the *Lactobacillus amylovorus* α- amylase gene. Appl. Environ. Microbiol. 60: 3529–3535

Fleischmann RD, Adams MD, White O, Clayton RA, Kirkness EF, Kerlavage AR, Bult CJ, Tomb J-F, Dougherty BA, Merrick JM, McKenney K, Sutton G, FitzHigh W, Fields C, Gocayne JD, Scott J, Shirley R, Liu L-I, Glodek A, Kelley JM, Weidman JF, Phillips CA, Spriggs T, Hedblom E, Cotton MD, Utterback TR, Hanna MC, Nguyen DT, Saudek DM, Brandon RC, Fine LD, Fritchman JL, Fuhrmann JL, Geoghagen NSM, Gnehm CL, McDonald LA, Small KV, Fraser CM, Smith HO & Venter JC (1995) Whole-genome random sequencing and assembly of *Haemophilus influenzae* Rd. Science 269: 496– 512

Forsman P & Alatossava T (1994) Repeated sequences and the sites of genome rearrangements in bacteriophages of *Lactobacillus delbrueckii* subsp. *lactis*. Arch. Virol. 137: 43–54

Fraser CM, Gocayne JD, White O, Adams MD, Clayton RA, Fleischmann RD, Bult CJ, Kerlavage AR, Sutton G, Kelley JM, Fritchman JL, Weidman JF, Small KV, Sandusky M, Fuhrmann J, Nguyen D, Utterback TR, Saudek DM, Phillips CA, Merrick JM, Tomb J-F, Dougherty BA, Bott KF, Hu P-C, Lucier TS, Peterson SN, Smith HO, Hutchison III CA & Venter JC (1995) The minimal gene complement of *Mycoplasma genitalium*. Science 270: 397–403

Fremaux C, De Antoni GL, Raya RR & Klaenhammer TR (1993) Genetic organization and sequence of the region encoding integrative functions from *Lactobacillus gasseri* temperate bacteriophage ϕadh. Gene 126: 61–66

Garmyn D, Ferain T, Bernard N, Hols P & Delcour J (1995a) Cloning, nucleotide sequence, and transcriptional analysis of the *Pediococcus acidilactici* l-(+)-lactate dehydrogenase gene. Appl. Environ. Microbiol. 61: 266–272

Garmyn D, Ferain T, Bernard N, Hols P, Delplace B & Delcour J (1995b) *Pediococcus acidilactici ldhD* gene: cloning, nucleotide sequence and structural analysis. J. Bacteriol. 177: 3427– 3437

Gasson MJ (1983) Plasmid complements of *Steptococcus lactis* NCDO 712 and other lactic streptococci after protoplast-induced curing. J. Bacteriol. 154: 1–9

Gasson MJ, Godon J-J, Pillidge C, Eaton TJ, Jury K & Shearman CA (1995) Characterization and exploitation of conjugation in *Lactococcus lactis*. Int. Dairy J. 5: 757–762

Gireesh T, Davidson BE & Hillier AJ (1992) Conjugal transfer in *Lactococcus lactis* of a 68-kilobase-pair chromosomal fragment containing the structural gene for the peptide bacteriocin nisin. Appl. Environ. Microbiol. 58: 1670–1676

Godon J-J, Chopin M-C & Ehrlich SD (1992a) Branched-chain amino acid biosynthesis genes in *Lactococcus lactis* subsp. *lactis*. J. Bacteriol. 174: 6580–6589

Godon J-J, Delorme C, Ehrlich SD & Renault P (1992b) Divergence of genomic sequences between *Lactococcus lactis* subsp.*lactis* and *Lactococcus lactis* subsp.*cremoris*. Appl. Environ. Microbiol. 58: 4045–4047

Godon J-J, Jury K, Shearman CA & Gasson MJ (1994) The *Lactococcus lactis* sex-factor aggregation gene *cluA*. Mol. Microbiol. 12: 655–663

Gonzalez CF & Kunka BS (1986) Evidence for plasmid linkage of raffinose utilization and associated α-galactosidase and sucrose hydrolase activity in *Pediococcus pentosaceus*. Appl. Environ. Microbiol. 51: 105–109

Guchte M van de, Daly C, Fitzgerald GF & Arendt EK (1994) Identification of *int* and *attP* on the genome of lactococcal bacteriophage Tuc2009 and their use for site-specific plasmid integration in the chromosome of Tuc2009-resistant *Lactococcus lactis* MG1363. Appl. Environ. Microbiol. 60: 2324–2329

Hayes F, Law J, Daly C & Fitzgerald GF (1990) Integration and excision of plasmid DNA in *Lactococcus lactis* subsp. *lactis*. Plasmid 24: 81–89

Herman RE & McKay LL (1985) Isolation and partial chracterization of plasmid DNA from *Streptococcus thermophilus*. Appl. Environ. Microbiol. 50: 1103–1106

Hill C (1993) Bacteriophage and bacteriophage resistance in lactic acid bacteria. FEMS Microbiol. Rev. 12: 87–108

Hill CW, Harvey S & Gray JA (1990) Recombination between rRNA genes in *Escherichia coli* and *Salmonella typhimurium*. In: Drlica K & Riley M (Eds) The bacterial chromosome (pp 335–340) Amer. Soc. Microbiol., Washington, DC

Hill CW & Harnish BW (1981) Inversions between ribosomal RNA genes of *Escherichia coli*. Proc. Natl. Acad. Sci. USA 78: 7069–7072

Hillier AJ & Davidson BE (1995) Pulsed-field electrophoresis. In: Howard J, Whitcombe DM & J.Howard and D.M.Whitcombe (Eds) Methods in Molecular Biology, Vol. 46: Diagnostic Bacteriology Protocols (pp 149–164) Humana Press Inc., Totowa, NJ

Hinnebusch J & Tilly K (1993) Linear plasmids and chromosomes in bacteria. Mol. Microbiol. 10: 917–922

Hols P, Ferain T, Garmyn D, Bernard N & Delcour J (1994) Use of homologous expression-secretion signals and vector-free stable chromosomal integration in engineering of *Lactobacillus plantarum* for α-amylase and levanase expression. Appl. Environ. Microbiol. 60: 1401–1413

Hugenholtz J (1993) Citrate metabolism in lactic acid bacteria. FEMS Microbiol. Rev. 12: 165–178

Jack RW, Tagg JR & Ray B (1995) Bacteriocins of Gram-positive bacteria. Microbiol. Rev. 59: 171–200

Jack RW, Wan J, Gordon J, Harmark K, Davidson BE, Hillier AJ, Wettenhall REH, Hickey MJ & Coventry MJ (1996) Characterization of the chemical and antimicrobial properties of piscicolin 126, a bacteriocin produced by *Carnobacterium piscicola* JG126. (In press)

Jahns A, Schäfer A, Geis A & Teuber M (1991) Identification, cloning and sequencing of the replication region of *Lactococcus lactis* ssp. *lactis* biovar. diacetylactis Bu2 citrate plasmid pSL2. FEMS Microbiol. Lett. 64: 253–258

Jarvis AW, Lubbers MW, Waterfield NR, Collins LJ & Polzin KM (1995) Sequencing and analysis of the genome of lactococcal phage c2. Int. Dairy J. 5: 963–976

Jarvis AW & Jarvis BDW (1981) Deoxyribonucleic acid homology among lactic streptococci. Appl. Environ. Microbiol. 41: 77– 83

Johansen E & Kibenich A (1992) Isolation and characterization of IS*1165*, an insertion sequence of *Leuconostoc mesenteroides* subsp. *cremoris* and other lactic acid bacteria. Plasmid 27: 200–206

Jumas-Bilak E, Maugard C, Michaux-Charachon S, Allardet- Servent A, Perrin A, O'Callaghan D & Ramuz M (1995) Study of the organization of the genomes of *Escherichia coli, Brucella melitensis* and *Agrobacterium tumefaciens* by insertion of a unique restriction site. Microbiology 140: 2425–2432

Kilpper-Bälz R, Fischer G & Schleifer KH (1982) Nucleic acid hybridization of group N and group D streptococci. Curr. Microbiol. 7: 245–250

Klaenhammer TR (1993) Genetics of bacteriocins produced by lactic acid bacteria. FEMS Microbiol. Rev. 12: 39–86

—(1995) Genetics of intestinal lactobacilli. Int. Dairy J. 5: 1019–1058

Klaenhammer TR & Fitzgerald GF (1994) Bacteriophages and bacteriophage resistance. In: Gasson MJ & Vos WM de (Eds) Genetics and biotechnology of lactic acid bacteria (pp 106–168) Blackie Academic and Professional, London

Kok J (1991) Special-purpose vectors for lactococci. In: Dunny GM, Cleary PP & McKay LL (Eds) Genetics and molecular biology of streptococci, lactococci and enterococci (pp 97–102) Amer. Soc. Microbiol., Washington, DC

Kordias N, Dobos M, Hillier AJ & Davidson BE (1993) Structure of the rRNA operons in *Lactococcus lactis* subsp. *lactis* MG1363. FEMS Microbiol. Rev. 12: P24

Kordias N, Baseggio N, Dobos M, Hillier AJ & Davidson BE (1996) Plasticity in the chromosome of *Lactococcus lactis*. J. Bacteriol. (In preparation)

Lamoureux M, Prévost H, Cavin JF & Diviès C (1993) Recognition of *Leuconostoc oenos* strains by the use of DNA restriction profiles. Appl. Microbiol. Biotechnol. 39: 547–552

Law J, Vos P, Hayes F, Daly C, Vos WM de & Fitzgerald G (1992) Cloning and partial sequencing of the proteinase gene complex from *Lactococcus lactis* subsp. *lactis* UC317. J. Gen. Microbiol. 138: 709–718

Law J, Buist G, Haandrikman A, Kok J, Venema G & Leenhouts K (1995) A system to generate chromosomal mutations in *Lactococcus lactis* which allows fast analysis of targeted genes. J. Bacteriol. 177: 7011–7018

Le Bourgeois P, Mata M & Ritzenthaler P (1989) Genome comparison of *Lactococcus* strains by pulsed-field gel electrophoresis. FEMS Microbiol. Lett. 59: 65–70

Le Bourgeois P, Lautier M, Mata M & Ritzenthaler P (1992a) Physical and genetic map of the chromosome of *Lactococcus lactis* subsp. *lactis* IL1403. J. Bacteriol. 174: 6752–6762

—(1992b) New tools for the physical and genetic mapping of *Lactococcus* strains. Gene 111: 109–114

Le Bourgeois P, Lautier M & Ritzenthaler P (1993) Chromosome mapping in lactic acid bacteria. FEMS Microbiol. Rev. 12: 109–124

Le Bourgeois P, Lautier M, Berghe L van den, Gasson MJ & Ritzenthaler P (1995) Physical and genetic map of the *Lactococcus lactis* subsp. *cremoris* MG1363 chromosome: comparison with that of *Lactococcus lactis* subsp. *lactis* IL 1403 reveals a large genome inversion. J. Bacteriol. 177: 2840–2850

Leenhouts KJ, Kok J & Venema G (1991a) Lactococcal plasmid pWVO1 as an integration vector for lactococci. Appl. Environ. Microbiol. 57: 2562–2567

Leenhouts KJ, Tolner B, Bron S, Kok J, Venema G & Seegers JFML (1991b) Nucleotide sequence and characterization of the broad-host-range lactococcal plasmid pWVO1. Plasmid 26: 55–66

Leenhouts KJ, Bolhuis AA, Kok JJ & Venema GG (1994) The sucrose and raffinose operons of *Pediococcus pentosaceus* PPE1.0. GenBank Accession No. L32093:

Leer RJ, Christiaens H, Verstraete W, Peters L, Posno M & Pouwels PH (1993) Gene disruption in *Lactobacillus plantarum* strain 80 by site-specific recombination: isolation of a mutant strain deficient in conjugated bile salt hydrolase activity. Mol. Gen. Genet. 239: 269–272

Leong-Morgenthaler P, Ruettener C, Mollet B, Hottinger H (1990) Construction of the physical map of *Lactobacillus bulgaricus*. Proc. Third Symp. Lactic Acid Bact. A28

Lillehaug D, Lindqvist BH & Birkeland NK (1991) Characterization of ϕLC3, a *Lactococcus lactis* subsp. *cremoris* temperate bacteriophage with cohesive single-stranded DNA ends. Appl. Environ. Microbiol. 57: 3206–3211

Lillehaug D & Birkeland N-K (1993) Characterization of genetic elements required for site-specific integration of the temperate lactococcal bacteriophage ϕLC3 and constructionof integration-negative ϕLC3 mutants. J. Bacteriol. 175: 1745–1755

Luchansky JB, Glass KA, Harsono KD, Degnan AJ, Faith NG, Cauvin B, Baccus-Taylor G, Arihara K, Bater B, Maurer AJ & Cassens RG (1992) Genomic analysis of *Pediococcus* starter cultures used to control *Listeria monocytogenes* in turkey summer sausage. Appl. Environ. Microbiol. 58: 3053–3059

Maguin E, Duwat P, Hege T, Ehrlich D & Gruss A (1992) New thermosensitive plasmid for Gram-positive bacteia. J. Bacteriol. 174: 5633–5638

Mahan MJ, Segall AM & Roth JR (1990) Recombination events that rearrange the chromosome: barriers to inversion. In: Drlica K & Riley M (Eds) The bacterial chromosome (pp 341–349) Amer. Soc. Microbiol., Washington, DC

Marugg JD, Gonzalez CF, Kunka BS, Lederboer AM, Pucci MJ, Toonen MY, Walker SA, Zoetmulder LCM & Vandenbergh PA (1992) Cloning, expression and nucleotide sequence of genes involved in production of pediocin PA–1, a bacteriocin from *Pediococcus acidilactici* PAC1.0. Appl. Environ. Microbiol. 58: 2360–2367

McKay LL (1983) Functional properties of plasmids in lactic streptococci. Antonie van Leewenhoek 49: 259–274

McKay LL & Baldwin KA (1990) Applications for biotechnology: present and future improvements in lactic acid bacteria. FEMS Microbiol. Rev. 87: 3–14

Mercenier A (1990) Molecular genetics of *Streptococcus thermophilus*. FEMS Microbiol. Rev. 87: 61–78

Mills DA, Choi CK, Dunny GM & McKay LL (1994) Genetic analysis of regions of the *Lactococcus lactis* subsp. *lactis* plasmid pRS01 involved in conjugative transfer. Appl. Environ. Microbiol. 60: 4413–4420

Mollet B, Knol J, Poolman B, Marciset O & Delley M (1993) Directed genomic integration, gene replacement, and integrative gene expression in *Streptococcus thermophilus*. J. Bacteriol. 175: 4315–4324

Mollet B & Delley M (1990) Spontaneous deletion formation within the β-galactosidase gene of *Lactobacillus bulgaricus*. J. Bacteriol. 172: 5670–5676

Motlagh A, Bukhtiyarova M & Ray B (1994) Complete nucleotide sequence of pSMB 74, a plasmid encoding the production of pediocin AcH in *Pediococcus acidilactici*. Lett. Appl. Microbiol. 18: 305–312

Nauta A, Sinderen D van, Karsens H, Smit E, Venema G & Kok J (1996) Inducible gene expression mediated by a repressor-operator system isolated from *Lactococcus lactis* bacteriophage r1t. Mol. Microbiol. (in press)

Neve H, Geis A & Teuber M (1988) Plasmid-encoded functions of ropy lactic acid streptococcal strains from Scandinavian fermented milk. Biochimie 70: 437–442

Nilsson D & Johansen E (1994) A conserved sequence in tRNA and rRNA promoters of *Lactococcus lactis*. Biochim. Biophys. Acta 1219: 141–144

Otto R, Vos WM de & Gavrieli J (1982) Plasmid DNA in *Streptococcus cremoris* Wg2: influence of pH on selection in chemostats of a variant lacking a protease plasmid. Appl. Environ. Microbiol. 43: 1272–1277

Pèbay M, Colmin C, Guèdon G, De Gaspèri C, Decaris B & Simonet JM (1992) Detection of intraspecific DNA polymorphism in *Streptococcus salivarius* subsp. *thermophilus* by a homologous rDNA probe. Res. Microbiol. 143: 37–46

Pilet M-F, Dousset X, Barré R, Novel G, Desmazeaud M & Piard J-C (1995) Evidence for two bacteriocins produced by *Carnobacterium piscicola* and *Carnobacterium divergens* isolated from fish and active against *Listeria monocytogenes*. J. Food Prot. 58: 256–262

Polzin KM & McKay LL (1991) Identification, DNA sequence, and distribution of IS*981*, a new, high-copy-number insertion sequence in lactococci. Appl. Environ. Microbiol. 57: 734–743

Polzin KM & McKay LL (1992) Development of a lactococcal integration vector by using IS*981* and a temperature-sensitive lactococcal replication region. Appl. Environ. Microbiol. 58: 476–484

Polzin KM & Shimizu-Kadota M (1987) Identification of a new insertion element, similar to gram-negative IS*26*, on the lactose plasmid of *Streptococcus lactis* ML3. J. Bacteriol. 169: 5481–5488

Pouwels PH & Leer RJ (1993) Genetics of lactobacilli: plasmids and gene expression. Antonie van Leewenhoek 64: 85–107

Quadri LE, Sailer M, Roy KL, Vederas JC & Stiles ME (1994) Chemical and genetic characterization of bacteriocins produced by *Carnobacterium piscicola* LV17B. J. Biol. Chem. 269: 12204–12211

Rauch PJG & Vos WM de (1992) Characterization of the novel nisin-sucrose conjugative transposon Tn*5276* and its insertion in *Lactococcus lactis*. J. Bacteriol. 174: 1280–1287

Raya RR, Fremaux C, De Antoni GL & Klaenhammer TR (1992) Site-specific integration of the temperate bacteriophage ϕadh into the *Lactobacillus gasseri* chromosome and molecular characterization of the phage *(attP)* and bacterial (*attB*) attachment sites. J. Bacteriol. 174: 5584–5592

Romero DA & Klaenhammer TR (1992) IS*946*-mediated integration of heterologous DNA into the genome of *Lactococcus lactis* subsp. *lactis*. Appl. Environ. Microbiol. 58: 699–702

—(1993) Transposable elements in lactococci: a review. J. Dairy Sci. 76: 1–19

Roussel Y, Colmin C, Simonet JM & Decaris B (1993) Strain characterization, genome size and plasmid content in the *Lactobacillus acidophilus* group (Hansen and Mocquot). J. Appl. Bacteriol. 74: 549–556

Roussel Y, Pébay M, Guedon G, Simonet J-M & Decaris B (1994) Physical and genetic map of *Streptococcus thermophilus* A054. J. Bacteriol. 176: 7413–7422

Salama M, Sandine W & Giovannoni S (1991) Development and application of oligonucleotide probes for identification of *Lactococcus lactis* subsp. *cremoris*. Appl. Environ. Microbiol. 57: 1313–1318

Salzano G, Moschetti G, Villani F & Coppola S (1993) Biotyping of *Streptococcus thermophilus* strains. Res. Microbiol. 144: 381–387

Schäfer A, Jahns A, Geis A & Teuber M (1991) Distribution of the IS elements IS*S1* and IS*904* in lactococci. FEMS Microbiol. Lett. 64: 311–317

Schleifer KH, Kraus J, Dvorak C, Kilpper-Bälz R, Collins MD & Fischer W (1985) Transfer of *Streptococcus lactis* and related streptococci to the genus *Lactococcus* gen. nov. System. Appl. Microbiol. 6: 183–195

Seegers JFML, Bron S, Franke CM, Venema G & Kiewiet R (1994) The majority of lactococcal plasmids carry a highly related replicon. Microbiol. 140: 1291–1300

Séchaud L, Cluzel P-J, Rousseau M, Baumgartner A & Accolas J-P (1988) Bacteriophages of lactobacilli. Biochimie 70: 401–410

Shimizu-Kadota M, Kiwaki M, Hirokawa H & Tsuchida N (1985) ISL*1*: a new transposable element in *Lactobacillus casei*. Mol. Gen. Genet. 200: 193–198

Sinderen D van, Karsens HA, Kok J, Terpstra P, Ruiters MHJ, Venema G & Nauta A (1996) Sequence analysis and molecular chracterization of the temperate lactococcal bacteriophage r1t. Mol. Microbiol. (in press)

Skaugen M & Nes IF (1994) Transposition in *Lactobacillus sake* and its abolition of lactocin S production by insertion of IS*1163*, a new member of the IS*3* family. Appl. Environ. Microbiol. 60: 2818–2825

Sneath PHA, Mair NS, Sharpe ME & Holt JG (ed.) (1989) Bergey's manual of systematic bacteriology, Vol 2 (p 1223) Williams & Wilkins, London

Solaiman DKY & Somkuti GA (1993) Shuttle vectors developed from a *Streptococcus thermophilus* native plasmid. Plasmid 30: 67–78

Steen MT, Chung YJ & Hansen JN (1991) Characterization of the nisin gene as part of a polycistronic operon in the chromosome of *Lactococcus lactis* ATCC 11454. Appl. Environ. Microbiol. 57: 1181–1188

Stiles ME (1994) Bacteriocins produced by *Leuconostoc* species. J. Dairy Sci. 77: 2718–2724

Stoffels G, Sahl H-G & Guòmundsdóttir A (1992) Isolation and properties of a bacteriocin-producing *Carnobacterium piscicola* isolated from fish. J. Appl. Bacteriol. 73: 309–316

Tailliez P, Ehrlich SD & Chopin M-C (1994) Characterization of IS*1201*, an insertion sequence isolated from *Lactobacillus helveticus*. Gene 145: 75–79

Tan PST, Poolman B & Konings WN (1993) Proteolytic enzymes of *Lactococcus lactis*. J. Dairy Res. 60: 269–286

Tanskanen EI, Tulloch DL, Hillier AJ & Davidson BE (1990) Pulsed-field gel electrophoresis of *Sma*I digests of lactococcal genomic DNA, a novel method of strain identification. Appl. Environ. Microbiol. 56: 3105–3111

Tenreiro R, Santos MA, Paveia H & Vieira G (1994) Inter-strain relationships among wine leuconostocs and their divergence from other *Leuconostoc* species, as revealed by low frequency restriction fragment analysis of genomic DNA. J. Appl. Bacteriol. 77: 271–280

Tulloch DL, Finch LR, Hillier AJ & Davidson BE (1991) Physical map of the chromosome of *Lactococcus lactis* subsp. *lactis* DL11 and localization of six putative rRNA operons. J. Bacteriol. 173: 2768–2775

Vaughan EE & Vos WM de (1995) Identification and characterization of the insertion element IS*1070* from *Leuconostoc lactis* NZ6009. Gene 155: 95–100

Vedamuthu ER (1994) The dairy *Leuconostoc*: use in dairy products. J. Dairy Sci. 77: 2725–2737

Volff JN, Vandewiele D, Simonet J & Decaris B (1993) Ultraviolet light, mitomycin C and nitrous acid induce genetic instability in *Streptomyces ambofaciens* ATCC 23877. Mut. Research 287: 141–156

Walker DC & Klaenhammer TR (1994) Isolation of a novel IS*3* group insertion element and construction of an integration vector for *Lactobacillus* spp. J. Bacteriol. 176: 5330–5340

Wallbanks S, Martinez-Murcia AJ, Fryer JL, Phillips BA & Collins MD (1990) 16S rRNA sequence determination for members of the genus *Carnobacterium* and related lactic acid bacteria and description of *Vagococcus salmoninarum* sp. nov. Int. J. Syst. Bacteriol. 40: 224–230

Williams AM, Fryer JL & Collins MD (1990) *Lactococcus piscium* sp. nov. a new *Lactococcus* species from salmonid fish. FEMS Microbiol. Lett. 68: 109–114

Worobo RW, Henkel T, Sailer M, Roy KL, Vederas JC & Stiles ME (1994) Characteristics and genetic determinant of a hydrophobic peptide bacteriocin, carnobacteriocin A, produced by *Carnobacterium piscicola*. Microbiology 140: 517–526

Yu W, Mierau I, Mars A, Johnson E, Dunny G & McKay LL (1995) Novel insertion sequence-like element IS*982* in lactococci. Plasmid 33: 218–225

Zwahlen M-C & Mollet B (1994) ISL*2*, a new mobile genetic element in *Lactobacillus helveticus*. Mol. Gen. Genet. 245: 334–338

METABOLISM

Antonie van Leeuwenhoek **70:** 187–221, 1996.

The proteolytic systems of lactic acid bacteria

Edmund R.S. Kunji[1], Igor Mierau[2], Anja Hagting[1], Bert Poolman[1] & Wil N. Konings[1]*
Departments of [1] Microbiology and [2] Genetics, Groningen Biomolecular Sciences and Biotechnology Institute, University of Groningen, Kerklaan 30, 9751 NN Haren, The Netherlands; (author for correspondence)*

Key words: proteolysis, proteinase, peptidase, peptide transport, lactococci, lactobacilli, casein hydrolysis

Abstract

Proteolysis in dairy lactic acid bacteria has been studied in great detail by genetic, biochemical and ultrastructural methods. From these studies the picture emerges that the proteolytic systems of lactococci and lactobacilli are remarkably similar in their components and mode of action. The proteolytic system consists of an extracellularly located serine-proteinase, transport systems specific for di-tripeptides and oligopeptides (> 3 residues), and a multitude of intracellular peptidases. This review describes the properties and regulation of individual components as well as studies that have led to identification of their cellular localization. Targeted mutational techniques developed in recent years have made it possible to investigate the role of individual and combinations of enzymes *in vivo*. Based on these results as well as *in vitro* studies of the enzymes and transporters, a model for the proteolytic pathway is proposed. The main features are: (i) proteinases have a broad specificity and are capable of releasing a large number of different oligopeptides, of which a large fraction falls in the range of 4 to 8 amino acid residues; (ii) oligopeptide transport is the main route for nitrogen entry into the cell; (iii) all peptidases are located intracellularly and concerted action of peptidases is required for complete degradation of accumulated peptides.

Introduction

Lactic acid bacteria are used in the production of a wide range of dairy products such as cheeses and yoghurts. Several metabolic properties of lactic acid bacteria serve special functions which directly or indirectly have impact on processes such as flavour development and ripening of dairy products. The main functions are (i) fermentation and depletion of the milk sugar lactose; (ii) reduction of the redox potential; (iii) citrate fermentation and (iv) degradation of casein (Olsen, 1990). The degradation of caseins plays a crucial role in the development of texture and flavour. Certain peptides contribute to the formation of flavour, whereas others, undesirable bitter-tasting peptides, can lead to off-flavour. Detailed understanding of these processes may lead to engineered lactic acid bacteria with improved proteolytic properties.

It has been well-established that many lactic acid bacteria, isolated from milk products, are multiple amino acid auxotroph (Chopin, 1993). The requirement for amino acids is strain dependent and can vary from 4 up to 14 different amino acids. In milk, the amounts of free amino acids and peptides are very low. Lactic acid bacteria, therefore, depend for growth in milk on a proteolytic system that allows degradation of milk proteins (caseins) (Mills & Thomas, 1981; Juillard et al., 1995b). Caseins constitute about 80% of all proteins present in bovine milk. The four different types of caseins found in milk, α_{S1}-, α_{S2}-, β- and κ-casein, are organized in micelles to form soluble complexes (Schmidt, 1982). In free solution, caseins behave as non-compact and largely flexible molecules with a high proportion of residues accessible to the solvent, i.e., like random coil-type proteins (Holt & Sawyer, 1988). Caseins contain all amino acids necessary for growth of lactic acid bacteria in milk to high cell density, but it can be calculated that only a minor fraction of the total is actually needed (less than 1%).

The structural components of the proteolytic systems of lactic acid bacteria can be divided into three groups on the basis of their function: (i) proteinases that breakdown caseins to peptides, (ii) peptidases that degrade peptides, and (iii) transport systems that translocate the breakdown products across the cytoplasmic membrane.

The proteinase is clearly involved in the initial degradation of caseins, yielding a large number of different oligopeptides. The initial analyses of the casein breakdown products liberated by the proteinases have indicated that, with a few exceptions, only large peptides are formed (Monnet et al., 1986; Visser et al., 1988; Monnet et al., 1989; Reid et al., 1991b; Pritchard & Coolbear, 1993). Consequently, further breakdown by extracellular peptidases was considered to be critical to fulfill the needs for essential and growth-stimulating amino acids. The external localization of proteinases is consistent with the finding that these are synthesized with a typical signal peptide sequence, but this property has not been found in any of the peptidases analysed so far (Kok & De Vos, 1994; Poolman et al., 1995). These findings are supported by biochemical and immunological data which indicate that the proteinases are present outside the cell, whereas most, if not all, peptidases are found in the cytoplasm.

These apparent discrepancies could be explained by: (i) the existence of extracellular peptidases which sofar have remained uncharacterized, possibly because most work has focussed on purification of soluble peptidases; (ii) a less restricted specificity of the cell envelope located proteinase; and/or (iii) the activity of membrane carriers capable of facilitating transport of peptides greater than 5–6 amino acid residues (Pritchard & Coolbear, 1993).

In this review, the list of putative components of the proteolytic pathway is updated and attempts are made to assign a physiological role to each of the enzymes. The main focus will be on the substrate specificity, expression/regulation and localization of the components. For detailed information on the biochemical properties of the enzymes, the cloning strategies and organization of the genes, the reader is referred to a series of reviews and the original references cited there (Tan et al., 1993a; Kok and De Vos, 1994; De Vos & Siezen, 1994; Poolman et al., 1995).

Most attention will be paid to the proteolytic system of lactococci, which is by far the best documented. The majority, if not all, of the enzymes necessary for degradation of caseins and transport of degradation products have been described. The second-best unravelled proteolytic systems are those of *Lactobacillus* (*Lb*) species, most notably *Lb. helveticus*, *Lb. delbrückii* and *Lb. casei*. Unfortunately, very little information is available on transport of casein breakdown products in these organisms. Whenever possible, similarities and differences between the proteolytic systems of lactococci and lactobacilli will be pointed out. Although work is advancing for other lactic actid bacteria, such as *Streptococcus, Pediococcus, Leuconostoc* and *Micrococcus* species, these data will not be discussed here.

In recent years, a series of elegant genetic tools have been developed for targeted inactivation of chromosomally located genes in lactococci and lactobacilli (Leenhouts 1991; Bhowmik et al., 1993; Leenhouts et al., 1996). These methods allowed, for the first time, the analysis of enzymes *in vivo* and have lead to a better understanding of the proteolytic pathway as a whole. Based on these and other results a model of the proteolytic pathway of *L. lactis* is presented, which accommodates most of the available data.

The proteinases of lactic acid bacteria

It has been well-established that degradation of caseins is initiated by a single cell wall-bound extracellular proteinase (PrtP) (Smid et al., 1991; Tan et al., 1993a; Pritchard & Coolbear, 1993; Kok & De Vos, 1994; De Vos & Siezen, 1994). The proteinases of many different lactic acid bacteria have been identified and characterized biochemically (Table 1). The biochemical and genetic properties, localization and specificity of the enzymes will be discussed in the following sections.

Genetic and biochemical properties of proteinases

The mature proteinase is a monomeric serine-proteinase with a molecular mass between 180–190 kDa, although breakdown products of smaller sizes are usually found upon isolation of the enzyme (Laan & Konings, 1989). The gene encoding PrtP has been cloned and sequenced for a number of *L. lactis* strains (Kok et al., 1988; Vos et al., 1989a; Kiwaki et al. 1989; Exterkate et al., 1993), *Lb. paracasei* (Holck & Næs, 1992) & *Lb. delbrückii* subsp. *bulgaricus* (Gilbert et al., 1996). The unprocessed proteinases of *Lb. paracasei* NCDO151, *L. lactis* Wg2 and NCDO763 consist of, 1902 amino acid residues, which compares well with the, 1946 residues of the *Lb. delbrückii* enzyme and the, 1962 residues for PrtP of *L. lactis* SK1, which have a duplication at the C-terminus (40 and 60 amino acids, respectively). The primary sequences of the lactococcal enzymes are more than 98% identical and more than 95% when compared to

Table 1. Proteinases of dairy lactic acid bacteria

Strain	Mw[a] (kDa)	Substrate[b]	Type[c]	pH[d]	Localization	Reference[g]
Lactococcus lactis subsp. *cremoris* WG2		κ-, β-casein	S		cell-wall[f]	Hugenholtz et al., 1987
	181				cell-wall[e]	Kok et al., 1988
Lactococcus lactis subsp. *cremoris* HP		κ-, β-casein	S	6.4	cell-wall[f]	Exterkate & De Veer, 1987a
Lactococcus lactis subsp. *cremoris* SK11	187	α_{s1}-, κ-, β-casein	S		cell-wall[e,f]	Vos et al., 1989a
Lactococcus lactis subsp. *cremoris* AC1		α_{s1}, κ-, β-casein			cell-wall[f]	Bockelmann et al., 1989
Lactococcus lactis subsp. *cremoris* AM1		α_{S1}-, κ-, β-casein			cell-wall[f]	Visser et al., 1991
Lactococcus lactis subsp. *cremoris* H2	180+	κ-, β-casein	S		cell-wall[f]	Coolbear et al., 1992
Lactococcus lactis subsp. *cremoris* NCDO763		α_{s1}-, κ-, β-casein	S	6.0	cell-wall[f]	Monnet et al., 1987
	181		S		cell-wall[e]	Kiwaki et al., 1989
Lactobacillus casei subsp. *casei* HN1		β-casein	S		cell-wall[f]	Kojic et al., 1991
Lactobacillus casei subsp. *casei* NCDO 151			S	6.5	cell-wall[f]	Næs & Nissen-Meyer, 1992
	181				cell-wall[e]	Holck & Næs, 1992
Lactobacillus delbrückii subsp. *bulgaricus* CNRZ 397	170+	α_{s1}-, β-casein	S[h]	5.5	cell-wall[f]	Laloi et al., 1991
			S		cell-wall[e]	Gilbert et al., 1996
Lactobacillus helveticus CNRZ 303		α_{s1}-, β-casein	S	7.5	cell-wall[f]	Zevaco & Gripon, 1988
Lactobacillus helveticus CP790	45+	α_{s1}-, β-casein	S	6.5	cell-wall[f]	Yamamoto et al., 1993
Lactobacillus helveticus L89	180+	α_{s1}-, β-casein	S	7.0	cell-wall[f]	Martín-Hernández et al., 1994

[a] If available, the molecular weight of the mature proteinase was calculated from the derived amino acid sequence of the cloned gene, otherwise as determined by gel filtration* or SDS-PAGE+. [b] Substrates degraded by the proteinase when tested. [c] Type of enzyme; S Serine-proteinase. [d] pH optimum of activity. [e] Localization as predicted from the presence of a signal or membrane anchor sequence. [f] Localization as predicted from fractionation studies and immuno-gold labeling. [g] Key references are only cited. [h] 45% of activity was recovered after incubation with 1 mM phenylmethylsulphonyl fluoride.

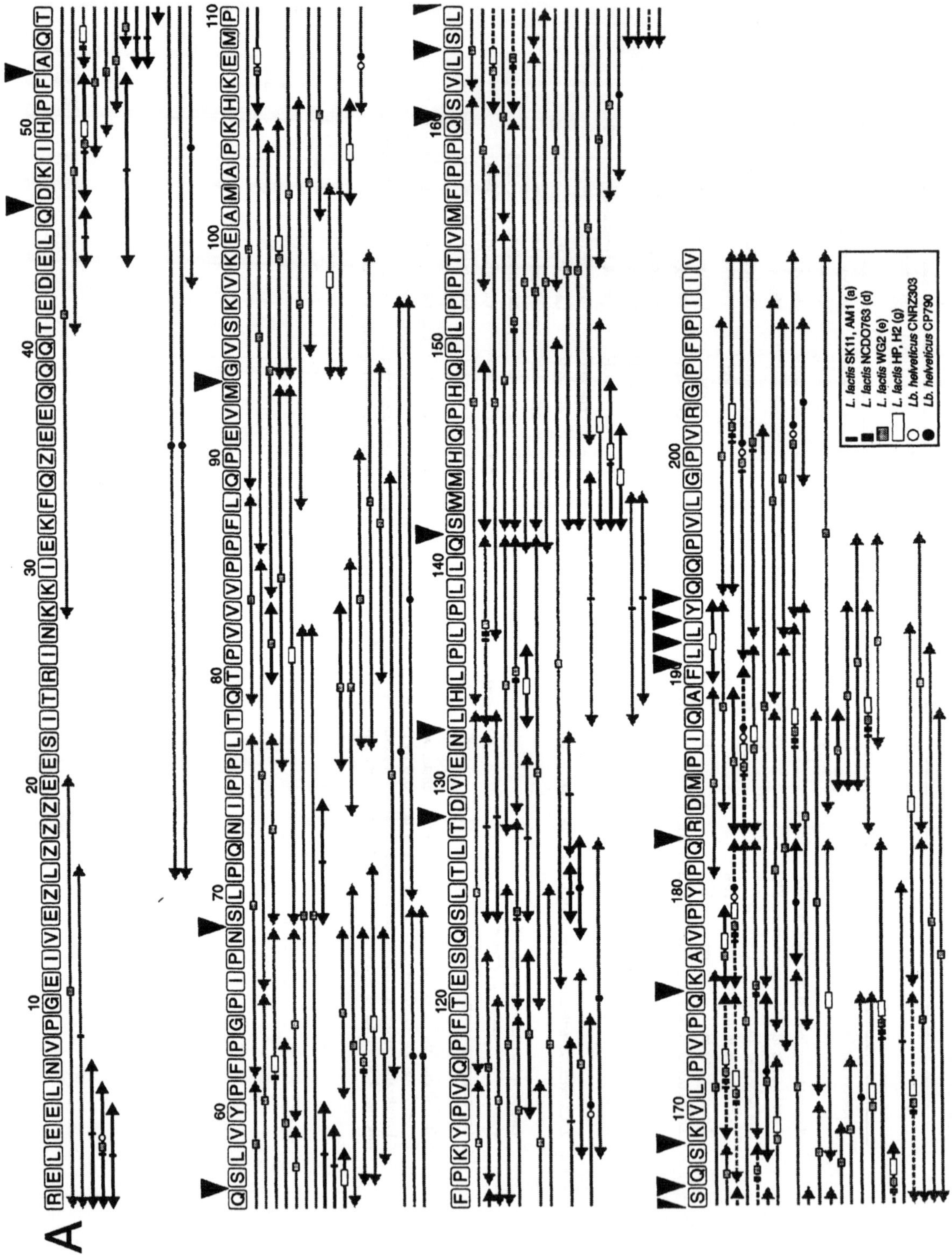
A
L. lactis SK11, AM1 (a)
L. lactis NCDO763 (d)
L. lactis WG2 (e)
L. lactis HP, H2 (g)
Lb. helveticus CNRZ303
Lb. helveticus CP790

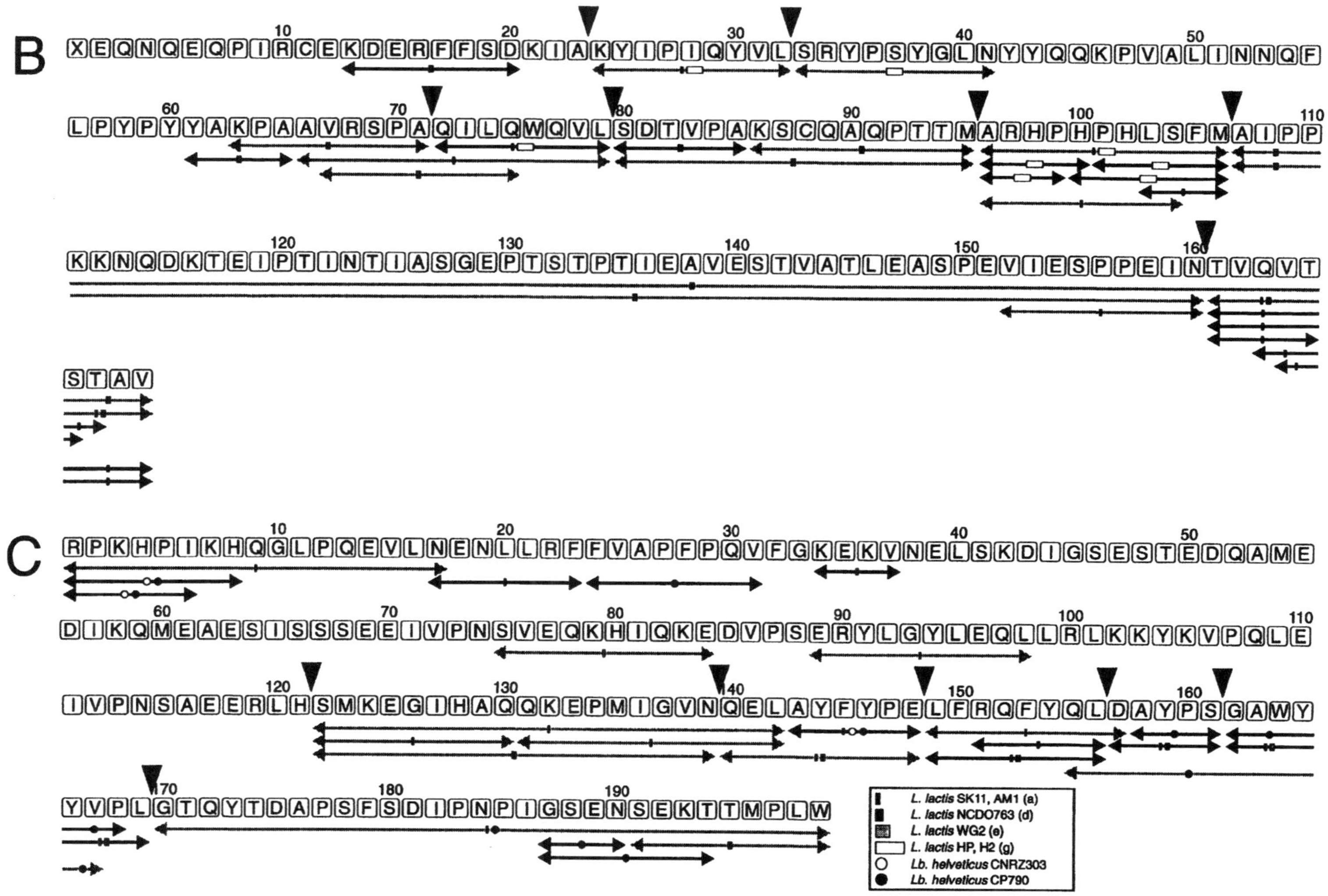

Figure 1. Primary sequences of β- (A), κ- (B) and α_{S1}-casein (C), and the peptides released by the activity of purified proteinases. Black arrows indicate degradation fragments of lengths of up to 8 amino acid residues (putative substrates of the oligopeptide transport system), while longer peptides are colored gray. The symbols relate to the strain or organism, in which the PrtP hydrolysate the particular peptide has been identified (shown in the legend of the figure). Black triangles show cleavage sites cut by all lactococcal proteinases and gray triangles indicate bonds hydrolyzed by the majority of the enzymes studied. (Sources: Monnet et al., 1986; Zevaco & Gripon, 1988; Visser et al., 1988; Monnet et al., 1989; Reid et al., 1991a; Reid et al., 1991b; Monnet et al., 1992; Yamamoto et al., 1993; Reid et al. 1994; Visser et al., 1994; Juillard et al., 1995a). Dashed arrows indicate peptides that are likely transported by the oligopeptide

the *Lb. paracasei* enzyme. The *Lb. delbrückii* enzyme shows up to 40% identity over the first 820 residues when compared to the lactococcal enzymes, while the C-terminal part does not share any homology. In *Lb. helveticus* CP790, a 45 kDa serine-proteinase has been identified, but sequence data are lacking to exclude the possibility that the isolated enzyme is an autoproteolytic product (Yamamoto et al., 1993). In another strain of this organism (L89), a proteinase of 180 kDa is found (Martín-Hernández et al., 1994).

Sequence alignments show that proteinases are related to subtilisins, which are serine-proteinases with similar catalytic domains (De Vos & Siezen, 1994). Comparison of the N-terminal sequence of the mature proteinase with that deduced from the nucleotide sequence revealed a typical signal sequence for Sec-dependent translocation and a prosequence, which both are removed by post-translational processing (Kok et al., 1988; Vos et al., 1989a; Kiwaki et al., 1989; Holck & Næs, 1992; Gilbert et al., 1996). The N-terminal part of the mature enzyme constitutes the catalytic domain (see below) and contains several conserved residues which are involved in catalysis and substrate positioning (Ser 433, His94, Asp30, Asn196 in the Sk11 enzyme) (De Vos & Siezen, 1994). The spacer that follows shows no homology to proteins of known function, but likely exposes the catalytic domain outside the cell-wall. The outermost C-terminal part is conserved in many surface proteins of Gram-positive bacteria and carries a sorting signal, typically LPXTG, followed by a putative membrane spanning α-helix and a small charged tail (Navarre & Schneewind, 1994). After translocation, the LPXT↓G sequence is cleaved (at the position indicated by the arrow) and the carboxyl group of threonine is probably covalently linked to a N-terminal glycine that is part of the cross-bridges in the peptidoglycan layer (Navarre & Schneewind, 1994).

In *L. lactis* Wg2, SK11 and *Lb. paracasei* NCDO 151 an enzyme has been identified which is involved in the maturation of the proteinase (PrtM) (Haandrikman et al., 1989; Vos et al., 1989b; Holck & Næs, 1992). The *prtM* gene product is also preceded by a signal sequence and has a consensus lipomodification site (Haandrikman et al., 1989; Vos et al. 1989b; Haandrikman et al., 1991; Holck & Næs, 1992; Sankaran & Wu, 1994). PrtM bears 30% amino acid sequence identity to PrsA of *Bacillus subtilis*, and these enzymes may function as extracellular chaperones (Kontinen et al., 1991). To date, a homolog of PrtM has not been identified in *Lb. delbrückii* subsp. *bulgaricus* (Gilbert et al., 1996).

Localization of the proteinases

The extracellular location of PrtP is supported by various kinds of data. Firstly, the proteinase can be liberated from the cell-wall with minimal lysis by treating cells with Ca^{2+}-free buffers (Mills & Thomas, 1981) or lysozyme (Laloi et al., 1991; Coolbear et al., 1992). The largest size of lactococcal PrtP, detected after release in Ca^{2+}-free buffer, is 165 kDa, which is believed to be the product of an intramolecular autoproteolytic event (Laan & Konings, 1989). Treatment with lysozyme yields a product of 180 kDa, which is close to the predicted size of the mature proteinase deduced from the primary sequence. Secondly, electron microscopy of immuno-gold labelled PrtP has confirmed a localization of the proteinase in the cell wall (Hugenholtz et al., 1987). Thirdly, without exception the genes encoding proteinases specify a typical N-terminal signal sequence, which targets the protein to the outside of the cell (see above) (Kok & De Vos, 1994; De Vos & Siezen 1994).

Specificity classes of proteinases

On the basis of degradation patterns of α_{S1}-, β- and κ-caseins, two proteinase specificity-classes have initially been described in lactococci, which are generally indicated as P_I and P_{III} (Visser et al., 1986). The primary substrates of P_I-type enzymes are β-casein and, to lesser extent, κ-casein, while P_{III}-type enzymes degrade α_{S1}-, β- and κ-caseins (Pritchard & Coolbear, 1993). Since P_I- and P_{III}-type proteinases from *Lactococcus* species are more than 98% identical, hybrid proteins could be constructed by swapping regions of the P_I-type proteinase of *L. lactis* subsp. *cremoris* Wg2 and the P_{III}-type enzyme of strain SK11. These studies limited the differences in the proteinase specificities to two regions (Vos et al., 1991). Substrate binding studies and computer modelling, based on the three-dimensional structure of subtilisins, suggest that the first region, with residues 131, 138, 142, 144 and 166 of SK11, is part of the substrate binding pocket of PrtP. The second region, in which particularly residues 747 and 748 are important, might be involved in electrostatic interactions with caseins.

The specificity of proteinases has been analysed further using the enzymes isolated from 16 different *L. lactis* strains and fragment 1-23 of α_{S1}-casein as substrate (Exterkate et al., 1993). On the basis of these

Table 2. Classification of proteinases according to the specificities toward α_{S1}-casein fragment 1–23

Group	Strains	Substr.	Cleavage sites in α_{s1}-casein fragment 1-23 [a] (5, 10, 15, 20) RPKHPIKHQGLPQEVLNENLLRF 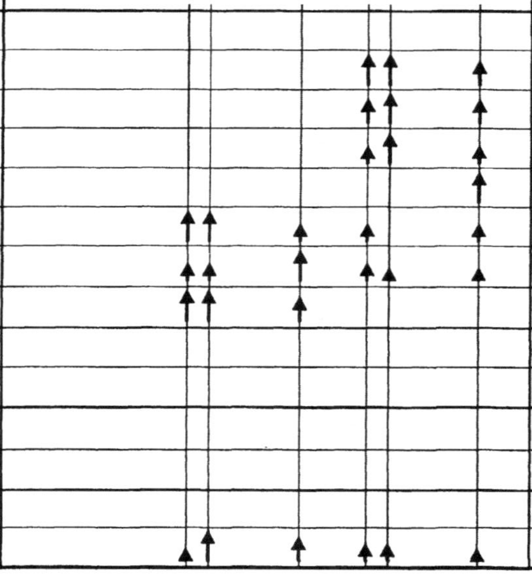	Amino acid substitutions at positions relevant for substrate binding [b]								
				131	138	142	144	166	177	747	748	763
	L. lactis											
a	AM1, SK11, US3	α_{s1}, β, κ		Ser	Lys	Ala	Val	Asn	Leu	Arg	Lys	Asn
b	AM2			Thr	Thr	Ala	Leu	Asp	Leu	Arg	Lys	Asn
c	E8	α_{s1}, β, κ		Thr	Thr	Ala	Leu	Asp	Ile	Arg	Lys	Asn
d	NCDO763, UC317	α_{s1}, β, κ		Thr	Thr	Ala	Leu	Asp	Leu	Arg	Lys	His
e	WG2, C13, KH	β, κ		Thr	Thr	Ser	Leu	Asp	Leu	Leu	Thr	Asn
f	Z8, H61, TR, FD27	α_{s1}, β, κ		Thr	Thr	Ala	Leu	Asp	Leu	Leu	Thr	His
g	HP	β, κ		Thr	Thr	Asp	Leu	Asp	Ile	Leu	Thr	His
	Lb. paracasei											
	NCDO151			Thr	Thr	Ala	Leu	Asp	Leu	Gln	Thr	Asn
	Lb. bulgaricus											
	NCDO1489			Ser	Gly	Asp	Ile	Val		Gly	Thr	
	Lb. helveticus											
	L89											

[a] The main cleavage sites are indicated by arrows; the sizes of the arrows are related to the relative cleavage rates. [b] Numbering is according to the sequence of the SK11 proteinase (Vos et al., 1989a). Amino acid substitutions at positions relevant for substrate binding (Vos et al., 1991) are indicated. (Data according to Holck & Næs, 1992; Exterkate et al. 1993; Martín-Hernández et al., 1994; Gilbert et al., 1996.)

studies, the lactococcal proteinases were classified into seven groups, which displayed a whole range of different specificities rather than two extremes, i.e., a P_I- and P_{III}-type. Table 2 presents the identified specificity classes of lactococcal proteinases and the amino acid substitutions that are thought to be responsible for the observed phenotypes. It is evident that the subtle changes in specificity do occur as a result of minor genetic variations in the structural gene of PrtP. The catalytic domain of PrtP is not only highly conserved among lactococcal species, but also when compared to homologs in lactobacilli. The *Lb. paracasei* enzyme differs from the *L. lactis* Wg2 proteinase in only two positions, that are regarded as important for substrate specificity and binding (Holck & Næs, 1992) (Table 2). The *Lb. delbrückii* proteinase has distinct substitutions at positions 138 (Gly), 166 (Val) and 747 (Gly), and similar residues at others. Finally, the 180 kDa serine-proteinase of *Lb. helveticus* L89 essentially cleaves the same bonds as the lactococcal proteinases, but the relative amounts of fragments are clearly different (Martín-Hernández et al., 1994).

Casein degradation products

β-casein. The products resulting from the action of proteinases on β-casein have been analysed *in vitro* using purified enzymes of different *L. lactis* (Monnet et al. 1986; Visser et al., 1988; Monnet et al., 1989; Reid et al. 1991b) and *Lb. helveticus* strains (Zevaco & Gripon, 1988; Yamamoto et al., 1993) (Figure 1A). After separation of the proteolytic products by liquid chromatography, the different peptides were collected, purified further when necessary, and identified by Edman degradation and/or amino acid composition analysis. In some cases, additional information was obtained from mass-spectrometrical analysis of the purified peptide. These studies indicate that only part of β-casein is degraded and that relatively large fragments – only a few contain less than 8 amino acid residues – are formed. However, inspection of the HPLC-profiles shows that only the most abundant peptides have been analysed. Recently, more than 95% of the peptides formed by the action of the proteinase of *L. lactis* subsp. *cremoris* Wg2 on β-casein have been recovered using liquid chromatography in combination with on-line ion-spray mass spectrometry (Juillard et al., 1995a). The results show that β-casein is degraded by PrtP into more than hundred different oligopeptides ranging from 4 to 30 residues, of which a major fraction falls in the range of 4–10 residues (Figure 1A). The proteinase activity does not yield detectable amounts of di- and tripeptides, and only traces of phenylalanine were measured. More than 50% of the peptides originate from the C-terminal part of β-casein, while about half of the remaining peptides are derived from the 60–105 region.

The peptides which are liberated from β-casein by the proteinase of *L. lactis* SK11 and AM1 (P_{III}-type), NCDO763 and Wg2 (intermediate-types) and HP, H2 (P_I-type) are indicated in Figure 1A. In total, thirteen bonds are cleaved systematically by all lactococcal enzymes studied to date (indicated as black triangles in Figure 1A), and an additional six bonds are cleaved by most enzymes (indicated as gray triangles in Figure 1A). The majority of these bonds are located in the C-terminal part of β-casein. The peptides liberated by hydrolysis of these bonds constitute the major fraction in hydrolysates and are present during the earliest times of degradation (Juillard et al., 1995a; Fang & Kunji unpublished results). These peptides are likely to be the main suppliers of amino acids during growth on β-casein (see below). In addition to these fragments, all types of enzymes produce a large number of different small oligopeptides (black arrows in Figure 1).

All bonds cleaved by the action of the proteinase of *Lb. helveticus* CNRZ 303 on β-casein are also hydrolysed by the lactococcal proteinases (Zevaco & Gripon, 1988). Again, the major products are derived from the C-terminal region of β-casein. In *Lb. helveticus* CP790 a considerably smaller cell-wall bound proteinase has been identified, i.e. 45 versus 180 kDa (Yamamoto et al., 1993), which is claimed to be a complete serine proteinase and not the product of an autoproteolytic event. Nonetheless, the major degradation products of this enzyme are virtually identical to that of the CRNZ303 enzyme and the lactococcal proteinases with respect to peptides of small molecular weights (Figure 1A).

κ-casein. The product formation from κ-casein has been studied for the proteinases of *L. lactis* NCDO763, SK11, H2 and AM1, but only the major products have been identified sofar (Monnet et al., 1992; Reid et al. 1994; Visser et al., 1994). This milk protein is hydrolysed by each of the enzymes, albeit with different degradation patterns even after 24 hours of incubation (Reid et al. 1994). Many bonds are systematically hydrolysed by all types, but also several type-dependent cleavage sites were reported (Figure 1B). However, inspection of the HPLC profiles of the original papers suggests that the same peptides might be present in each hydrolysate, although in different amounts. The degradation of κ-casein yields a large number of small oligopeptides, which originate mainly from region 96–106 and the C-terminal part.

α_{S1}-casein. Degradation of α_{S1} and α_{S2}-casein is confined to P_{III}-type and intermediate-type proteinases, while P_I-type proteinases cannot hydrolyse this substrate (Visser et al. 1986; Bockelmann et al., 1989). However, this conclusion is based on SDS-PAGE and not on the more sensitive HPLC analysis that has been used to characterize the degradation of β- and κ-casein. About 25 major oligopeptides were identified in the product formation of α_{S1}-casein by several proteinases, of which about half originate from the C-terminal region (Figure 1C) (Reid et al., 1991a; Monnet et al., 1992). Again, several different small oligopeptides are found in the hydrolysates of the various enzymes, of which several are bordered by preferential cleavage sites.

Cleavage sites. Various researchers have attempted to define specificity rules for the proteinase based on statistical analysis of cleavage site residues and putative interaction of these residues with the PrtP binding pocket (Monnet et al. 1992; Vos et al., 1991; De Vos & Siezen, 1994; Juillard et al., 1995a). It is apparent from inspection of Figure 1 that the proteinases have a very broad substrate specificity. On the basis of our recent studies, in which more than 95% of the β-casein degradation products have been identified, it is not possible to define unequivocally a consensus cleavage site. However, it is apparent that particular bonds are preferentially hydrolysed if the peptide product formation of different proteinases is compared. In addition to the specificity of PrtP *per se*, intrinsic properties of caseins might also play a role in degradation. Caseins have been described to behave, in free solution, as non-compact and largely flexible molecules with a high proportion of residues accessible to the solvent (Holt & Sawyer, 1988). Nonetheless, circular dichroism and Raman spectral studies have shown that β-casein contains about 12% α-helical structure (Swaisgood, 1993). Secondary structure predictions indicate that region 21–38 of β-casein has the potential to form α-helices, i.e. in the proximity of the phosphorylation sites. The same region has high sequence identity to the C-terminal part of hen egg-white riboflavin-binding protein (Holt & Sawyer, 1988). Possibly, the inability of the proteinases to degrade this part of the N-terminal region is related to the presence of α-helical structure.

It is striking that in several regions almost each peptide bond is cleaved by all types of proteinases, such as in regions 160–170 and, 190–195 in β-casein, while in other regions the bonds are only incidently cut. This could reflect the broad specificity of PrtP, but can also be attributed to a better accessibility of particular loops for degradation. Furthermore, the hydrolysis of casein by PrtP is a complex process, in which parameters such as refolding of the substrate after cleavage and aggregation of casein molecules may also play a role.

In conclusion, the biochemical properties of the proteinases of the various lactic acid bacteria are very similar, i.e. most enzymes, if not all, are serine-proteinases of similar size (Table 1). Sequence comparisons reveal a remarkably high degree of identity, even when proteinases of different specificity classes or organisms are compared. Also the product formation seems to cross borders of specificity classes and species, which is apparent from the large number of preferential cleavage sites. Many different small oligopeptides (4–8 amino acid residues) are formed, which contain all essential and growth-stimulating amino acids, and many of these peptides are produced in high amounts. To date no significant amounts of free amino acids and di/tripeptides have been detected in hydrolysates formed by the various proteinases.

Amino acid and peptide transport systems

To utilize amino acids for biosynthesis, degradation products derived from casein have to traverse the membrane at one stage or another. In the following sections, the transport processes are reviewed with the emphasis on peptide transport systems.

Lactococcal amino acid transport systems
Lactococci possess at least 10 amino acid transport systems which have a high specificity for structurally similar amino acids, e.g. Glu/Gln, Leu/Ile/Val, Ser/Thr, Ala/Gly, Lys/Arg/Orn (Konings et al., 1989). Several amino acid transport systems were characterized as being driven by hydrolysis of ATP, i.e. those for Glu/Gln, Asn and Pro/Glycine-Betaine (Konings et al., 1989; Poolman 1993; Molenaar et al., 1993). The amino acid transport systems for Leu/Val/Ile, Ala/Gly, Ser/Thr and Met are driven by the proton motive force, whereas the Arg/Orn antiporter is driven by the concentration gradient of both solutes (Konings et al., 1989). To date, genes encoding lactococcal amino acid transport systems have not been cloned.

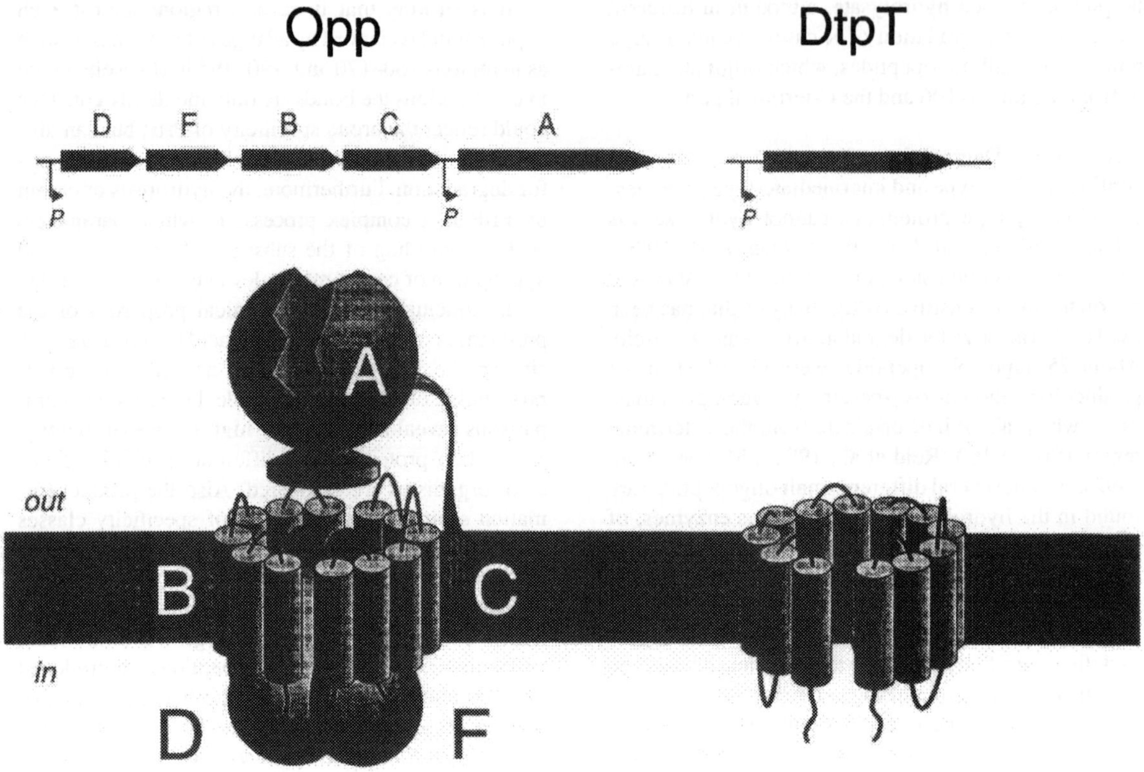

Figure 2. Model and gene organization of the oligopeptide transport system (Opp) and di/tripeptide transporter (DtpT) of *Lactococcus lactis*. (Data from Tynkkynen et al., 1993; Hagting et al., 1994.)

The lactococcal di/tripeptide transporter for hydrophilic substrates (DtpT)

Transport studies in peptidase-free membrane vesicles of *L. lactis* have shown that relatively hydrophilic di- and tripeptides are transported by a proton motive force-driven transport mechanism (Smid et al., 1989). The gene (*dtpT*) encoding this di-tripeptide transport protein has been cloned by complementation and functionally expressed in *E. coli* and *L. lactis* (Hagting et al., 1994). The amino acid sequence deduced from the nucleotide sequence of *dtpT* shows no significant similarity to other known bacterial peptide transport systems, which all belong to the ABC superfamily (see below) and couple transport to the hydrolysis of ATP (Higgins, 1992). In fact, DtpT belongs to a new family of proton motive force-driven peptide transport systems, called the PRT-family, which sofar has only eukaryotic counterparts of DtpT (Steiner et al., 1995). Recent database searches have revealed bacterial open reading frames (ORFs) with significant similarity to DtpT (unpublished results). One of these ORFs correspond to a partial sequence located downstream of aminopeptidase N gene of *Lb. helveticus* (Christensen et al., 1995). The secondary structure of DtpT is similar to that of proton motive force driven secondary transport proteins and consists of twelve putative membrane-spanning α-helices (Figure 2). Using flanking regions of the gene, *dtpT* has been deleted from the chromosome via homologous recombination. Characterisation of the Δ*dtpT* strain indicates that DtpT is the only transport protein for hydrophilic di- and tripeptides in *L. lactis*. Its role in the uptake of casein degradation products is discussed below (section: The role of peptide transport systems *in vivo*).

The lactococcal oligopeptide transport system (Opp)

From mutant analysis it has become apparent that *L. lactis* also possesses a transporter that is specific for oligopeptides (Opp) (Kunji et al., 1993). Growth experiments in chemically defined media containing peptides of varying length have suggested that Opp transports peptides up to lengths of 8 residues (Tynkkynen et al., 1993). Preliminary experiments, in which translocation of peptides formed by the action of PrtP on β-casein was analysed, indicate that oligopeptides consisting of up to 10 amino acids may be transported (Fang & Kunji, unpublished results).

On the basis of metabolic inhibitor studies it has been concluded that oligopeptide transport is driven by ATP rather than the proton motive force (Kunji et al., 1993). The genes encoding the oligopeptide transport protein have been cloned, sequenced and functionally expressed in strains of *L. lactis* (Tynkkynen et al., 1993). Five open reading frames correspond to polypeptides that are typical components of binding protein-dependent transport systems (OppDFBCA) (Figure 2), and on the basis of sequence comparisons the system has been classified as a member of the Binding Cassette (ABC) superfamily (Higgins, 1992). The sixth gene of the operon, that is located 5′ of *oppA*, encodes the endopeptidase PepO (Mierau et al., 1993; Tynkkynen et al., 1993). The five subunits of the oligopeptide transport system include a peptide binding protein (OppA), two integral membrane proteins (OppB and OppC), and two ATP-binding proteins (OppD and OppF) (Figure 2). The derived amino acid sequence of the *oppA* gene has a consensus prolipoprotein cleavage site Leu-Ser-Ala↓Cys (Tynkkynen et al., 1993), which, through fatty acid modification of the N-terminal cysteine, anchors the mature protein at the outer surface of the cytoplasmic membrane (Von Heijne, 1989; Sankaran and Wu, 1994; Detmers & Kunji unpublished results).

OppA serves as the receptor protein that delivers peptides to the membrane-bound translocator complex. Elucidation of the tertiary structure of OppA of *Salmonella typhimurium* indicated that the protein is composed of three domains: Domains I and III are involved in the binding of oligopeptides in the manner of a Venus fly-trap; and domain II is typical for peptide binding proteins but its function has not been established (Figure 3) (Tame et al., 1994). The structure has shed light on the intriguing observation that OppA is able to bind peptides with high affinity, which differ in size and amino acid composition (Tame et al., 1994). It turns out that the peptide backbone is bound to the binding pocket of OppA through salt bridges and hydrogen bonds, while the side chains of the peptide are accommodated in large pockets with minimal interaction with OppA. The N-terminal end of the peptide is bound to Asp419, while the C-terminal end is bound to Arg413, His371 or Lys307 in case of a tri-, tetra- or pentapeptide, respectively (Figure 3). The same principal has been observed in the binding of dipeptides to the dipeptide binding protein of *E. coli* (Dunten & Mowbray, 1995). OppA of *S. typhimurium* and *L. lactis* are homologous, but the identity between the two genes is only 25%. The lactococcal OppA is about 60 amino acids larger and might have an extra loop at the N-terminus which could function as a spacer to connect the protein to its membrane anchor (Figure 3).

OppB and OppC are highly hydrophobic proteins that, on the basis of hydropathy profiling, are able to span the cytoplasmic membrane in α-helical configuration six times (Figure 2). These proteins are likely to constitute the pathway that facilitates the translocation of oligopeptides across the membrane. OppD and OppF are homologous to the ATP binding protein(s) (domains) of the ABC-transporter superfamily (Higgins, 1992). These proteins most likely couple the hydrolysis of ATP to conformational changes in OppB/C that allow passage of the peptides across the membrane (Figure 2).

To discriminate between OppDFBCA and PepO as essential components of the proteolytic pathway of *L. lactis*, two integration mutants have been constructed, one defective in OppA and the other in PepO. Growth of these mutants in milk and in a chemically defined medium with oligopeptides has shown that the OppDFBCA system, but not the endopeptidase, is essential for the utilization of milk proteins and oligopeptides (Tynkkynen et al., 1993).

In *E. coli* and *S. typhimurium* peptides of 3 to 6 amino acids are transported by Opp. Translocation of longer peptides in Gram-negative bacteria may be restricted by the upper size exclusion limits of the outer membrane pores rather than the transporter (Payne & Smith, 1994). Moreover, the upper size limits of Opp have not been studied systematically. Since the number of amino acid combinations increases rapidly with the size of the peptide, it is difficult to explore the upper size exclusion range without taking into account the specificity. One approach to solve this problem is to use random libraries of peptides, which contain all amino acids in random order but do have a specified size (Momburg et al., 1994). The other approach we are currently taking is the investigation of transport of natural substrates, i.e. fragments of various lengths liberated from β-casein by the action of the proteinase. This is done by following the uptake of casein degradation products *in vivo* by ion-spray mass spectrometry

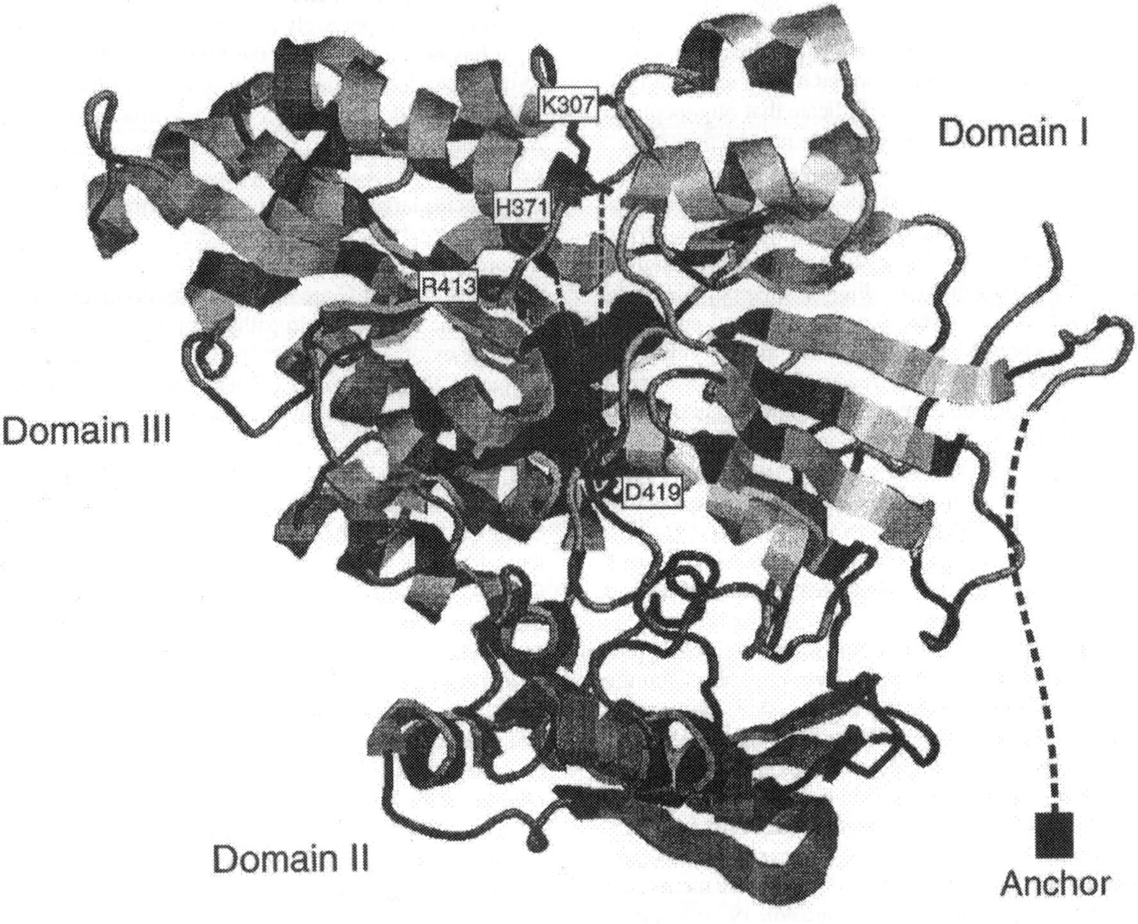

Figure 3. Tertiary structure of the oligopeptide binding protein (OppA) of *Salmonella typhimurium*. The identical residues of OppA of *Lactococcus lactis* are colored black and Van der Waals spheres depict the bound substrate tri-lysine. The amino acid residues, essential for binding of the peptide, are indicated in boxes and the direction of the substrate backbone is depicted by a dotted line. Also indicated is the position of the putative spacer which connects the membrane anchor to OppA of *L. lactis*. (Data from Tynkkynen et al., 1993; Tame et al., 1994.)

and by analysing the transport of chemically synthesized peptides *in vitro*.

The lactococcal di/tripeptide transport system for hydrophobic substrates (DtpP)

Mutants defective in Opp and DtpT are impaired in their ability to utilize hydrophilic di/tripeptides and oligopeptides, but are still capable of using a wide range of hydrophobic di- and tripeptides. This observation led to the discovery of a third peptide transport system, designated DtpP (Foucaud et al., 1995). Inhibitor studies have indicated that this system is driven by hydrolysis of ATP, which is the same mechanism of energy coupling as described for Opp. The genes encoding DtpP have not been cloned, but inhibition by *ortho*-vanadate suggests that it also belongs to the family of ATP-dependent transporters (Higgins, 1992).

Amino acid and peptide transport systems in lactobacilli

While a lot is known about amino acid and peptide transport in lactococci, virtually no information is available on similar systems in lactobacilli. Preliminary results suggest that the amino acid transport systems of *Lb. helveticus* are similar to those of *L. lac-*

tis, because the mechanism of uptake is identical for the same amino acids (Nakajima, unpublished results). The gene coding for a branched chain amino acid carrier (*brnQ*) of *Lb. delbrückii* subsp. *lactis* has been cloned and sequenced (Stucky et al., 1995a). Like its lactococcal counterpart, this branched chain amino acid carrier is also driven by the proton motive force. Recently, the genes encoding a transporter specific for aromatic amino acids (*aroP*) and one for dipeptides (*dppE*) have been cloned and sequenced in this organism (Vongerichten and Krüger, unpublished results).

As indicated above, a homolog of DtpT is specified by a sequence located downstream of *pepN* in *Lb. helveticus* (Christensen et al., 1995). Transport experiments have shown that substrates, typical for the lactococcal DtpT, are indeed transported by this organism. Preliminary experiments also indicate that an oligopeptide transport system is present in *Lb. helveticus* (Nakajima, unpublished results).

Peptidases of lactic acid bacteria

Following breakdown by PrtP and/or uptake, the casein-derived peptides need to be hydrolysed further by peptidases. The specificity of individual peptidases, their cellular location and their role in the utilization of caseins and peptides will be discussed in the following sections. The discussion of the peptidase specificities is confined to those enzymes which have been purified to homogeneity and characterized biochemically. The genes specifying most of these enzymes are available and biochemical data, sequence alignments, substrate specificities and immunological data are used to compare enzymes of different organisms.

A true comparison of the specificities of peptidases is only possible when well-designed sets of peptides are used. Unfortunately, most studies are inspired by the availability of peptides in the laboratory freezers and a systematic approach has been chosen in only a few cases. Nonetheless, some generalizations can be made, but the set of data is insufficient for true statistical evaluation. Many authors have claimed differences in specificity, but these could well be minor variations on a general theme. Tables 3 and 4 summarize general biochemical and genetic properties of the best characterized peptidases of lactococci and lactobacilli.

General aminopeptidases

Aminopeptidase N. For aminopeptidase N (PepN), there is a clear consensus about the biochemical properties of the enzyme. In most, if not all, organisms studied, the enzyme is a monomeric metallopeptidase of about 95 kDa. Sequence alignments have shown that the gene is conserved among dairy lactic acid bacteria; the primary sequence of PepN of *Lb. helveticus* is 72% and 49% identical to the enzyme of *Lb. delbrückii* subsp. *lactis* and *L. lactis* subsp. *cremoris*, respectively (Tan et al. 1992a; Strøman, 1992; Klein et al., 1993; Christensen et al., 1995). In addition, the primary sequence of *pepN* is homologous to the mammalian aminopeptidase N. The conserved signature sequence F–GAMEN-G indicates that PepN belongs to the subclass of zinc-dependent metallo-peptidases.

PepN is capable of cleaving N-terminal amino acids from a wide range of peptides differing both in size and composition. In most papers, the specificity of the enzyme for di- and tripeptides has been investigated, but as will be shown below, PepN activity is also directed towards oligopeptides (Tan et al., 1990; Miyakawa et al. 1992; Baankreis, 1992; Arora & Lee, 1992; Tan et al. 1993b; Niven et al., 1995; Sasaki et al., 1996a). Generally, dipeptides containing Pro in either two positions are not cleaved, while tripeptides which contain Pro in either first or second position are hydrolysed (Tan et al., 1990; Miyakawa et al., 1992; Arora & Lee, 1992; Tan et al. 1993b). The enzyme of *L. lactis* shows a marked preference for dipeptides containing Arg as the N-terminal residue, but, to a lesser extent, is also capable of cleaving other residues such as Lys & Leu (Niven et al., 1995). An increase in activity is observed with increasing hydrophobicity of the C-terminal residue of the dipeptide Arg-X (Niven et al., 1995). A similar relationship is observed for the hydrolysis of Ala-X and Leu-X peptides by PepN of *Lb. helveticus* (Miyakawa et al., 1992). Tripeptides are also readily cleaved by PepN, but information about preferred peptides is not available.

Three different studies have reported the hydrolysis of oligopeptides by PepN of *L. lactis* (Baankreis, 1992; Tan et al., 1993b; Niven et al., 1995). First, in a study in which a tryptic digest of β-casein was incubated with purified PepN, oligopeptides ranging from 4 to 12 acid residues were hydrolysed (Tan et al., 1993b); several of these fragments contained Pro and even Pro-X-Pro sequences. Second, a *PepN* mutant was found to be impaired in its ability to hydrolyse peptides like $(Lys)_4$, $(Lys)_5$ and $(Lys)_3$ $Trp(Lys)_3$, while di- and tripeptide hydrolysis was unaffected due to activity of other peptidases (Baankreis, 1992). Apparently, these oligopeptides are not hydrolysed by endopeptidases present in this organism. Third, several peptides with

Table 3. General aminopeptidases of dairy lactic acid bacteria

	Name	Substrate n = 1, 2, 3 ...	Strain	Mw[a] kDa	Quat. struct.	Type[b] e[b]	pH Optimum [c]	Localization	Reference
Aminopeptidase N	PepN	X↓(X)$_n$	*Lactococcus lactis* subsp. *cremoris* Wg2	95*	mono	M	7		Tan & Konings, 1990
			Lactococcus lactis subsp. *cremoris* MG1363	95*				intracellular[d,e]	Van Alen-Boerighter et al., 1991
			Lactococcus lactis subsp. *cremoris* HP	95*				intracellular[e]	Exterkate et al., 1992
			Lactococcus lactis subsp. *cremoris* Wg2	95				intracellular[d]	Strøman, 1992
			Lactococcus lactis subsp. *cremoris* Wg2	95				intracellular[d]	Tan et al., 1992a
			Lactococcus lactis subsp. *cremoris* Wg2					intracellular[e]	Tan et al., 1992b
			Lactobacillus casei subsp. *casei* LGG	87+	mono	M	7		Arora & Lee, 1992
			Lactobacillus delbrückii subsp. *lactis* DSM7290	95	mono	M	7	intracellular[d]	Klein et al., 1993
			Lactobacillus delbrückii subsp. *bulgaricus* B14	95*	mono	M	7		Bockelmann et al., 1992
			Lactobacillus helveticus LHE511	92*	mono	M	7	cell-wall[e]	Miyakawa et al., 1992
			Lactobacillus helveticus CNRZ32	97*	mono		6.5		Khalid & Marth, 1990a
			Lactobacillus helveticus ITGL1	97*	mono	M	6.5	cell-wall[e]	Blanc et al., 1993
			Lactobacillus helveticus CNRZ32	97				intracellular[d]	Varmanen et al., 1994
			Lactobacillus helveticus CNRZ32	97				intracellular[d]	Christensen et al., 1995
			Lactobacillus helveticus SBT2171	95*	mono	M	7		Sasaki et al., 1996a
			Lactobacillus helveticus SBT2171					intracellular[e]	Bosman et al., 1996
Aminopeptidase C	PepC	X↓(X)$_n$	*Lactococcus lactis* subsp. *cremoris* AM2	50*	hexa	T	7	intracellular[e]	Neviani et al., 1989
			Lactococcus lactis subsp. *cremoris* Wg2					intracellular[e]	Tan et al., 1992b
			Lactococcus lactis subsp. *cremoris* AM2	50				intracellular[d]	Chapot-Chartier et al., 1993
			Lactobacillus delbrückii subsp. *lactis* DSM7290	51		T		intracellular[d]	Klein et al., 1994a
			Lactobacillus delbrückii subsp. *bulgaricus* B14	54*	tetra	T	7		Wohlrab & Bockelmann, 1993
			Lactobacillus helveticus CNRZ32	50		T		intracellular[d]	Fernández et al., 1994
			Lactobacillus helveticus CNRZ32	51		T		intracellular[d]	Vesanto et al., 1994
Tripeptidase	PepT	X↓X–X	*Lactococcus lactis* subsp. *cremoris* Wg2	52*	di	M	7.5		Bosman et al., 1990
			Lactococcus lactis subsp. *cremoris* Wg2					intracellular[e]	Tan et al., 1992b
			Lactococcus lactis subsp. *cremoris* AM2	52*	di	M	7.5	intracellular[e]	Bacon et al., 1993
			Lactococcus lactis subsp. *cremoris* MG1363	46				intracellular[d]	Mierau et al., 1994
	53	X↓X–X	*Lactococcus lactis* subsp. *cremoris* IMN-C12	23+	tri	T	5.8	cell-wall[e]	Sahlstrøm et al., 1993

Table 3. Continued

	Name	Substrate n = 1, 2, 3 . . .	Strain	Mw[a] kDa	Quat. struct.	Type[b] e[b]	pH Optimum [c]	Localization	Reference
Dipeptidase	PepV	X↓X	*Lactococcus lactis* subsp. *cremoris* H61	50*	di	M	8	intracellular[e]	Hwang et al., 1981
			Lactococcus lactis subsp. *cremoris* Wg2	49*	mono		8	intracellular[e]	Van Bven et al., 1988
			Lactococcus lactis subsp. *cremoris* MG1363	51				intracellular[d]	Fayard and Mierau, 1996
			Lactobacillus delbrückii subsp. *lactis* DSM7290	52	–	M		intracellular[d]	Vongerichten et al., 1994
			Lactobacillus delbrückii subsp. *bulgaricus* B14	51	mono	M	7.5		Wohlrab & Bockelmann, 1992
			Lactobacillus helveticus SBT2171	50*	mono	M	8		Tan et al., 1995
			Lactobacillus helveticus SBT2171					intracellular[e]	Bosman et al., 1996
			Lactobacillus helveticus CNRZ32	50	mono	M		intracellular[d]	Shao et al., 1996
	PepD	X↓X	*Lactobacillus helveticus* CNRZ32	54				intracellular[d]	Dudley et al., 1996
			Lactobacillus helveticus 53/7	54	octo	T	6	intracellular[d]	Vesanto et al., 1996

[a] If available, the molecular weight of the monomer was calculated from the derived amino acid sequence of the cloned gene, otherwise as determined by gel filtration* or SDS-PAGE+. [b] Type of enzyme; M Metallo-peptidase; S Serine-peptidase; T Thiol-peptidase. [c] pH optimum of activity. [d] Localization as predicted from the absence of a signal or membrane anchor sequence. [e] Localization as predicted from fractionation studies and immuno-gold labeling.

the formula Lys-Phe-$(Gly)_n$ were tested for hydrolysis by purified PepN. Using V_{max}/K_m as measure of the overall efficiency of degradation, it was shown that a hexamer is the optimal substrate for PepN (Niven et al., 1995). A recently characterized PepN of *Lb. helveticus* was also able to hydrolyse peptides of lengths up to 10 amino acid residues, even those containing Pro in first position (Sasaki et al., 1996a).

Aminopeptidase C. The general aminopeptidase C (PepC) is a thiol-peptidase of about 50 kDa in all organisms studied to date (Neviani et al., 1989; Chapot-Chartier et al., 1993; Wohlrab & Bockelmann, 1993; Klein et al., 1994a; Fernández et al., 1994; Vesanto et al., 1994). However, differences with respect to the quaternary structure have been reported, i.e. tetramer vs hexamer (Table 3). PepC is also highly conserved among dairy lactic acid bacteria. The amino acid identity of *Lb. helveticus* PepC is 48% and 73% compared to the enzymes of *L. lactis* subsp. *cremoris* and *Lb. delbrückii* subsp. *lactis*, respectively (Chapot-Chartier et al., 1993; Fernández et al., 1994; Klein et al., 1994a; Vesanto et al., 1994). PepC shows some similarity to mammalian bleomycin hydrolase, which is putatively involved in the degradation of the glycopeptide antibiotic bleomycin. PepC has recently been crystallized for structural determinations (Mistou et al., 1994).

Characterization of the substrate specificity of PepC has been largely confined to *para*-nitro-anilide or β-naphthylamide derivatives. In the case of *Lb. delbrückii* subsp. *lactis*, PepC has similar activities as PepN for substrates like Leu-β-NA and His-β-NA, but differs from PepN in its ability to hydrolyse Gly-β-NA, Asp-β-NA, Gly-Ala-β-NA and Gly-Phe-β-N (Klein et al., 1994a). This indicates that PepN and PepC have distinct as well as overlapping specificities. PepC of *Lb. delbrückii* subsp. *bulgaricus* hydrolyses a broad range of di- and tripeptides (Wohlrab & Bockelmann, 1993). Similar to PepN, dipeptides containing Pro are not cleaved, but Pro-containing tripeptides are to some extent. Some preference for dipeptides with Ala, Leu or Lys in N-terminal position has been observed. In the same study, a hexa- and pentapeptide are not hydrolysed by PepC, but the upper size limits of this enzyme have not been explored systematically.

Tripeptidases. Two peptidases capable of cleaving tripeptides have been identified in strains of *Lactococcus lactis* subsp. *cremoris*, but not in other dairy lactic acid bacteria. One tripeptidase (PepT) has been characterized in two different studies and the gene has been cloned and sequenced (Bosman et al., 1990; Bacon et al., 1993; Mierau et al., 1994). PepT is a metalloenzyme and a homodimer of about 105 kDa. Primary sequence alignments revealed that PepT is 47% identical to PepT of *S. typhimurium*. The other tripeptidase, designated peptidase 53, is a trimer of 72 kDa which has been claimed to be a thiol-enzyme (Sahlstrøm et al., 1993). Distinct pH optima for Leu-Leu-Leu hydrolysis by the two enzymes have been found, i.e. 7.5 and 5.8 for PepT and peptidase 53, respectively.

The specificity of PepT is strictly limited to tripeptides as no detectable activity for di- and oligopeptides is observed (Bosman et al., 1990; Bacon et al., 1993). PepT was characterized as having a broad specificity; even substrates such as tri-ornithine were degraded. Although not systematically investigated, PepT was found to hydrolyse tripeptides with Pro in first and third position, but is unable to cleave peptides having Pro in second position. With respect to substrate specificity, peptidase 53 is clearly distinct from PepT (Sahlstrøm et al., 1993). Peptidase 53 has a preference towards tripeptides, but is also able to cleave di- and oligopeptides, albeit with lower activity. Hydrolysis of Pro-containing peptides by peptidase 53 has not been investigated.

Dipeptidases. The general dipeptidase (PepV) is a monomeric metallopeptidase of about 51 kDa, although the enzyme isolated from *L. lactis* subsp. *cremoris* H61 has been reported to have a homodimeric quaternary structure (Hwang et al., 1981). The primary sequence of the dipeptidase of *L. lactis* subsp. *cremoris* and *Lb. delbrückii* subsp. *lactis* do not resemble the general dipeptidase PepD of *E. coli* (Vongerichten et al., 1994; Fayard & Mierau, 1996). Recently, the gene coding for a totally distinct dipeptidase, designated PepD or PepDA, has been cloned in two strains of *Lb. helveticus* (Dudley et al., 1996; Vesanto et al., 1996). Although similar in subunit size and specificity compared to PepV, PepD/PepDA has a distinct pH optimum (6 vs 8), quaternary structure (octomer vs monomer) and catalytic properties (thiol-peptidase vs metallopeptidase). In addition to PepD, *Lb. helveticus* strains also have a PepV-like enzyme, which has been purified and characterized (Tan et al., 1995). The deduced primary sequence of the recently cloned and sequenced *pepV* gene of *Lb. helveticus* CNRZ32 has 69% identity with the one from *Lb. delbrückii* (Shao et al., 1996).

The specificity of dipeptidases is really confined to dipeptides, since none of the tested tri- and oligopep-

Table 4. Proline-specific peptidases of dairy lactic acid bacteria

Enzyme	Name	Substrate n = 1, 2, 3 ...	Strain	Mw[a] kDa	Quat. struct.	Type [b]	pH Opt [c]	Localization	Reference
Prolidase	PepQ	X↓Pro	*Lactococcus lactis* subsp. *cremoris*H61	43*	mono	M	7		Kaminogawa et al., 1984
			Lactococcus lactis subsp. *cremoris* AM2	42*	mono	M	8	intracellular[e]	Booth et al., 1990b
			Lactobacillus delbrückii subsp. *lactis* DSM7290	41	–	M	–	intracellular[d]	Stucky et al., 1995b
Aminopeptidase P	PepP	X↓Pro-$(X)_n$	*Lactococcus lactis* subsp. *cremoris* NCDO763	43*	mono	M	8	intracellular[e]	Mars & Monnet, 1995
X-prolyl-dipeptidyl aminopeptidase	PepX	X-Pro↓$(X)_n$	*Lactococcus lactis* subsp. *lactis* H1	83+	di	S		intracellular[d]	Lloyd and Pritchard, 1991
			Lactococcus lactis subsp. *cremoris* P8-2-47	90*	mono	S	7	cell-wall[e]	Kiefer-Partsch et al., 1989
			Lactococcus lactis subsp. *cremoris* NCDO763	88*	di	S	8.5	intracellular[e]	Zevaco et al., 1990
			Lactococcus lactis subsp. *cremoris* AM2	59*	di	S	7.5	intracellular[e]	Booth et al., 1990a
			Lactococcus lactis subsp. *cremoris* P8-2-47	88		S		intracellular[d]	Mayo et al., 1991
			Lactococcus lactis subsp. *cremoris* NCDO763	88				intracellular[d]	Nardi et al., 1991
			Lactococcus lactis subsp. *cremoris* nTR	88*	di	S	7.5	intracellular[e]	Yan et al., 1992
			Lactococcus lactis subsp. *cremoris* Wg2					intracellular[e]	Tan et al., 1992b
			Lactobacillus casei subsp. *casei* LLG	79+	mono	S	8		Habibi-Najafi & Lee, 1994
			Lactobacillus delbrückii subsp. *lactis* DSM7290	88	mono	S	7	intracellular[d]	Meyer-Barton et al., 1993
			Lactobacillus delbrückii subsp. *bulgaricus* CNRZ397	82*	di	S	7	intracellular[e]	Atlan et al., 1990
			Lactobacillus delbrückii subsp. *bulgaricus* B14	95+	di	S	6.5		Bockelmann et al., 1991
			Lactobacillus helveticus CNRZ32	72+	mono	S	7		Khalid & Marth, 1990b
			Lactobacillus helveticus 53/7	91	di	S	6.5	intracellular[d]	Vesanto et al., 1995a
			Lactobacillus helveticus CNRZ32	88		S		intracellular[d]	Yüksel & Steele, 1995
			Lactobacillus helveticus SBT2171	95*	di	(S)[f]	6.5		Sasaki et al., 1996b
			Lactobacillus helveticus SBT2171					intracellular[e]	Bosman et al., 1996
Prolinase	PepR	Pro↓X	*Lactobacillus helveticus* CNRZ32	35				intracellular[d]	Dudley & Steele, 1994
			Lactobacillus helveticus 53/7	35			7.5		Varmanen et al., 1996b
Proline iminopeptidase	PepI	Pro↓X–$(X)_n$	*Lactococcus lactis* subsp. *cremoris* HP	50+		M	8.5	intracellular[e]	Baankreis & Exterkate, 1991
			Lactobacillus delbrückii subsp. *lactis* DSM7290	33		S		intracellular[d]	Klein et al., 1994b
			Lactobacillus delbrückii subsp. *bulgaricus* CNRZ397	33		S		intracellular[d]	Atlan et al., 1994
			Lactobacillus delbrückii subsp. *bulgaricus* CNRZ397		tri		6.5	cell-wall[e]	Gilbert et al., 1994
			Lactobacillus helveticus 53/7	34	di	S		intracellular[d]	Varmanen et al., 1996a

[a] If available, the molecular weight of the monomer was calculated from the derived amino acid sequence of the cloned gene, otherwise as determined by gel filtration* or SDS-PAGE+. [b] Type of enzyme; M Metallo-protease; S Serine-protease. [c] pH optimum of activity. [d] Localization as predicted from the absence of a signal sequence or membrane anchor sequence. [e] Localization as predicted from fractionation studies and immuno-gold labeling. [f] 50% inhibition by phenylmethylsulfonyl fluoride.

tides are hydrolysed (Hwang et al., 1981; Van Boven et al. 1988; Wohlrab & Bockelmann, 1992; Vongerichten et al. 1994; Tan et al., 1995). Of the 400 possible dipeptides about 80 have been tested, but only a minor fraction has been used for systematic studies. Comparison is therefore difficult, but again some generalizations can be made. Some preference towards N-terminal hydrophobic residues has been observed, whereas specificity for Pro-, Glu- or Asp-containing dipeptides is absent or very low (Van Boven et al., 1988; Tan et al., 1995). Small differences in specificity between PepVs of different organisms have also been noticed; the lactococcal dipeptidase hydrolyses Ala-Ala and Ala-Gly, whereas the dipeptidase of *Lb. helveticus* does not; the reverse is true for His-Leu (Tan et al., 1995).

Peptidases involved in the hydrolysis of Pro-containing peptides

Specialized peptidases capable of hydrolysing Pro-containing sequences have been postulated to be important for the degradation of casein-derived peptides because of the high content of proline in these molecules. General peptidases, such as PepN, PepC, and PepT, are also able to cleave Pro-containing tri- and oligopeptides *in vitro*, but activities observed are usually low (see above). To our knowledge, no significant hydrolysing activities have been found for Pro-containing dipeptides other than that of prolinases and prolidases (Table 4). By definition, the only difference between the specificity of a prolidase and aminopeptidase P, and between proline iminopeptidase and prolinase is the size of the substrate hydrolysed. Although distinct in substrate size limits, the biochemical and genetic properties of the enzymes are very similar (Table 4). Therefore, a thorough investigation of the upper substrate size-limits is required to correctly classify these enzymes.

Aminopeptidase P and prolidase. One aminopeptidase P (PepP) (Mars & Monnet, 1995) and two prolidases (PepQ) (Booth et al., 1990b; Kaminogawa et al., 1984) have been identified in different *L. lactis* subsp. *cremoris* strains. Recently, the gene encoding a prolidase (PepQ) from *Lb. delbrückii* subsp. *lactis* has been cloned and sequenced, and the deduced polypeptide was found to be homologous to PepP and PepQ of *E. coli* (Stucky et al., 1995b). PepP and PepQ are both monomeric metallo-peptidases of about 42 kDa. In agreement with its classification, PepP exclusively hydrolyses oligopeptides of up to 10 residues containing X↓Pro-Pro-$(X)_n$ or X↓Pro-$(X)_n$ sequences at the N-terminus (Mars & Monnet, 1995). No hydrolysis of di- and tripeptides with similar sequences occurs. The specificities of the three putative prolidases have been studied, but the upper size limits have not been systematically explored. One prolidase exclusively hydrolyses X-Pro dipeptides (Kaminogawa et al., 1984), while the other two also cleave dipeptides and tripeptides that do not even contain Pro (Booth et al. 1990b; Stucky et al., 1995b). The latter two enzymes also cleave Pro-X dipeptides, which is a typical activity for a prolinase.

Proline iminopeptidase and prolinase. By definition, a proline iminopeptidase recognizes tri- and oligopeptides which contain Pro-X sequences at the N-terminus, while a prolinase only cleaves Pro-X dipeptides. The proline iminopeptidases of four organisms have been characterized, and designated PepI (or PepIP) (Baankreis & Exterkate, 1991; Klein et al., 1994b; Atlan et al., 1994; Gilbert et al., 1994; Varmanen et al., 1996a). The primary sequence of PepI of *Lb. helveticus* shares 75% identity with that of *Lb. delbrückii* subspecies *lactis* and *bulgaricus* (Klein et al. 1994b; Atlan et al., 1994; Varmanen et al., 1996a). Two proline iminopeptidases are proteases with a subunit size of about 33 kDa, with the possible exception of the one characterized by Baankreis & Exterkate (1991) which is 50 kDa based on SDS-PAGE (or 36 kDa based on 110 kDa by gel-filtration and assumption of a trimeric quaternary structure). Sequence comparisons have revealed a structural motif GQSWGG indicative of a serine-active site, and inhibitor studies confirmed that the enzyme is a (metal-independent) serine peptidase (Atlan et al., 1994; Gilbert et al., 1994; Klein et al., 1994b). The lactococcal enzyme is the only proline iminopeptidase, which has been characterized as a metal-dependent enzyme (Baankreis & Exterkate, 1991). In view of the high similarity/identity of the primary sequences, it is puzzling that the *Lb. helveticus* enzyme is described as a dimer (Varmanen et al., 1996), while the *Lb. delbrückii* enzyme is characterized to be a trimer (Gilbert et al., 1994). The proline iminopeptidase of *Lb. delbrückii* subsp. *bulgaricus* hydrolyses preferentially Pro-X dipeptides and Pro-Gly-Gly, but also cleaves some Ala-X, Gly-X and Leu-X sequences, albeit with lower activities (Gilbert et al., 1994). Similar observations were made for the proline iminopeptidase of *L. lactis* subsp. *cremoris*. Two pentapeptides with Pro in first position were not hydrolysed, while a tetrapeptide with Pro in first position was degraded with low activity (Baankreis & Exterkate, 1991). However, the upper size limits of the proline iminopeptidases have not been explored systematically.

Recently, a prolinase (PepR) has been found in two different strains of *Lb. helveticus*, and its gene has been cloned and sequenced (Dudley & Steele, 1994; Varmanen et al., 1996b). The amino acid sequence bears significant similarity with PepI of *Lb. delbrückii* (35%). The prolinase of *Lb. helveticus* has a subunit size of 35 kDa, which is similar to that of PepI and the same signature sequence (GQSWGG) is found. In agreement with its denomination, PepR predominantly hydrolyses Pro-X dipeptides, but also shows some activity towards tripeptides and Pro-lacking dipeptides.

X-prolyl-dipeptidyl aminopeptidase. Of all peptidases able to cleave Pro-containing sequences, the X-prolyl-dipeptidyl aminopeptidase (PepX) has received most attention. The PepX enzymes have been characterized as serine-peptidases. A number of *pepX* genes have been cloned from different dairy lactic acid bacteria (Mayo et al., 1991; Nardi et al., 1991; Meyer-Barton et al., 1993; Vesanto et al., 1995a, b; Yüksel & Steele, 1995). The primary sequence of PepX from *Lb. helveticus* is 70%, 37% and 37% identical to the enzymes of *Lb. delbrückii* subsp. *lactis*, *L. lactis* subsp. *cremoris* and *L. lactis* subsp. *lactis*, respectively. Some variations in the molecular mass of the monomer and the quaternary structure have been reported, but these differences may largely be due to the methods chosen for size determination (Table 4).

The substrate specificity of PepX has mainly been inferred from chromogenic substrates as X-Pro-*p*-nitroanilides or X-Pro-aminomethylcoumarins. Usually, PepX cleaves N-terminal X-Pro dieptides from tri- and oligopeptides (up to eleven residues) as, for instance, can be inferred from its activity towards Tyr-Pro-Phe, Tyr-Pro-Phe-Pro and β-casein fragment 60–66, Tyr-Pro-Phe-Pro-Gly-Pro-Ile (Kiefer-Partsch et al., 1989; Booth et al., 1990a; Lloyd & Pritchard, 1991; Yan et al. 1992; Sasaki et al., 1996b). The specificity is, however, not limited to X-Pro containing sequences, because Lys-Ala↓Val-Pro↓Tyr-Pro↓Gln (β-casein fragment 176–182) (Lloyd & Pritchard, 1991), and Tyr-Gly↓Gly-Phe-Met (Mayo et al., 1993) are cleaved at positions indicated by the arrows. On the other hand, Arg-Lys-Asp-Val, Arg-Tyr-Leu-Gly-Tyr-Leu and generally X-Pro-Pro-$(X)_n$ sequences are not hydrolysed (Booth et al., 1990a; Lloyd & Pritchard, 1991; Sasaki et al., 1996b).

Unique aminopeptidases

In addition to PepP, peptidase 53 and PepT of *L. lactis*, and PepD and PepR of *Lb. helveticus*, a number of other aminopeptidases have been identified, which have no counterpart in other organisms sofar.

Glutamyl aminopeptidase. In *L. lactis* subspecies, a glutamyl aminopeptidase (PepA) has been identified, which liberates N-terminal Glu and Asp residues from di-, tripeptides and oligopeptides consisting of up to ten amino acid residues (Exterkate & De Veer, 1987c; Niven, 1991; Bacon et al. 1994). This enzyme might well complement the inability of dipeptidases or other enzymes to release N-terminal acidic residues. Because glutamate is abundantly present as free amino acid in milk, PepA is probably not important for the liberation of Glu *per se*, but for the continued degradation of the peptides by other enzymes. The gene encoding this enzyme has been cloned and sequenced (l'Anson et al., 1995); and the deduced primary sequence is found to be homologous to endoglucanase from *Clostridium thermocellum* (30% identity). PepA is a metallo-peptidase with a subunit size of 38 kDa and probably has a hexameric quarternary structure (Niven, 1991; Bacon et al., 1994), although a trimer has also been reported (Exterkate & De Veer, 1987c).

Pyrrolidone carboxylyl peptidase. The pyrrolidone carboxylyl peptidase (Pcp), identified in *L. lactis* subsp. *cremoris*, specifically cleaves N-terminal pyrrolidone carboxylyl residues of peptides and proteins (Exterkate, 1977). The primary sequence encodes a protein of 25 kDa, which shows significant similarity to analogous enzymes from *B. subtilis* and *Streptococcus pneumonia* (Haandrikman, unpublished results).

Leucyl aminopeptidase. Recently, the gene specifying a leucyl aminopeptidase (PepL) from *Lb. delbrückii* subsp. *lactis* has been cloned and sequenced (Klein et al., 1995). The gene product is predicted to be 35 kDa and can be classified as a serine-peptidase. This enzyme hydrolyses preferentially N-terminal Leu-containing dipeptides and some tripeptides. Another distinct aminopeptidase has been identified in *Lb. delbrückii* subsp. *bulgaricus* (Wohlrab & Bockelmann, 1994). It consists of eight subunits of about 32 kDa and might resemble the AC1-aminopeptidase identified in *L. lactis* subsp. *cremoris* (Geis et al., 1985). The enzyme is a metallo-enzyme and hydrolyzes di- and tripeptides and possibly also tetrapeptides. The highest hydrolysis rates were obtained with peptides carrying a N-terminal Leu or Lys, but activities towards Pro-containing di- and tripeptides were also measured.

Endopeptidases
Two different endopeptidases have been identified in *L. lactis* subsp. *cremoris*. One is a 70 kDa monomeric metallopeptidase capable of hydrolysing oligopeptides, but unable to hydrolyse caseins (PepO) (Tan et al., 1991). The gene encoding this enzyme has been cloned and sequenced, and the gene product exhibits significant identity to mammalian enkephalinases (27% amino acid identity) (Mierau et al., 1993). Study of the primary sequence revealed the presence of a consensus His-Glu-X-X-His sequence, typical for Zn-dependent metallopeptidases. Interestingly, the *pepO* gene is located immediately downstream of the genes of the oligopeptide transport system (*oppDFBCA*) (see above) (Tynkkynen et al., 1993). Following gene inactivation of *pepO*, another enzyme with similar activity, molecular mass and immunological properties was identified in this organism (Hellendoorn & Mierau, 1996). Cloning and nucleotide sequencing of this gene revealed that the amino acid sequence of PepO2 is 88% identical to that of PepO.

An additional endopeptidase activity with a specificity that is distinct from that of the PepO enzymes was purified by Monnet et al. (1994). This oligopeptidase was designated PepF; the *pepF* gene was cloned and sequenced, and was shown to specify a 70 kDa monomeric metallopeptidase. Sequence data show that this enzyme is different from the PepO enzymes, and that it resembles mammalian thimet oligopeptidases. Also this enzyme contains a typical Zn-binding site characteristic for Zn-dependent metallopeptidases. PepF hydrolyses peptides containing between 7 and 17 amino acids with a rather broad specificity. In contrast to PepO enzymes, this peptidase cannot degrade Met-enkephalin. Recently, another copy of the gene (80% identity) was found on a plasmid, which also carries the lactose utilization and proteinase genes, and the corresponding enzyme was designated PepF2 (Nardi et al., 1995). Similar enzymes as the ones described above have been purified from other strains of *L. lactis* (Muset et al., 1989; Baankreis 1992; Pritchard et al., 1994).

Recently, an endopeptidase of *Lb. helveticus* has been purified to homogeneity (Sasaki et al., 1996c). Like its lactococcal counterparts, it is a 70 kDa monomeric metallo-enzyme. The *Lb. helveticus* enzyme has a broad specificity for peptides of 3 up to 34 amino acid residues. It remains to be established whether this enzyme is PepO or PepF-like, but immunoblotting showed that it is immunologically distinct from the lactococcal enzymes. The genes coding for two endopeptidases have been cloned from another strain of *Lb. helveticus* (Fenster et al., 1996). One has a high amino acid identity to PepO of *L. lactis*, about the same size, and also contains a typical Zn-peptidase motif. The genes adjacent to this gene are not coding for the oligopeptide transport system like in *L. lactis*. The other putative endopeptidase is smaller (about 440 amino acids) and the primary sequence bears a cysteine peptidase motif. Protein homology searches revealed about 40% amino acid identity to PepC enzymes of lactic acid bacteria. This enzyme has been purified to homogeneity and appears to be able to hydrolyse Met-enkephalin in a PepO-like manner.

Carboxypeptidases
Until now no carboxypeptidases have been characterized in lactic acid bacteria.

Localization of peptidases

While the cellular location of the proteinase is undisputed, the localization of peptidases has been subject to controversies. Fractionation studies have suggested that some peptidases are present in cell-wall fractions, and on the basis of the assumption that PrtP-generated casein degradation products are too big for transport, extracellular peptidases have been implicated in the proteolytic pathway (Geis et al., 1985; Kiefer-Partsch et al., 1989; Exterkate & De Veer, 1987c; Blanc et al., 1993; Gilbert et al., 1994; Law, 1979; Law & Kolstad, 1983; Smid et al., 1989). Most biochemical, genetic and immunological data, however, suggest an intracellular location for most, if not all, enzymes studied to date (Tables 3 and 4). Below, the arguments and methods are listed that have been used to assign a cellular location to the peptidases. To our opinion, the experimental data described in the literature are often misinterpreted.

First, in most enzyme purification schemes, total cell extracts are used as starting material. Very rarely, cell fractionation of some sort is carried out as a first purification step (Sahlstrøm et al., 1993; Blanc et al., 1993; Gilbert et al., 1994), and only in isolated instances peptidase activity has been recovered from the growth medium (Law, 1977). For none of the enzymes isolated and purified, detergents were used (or needed) to solubilize the proteins from the membrane. In all cases the purification methodology indicates that the isolated enzymes are readily soluble in water.

Second, the pH optimum of the enzyme may indicate where the enzyme activity would be maximal, but this information provides no argument for a possible cellular location.

Third, cell-fractionation studies are only useful when performed carefully and with the appropriate controls. For example, on the basis of fractionation studies the localization of the X-prolyl dipeptidyl aminopeptidase of *L. lactis* subsp. *cremoris* has been reported to be cell-wall bound (Kiefer-Partsch et al., 1989), inner-membrane bound (Yan et al., 1992) and cytoplasmic (Booth et al., 1990a). Critical for a correct interpretation of the results is the use of marker enzymes, both negative and positive for a given cell fraction. Many authors have isolated putatively extracellular enzymes after washing the cells several times with water or slightly alkaline buffers, or after incubation with lysozyme (Law, 1977; Geis et al. 1985; Kiefer-Partsch et al., 1989; Miyakawa et al., 1992). Such incubations may lead to significant cell lysis and, particularly, when enzymes are present in high amounts and/or high activities, the cellular location is easily erroneously assigned. To complicate things even further, many strains contain autolysins which render the cells variably susceptible to lysis (Coolbear et al., 1994; Chapot-Chartier et al., 1994).

Various enzymes such as β-galactosidase (Blanc et al., 1993; Atlan et al., 1994), glucose-6-phosphate dehydrogenase (Gilbert et al., 1994), aldolase and malate dehydrogenase (Sahlstrøm et al., 1993) have been used as cytoplasmic markers, but high activities are often associated with membrane preparations (Foucaud and Poolman, 1992; Poolman et al., 1991; Kunji unpublished results). Such activities can only be removed by appropriate procedures and by further purification of membrane fractions by density gradient centrifugation. A gentle shock of osmotically stabilized protoplasts with distilled water (Blanc et al., 1993) is inappropriate to isolate peripheral (external) membrane associated proteins. Such a treatment combined with extensive centrifugation is far more likely causing lysis of protoplasts and subsequent release of cytoplasmic enzymes. The variations in the extent of lysis can be dramatic as in one and the same study under similar conditions, values ranging from 0.1 to 77% have been observed (Blanc et al., 1993), which prevents any conclusions to be made. The use of antibodies as marker for fractionation studies has also to be taken with caution. Titers of the antibodies used may differ considerably and quantification is complicated by non-linearity between signal and protein amounts.

Fourth, immunogold labeling experiments have been performed on whole cells after fixation by aldehydes and embedding in Lowicryl K_4M (polymerization is induced by UV at -35 °C). Ultrathin sections are incubated with specific antibodies and labeled with protein A gold particles (Tan et al., 1992). Electronmicrographs reveal subsequently the localization of the gold-particles. Eminent to success of such an approach is the fixation of whole cells prior to incubation with specific antibodies. Any disruption of the cell membrane prior to or during fixation might lead to loss of enzymes, which may even be 'captured' by an undisrupted cell-wall and lead to misinterpretation of the data.

Fifth, the deduced primary sequence of a gene can give information on the presence or absence of a signal sequence required for translocation of the protein by the general secretory machinery (Sec pathway) (Driessen, 1994). Since signal sequences are not found in the peptidases studied to date, many authors have suggested the existence of dedicated secretory systems involved in the secretion of peptidases without cleavable signal sequence (Tan et al., 1992; Sahlstrøm et al. 1993). Although such systems have been described (Driessen, 1994), there is yet no evidence for signal sequence independent excretion of peptidases in lactic acid bacteria.

Sixth, as an adaptation to the absence of an outer membrane, membrane-associated enzymes of Gram-positives have anchors which keep the proteins attached to the cell. Proteins, as diverse as the maturation protein (PrtM) (Haandrikman et al., 1991), the oligopeptide binding protein (OppA) (Tynkkynen et al., 1993) and the nisin immunity protein (NisI) (Kuipers et al., 1993), all contain a typical signal sequence and a consensus lipo-modification site (generally LAX↓C) (Sankaran & Wu, 1994). The cysteine that follows the cleavage site is modified with three fatty acid tails that anchors the mature protein to the external surface of the cell. In fact, even after extraction with urea/cholate, 80% of total OppA is still attached to membrane vesicles of *L. lactis* (Detmers & Kunji, unpublished results). Alternatively, other extracellularly located enzymes, such as the proteinase (PrtP) (Kok & De Vos, 1994) and the nisin maturation protein (NisB) (Kuipers et al., 1993), contain a signal sequence typical for the translocation by the Sec-dependent secretory pathway (Palmen et al., 1994) and a C-terminal anchor (see above) (Navarre & Schneewind, 1994). A putative anchor can be deduced from sequence comparisons, hydropathy profiling and secondary structure predictions.

Localization of individual peptidases in lactococci

Tables 3 and 4 list the general properties of most peptidases of dairy lactococci and lactobacilli characterized to date and their proposed localization. Assignment of the cellular localization is based on fractionation and immunogold labelling studies, and on the presence or absence of a membrane anchor and/or signal sequence as can be inferred from protein sequences. Listed is also the pH range in which the enzyme has optimum activity.

The genes of eleven peptidases of *L. lactis*, i.e. PepO, PepO2, PepF, PepF2, PepN, PepC, PepT, PepV, PepA, PepX and PcP, have been cloned and sequenced [most of them are reviewed in Kok & De Vos (1994) and Poolman et al. (1995); others are described in Monnet et al., 1994; l'Anson et al., 1995; Nardi et al., 1995; Hellendoorn & Mierau, 1996; & Faynard & Mierau 1996]. In none of the inferred amino acid sequences of the peptidase genes a typical signal sequence or membrane anchor is detected. Data obtained from fractionation and immunogold labelling studies point towards an intracellular localization for PepN, PepO, PepT, PepX and PepC (see Tables 3 and 4) or are conflicting, e.g. PepA (Exterkate & De Veer, 1987c; Bacon et al., 1994) and PepX (Kiefer-Partsch et al., 1989; Booth et al., 1990a; Yan et al., 1992). Sequence data are lacking for AC1-aminopeptidase (Geis et al., 1985), and peptidase 53 (Sahlstrøm et al., 1993).

The view that most, if not all, lactococcal peptidases are located inside the cell is also supported by growth and uptake experiments performed with isogenic peptide transport mutants. If a peptide transport mutant is unable to utilize a particular peptide, it must mean that peptidases involved in hydrolysis of the peptide are **not** located extracellularly, since extracellular breakdown products would otherwise have entered the cell via the amino acid and/or, in case of Opp substrates, via the di- and tripeptide transport systems. Substrates as diverse as Ala-Ala, Pro-Ala, Ala-Pro, Ala-Pro-Leu, Leu-Gly-Gly, Ala-Ala-Ala-Ala, Tyr-Gly-Gly-Phe-Leu and Val-His-Leu-Thr-Pro-Val-Gly-Lys are not hydrolysed extracellularly, providing compelling evidence for an intracellular location of PepV, PepR (sofar not characterized), PepQ, PepX, PepT, PepN, PepC, PepO and PepO2 (Tynkkynen et al., 1993; Hagting et al., 1994; Kunji et al., 1995; Foucaud et al., 1995; Kunji et al., 1996). Similar experiments with substrates typical for peptidase 53 (Sahlstrøm et al., 1993), such as Leu-Leu-Leu, have made it highly unlikely that indeed this enzyme is present extracellularly in *L. lactis* MG1363.

Moreover, in peptide transport assays in *L. lactis*, a wide variety of peptides (from 2 up to ten residues) have been used, and extracellular and intracellular fractions have been analysed by liquid chromatography (sometimes in combination with mass spectrometry) (Fang & Kunji, unpublished results). Under conditions that cell lysis is minimal, through the use of mutants in which an autolysin gene is inactivated (Buist et al., 1995), we have never detected any extracellular hydrolysis of peptides, not even when high cell densities were used. In fact, by comparing amino acid accumulation from β-casein degradation in wildtype cells and oligopeptide transport mutants, we could demonstrate that extracellular peptidase activity involved in degradation of this protein substrate is lacking in *L. lactis* (see below) (Kunji et al., 1995). In addition, recent studies, in which the product formation of the purified proteinase was compared to that of PrtP present on the cell surface of oligopeptide transport and autolysin deficient mutants, have shown that the casein-derived peptides are not significantly altered even after prolonged incubations (Kunji & Fang, unpublished results).

Localization of individual peptidases in lactobacilli

Sequence data and fractionation studies indicate that the majority of peptidases in lactobacilli is located in the cytoplasm (Tables 3 and 4), but conflicting data have also been reported. In *Lb. helveticus* two PepN-like aminopeptidases have been described to be cell-wall associated (Miyakawa et al., 1992; Blanc et al., 1993). Recent immunological studies have indicated that PepN as well as PepV, PepX and PepO of *Lb. helveticus* are located intracellularly (Bosman et al., 1996). In addition, the gene encoding PepN does not contain a signal sequence or membrane anchor (Christensen et al., 1995).

The prolyl iminopeptidase of *Lb. delbrückii* subsp. *bulgaricus* is sofar the only proline-specific peptidase which has been detected in cell-wall fractions (Gilbert et al., 1994), but the deduced amino acid sequence does not contain a signal sequence or membrane anchor (Atlan et al., 1994). The observed periplasmic location in *E. coli* of 45% of total prolyl iminopeptidase activity is probably an artifact of heterologous expression or isolation procedure (Gilbert et al., 1994). Hydropathy profiling of the deduced protein indicated that two short hydrophobic sequences are present, but this is not uncommon for interior stretches of soluble proteins.

Concluding remarks

In conclusion, it strikes as somewhat odd that peptidases, which are so similar in primary sequence and function, would be located at different places in different lactic acid bacteria. Consensus about the location for all peptidases of the same class would be expected, whether inside or outside. The localization of peptidases has also been studied in detail in *E. coli* and *S. typhimurium*. In these organisms, peptidases with very similar biochemical characteristics, and in many cases homologous to those of lactic acid bacteria, have all been found intracellularly (Lazdunski, 1989). Extracellular degradation of peptides down to amino acids would lead to a substantial loss of amino acids through diffusion. In contrast, amino acids liberated from intracellularly accumulated peptides, can directly be used for biosynthesis or other metabolic activities. In view of the concept that concerted action of peptidases is required for complete degradation of peptides, it seems logical that peptidases are located in the same compartment, otherwise, extensive relocation of degradation products would be necessary to complete hydrolysis. The degradation is most efficient when the concentration of enzymes and substrates is highest, which is most easily achieved when these are gathered in the cytoplasm. A prerequisite for efficient degradation by cytoplasmic peptidases is that the PrtP-generated casein breakdown products can be translocated into the cell.

The role of peptide transport systems in growth on milk

In recent years, a number of directional mutagenesis techniques have been developed for inactivation of chromosomally located genes (Leenhouts, 1991; Bhowmik et al., 1993; Leenhouts et al., 1996). Mutants have been constructed which lack a functional di- and tripeptide transport system (DtpT) and/or oligopeptide transport system (Opp), but do express the P_I-type proteinase (Tynkkynen et al., 1993; Hagting et al., 1994; Kunji et al. 1995). Mutants which lack a functional Opp system are unable to grow on milk (Tynkkynen et al., 1993), while growth of mutants lacking DtpT is unaffected in this medium (Kunji et al., 1995). This observation indicates that one or more essential amino acids enter the cell via uptake through Opp. To circumstantiate this finding, cells were incubated with β-casein in the presence of glucose and chloramphenicol. The wildtype strain and the DtpT$^-$ mutant accumulate all amino acids present in β-casein (Figure 4A), whereas amino acids are not accumulated significantly inside the cells of Opp$^-$ and DtpT$^-$ Opp$^-$ mutants (Figure 4B). When cells are incubated with a mixture of amino acids mimicking the composition of β-casein, the amino acids are taken up to the same extent in all four strains.

These and other experiments have revealed a number of important properties of the proteolytic pathway of *L. lactis*. First, all the essential and growth-stimulating amino acids can be released from β-casein by the action of the proteinase PrtP in a form that can be transported by the oligopeptide transport system exclusively. When a functional oligopeptide transport system is absent no significant intracellular accumulation of amino acids is observed. Second, consistent with the observation that PrtP does not release significant amounts of di- and tripeptides from β-casein (Juillard et al., 1995a), inactivation of the di-tripeptide transport system has no effect on the utilization of this protein substrate. Since di-tripeptide transport mutants selected on the basis of resistance towards L-Ala-β-chloro-L-Ala are affected in their ability to grow on a mixture of caseins (Smid et al., 1989), we speculate that this phenotype is due to secondary mutations or to the inability to transport essential amino acids (most likely His and/or Leu) in the form of small peptides that are released from proteins other than β-casein. In fact, growth of *L. lactis* HP has been shown to be dependent on the utilization of both β- and κ-casein (Exterkate & De Veer, 1987b). Third, the observation that a single mutation, abolishing oligopeptide transport activity, results in a defect to accumulate amino acids argues strongly against the involvement of extracellular peptidases in the degradation of β-casein. If peptidases would have been present externally, amino acids and di- and tripeptides would have been formed and subsequently been taken up by the corresponding transport systems.

Preliminary results show that peptides from the C-terminal end of β-casein are transported by Opp (indicated as dotted arrows in Figure 1A). Most of these peptides are in the range of 4 to 8 residues, but transport of at least one nonamer and one decamer has been observed. Transport of these peptides into the

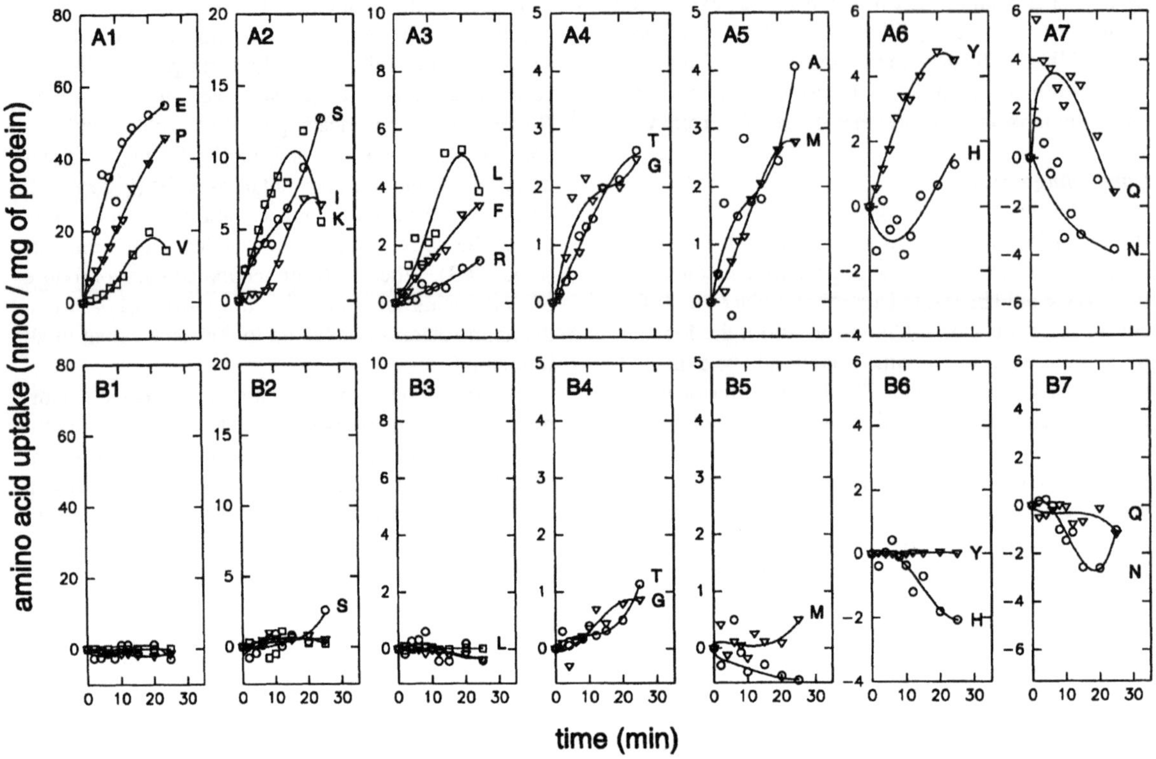

Figure 4. Time dependence of intracellular amino acid accumulation by *L. lactis* wildtype (A) and oligopeptide transport mutant (B). β-casein was added to glycolyzing chloramphenicol-treated cells at time zero (min). The amino acids are indicated by their 1 letter code. (Data from Kunji et al., 1995.)

cell largely explains the intracellular amino acid accumulation upon addition of β-casein to wildtype cells. Amino acids that are rare in β-casein, such as Ala and Tyr, are readily accumulated and these residues are indeed present in the C-terminal fragments. As stated before, the C-terminal peptides are flanked by preferential cleavage sites and are present in relatively large amounts in the hydrolysates, already at the earliest times of degradation. Transport of these peptides would supply the cell with all essential and growth stimulating amino acids with the exception of His. Even though Glu and Asn are not present in these fragments, Gln and Asp can be converted into these amino acids. Growth experiments with β-casein have shown that His and Leu are not liberated from β-casein at rates high enough to meet the growth requirements of *L. lactis* (Kunji et al., 1995). In hydrolysates of κ-casein, peptides are present which fall within the size exclusion limits of Opp and contain His and Leu. The degradation of α-caseins may also contribute significantly to supply of amino acids such as Leu, i.e. when the appropriate proteinase activity is present.

The role of peptidases in vivo

In recent years a number of single peptidase mutants have been constructed by targeted deletion or disruption of the corresponding genes. Lactococcal mutants lacking either PepX, PepN, PepO, PepT, PepF, PepC and PepA are not or only slightly affected in their ability to grow in milk (Mayo et al., 1993; Baankreis, 1992; Mierau et al., 1993; Mierau et al., 1994; Monnet et al., 1994; Erra-Pujada et al. 1995; l'Anson et al., 1995). Similar results have been obtained when single mutations of PepC, PepN and PepX were generated in *Lb. helveticus* (Christensen and Steele, 1996). The observation that single peptidases are not essential can have several reasons: (i) some peptidases might not be involved in casein degradation at all, (ii) peptidases with overlapping specificities might be present and/or

(iii) alternative peptides, whose degradation is undisturbed by the mutation, might be used to supply the cell with free amino acids.

Recently, a set of sixteen single and multiple peptidase deletion mutants has been constructed in *L. lactis*, in which combinations of up to all five of the following peptidase genes were inactivated: *pepO, pepN, pepC, pepX* and *pepT* (Mierau et al., 1996). When the ability of these strains to grow in milk was tested, it was observed that an increasing number of peptidase mutations leads to an increasing growth defect (Mierau et al., 1996) (Figure 5). Ultimately, the fivefold mutant grows more than 10 times slower in milk than the wildtype strain. Similar results were obtained with *Lb. helveticus* mutants, in which combinations of *pepC, pepN* and *pepX* were made (Christensen & Steele, 1996). The main exception being that PepX appears to play a more important role in this organism than in *L. lactis*.

In *L. lactis*, the growth of the manifold mutants in complex broth was identical to that of the wildtype, indicating that the lower growth rates in milk are not due to a general defect in proteolytic housekeeping functions of the cell. The phenotype of the mutants is also not caused by a decreased expression or activity of the other components of the proteolytic pathway, since neither PrtP, Opp nor other peptidase activities were seriously affected by the peptidase mutations (see below).

The slowest growing mutants, i.e. $[XTNC]^-$ and [XTOCN], had much lower intracellular amino acid pools than the wild-type, and peptides were accumulated inside the cell of these mutants. Thus, the lower growth rates can directly be attributed to the inability of the mutants to breakdown casein-derived peptides, providing, for the first time, direct evidence for the functioning of lactococcal peptidases in the degradation of milk proteins. The observation that the five-fold mutant has growth rates close to zero, also indicates that PepN, PepC, PepO, PepT and (to a lesser extent) PepX are crucial for the degradation of casein-derived peptides.

The complexity of the peptide mixture in milk makes it difficult to trace the fate of individual molecules and to assign a particular role to individual peptidases in the proteolytic pathway. Therefore, the same set of peptidase mutants of *L. lactis* has been used to study the fate of single peptides in growth experiments and by chromatographic analysis of intracellular fractions (Kunji et al., 1996). Several multiple peptidase mutants were unable to utilize particular peptides and, as a consequence, accumulated these peptides intracellularly. Apparently, the failure to grow relates to the inability of the cells to hydrolyse the peptides. The observation that peptide transport mutants are impaired in their ability to accumulate peptides, combined with the finding that peptides are translocated into the cell as whole entities, proves unequivocally that transport precedes degradation (Kunji et al., 1996).

Mutants lacking PepN, PepC, PepT plus PepX cannot utilize peptides such as Leu-Gly-Gly, Gly-Phe-Leu, Leu-Gly-Pro, Ala-Pro-Leu and Gly-Leu-Gly-Leu, indicating that no other peptidases are present in *L. lactis* MG1363 to hydrolyse these molecules. The fivefold mutant $[XTOCN]^-$ still grows on Gly-Leu and Tyr-Gly-Gly-Phe-Leu, confirming the presence of a dipeptidase and another endopeptidase (e.g. PepO2). The general aminopeptidases PepN, PepC and PepT have overlapping, but not identical specificities and differ in their overall activity towards individual peptides (Kunji et al., 1996). In contrast, PepX has a unique specificity, because it is the only enzyme which degrades Ala-Pro-Leu efficiently. Certain peptides can only be broken down by the concerted action of different peptidases, e.g. in a *pepN* background, Leu can only be liberated from Gly-Leu-Gly-Leu if PepC plus PepT are present.

Regulation of expression of components of the proteolytic pathway

The regulation of expression of the various components of the proteolytic pathway is a still largely unexplored area of research. Few studies have been performed in which the expression of enzymes has been studied in different media, including whey permeate, milk, complex media, and chemically defined media (CDM) with either amino acids, peptides or caseins. Published data cannot always be compared directly due to differences in strains, variations in media, etc. In general, the expression of components of the proteolytic pathway of *L. lactis* is highest in media containing amino acids only, while peptides generally down-regulate expression.

In *L. lactis* AM1, synthesis of the proteinase is repressed when the growth medium contains casitone, an enzymatic digest of casein that mainly consists of peptides (Exterkate, 1985). The production of the proteinase of *L. lactis* Wg2 is also inhibited in media containing casein or a tryptic digest of casein (Laan et al., 1993). Moreover, the addition of the dipeptide Leu-Pro to chemically defined medium leads to a decrease of the proteinase production by this strain (Laan et al.,

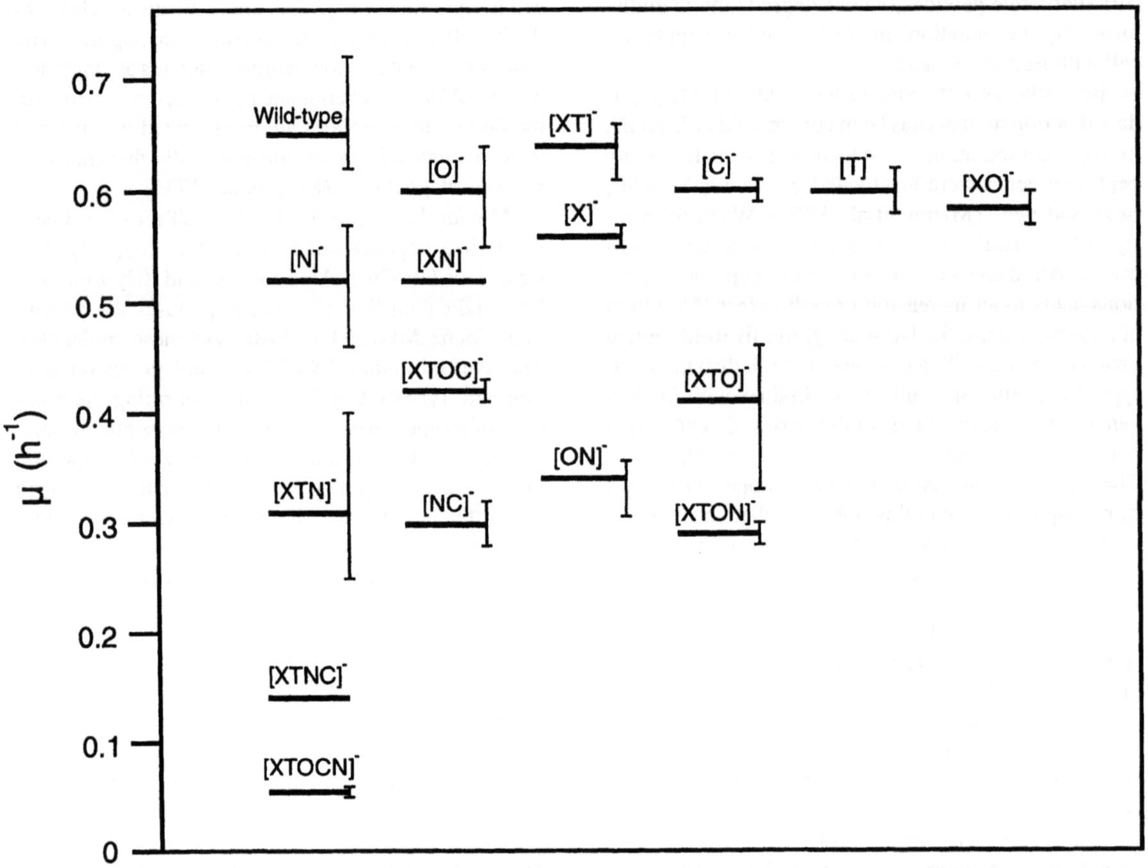

Figure 5. Growth rates of *Lactococcus lactis* MG1363 and peptidase mutants in milk. Horizontal bars indicate the growth rates and vertical bars mark the variation of the data by showing the range of the highest and the lowest values measured for a particular strain. X, T, O, C and N refer to deletions of *pepX, pepT, pepO, pepC* and *pepN* genes, respectively. (Data from Mierau et al., 1996.)

1993). By fusing the promoter regions of *prtP* and *prtM* to the reporter gene *gusA*, it was possible to follow the transcriptional regulation of the proteinase expression directly (Marugg et al., 1995). The GusA activities were highest in media containing no or low amounts of peptides, and repression is observed in peptide-rich media. Again, addition of Leu-Pro and Pro-Leu were found to down-regulate the expression of PrtP in wild-type cells. Importantly, in mutants, lacking DtpT, the synthesis of GusA (driven by the PrtP and PrtM promoters) was not affected by these dipeptides. In agreement with these results, significant higher expression of PrtP was observed when mutants lacking DtpT were grown in a complex medium as compared to the wild-type (Kunji et al., 1995).

Also the expression of the peptide transport systems of *L. lactis* is (moderately) affected by the composition of the growth medium. Expression of DtpT is five-fold higher when cells are grown on chemically defined media containing only amino acids as compared to complex media containing both peptides and amino acids as source of nitrogen (Hagting, unpublished results). Likewise, the expression of OppA is increased more than 10-fold when cells are grown on chemically defined media containing amino acids rather than complex broth (Detmers and Kunji, unpublished results). The expression of DtpP, increases two to three-fold when CDM is used instead of a complex medium to culture the cells (Foucaud et al., 1995). A similar increase in transport activity is observed when Leu-Leu or Leu-Leu-Leu are added to CDM, suggesting that, in contrast to DtpT and Opp, these peptides may serve as inducers of the DtpP system.

The regulation of peptidase expression has not been studied in detail. Preliminary experiments suggest that expression of PepN in *L. lactis* is regulated in a similar manner as the proteinase (Meijer et al., 1995). Genes specifying putative regulator proteins have been found in the proximity of peptidase genes. A gene coding for a potential transcription regulator protein, designated PepR1, was identified by sequence comparisons and is located upstream of the prolidase gene *pepQ* of *Lb. delbrückii* subsp. *lactis* (Stucky et al., 1996). The deduced protein has a molecular mass of 37 kDa and shows significant similarity to catabolite control proteins of various organisms (Stucky et al., 1996). A hybrid of *pepQ* and the β-galactosidase reporter gene displays an enhanced expression when co-expressed with *pepR1* in *E. coli*.

Despite the observations that synthesis of components of the proteolytic system of *L. lactis* is affected by peptides, no major changes were observed in the specific activities of PrtP, Opp and the remaining peptidases in the peptidase mutants when the cells were grown in milk (Mierau et al., 1996). This is surprising, because one would expect that under the nitrogen starvation conditions prevailing in the multiple peptidase mutants, the cells would want to compensate for the peptidase deficiency by increasing the expression levels of other components of the proteolytic system. Thus, the regulation of expression through changes in amino acid and/or peptide levels in the cell appear to be minimal when *L. lactis* is grown in milk.

The proteolytic pathway of Lactococcus lactis*: a model*

Gene inactivation experiments have demonstrated that PrtP and Opp are essential for growth of *L. lactis* on media containing casein(s) as sole source of amino acids (Kok & De Vos, 1994; Tynkkynen et al., 1993; Kunji et al., 1995; Juillard et al., 1995b). Various lines of evidence indicate that the peptidases are located intracellularly and, consequently, cannot play a role in extracellular hydrolysis of caseins or casein-derived peptides (unless cell lysis occurs). These observations, together with the evidence that the Opp system transports peptides with a length of 4 up to at least 8 residues (maybe 10), suggest that a major fraction of essential and growth-stimulating amino acids enters the cell in the form of oligopeptides (Figure 6). A large number of peptides are released from β-casein by the activity of the proteinase, and a major fraction of the β-casein derived peptides falls in the range of 4 to 10 residues (Juillard et al., 1995a). Several of these peptides are released from the C-terminal part of β-casein in high amounts, irrespective of the type of proteinase (Figure 1A). Figure 1B shows the identified peptides released from activity of the various proteinases on β-casein. The β- and κ-casein-derived oligopeptides together contain all amino acids necessary for growth of *L. lactis*.

From the analysis of the extracellular and the intracellular fractions it can be concluded that only a few of the peptides generated by the proteinase are actually utilized by *L. lactis* (Kunji et al., 1995; Fang & Kunji unpublished results). The fact that large peptides do accumulate in the medium despite a functional Opp system is a consequence of the size-exclusion limits of the oligopeptide transporter. Peptides up to a length of 30 amino acids are formed by PrtP, and, although the upper size exclusion limits of Opp have not been established unequivocally, most of these large fragments will not be transported (Juillard et al., 1995a). Furthermore, although the lactococcal oligopeptide transport system has a broad substrate specificity, certain peptides may not be transported due to competition of peptides for the oligopeptide binding protein. In addition, a part of the peptide pool may also be taken up with a rate that is lower than the production rate by the proteinase.

Since transport precedes degradation (see above), the specificity of the Opp system will largely determine the oligopeptides which enter the peptidolytic pathway during growth in milk. These peptides are degraded efficiently by a multitude of peptidases, because all peptidases which are needed for degradation are present in the cytoplasm (Figure 6).

The peptidases can be divided into two classes: (1) *primary peptidases*, that generate free amino acids from oligopeptides directly (PepN, PepA, PepI, PepP and possibly PepC), and (ii) *secondary peptidases*, that need degradation by other peptidases to complete hydrolysis to the level of free amino acids (PepO, PepO2, PepF, PepF2, PepX, PepR, PepQ, PepT and PepV) (Figure 6). Endopeptidases (and PepX) require further degradation of the breakdown products by aminopeptidases, tri- and dipeptidases, while di- and tripeptidases require the initial activity of oligo- and/or aminopeptidases capable of hydrolysing oligopeptides. From inactivation studies it has become apparent that PepN, PepC, PepT, PepX and PepO activities are crucial in the peptidolytic pathway. Inactivation of the

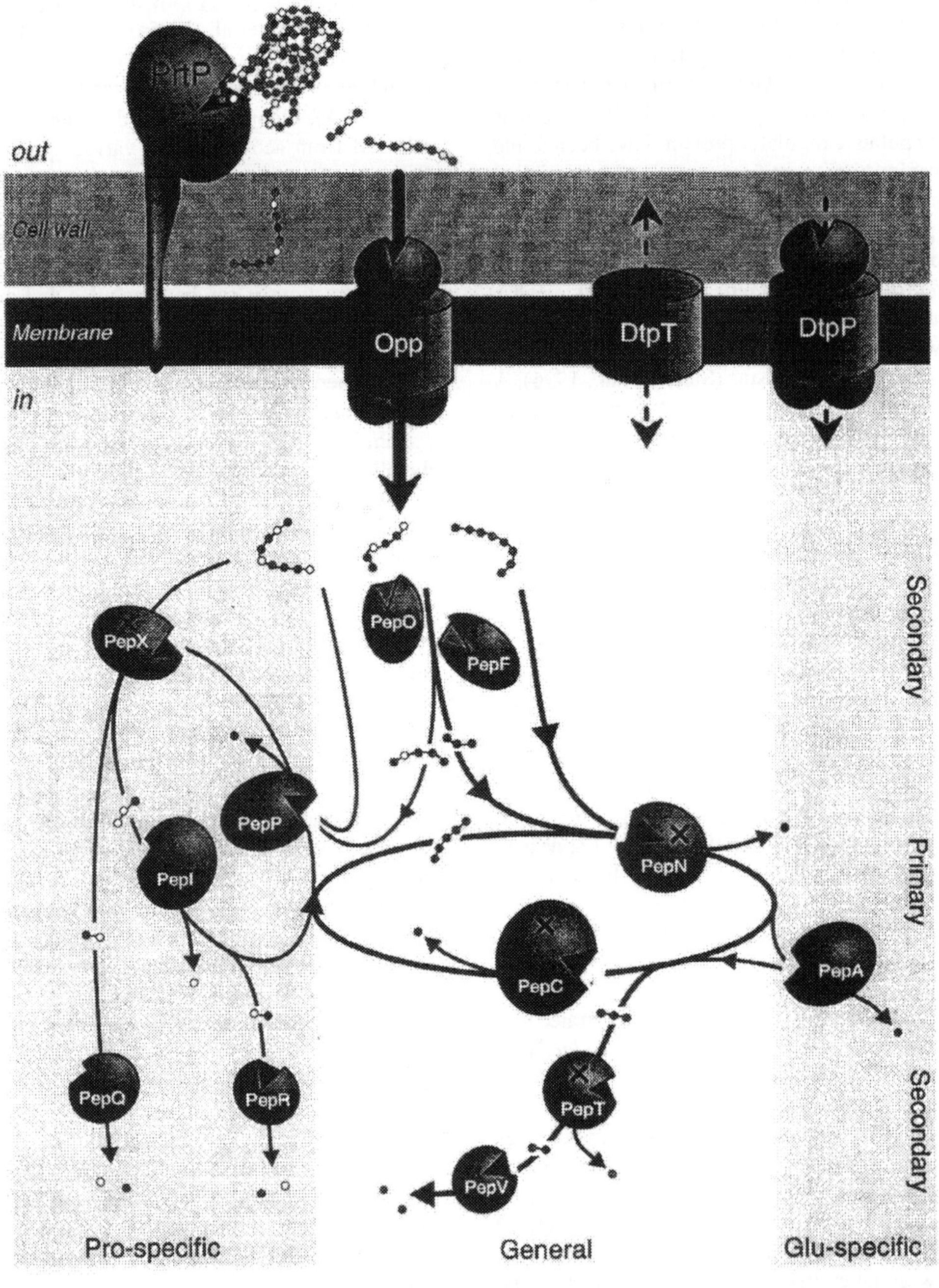

Figure 6. Proposed model for the proteolytic system of *Lactococcus lactis*. The role of the *primary* (PepN, PepC, PepI, PepP and PepA) and *secondary* enzymes (PepO, PepF, PepX, PepQ, PepR, PepV and PepT) and peptidolytic cycles is depicted schematically (various alternative routes of breakdown are possible for most peptides). For simplicity, PepO2 and PepF2 are not depicted. The peptidases inactivated in the five-fold mutant (see text) are indicated by X. Proline residues of casein or casein-derived peptides are depicted as white circles and Glu/Asp as black.

primary peptidases PepN and PepC will directly affect the release of amino acids from oligopeptides (Mierau et al., 1996; Christensen & Steele, 1996). The five-fold peptidase mutant has growth rates in milk close to zero, even though it still contains all the dipeptidases and endopeptidases (only PepO is missing), and several specific aminopeptidases (PepA, PepI and PepP). This indicates that degradation to the levels of dipeptides is severely impaired in this mutant and that the proteinase does not produce dipeptides in large amounts. Moreover, the five-fold mutant provides evidence for the notion that the endopeptidases, PepO2, PepF and PepF2, do not form significant amounts of dipeptides, and dipeptide formation by endopeptidases has thus become dependent on other activities which are inactivated in the multiple peptidase mutant. The Pro- and Glu/Asp-specific peptidases capable of oligopeptide degradation, i.e. PepI, PepP and PepA, are clearly not active enough and/or too specific to complement the peptidase deficiencies in the five-fold mutant.

Concluding remarks and prospects

The model proposed for *L. lactis* is compatible with most studies on specificity, location and role of the various enzymes in proteolysis. However, important questions remain to be answered. Are the casein degradation experiments *in vitro* a true reflection of the situation in milk? In the *in vitro* experiments, purified casein preparations are used, while in milk, caseins are in part organized in micelles. In addition, other proteolytic activities naturally present in milk, such as that of plasmins, might alter the product formation on caseins considerably and thereby influence the growth of lactic acid bacteria (Grufferty & Fox, 1988). Recently, the plasmin precursor plasminogen has been isolated from bovine milk and characterized (Benfeldt et al., 1994). Studies on the product formation of plasmins on caseins have been intiated in the laboratory of Petersen.

Does a purified enzyme yield the same product formation as a cell wall associated enzyme? Preliminary data indicate that peptides which are thought to be important for growth are released by the purified as well as the cell-wall associated enzymes (Fang & Kunji, unpublished results). How is casein-degradation influenced by the decreasing pH values prevailing during growth in milk? Does the degradation of casein by other proteinases indeed proceed as far as reported for the *L. lactis* Wg2 proteinase (Juillard et al., 1995a)?

Why are amino acid transport systems present in lactic acid bacteria? During initial growth of *L. lactis* in milk, amino acids are used as sources of nitrogen, but total consumption of free amino acids is very low (Juillard et al., 1995b). Peptide transport mutants, which are only able to utilize free amino acids, grow only to very low cell densities. Based on these studies it could be estimated that free amino acids contribute less than 2% to growth of the organism in milk. Alternatively, the amino acid transport systems might play a role in maintaining a balanced amino acid pool in the following way. The composition of transported oligopeptides will not correspond to the amino acid needs of the organism. If the concentration gradients of the amino acids exceed the driving force imposed by the amino acid transporter, the residues may leave the cell by facilitated diffusion. If efflux of amino acids via the transport proteins is coupled to proton extrusion, a proton motive force will be generated. In fact, if the total energy costs of oligopeptide uptake are lower than the amount of energy generated by amino acid residue driven efflux of protons, metabolic energy is generated. Such mechanisms might be important for the survival of lactic acid bacteria in dairy products when milk sugars have been depleted.

What is the function of DtpT and DtpP? Inactivation of the di/tripeptide transport systems does not result in a growth defect in milk (Kunji et al., 1995; Juillard unpublished results). Since DtpT is just like most amino acid transport systems a reversible enzyme, it is possible that DtpT can also function as an efflux system for partial degradation products.

What role has the putatively extracellular peptidase 53 in the proteolytic system (Sahlstrøm et al., 1993)? Our results indicate that mutations in *pepT, pepX, pepN* plus *pepC* remove all tripeptidase activity in *L. lactis* MG1363 (Kunji et al., 1996). What is the role of proline- and glutamate/aspartate-specific peptidases and of the PepF enzymes in the degradation of casein-derived peptides? It might well be that some of these peptidases also have roles in intrinsic protein turnover.

The model we propose for *L. lactis* may be extrapolated to the proteolytic systems of *Lactobacillus* species, but even more questions remain to be answered because several components of the proteolytic pathway of these bacteria have been studied in much less detail. Anyhow, as far as a comparison can be made, the proteolytic systems of lactococci and lactobacilli appear to be very similar. Since these organisms are faced with similar challenges during growth in milk, profound differences in the general scheme of the proteolytic pathway are not to be expected. All organisms require efficient degradation of caseins for optimal growth. The amount of amino acids necessary for growth is strain dependent, but caseins provide enough amino acids to meet any demand. For initial degradation, most lactic acid bacteria possess a single extracellularly located proteinase, which degrades caseins into oligopeptides. These enzymes are subject to only little genetic variation. Since the casein-derived peptides are variable in composition and size, concerted action of peptidases is required to complete degradation and this is most efficiently done when all enzymes are located at the same location. Although peptide transport systems are still poorly characterized in lactobacilli, preliminary experiments suggest that activities similar to those in *L. lactis* are indeed present.

If the overall strategy of dairy lactic acid bacteria to grow in milk is the same, how is it possible that these organisms yield such a variety of different dairy products? A recent comparison of proteolytic activities of different lactococci and lactobacilli has revealed that overall activities of enzymes may vary considerably (Crow et al., 1994; Sasaki et al., 1995). Such variations are not important for growth of the organism (the amounts of amino acids produced from caseins are always in excess), but might have a profound effect on flavour and texture of the final food product. Additionally, through acquired mutations, the enzymes might have an altered sensitivity towards pH values, salt concentrations, temperature and prolonged periods of incubation, which are conditions prevailing during ripening of dairy products. In addition to differences in activity of the enzymes, particular activities seem to be confined to certain *Lactobacillus* and *Lactococcus* strains. For instance, significant PepA-like activity has only been detected in lactococci and not in lactobacilli (Sasaki et al., 1995).

Finally, since the peptidases are located intracellularly, differences in susceptibility to lysis might play an important role in ripening of dairy products (Chapot-Chartier et al., 1994; Coolbear et al., 1994; Buist et al., 1995) and offers a means to manipulate proteolysis.

The proteolytic system of lactic acid bacteria has become the paradigm of research on proteolysis in bacteria. With the genetic and biochemical tools available, it has now become possible to manipulate the pathways of protein and peptide degradation, and amino acid and peptide transport. These developments have paved the way to new, more economical and better quality food products.

Acknowledgements

The authors would like to acknowledge, Danièle Atlan, Frank Detmers, Gang Fang, Taisuke Iwasaki, Vincent Juillard, Jürgen Klein, Veronique Monnet, Hajime Nakajima, Airi Palva, Tjwan Tan, Masahiro Sasaki, Roland Siezen & Jim Steele for providing us with information prior to publication.

References

Arora G & Lee BH (1992) Purification and characterization of aminopeptidase from *Lactobacillus casei* subsp. *casei* LLG. J. Dairy Sci. 75: 700–710

Atlan D, Gilbert C, Blanc B & Portalier R (1994) Cloning, sequencing and characterizatioon of the *pepIP* gene encoding a proline iminopeptidase from *Lactobacillus delbrückii* subsp. *bulgaricus* CNRZ397. Microbiology 140: 527–535

Atlan D, Laloi P & Portalier R (1990) X-prolyl-dipeptidyl aminopeptidase of *Lactobacillus delbrückii* subsp. *bulgaricus*: Characterization of the enzyme and isolation of deficient mutants. Appl. Environ. Microbiol. 56: 2174–2179

Baankreis R (1992) The role of lactococcal peptidases in cheese ripening. University of Amsterdam, Ph.D. Thesis

Baankreis R & Exterkate R (1991) Characterization of a peptidase from *Lactococcus lactis* ssp. *cremoris* HP that hydrolyses di- and tripeptides containing proline or hydrophobic residues as the aminoterminal amino acid. Syst. Appl. Microbiol. 14: 317–323

Bacon CL, Jennings PV, Fhaolain IN & O'Cuinn G (1994) Purification and characterization in an aminopeptidase A from cytoplasm of *Lactococcus lactis* subsp. *cremoris* AM2. Int. Dairy J. 4: 503–519

Bacon CL, Wilkinson M, Jennings PV, Fhaoláin IN & O'Cuinn G (1993) Purification and characterization of an aminotripeptidase from cytoplasm of *Lactococcus lactis* subsp. *cremoris* AM2. Int. Dairy J. 3: 163–177

Benfeldt C, Larsen LB, Rasmussen JT, Andreasen PA & Petersen TE (1994) Isolation and characterization of plasminogen and fragments of plasminogen from bovine milk. Int. Dairy J. 4

Blanc B, Laloi P, Atlan D, Gilbert C & Portalier R (1993) Two cell-wall-associated aminopeptidases from *Lactobacillus helveticus* and the purification and characterization of APII from strain ITGL1. J. Gen. Microbiol. 139: 1441–1448

Bockelmann W, Fobker M & Teuber M (1991) Purification and characterization of the X-prolyl-dipeptidyl-aminopeptidase from *Lactobacillus delbrückii* subsp. *bulgaricus* and *Lactobacillus acidophilus*. Int. Dairy J. 1: 51–66

Bockelmann W, Monnet V, Geis A, Teuber M & Gripon J-C (1989) Comparison of cell wall proteinases from *Lactococcus lactis* subsp. *cremoris* AC1 and *Lactococcus lactis* subsp. *lactis* NCDO 763. Appl. Microbiol. Biotechnol. 31: 278–282

Bockelmann W, Schulz Y & Teuber M (1992) Purification and characterization of an aminopeptidase from *Lactobacillus delbrückii* subsp. *bulgaricus*. Int. Dairy J. 2: 95–107

Booth M, Phaoláin IN, Jennings PV & O'Cuinn G (1990a) Purification and characterization of a post-proline dipeptidyl aminopeptidase from *Streptococcus cremoris* AM2. J Dairy Res. 57: 89–99

Booth M, Jennings PV, Fhaoláin IN & O'Cuinn G (1990b) Prolidase activity of *Lactococcus lactis* subsp. *cremoris* AM2: partial purification and characterization. J. Dairy Res. 57: 245–254

Bosman BW, Sasaki M, Iwasaki T & Tan PST (1996) Localization of proteolytic enzymes in *Lactobacillus helveticus* SBT 2171. Manuscript in preparation

Bosman BW, Tan PST & Konings WN (1990) Purification and characterization in a tripeptidase from *Lactococcus lactis* subsp. *cremoris* Wg2. Appl. Environ. Microbiol. 56: 1839–1843

Bhowmik T, Fernández L & Steele JL (1993) Gene replacement in *Lactobacillus helveticus*. J. Bacteriol. 175: 6341–6344

Buist G, Kok J, Leenhouts KJ, Dabrowska M, Venema G & Haandrikman AJ (1995) Molecular cloning and nucleotide sequence of the gene encoding the major peptidoglycan hydrolase of *Lactococcus lactis*, a muramidase needed for cell separation. J. Bacteriol. 177: 1554–1563

Chapot-Chartier M-P, Deniel C, Rousseau M, Vassal L & Gripon J-C (1994) Autolysis of two strains of *Lactococcus lactis* during cheese ripening. Int. Dairy J. 4: 251–269

Chapot-Chartier M-P, Nardi M, Chopin M-C, Chopin A & Gripon J-C (1993) Cloning and sequencing of *pepC*, a cysteine aminopeptidase gene from *Lactococcus lactis* subsp. *cremoris* AM2. Appl. Environ. Microbiol. 59: 330–333

Chopin A (1993) Organization and regulation of genes for amino acid biosynthesis in lactic acid bacteria. FEMS Microbiol. Rev. 12: 21–38

Christensen JE, Lin D, Palva A & Steele JL (1995) Sequence analysis, distribution and expression of an aminopeptidase N-encoding gene from *Lactobacillus helveticus* CNRZ32. Gene 155: 89–93

Christensen JE & Steele JL (1996) Characterization of peptidase-deficient *Lactobacillus helveticus* derivatives. Unpublished results

Coolbear T, Pillidge CJ & Crow VL (1994) The diversity of potential cheese ripening characteristics of lactic acid starter bacteria: 1. resistance to cell lysis and levels and cellular distribution of proteinase activities. Int. Dairy J. 4: 697–721

Coolbear T, Reid JR & Pritchard GG (1992) Stability and specificity of the cell wall-associated proteinase from *Lactococcus lactis* subsp. *cremoris* H2 released by treatment with lysozyme in the presence of calcium ions. Appl. Environ. Microbiol. 58: 3263–3270

Crow VL, Holland R, Pritchard GG & Coolbear T (1994) The diversity of potential cheese ripening characteristics of lactic acid starter bacteria: 2. The elvels and subcellular distribution of peptidase and esterase activities. Int. Dairy J. 4: 723–742

Detmers & Kunji, unpublished results

De Vos WM & Siezen RJ (1994) Engineering pivotal proteins for lactococcal proteolysis. In: Andrews AT & Varley J (Eds) Biochemistry of Milk Products, pp 56–71. Royal Society of Chemistry, Cambridge, England

Driessen AJM (1994) How proteins cross the bacterial cytoplasmic membrane. J. Membr. Biol. 142: 145–159

Dudley EG, Husgen AC, He W & Steele JL (1996) Sequencing, distribution, and inactivation of the dipeptidase A gene (pepDA) from *Lactobacillus helveticus* CNRZ32. J. Bacteriol. in press

Dudley EG & Steele JL (1994) Nucleotide sequence and distribution of the *pepPN* gene from *Lactobacillus helveticus* CNRZ32. FEMS Microbiol. Lett. 119: 41–46

Dunten P & Mowbray SL (1995) Crystal structure of the dipeptide binding protein from *Escherichia coli* involved in active transport and chemotaxis. Prot. Sci. 4: 2327–2334

Erra-Pujada M, Mistou MY & Gripon J-C (1995) Construction et étude d'une souche de *Lactococcus lactis* dont le gène *pepC* n'est plus fonctionnel. 7ème Colloque du Club des Bactéries Lactiques. p 8

Exterkate FA (1977) Pyrrolidone carboxylyl peptidase in *Streptococcus cremoris*: dependence on an interaction with membrane components. J. Bacteriol. 129: 1281–1288

— (1984) Location of peptidases outside and inside the membrane of *Streptococcus cremoris*. Appl. Environ. Microbiol. 47: 177–183

— (1985) A dual-directed control of cell wall proteinase production in *Streptococcus cremoris* AM1: a possible mechanism of regulation during growth in milk. J Dairy Sci. 68: 562–571

Exterkate FA, Alting AC & Bruinenberg PG (1993) Diversity of cell envelope proteinase specificity among strains of *Lactococcus lactis* and its relationship to charge characteristics of the substrate-binding region. Appl. Environ. Microbiol. 59: 3640–3647

Exterkate FA, De Jong M, De Veer GJCM & Baankreis R (1992) Location and characterization of aminopeptidase N in *Lactococcus lactis* subsp. *cremoris* HP. Appl. Microbiol. Biotechnol. 37: 46–54

Exterkate FA & De Veer GJCM (1987a) Complexity of the native cell wall proteinase of *Lactococcus lactis* subsp. *cremoris* HP and purification of the enzyme. System. Appl. Microbiol. 9: 183–191

— (1987b) Optimal growth of *Streptococcus cremoris* HP in milk is related to β- and κ-casein degradation. Appl. Microbiol. Biotechnol. 25: 471–475

— (1987c) Purification and properties of a membrane- bound aminopeptidase A from *Streptococcus cremoris*. Appl. Environ. Microbiol. 53: 577–583

Fang G & Kunji ERS (1996) Unpublished results

Fayard B & Mierau I (1996) Dipeptidase (pepV) from *Lactococcus lactis* subsp. *cremoris*. Unpublished results

Fenster KM, Chen YS & Steele JL (1996) Endopeptidases from *Lactobacillus helveticus* CNRZ32. Unpublished results

Fernández L, Bhowmik T & Steele JL (1994) Characterization of the *Lactobacillus helveticus* CNRZ32 *pepC* gene. Appl. Environ. Microbiol. 60: 333–336

Foucaud C, Kunji ERS, Hagting A, Richard J, Konings WN, Desmazeaud M & Poolman B (1995) Specificity of peptide transport systems in *Lactococcus lactis*: Evidence for a third system which transports hydrophobic di- and tripeptides. J. Bacteriol. 177: 4652–4657

Foucaud C & Poolman B (1992) Lactose transport protein of *Streptococcus thermophilus*. Functional reconstitution of the protein and characterization of the kinetic mechanism of transport. J. Biol. Chem. 267: 22087–22094

Geis A, Bockelmann W & Teuber T (1985) Simultaneous extraction and purification of a cell wall-associated peptidase and β-casein specific protease from *Streptococcus cremoris* AC1. Appl. Microbiol. Biotechnol. 23: 79–84

Gilbert C, Atlan D, Blanc B & Portalier R (1994) Proline iminopeptidase from *Lactobacillus delbrückii* subsp. *bulgaricus* CNRZ397: purification and characterization. Microbiology 140: 537–542

Gilbert C, Atlan D, Blanc B, Portalier R, Germond GJ, Lapierre L & Mollet B (1996) A new cell surface proteinase: sequencing and analysis of the prtB gene from *Lactobacillus delbrückii* subsp. *bulgaricus*. Submitted for publication

Grufferty MB & Fox PF (1988) Milk alkaline proteinase. J. Dairy Res. 4: 609–630

Haandrikman AJ (1994) Pyrrolidone carboxylyl peptidase (Pcp) in *L. lactis* subsp. *cremoris*. Unpublished results

Haandrikman AJ, Kok J, Laan H, Soemitro S, Ledeboer AM, Konings WN & Venema G (1989) Identification of a gene required for the maturation of an extracellular serine proteinase. J. Bacteriol. 171: 2789–2794

Haandrikman AJ, Kok J & Venema G (1991) Lactococcal proteinase maturation protein PrtM is a lipoprotein. J. Bacteriol. 173: 4517–4525

Habibi-Najafi MB & Lee BH (1994) Purification and characterization of X-prolyl dipeptidyl peptidase from *Lactobacillus casei* subsp. *casei* LLG. Appl. Microbiol. Biotechnol. 42: 280–286

Hagting A, Kunji ERS, Leenhouts KJ, Poolman B & Konings WN (1994) The di- and tripeptide transport protein of *Lactococcus lactis*. J. Biol. Chem. 269: 11391–11399

Hagting A. Unpublished results

Hellendoorn MA & Mierau I (1996) PepO2 of *Lactococcus lactis* subsp. *cremoris*. Unpublished results

Higgins CF (1992) ABC transporters: from microorganisms to man. Ann. Rev. Cell. Biol. 8: 67–113

Holck A & Næs H (1992) Cloning, sequencing and expression of the gene encoding the cell-envelope-associated proteinase from *Lactobacillus paracasei* subsp. *paracasei* NCDO151. J. Gen. Microbiol. 138: 1353–1364

Holt C & Sawyer L (1988) Primary and predicted secondary structures of the caseins in relation to their biological function. Protein Engineer. 2: 251–259

Hugenholtz H, Van Sinderen D, Kok J & Konings WN (1987) Cell wall-associated proteases of *Streptococcus cremoris* Wg2. Appl. Environ. Microbiol. 53: 853–859

Hwang I-K, Kaminogawa S & Yamauchi K (1981) Purification and properties of dipeptidase from *Streptococcus cremoris*. Agric. Biol. Chem. 45: 159–165

I'Anson K, Movahedi S, Griffin HG, Gasson MJ & Mulholland F (1995) A non-essential glutamyl aminopeptidase is required for optimal growth of *Lactococcus lactis* MG1363 in milk. Microbiology 141: 2873–2881

Juillard V, Laan H, Kunji ERS, Jeronimus-Stratingh CM, Bruins AP & Konings WN (1995a) The extracellular P_I-type proteinase of *Lactococcus lactis* hydrolyzes β-casein into more than one hundred different oligopeptides. J. Bacteriol. 177: 3472–3478

Juillard V, Le Bars D, Kunji ERS, Konings WN, Gripon J-C & Richard J (1995b) Oligopeptides are the main source of nitrogen for *Lactococcus lactis* during growth in milk. Appl. Environ. Microbiol. 61: 3024–3030

Kaminogawa S, Azuma N, Hwang I-K, Suzuki Y & Yamauchi K (1984) Isolation and characterization of a prolidase from *Streptococcus cremoris* – H61. Agric. Biol. Chem. 48: 3035–3040

Khalid NM & Marth EH (1990a) Partial purification and characterization of an aminopeptidase from *Lactobacillus helveticus* CNRZ32. System. Appl. Microbiol. 13: 311–319

— (1990b) Purification and partial characterization of a prolyl-dipeptidyl aminopeptidase from *Lactobacillus helveticus* CNRZ32. Appl. Environ. Microbiol. 56: 381–388

Kiefer-Partsch B, Bockelmann W, Geis A & Teuber M (1989) Purification of an X-prolyl-dipeptidyl aminopeptidase from the cell wall proteolytic system of *Lactococcus lactis* subsp. *cremoris*. Appl. Microbiol. Biotechnol. 31: 75–78

Kiwaki M, Ikemura H, Shimizu-Kadota M & Hirashima A (1989) Molecular characterization of a cell wall-associated proteinase gene from *Streptococcus lactis* NCDO763. Mol. Microbiol. 3: 359–369

Klein JR, Dick A, Schick J, Matern HT, Henrich B & Plapp R (1995) Molecular cloning and DNA sequence analysis of *pepL*, a leucyl aminopeptidase gene from *Lactobacillus delbrückii* subsp. *lactis* DSM7290. Eur. J. Biochem. 228: 570–578

Klein J-R, Henrich B & Plapp R (1994a) Cloning and nucleotide sequence analysis of the *Lactobacillus delbrückii* ssp. *lactis* DSM7290 cysteine aminopeptidase gene *pepC*. FEMS Microbiol. Lett. 124: 291–300

Klein J-R, Klein U, Schad M & Plapp R (1993) Cloning, DNA sequence analysis and partial characterization of *pepN*, a lysyl aminopeptidase from *Lactobacillus delbrückii* subsp. *lactis* DSM7290. Eur. J. Biochem. 217: 105–114

Klein J-R, Schmidt U & Plapp R (1994b) Cloning, heterologous expression, and sequencing of a novel proline iminopeptidase gene, *pepI*, from *Lactobacillus delbrückii* subsp. *lactis* DSM7290. Microbiology 140: 1133–1139

Kojic M, Fira D, Banina A & Topisirovic L (1991) Characterization of the cell wall-bound proteinase of *Lactobacillus casei* HN14. Appl. Environ. Microbiol. 57: 1753–1757

Kok J (1990) Genetics of the proteolytic system of lactic acid bacteria. FEMS Microbiol. Rev. 87: 15–42

Kok J & De Vos WM (1994) The proteolytic system of lactic acid bacteria. In: Gasson M & De Vos W (Eds) Genetics and Biotechnology of Lactic Acid Bacteria. pp 169–210. Blackie and Professional, London, England

Kok J, Leenhouts KJ, Haandrikman AJ, Ledeboer AM & Venema G (1988) Nucleotide sequence of the cell wall-associated proteinase gene of *Streptococcus cremoris* Wg2. Appl. Environ. Microbiol. 54: 231–238

Konings WN, Poolman B & Driessen AJM (1989) Bioenergetics and solute transport in Lactococci. CRC Crit. Rev. Microbiol. 16: 419–476

Kontinen VP, Saris P & Sarvas M (1991) A gene (prsA) of *Bacillus subtilis* is involved in a novel, late stage of protein export. Mol. Microbiol. 5: 1273–1283

Kuipers OP, Rollema HS, De Vos WM & Siezen RJ (1993) Characterization of the nisin gene cluster *nisABTCIPR* of *Lactococcus lactis*; requirement of expression of the *nisA* and *nisI* genes for producer immunity. Eur. J. Biochem. 216: 281–291

Kunji ERS, Hagting A, De Vries CJ, Juillard V, Haandrikman AJ, Poolman B & Konings WN (1995) Transport of β-casein-derived peptides by the oligopeptide transport system is a crucial step in the proteolytic pathway of *Lactococcus lactis*. J. Biol. Chem. 270: 1569–1574

Kunji ERS, Mierau I, Poolman B, Konings WN, Venema G & Kok J (1996) Fate of peptides in peptidase mutants of *Lactococcus lactis*. Mol. Microbiol. (submitted)

Kunji ERS, Smid EJ, Plapp R, Poolman B & Konings WN (1993) Di-tripeptides and oligopeptides are taken up via distinct transport mechanisms in *Lactococcus lactis*. J. Bacteriol. 175: 2052–2059

Laan H, Bolhuis H, Poolman B, Abee T & Konings WN (1993) Regulation of proteinase synthesis in *Lactococcus lactis*. Acta Biotechnol. 13: 95–101

Laan H & Konings WN (1989) The mechanism of proteinase release from *Lactococcus lactis* subspecies *cremoris* Wg2. Appl. Environ. Microbiol. 55: 3103–3106

Laloi P, Atlan D, Blanc B, Gilbert C & Portalier R (1991) Cell-wall-associated proteinase of *Lactobacillus delbrückii* subsp. *bulgaricus* CNRZ397: differential extraction, purification and properties of the enzyme. Appl. Microbiol. Biotechnol. 36:, 196–204

Law BA (1977) Dipeptide utilization by starter streptococci. J. Dairy Res. 44: 309–317

— (1979) Extracellular peptidases in group N streptococci used as cheese starters. J. Appl. Bacteriol. 46: 455–463

Law BA & Kolstad J (1983) Proteolytic systems in lactic acid bacteria. Antonie van Leeuwenhoek 49: 225–245

Lazdunski AM (1989) Peptidases and proteases of *Escherichia coli* and *Salmonella typhimurium*. FEMS Microbiol. Rev. 63: 265–276

Leenhouts (1991) Ph.D. Thesis, University of Groningen

Leenhouts KJ, Buist G, Bolhuis A, Ten Berge A, Kiel J, Mierau I, Dabrowska M, Venema G & Kok J (1996) A general system for generating unlabelled gene-replacements in the bacterial chromosome. (Manuscript in preparation)

Lloyd RJ & Pritchard GG (1991) Characterization of X-prolyl dipeptidyl aminopeptidase from *Lactococcus lactis* subsp. *lactis* H1. J. Gen. Microbiol. 137: 49–55

Mars I & Monnet V (1995) An aminopeptidase P from *Lactococcus lactis* with original specificity. Biochim. Biophys. Acta 1243: 209–215

Martín-Hernández MC, Alting AC & Exterkate FA (1994) Purification and characterization of the mature, membrane-associated cell-envelope proteinase of *Lactobacillus helveticus* L89. Appl. Microbiol. Biotechnol. 40: 828–834

Marugg JD, Meijer W, Van Kranenburg R, Laverman P, Bruinenberg PG & De Vos WM (1995) Medium-dependent regulation of proteinase gene expression in *Lactococcus lactis*: Control of transcription by specific dipeptides. J. Bacteriol. 177: 2982–2989

Mayo B, Kok J, Bockelman W, Haandrikman A, Leenhouts KJ & Venema G (1993) Effect of X-prolyl dipeptidyl aminopeptidase deficiency on *Lactococcus lactis*. Appl. Environ. Microbiol. 5: 2049–2055

Mayo B, Kok J, Venema K, Bockelman W, Teuber M, Reinke H & Venema G (1991) Molecular cloning and sequencing analysis of the X-prolyl dipeptidyl aminopeptidase gene from *Lactococcus lactis* subsp. *cremoris*. Appl. Environ. Microbiol. 57: 38–44

Meijer WC, Looijestijn E, Marugg JD & Hugenholtz J (1995) Expression and release of proteolytic starter enzymes during cheese ripening. 7th European Congress on Biotechnology. Poster MEP23

Meyer-Barton EC, Klein JR, Imam M & Plapp R (1993) Cloning and sequence analysis of the X-prolyl-dipeptidyl-aminopeptidase gene (*pepX*) from *Lactobacillus delbrückii* ssp. *lactis* DSM7290. Appl. Microbiol. Biotechnol. 40: 82–89

Mierau I, Haandrikman AJ, Velterop O, Tan PST, Leenhouts KL, Konings WN, Venema G & Kok J (1994) Tripeptidases gene (*pepT*) of *Lactococcus lactis*: Molecular cloning and nucleotide sequencing of *pepT* and construction of a chromosomal deletion mutant. J. Bacteriol. 176: 2854–2861

Mierau I, Kunji ERS, Leenhouts KJ, Hellendoorn MA, Haandrikman AJ, Poolman B, Konings WN, Venema G & Kok J (1996) Multiple peptidase mutants of *Lactococcus lactis* are severely impaired in their ability to grow in milk. (Submitted)

Mierau I, Tan PST, Haandrikman AJ, Kok J, Leenhouts KJ, Konings WN & Venema G (1993) Cloning and sequencing of the gene for a lactococcal endopeptidase, an enzyme with sequence similarity to mammalian enkephalinase. J. Bacteriol. 175: 2087–2096

Mills OE & Thomas TD (1981) Nitrogen sources for growth of lactic streptococci in milk. N.Z. J. Dairy Sci. Technol. 16: 43–55

Mistou M-Y, Rigolet P, Chapot-Chartier M-P, Nardi M, Gripon J-C & Brunie S (1994) Crystallization and preliminary X-ray analysis of PepC, a thiol aminopeptidase from *Lactococcus lactis* homologous to bleomycin hydrolase. J. Mol. Biol. 237: 160–162

Miyakawa H, Kobayashi S, Shimamura S & Tomita M (1992) Purification and characterization of an aminopeptidase from *Lactobacillus helveticus* LHE-511. J. Dairy Sci. 75: 27–35

Molenaar D, Hagting A, Alkema H, Driessen AJM & Konings WN (1993) Characteristics and osmoregulatory roles of uptake systems for proline and glycne betaine in *Lactococcus lactis*. J. Bacteriol. 175: 5438–5444

Momburg F, Roelse J, Howard JC, Butcher GW, Hämmerling GJ & Neefjes JJ (1994) Selectivity of MHC-encoded peptide transporters from human, mouse and rat. Nature 367: 648–651

Monnet V, Bockelman W, Gripon J-C & Teuber M (1989) Comparison of cell wall proteinases from *Lactococcus lactis* subsp. *cremoris* AC1 and *Lactococcus lactis* subsp. *lactis* NCDO 763. Appl. Microbiol. Biotechnol. 31: 112–118

Monnet V, LeBars D & Gripon J-C (1986) Specifity of a cell wall proteinase from *Streptococcus lactis* NCDO 763 towards bovine β-casein. FEMS Microbiol. Lett. 36: 127–131

— (1987) Purification and characterization of a cell wall proteinase from *Streptococcus lactis* NCDO 763. J. Dairy Res. 54: 247–255

Monnet V, Ley JP & Gonzalez S (1992) Substrate specificity of the cell envelope-located proteinase of *Lactococcus lactis* subsp. *lactis* NCDO 763. Int. J. Biochem. 24: 707–718

Monnet V, Nardi M, Chopin A, Chopin M-C & Gripon J-C (1994) Biochemical and genetic characterization of PepF, an oligopeptidase from *Lactococcus lactis*. J. Biol. Chem. 269: 32070–32076

Muset G, Monnet V & Gripon J-C (1989) Intracellular proteinase of *Lactococcus lactis* subsp. *lactis* NCDO 763. J. Dairy Res. 56: 765–778

Nakajima H (1996) Amino acid and peptide transport systems. Unpublished results

Nardi M, Chopin M-C, Chopin A, Cals M-M & Gripon J-C (1991) Cloning and DNA sequence analysis of an X-prolyl dipeptidyl aminopeptidase gene from *Lactococcus lactis* subsp. *lactis* NDCO763. Appl. Environ. Microbiol. 57: 45–50

Nardi M, Renault P, Gripon J-C & Monnet V (1995) Duplication d'un gene *pepF* codant une oligopeptidase chez *Lactococcus lactis*. 7ème Colloque du Club des Bactéries Lactiques. p 7

Næs H & Nissen-Meyer J (1992) Purification and N-terminal amino acid determination of the cell wall bound proteinase from *Lactobacillus paracasei* subsp. *paracasei*. J. Gen. Microbiol. 138: 313–318

Navarre WW & Schneewind O (1994) Proteolytic cleavage and cell wall anchoring at the LPXTG motif of surface proteins in Gram-positive bacteria. Mol. Microbiol. 14: 115–121

Neviani E, Boquien CY, Monnet V, Phan Thanh L & Gripon J-C (1989) Purification and characterization of an aminopeptidase from *Lactococcus lactis* subsp. *cremoris* AM2. Appl. Environ. Microbiol. 55: 2308–2314

Niven GW (1991) Purification and characterization of aminopeptidase A from *Lactococcus lactis* subsp. *lactis* NCDO712. J. Gen. Microbiol. 137: 1207–1212

Niven GW, Holder SA & Strøman P (1995) A study of the substrate specificity of aminopeptidase N from *Lactococcus lactis* subsp. *cremoris*. Appl. Microbiol. Biotechnol. 43

Olson NF (1990) The impact of lactic acid bacteria on cheese flavor. FEMS Microbiol. Rev. 87: 131–148

Payne JW & Smith MW (1994) Peptide transport by microorganisms. In: Rose AH & Tempest DW (Eds) Advances in Microbial Physiology. Vol. 36, pp 2–80. Academic Press, London

Poolman B (1993) Energy transduction in lactic acid bacteria. FEMS Microbiol. Rev. 12: 125–148

Poolman B, Molenaar D, Smid EJ, Ubbink T, Abee T, Renault PP & Konings WN (1991) Malolactic fermentation: electrogenic malate uptake and malate/lactate antiport generate metabolic energy. J. Bacteriol. 173: 6030–6037

Poolman B, Kunji ERS, Hagting A, Juillard V & Konings WN (1995) The proteolytic pathway of *Lactococcus lactis*. J. Appl. Bacteriol. Symp. Suppl. 79: 65–75

Pritchard GG & Coolbear T (1993) The physiology and biochemistry of the proteolytic system in lactic acid bacteria. FEMS Microbiol. Rev. 12: 179–206

Pritchard GG, Freebairn AD & Coolbear T (1994) Purification and characterization of an endopeptidase from *Lactococcus lactis* subsp. *cremoris* SK11. Microbiology 140: 923–930

Reid JR, Coolbear T, Pillidge CJ & Pritchard GG (1994) Specificity of hydrolysis of bovine κ-casein by cell envelope-associated proteinases from *Lactococcus lactis* strains. Appl. Environ. Microbiol. 60: 801–806

Reid JR, Moore CH, Midwinter GG & Pritchard GG (1991a) Action of a cell wall proteinase from *Lactococcus lactis* subsp. *cremoris* SK11 on bovine α_{S1}-casein. Appl. Microbiol. Biotechnol. 35: 222–227

Reid JR, Ng KH, Moore CH, Coolbear T & Pritchard GG (1991b) Comparison of bovine β-casein hydrolysis by P_I and P_{III}-type proteinases from *Lactococcus lactis* subsp. *cremoris*. Appl. Microbiol. Biotechnol. 35: 477–483

Sahlstrøm S, Chrzanowska J & Sørhaug T (1993) Purification and characterization of a cell wall peptidase from *Lactococcus lactis* subsp. *cremoris* IMN-C12. Appl. Environ. Microbiol. 59: 3076–3082

Sankaran K & Wu HC (1994) Lipid modification of bacterial prolipoprotein; transfer of diacylglyceryl moiety from phosphatidylglycerol. J. Biol. Chem. 269: 19701–19706

Sasaki M, Bosman BW & Tan PST (1995) Comparison of proteolytic activities in various lactobacilli. J. Diary Res. 62: 601–610

— (1996a) Characterization of a new, broad substrate specificity aminopeptidase from the dairy organism *Lactobacillus helveticus* SBT 2171. Microbiol. in press

Sasaki M, Bosman BW, Iwasaki T & Tan PST (1996b) The purification and characterization of a 95 kDa X-prolyl-dipeptidyl aminopeptidase from *Lactobacillus helveticus* SBT 2171. (Submitted)

— (1996c) The purification and characterization of a new oligopeptidase from *Lactobacillus helveticus* SBT2171. (Submitted)

Schmidt DG (1982) Association of caseins and casein micelle structure. In: Fox PF (Ed) Developments in Dairy Chemistry, vol. 1, pp 61–68. Elsevier, London, U.K.

Shao W, Parkin KL & Steele JL (1996) Characterization of two dipeptidases from *Lactobacillus helveticus*. (Submitted)

Smid EJ, Driessen AJM & Konings WN (1989) Mechanism and energetics of dipeptide transport in membrane vesicles of *Lactococcus lactis*. J. Bacteriol. 171: 292–298

Smid EJ, Poolman B & Konings WN (1991) Casein utilization by lactococci. Appl. Environ. Microbiol. 57: 2447–2452

Steiner H-Y, Naider F & Becker JM (1995) The PRT family: a new group of peptide transporters. Mol. Microbiol. 16: 825–834

Strøman P (1992) Sequence of a gene (*lap*) encoding a 95.3-kDa aminopeptidase from *Lactococcus lactis* ssp. *cremoris* Wg2. Gene 113: 107–112

Stucky K, Hagting A, Klein JR, Matern H, Henrich B, Konings WN & Plapp R (1995a) Cloning and characterization of *brnQ*; a gene encoding a low affinity branched chain amino acid carrier of *Lactobacillus delbrückii* subsp. *lactis*. Mol. Gen. Genet. 249: 682–690

Stucky K, Klein JR, Schüller A, Matern H, Henrich B & Plapp R (1995b) Cloning and DNA sequence analysis of *pepQ*, a prolidase gene from *Lactobacillus delbrückii* subsp. *lactis* DSM7290 and partial characterization of its product. Mol. Gen. Genet. 247: 494–500

Stucky K, Schick J, Klein JR, Henrich B & Plapp R (1996) Characterization of *pepR1*, a gene coding for a potential transcriptional regulator of *Lactobacillus delbrückii* subsp. *lactis* DSM7290. FEMS Microbiol. Lett. (in press)

Swaisgood HE (1993) Symposium: genetic perspectives on milk proteins: comparative studies and nomenclature. J. Dairy Sci. 76: 3054–3061

Tame JRH, Murshudov GN, Dodson EJ, Neil TK, Dodson GG, Higgins CF & Wilkinson AJ (1994) The structural basis of sequence-independent peptide binding by OppA protein. Science 264: 1578–1581

Tan PST, Chapot-Chartier M-P, Pos KM, Rousseaud M, Boquien C-Y, Gripon J-C & Konings WN (1992b) Localization of peptidases in Lactococci. Appl. Environ. Microbiol. 58: 285–290

Tan PST & Konings WN (1990) Purification and characterization of an aminopeptidase from *Lactococcus lactis* subsp. *cremoris* Wg2. Appl. Environ. Microbiol. 56: 526–532

Tan PST, Poolman B & Konings WN (1993a) Proteolytic enzymes of *Lactococcus lactis*. J. Dairy Res. 60: 269–286

Tan PST, Pos KM & Konings WN (1991) Purification and characterization of an endopeptidase from *Lactococcus lactis* subsp. *cremoris* Wg2. Appl. Environ. Microbiol. 57: 3593–3599

Tan PST, Sasaki M, Bosman BW & Iwasaki T (1995) Purification and characterization of a dipeptidase from *Lactobacillus helveticus* SBT 2171. Appl. Environ. Microbiol. 61: 3430–3435

Tan PST, Van Alen-Boerrigter IJ, Poolman B, Siezen RJ, De Vos WM & Konings WN (1992a) Characterization of the *Lactococcus lactis pepN* gene encoding an aminopeptidase homologous to mammalian aminopeptidase N. FEBS Lett. 306: 9–16

Tan PST, Van Kessel TAJM, Van de Veerdonk FLM, Zuurendonk PF, Bruins AP & Konings WN (1993b) Degradation and debittering of a tryptic digest from β-casein by aminopeptidase N from *Lactococus lactis* subsp. *cremoris* Wg2. Appl. Environ. Microbiol. 59: 1430–1436

Tynkkynen S, Buist G, Kunji E, Kok J, Poolman B, Venema G & Haandrikman AJ (1993) Genetic and biochemical characterization of the oligopeptide transport system of *Lactococcus lactis*. J. Bacteriol. 175: 7523–7532

Van Alen-Boerrigter IJ, Baankreis R & De Vos WM (1991) Characterization and overexpression of the *Lactococcus lactis pepN* gene and localization of its product, aminopeptidase N. Appl. Environ. Microbiol. 57: 2555–2561

Van Boven A, Tan PST & Konings WN (1988) Purification and characterization of a dipeptidase from *Streptococcus cremoris* Wg2. Appl. Environ. Microbiol. 54: 43–49

Varmanen P, Ranthanen T & Palva A (1996a) Characterization of a novel ABC transporter-proline iminopeptidase operon from *Lactobacillus helveticus*. (Submitted)

Varmanen P, Steele JL & Palva A (1996b) Characterization of a prolinase gene and its product, and an adjacent ABC transporter gene from *Lactobacillus helveticus*. Microbiology (in press)

Varmanen P, Vesanto E, Steele JL & Palva A (1994) Characterization and expression of the *pepN* gene encoding a general aminopeptidase from *Lactobacillus helveticus*. FEMS Microbiol. Lett. 124: 315–320

Vesanto E, Peltoniemi K, Purtsi T, Steele JL & Palva A (1996) Molecular characterization, overexpression and purification of a novel dipeptidase from *Lactobacillus helveticus*. Appl. Microbiol. Biotechnol. submitted

Vesanto E, Savijoki K, Rantanen T, Steele JL & Pavla A (1995a) An X-prolyl dipeptidyl iminopeptidase (pepX) gene from *Lactobacillus helveticus*. Microbiology 141: 3067–3075

— (1995b) Molecular characterization, heterologous expression and purification of an X-prolyl-dipeptidyl aminopeptidase gene from *Lactobacillus helveticus*. (Submitted)

Vesanto E, Varmanen P, Steele JL & Palva A (1994) Characterization and expression of the *Lactobacillus helveticus pepC* gene encoding a general aminopeptidase. Eur. J. Biochem. 224: 991–997

Visser S, Exterkate FA, Slangen CJ & De Veer GJCM (1986) Comparative study of action of cell wall proteinases from various strains of *Streptococcus cremoris* on bovine α_{S1}-, β- and κ-casein. Appl. Environ. Microbiol. 52: 1162–1166

Visser S, Robben AJPM & Slangen CJ (1991). Specificity of a cell-envelope located proteinase (P_{III}-type) from *Lactococcus lactis* subsp. *cremoris* AM1 in its action on bovine β-casein. Appl. Microbiol. Biotechnology 35: 477–483

Visser S, Slangen CJ, Exterkate FA & De Veer GJCM (1988) Action of a cell wall proteinase (P_I) from *Streptococcus cremoris* HP on bovine β-casein. Appl. Microbiol. Biotechnol. 29: 61–66

Visser S, Slangen CJ, Robben AJPM, Van Dongen WD, Heerma W & Haverkamp J (1994) Action of a cell-envelope proteinase (CEP_{III}-type) from *Lactococcus lactis* subsp. *cremoris* AM1 on bovine κ-casein. Appl. Microbiol. Biotechnol. 41: 644–651

Vongerichten KF, Klein JR, Matern H & Plapp R (1994) Cloning and nucleotide sequence analysis of *pepV*, a carnosinase gene from *Lactobacillus delbrückii* subsp. *lactis* DSM7290, and partial characterisation of the enzyme. Microbiol. 140: 2591–2600

Vongerichten KF & Krüger E. Unpublished results

Von Heijne G (1989) The structure of signal peptides from bacterial lipoproteins. Protein Engineer. 2: 531–534

Vos P, Boerrigter IJ, Buist G, Haandrikman AJ, Nijhuis M, De Reuver MB, Siezen RJ, Venema G, De Vos W & Kok J (1991) Engineering of the *Lactococcus lactis* serine proteinase by construction of hybrid enzymes. Protein Engineer. 4: 479–484

Vos P, Simons G, Siezen RJ & De Vos WM (1989a) Primary structure and organization of the gene for a prokaryotic cell envelope-located serine proteinase. J. Biol. Chem. 264: 14579–13585

Vos P, Van Asseldonk M, Van Jeveren F, Siezen R, Simons G & De Vos WM (1989b) A maturation protein is essential for the production of active forms of *Lactococcus lactis* SK11 serine proteinase located in or secreted from the cell. J. Bacteriol. 171: 2795–2802

Wohlrab Y & Bockelman W (1992) Purification and characterization of a dipeptidase from *Lactobacillus delbrückii* subsp. *bulgaricus*. Int. Dairy J. 2: 345–361

— (1993) Purification and characterization of a second aminopeptidase (PepC-like) from *Lactobacillus delbrückii* subsp. *bulgaricus* B14. Int. Dairy J. 3: 685–701

— (1994) Purification and characterization of a new aminopeptidase from *Lactobacillus delbrückii* subsp. *bulgaricus* B14. Int. Dairy J. 4: 409–427

Yamamoto N, Akino A & Takano T (1993) Purification and specificity of a cell-wall-associated proteinase from *Lactobacillus helveticus* CP790. J. Biochem. 114: 740–745

Yan T-R, Ho S-C & Hou C-L (1992) Catalytic properties of X-prolyl dipeptidyl aminopeptidase from *Lactococcus lactis* subsp. *cremoris* nTR. Biosci. Biotech. Biochem. 56: 704–707

Yüksel, GÜ & Steele JL (1995) DNA sequence analysis, expression, distribution, and physiological role of the X-prolyl dipeptidyl aminopeptidase (PepX) gene from *Lactobacillus helveticus* CNRZ 32. (Submitted)

Zevaco C, Monnet V & Gripon J-C (1990) Intracellular X-prolyl dipeptidyl peptidase from *Lactococcus lactis* subsp. *lactis*: purification and properties. J. Appl. Bacteriol. 68: 357–366

Zevaco C & Gripon J-C (1988) Properties and specificity of a cell-wall proteinase from *Lactobacillus helveticus*. Le Lait 68: 393–408

Antonie van Leeuwenhoek **70**: 223–242, 1996.

Metabolic engineering of sugar catabolism in lactic acid bacteria

Willem M. de Vos[1,2,*]
[1]*Department of Biophysical Chemistry, NIZO, P.O. Box 20, 6710 BA Ede, The Netherlands;* [2]*Department of Microbiology, Wageningen Agricultural University, The Netherlands; (*Requests for offprints: NIZO, Kernhemseweg 2, 6718 ZB Ede, The Netherlands)*

Key words: Lactococcus lactis, Streptococcus thermophilus, metabolism, oxygen, lactose, pyruvate, diacetyl, polysaccharides

Abstract

Lactic acid bacteria are characterized by a relatively simple sugar fermentation pathway that, by definition, results in the formation of lactic acid. The extensive knowledge of traditional pathways and the accumulating genetic information on these and novel ones, allows for the rerouting of metabolic processes in lactic acid bacteria by physiological approaches, genetic methods, or a combination of these two. This review will discuss past and present examples and future possibilities of metabolic engineering of lactic acid bacteria for the production of important compounds, including lactic and other acids, flavor compounds, and exopolysaccharides.

Introduction

Lactic acid bacteria are well-recognized for their capacity to convert a great number of substrates in industrial food fermentations. The end products of these anaerobic conver sions confer the necessary protection against spoilage, contribute to the generation of the desired flavors, and add to the texture of the final products. The many metabolic reactions that underlie these conversions vary widely and depend on the lactic acid bacteria used, the fermentation conditions, and the nature of the fermentation substrates that usually include unprocessed food stuffs, such as milk, meat or vegetables. In spite of the vast amount of organic molecules that collectively can be metabolized by lactic acid bacteria, most individual species or strains have the capacity to exploit only a limited number of substrates as sole energy source (London, 1990; Thompson & Gentry-Weeks, 1994). In many cases the only compounds that support growth are mono- and disaccharides, as is the case for the main species of the dairy lactic acid bacteria that have been adapted to continuous growth in milk (see Table 1).

Although in selected cases specific deficiencies in substrate transport or further degradation may account for a limited catabolic potential, the main factor causing the metabolic inflexibility of lactic acid bacteria is the absence of functional electron transport chains. This prevents the generation of energy by the reduction of external electron acceptors, thereby limiting the number of catabolic pathways that provide energy. Only two ways are known by which lactic acid bacteria generate metabolic energy (Poolman, 1993; Konings et al., 1995). The simplest and most direct one is substrate level phosphorylation that couples the dephosphorylation of a high energy phosphate group to ATP formation. Only a small number of these reactions that generate primary metabolic energy actually occur. The two substrate level phosphorylation reactions, involving phosphoglycerate kinase and pyruvate kinase, that operate in the glycolytic pathway are well-known and found in all lactic acid bacteria (Figure 1). Moreover, the acetate kinase pathway, generating ATP and acetate from acetyl-phosphate, is also widely distributed. In contrast, specific ATP-generating reactions, such as that in the arginine deiminase pathway, are limited to only few species (Driessen et al., 1987). A second, indirect way for the generation of metabolic energy in lactic acid bacteria may result from the conversion of a solute gradient into an electrochemical gradient of

Table 1. Sugars utilized by dairy lactic acid bacteria. The data have been obtained with *L. lactis* subsp. *cremoris* MG1363 harboring the mini-lactose plasmid pMG820 (de Vos & Gasson, 1989), *Leuconostoc lactis* NZ6009 (Vaughan et al., 1996), *S. thermophilus* NZ302G (Catzeddu et al., 1996) and *Lactobacillus delbrueckii* subsp. *bulgaricus* (further designated with its trivial name *Lactobacillus bulgaricus*) (NIZO collection).

	Lactococcus lactis	*Leuconostoc lactis*	*Streptococcus thermophilus*	*Lactobacillus bulgaricus*
Fructose	+	+	+	+
Glucose	+	+	+	+
Lactose	+	+	+	+
Galactose	+	+	+	-
Sucrose	-	-	+	-
Maltose	+	+	-	-
Mannose	+	+	-	-
Mellibiose	-	+	-	-
Ribose	-	+	-	-
Threalose	+	+	-	-
Cellobiose	+	-	-	-
Salicin	+	-	-	-
α-methylD-glycoside	+	+	-	-
N-acetylglucosamine	+	+	-	-

protons, usually by the efflux of weak acids or other charged products via secondary transporters. The thus formed proton gradient is used to generate ATP via the membrane-located ATPase and the overall process is now known as secondary metabolic energy generation (Konings et al., 1995). This system was first discovered for the excretion of lactic acid in *Lactococcus lactis* although the secondary transport system involved in this process has not yet been characterized (Otto et al., 1980). Pathways that allow growth on single substrates via this form of energy conservation include the malolactic and citrolactic fermentation (Poolman, 1993; Hugenholtz et al., 1993; Marty-Teysset et al., 1996).

Another factor contributing to the limited metabolic versatility of lactic acid bacteria is their usual growth under anaerobic conditions. Since lactic acid bacteria are not capable of producing hydrogen, they show classical fermentative growth on sugars that are both electron donor and acceptor. This considerably limits the possibilities for the formation of other end-products than lactic acid, acetate, ethanol and carbon dioxide. These products are formed via a limited number of well-established and mutually exclusive sugar fermentation pathways in lactic acid bacteria, the glycolytic (homolactic), phosphoketolase (heterolactic), and bifidobacterial pathways (Thompson & Gentry- Weeks, 1994).

Recent advances in the physiology and genetics of lactic acid bacteria have allowed for increasing the metabolic capacity of lactic acid bacteria not only by changing the fermentation conditions but also by applying genetic techniques that result in the inactivation of undesired genes or the overexpression of existing or novel ones. This approach is termed here metabolic engineering and involves the rerouting of metabolic fluxes to create alternative energy generating or consuming systems, change the concentration of existing intermediates or end-products, or produce completely novel products. In order to apply metabolic engineering in a rational way, it is essential to have an extensive knowledge of the pathways under investigation, including their fluxes, control and genetics. This also permits the application of the metabolic control theory that aims to quantify the fluxes in a pathway by calculating the control coefficients of the various metabolic reactions (Kell & Westerhoff, 1986; Jensen et al., 1995). In spite of the limited number of substrates that support growth of individual strains, lactic acid bacteria are ideally suited for metabolic engineering since the biosynthesis of cell components and sugar catabolism are two, almost completely separated, pathways. Most attention has been focused on metabolic engineering of dairy lactic acid bacteria that ferment a restricted number of sugars by well-established pathways and are among the best genetically characterized

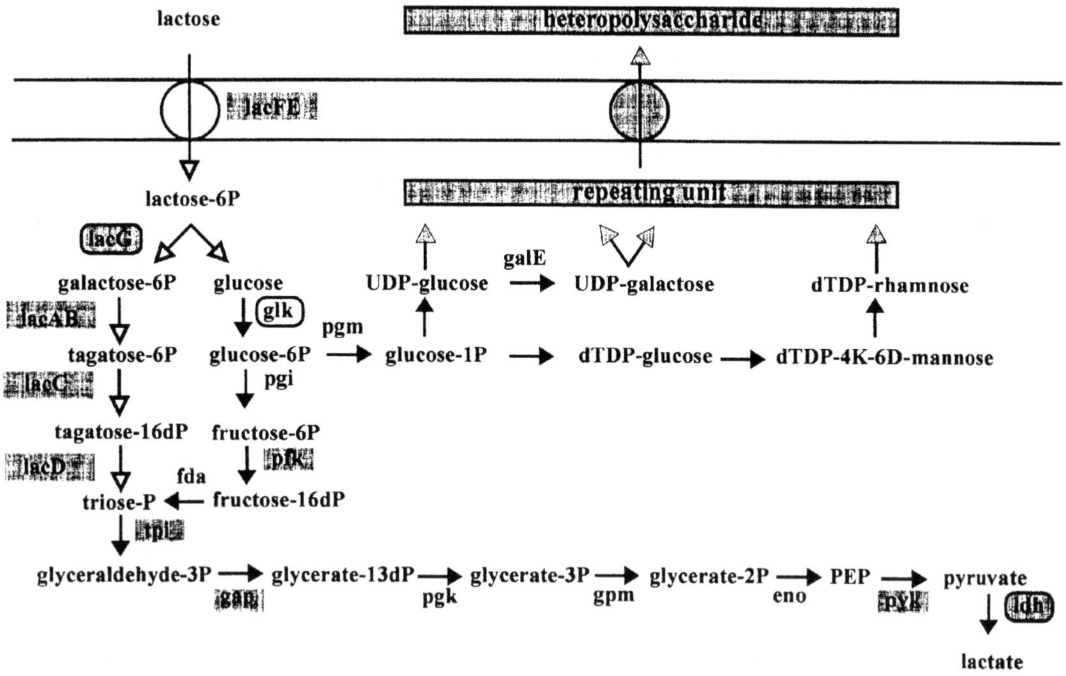

Figure 1. Schematic representation of pathways involved in lactose catabolism (top left), glycolysis (bottom) and exopolysaccharide production (top right) in *L. lactis*. Known enzymes are indicated using abbreviations based on established genetic nomenclature. Triose-P represents both glyceraldehyde–3P and dihydroxyacetone-P, from which the latter is converted into glyceraldehyde–3P as indicated. Enzymes for which the genes have been identified and characterized in lactococci are shaded; enzymes for which the genes have been subject to site-directed or classical mutagenesis are boxed. Reactions catalyzed by enzymes encoded by the lactose-plasmid pMG820 are indicated with white arrowheads, those encoded by the exopolysaccharide plasmid pNZ4000 are indicated with shaded arrowheads. The (multiple) reactions involved in production of the pNZ4000-encoded exopolysaccharide have not yet been established in *L. lactis* but are based on the exopolysaccharide structure and similar reactions in other bacteria; dTDP–4K–6D-mannose represent dTDP–4-keto–6-deoxy- mannose. See text for further explanation and references.

species (Gasson & de Vos, 1994). This review will discuss the present examples and future possibilities of metabolic engineering of the sugar catabolism for the production of important compounds by these and other lactic acid bacteria including lactic and other acids, flavor compounds, and polysaccharides.

Engineering sugar-specific transport and catabolism

In order to be able to utilize sugars for energy production, synthesis of cell constituents, or formation of desired endproducts, lactic acid bacteria should be able to transport these molecules and activate them into hexose-phosphates that may be degraded via well-established fermentation pathways. Three basically different systems for sugar transport are known in lactic acid bacteria viz. primary transport, secondary transport and group translocation systems. In primary transport systems ATP hydrolysis is directly coupled to sugar translocation via a sugar transport ATPase, an ATP-binding cassette protein (Fath & Kolter, 1994). In secondary transport systems sugar translocation is coupled to ions or other solutes (Poolman, 1993). The group-translocation, finally, occurs in lactic acid bacteria during sugar transport via the phosphoenolpyruvate (PEP)-dependent phosphotransferase system (PTS). In this system the sugar is phosphorylated during transport and the phosphate-group originates from the conversion of PEP into pyruvate and is translocated to the sugar via a cascade of reactions involving two general cytoplasmic phosphocarriers, Enzyme I and HPr (Postma et al., 1993).

All three known transport systems have sugar-specific components and differ in their mechanism, bioenergetics, and further catabolic pathways (Thompson, 1987; Poolman, 1993, de Vos & Vaughan, 1994). Moreover, the distribution of these transport systems varies among the different lactic acid bacteria and the PTS has been reported to be widespread in homolactic bacteria but absent from heterolactic bacteria, although recently *Lactobacillus brevis* was found to be an exception (Saier et al., 1996).

Because of their key function it is conceivable that sugar transport systems exert significant control of the flux in a catabolic pathway. It has recently been suggested for *E. coli* that flux control should rather reside in substrate transport than in the H^+-ATPase at the other end of the pathway (Jensen et al., 1993). Moreover, in this host the glucose PTS has found to play an important role in flux control (van Dam et al., 1994). As a consequence, it may be relevant for present and future metabolic engineering studies to summarize the physiological and genetic control of sugar transport and sugar-specific catabolic reactions.

Many transport systems have now been studied at the genetic level in various lactic acid bacteria (Table 2). In all cases investigated, the genes involved in sugar transport are located in clusters or operons that also code for sugar-specific catabolic reactions. Several of these are located on unstable elements, such as plasmids or transposons, and their curing or transfer may change the catabolic potential of the host strain. In many cases transcriptional regulators are also included in these gene clusters and the regulation of several operons has been studied in detail (Figure 2). Examples of the different transport systems will be summarized below with specific attention for engineering the control of these and, in selected cases, other sugar-specific genes.

Control of sugar transport and sugar-specific catabolism: primary and secondary transport systems

The best studied primary transport system is the *Streptococcus mutans msm* operon involved in transport of multiple sugars, including raffinose, melibiose, stachyose, sucrose, and isomaltosaccharides. This transport system is encoded by 4 genes and consists of the sugar- binding lipoprotein (*msmE*), two membrane components (*msmF* and *msmG*), and an ATP-binding protein (*msmK*) (Figure 2). These genes are part of an operon including 4 other genes, *msmR* coding for a divergently transcribed activator protein belonging to the AraC family of regulators, *aga* coding for α-galactosidase, *gtfA* coding for sucrose phosphorylase, and *dexB* coding for a dextran glucosidase (Russell et al., 1992). The recently reported regulation of the *msm* operon by the product of the *msmR* gene and the intracellular inducers, fructose, glucose–1-P or glucose–6-P, indicates that there are possibilities to increase the expression of this, and possibly other, primary transport systems (Tao et al., 1995).

All secondary sugar transporters that have so far been discovered in lactic acid bacteria belong to the galactoside-pentose-hexuronide group of translocators (Poolman et al., 1996; Table 2). The hybrid LacS protein of *Streptococcus thermophilus* is the prototype of these carriers that consist of a hydrophobic domain, essential and sufficient for sugar transport, and a carboxyterminal enzyme IIA domain (Poolman et al., 1989). Depending on the conditions, LacS can operate as a proton symport system or as a lactose-galactose exchanger (Poolman, 1993) (Figure 3). The enzyme IIA domain is a target for regulation of sugar transport as was shown by its deletion from LacS and site-directed mutagenesis of its phosphorylation site (Poolman et al., 1995). However, this domain is absent in the galactose and xylose transporters encoded by the *L. lactis galA* and *Lactobacillus pentosus xylP* genes, respectively, that also belong to the family of galactoside-pentose-hexuronide group of translocators (Table 2; Poolman et al., 1996). The *S. thermophilus lacS* gene is the first of the *lacSZ* operon which also codes for a ß-galactosidase (Poolman et al., 1989; Schroeder et al., 1991). Recent studies have shown that in *S. thermophilus* CNRZ302 the *lacSZ* operon preceded by the *gal* gene cluster of that codes for the Leloir enzymes galactokinase (*galK*), transferase (*galT*), epimerase (*galE*) and mutarotase (*galM*) (Poolman et al, 1990; Catzeddu et al., 1996) (Figure 3). The expression of the *gal* genes is controlled by the product of the divergently transcribed *galR* gene coding for a transcriptional activator, homologous to the *E. coli* GalR/LacI family of regulators, which probably requires galactose as an inducer. Recent deletion studies have indicated that expression of the *lacSZ* operon is in part also controlled by GalR (E.E. Vaughan & W.M. de Vos, unpublished data).

Control of sugar transport and sugar-specific catabolism: PTS and its components

Many sugars in lactic acid bacteria are transported via a sugar-specific PTS. The PTS is bioenergetically the most efficient transport system since the sugar is translocated and phosphorylated in a single

Table 2. Summary of genetically characterized components of sugar transport systems utilized by lactic acid bacteria. The designations of the genes are shown as is their location (X: chromosomal; P: plasmid; T: congugative transposon).

Sugars	Genes	Location	Organism	Reference
Primary Transport Systems				
ribose, raffinose sucrose, stachyose	*msmEFGK*	X	*Streptococcus mutans*	Russell et al., 1991
maltose	*malK*	X	*Lactococcus lactis*	Law et al., 1994
Secondary Transport Systems				
lactose	*lacS*	X	*Streptococcus thermophilus*	Poolman et al., 1989
	lacS	X	*Lactobacillus bulgaricus*	Leong-Morgenthaler et al., 1991
	lacS	P	*Leuconostoc lactis*	Vaughan et al., 1996
galactose	*galA*	X	*Lactococcus lactis*	B. Grossiord & W.M. de Vos, unpublished data
xylose	*xylP*	X	*Lactobacillus xylosus*	Lokman et al., 1994; Poolman et al., 1996
raffinose	*rafS*	P	*Pediococcus acidilactii*	K. Leenhouts, personal communication; Poolman et al., 1996
Phosphotransferase Systems				
all PTS sugars	*ptsHI*	X	*Streptococcus salivarius*	Gagnon et al., 1992, Gagnon et al., 1993
	ptsHI	X	*Staphylococcus carnosus*	Kohlbrecher et al., 1992
	ptsHI	X	*Streptococcus mutans*	Boyd et al., 1994
	ptsHI	X	*Lactococcus lactis*	E.J. Luesink, O.P.Kuipers & W.M. de Vos
lactose	*lacFE*	P	*Lactococcus lactis*	de Vos et al., 1990
	lacFE	P	*Lactobacillus casei*	Alpert & Chassy, 1990
	lacFE	X	*Streptococcus mutans*	Rosey & Stewart, 1993
sucrose	*scrA*	X	*Streptococcus mutans*	Sato et al., 1989
	sacA	X	*Staphylococcus carnosus*	Wagner et al., 1993
	sacB	T	*Lactococcus lactis*	de Vos et al., 1995; Rauch & de Vos, 1992
glucose	*ptsG*	X	*Staphylococcus carnosus*	W. Hengstenberg, personal communication

step at the cost of one ATP, which otherwise would have been gained during substrate level phosphorylation from phosphoenolpyruvate. The best characterized PTS in lactic acid bacteria is the *lac* operon of the *L. lactis* mini-lactose plasmid pMG820, which includes the genes for lactose transport (*lacEF*), lactose–6-P hydrolysis (*lacG*) and the tagatose–6-P pathway (*lacABCD*) (de Vos et al., 1990; van Rooijen et al., 1991) (Figure 1). Expression of this *lac* operon is controlled by the *lacR* gene product, a transcriptional repressor belonging to the DeoR family of repressors (van Rooijen & de Vos 1990). Expression of these *lac* genes is dependent the inducer tagatose–6-P, an intermediate of the tagatose–6-P pathway (van Rooijen et al., 1993).

Various elements of the PTS have now been genetically characterized in lactic acid bacteria (Table 2). These include the highly homologous *ptsH* and *ptsI* genes, usually organized as a *ptsHI* operon, coding for the general enzymes HPr and Enzyme I, respectively. It is well-known that HPr plays an important regulatory role in gram-positive bacteria with a low GC content. This phosphocarrier contains two phosphorylation sites, a histidine residue at the amino-terminal side of the carrier that is phosphorylated by Enzyme I and involved in phosphotransfer to the sugar permease, and a serine residue at the carboxy-terminal side of the protein that may be phosphorylated by an ATP-dependent, metabolite-activated protein kinase. Both phosphorylation sites are conserved in the presently known HPr sequences in lactic acid bacteria and the possible single and double phosphorylated HPr molecules have important functions. Next to a catalytic role in the PTS, the histidine-phosphorylated HPr (His-HPr) regulates non-PTS enzymes by phosphorylation in different ways since the glycerol kinase of *E. faecalis* is activated by this phosphorylation while the LacS of *S. thermophilus* shows a reduced transport activity upon

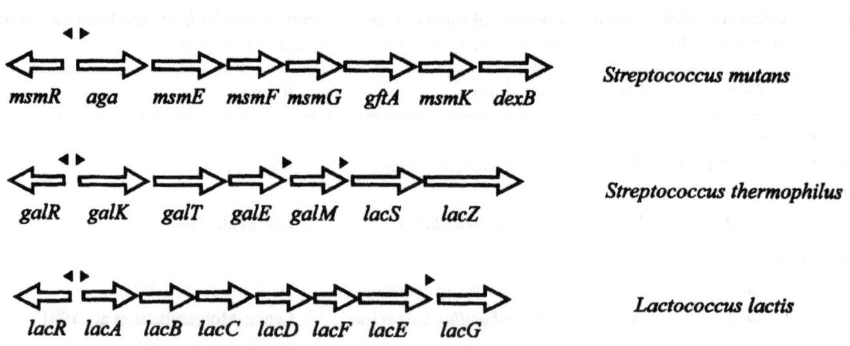

Figure 2. Schematic outline of the genetic organization of the *S. mutans msm* operon (Russell et al., 1992), the *S. thermophilus gal/lac* gene cluster (Poolman et al., 1987; Schroeder et al., 1991; Catzeddu et al., 1996) and the *Lactococcus lactis lac* operon (de Vos et al., 1990; van Rooijen et al., 1991; van Rooijen & de Vos, 1991). Arrows indicate mapped promoters.

phosphorylation of the enzyme IIA domain (Deutscher & Sauerwald, 1986; Poolman et al., 1995). Similarly, the serine-phosphorylated HPr, HPr(Ser-P), has several functions. One is the allosteric control by HPr(Ser-P) of the activity of both PTS and non-PTS lactose permeases as found in *L. lactis* and *Lactobacillus brevis*, respectively (Ye et al., 1994; Ye & Saier Jr. 1995). In addition, HPr(Ser-P) controls a sugar-phosphate phosphatase in *L. lactis*, resulting in inducer exclusion and expulsion, the rapid efflux of preaccumulated sugars or sugar metabolites (Reizer, 1989; Ye et al., 1994). Finally, HPr(Ser-P) also interacts with the catabolite control protein CcpA, which negatively controls the expression of a number of genes or operons that are under catabolite repression (Hueck & Hillen 1995). Sequences that conform to the *cis*-acting target site for CcpA, the catabolite responsive element, have been detected in various sugar catabolic operons in lactic acid bacteria (de Vos & Simons 1994). Recently, it was established that *ccpA*-like genes are widely distributed in lactic acid and other gram-positive bacteria (Kuster et al., 1996). Therefore, it is likely that mutations in *ptsH*, specifically those that affect serine phosphorylation, may have profound effects on the general sugar metabolism, as has been demonstrated in other gram-positive bacteria (Reizer et al., 1989; Deutscher et al., 1995). With the availability of the general and sugar-specific genes it may be expected that the physiological and genetic control of transport and degradation of PTS- and other sugars can now be studied effectively in lactic acid bacteria. In addition, it allows for the engineering of global regulatory systems such as catabolite repression since *ccpA*-like genes have been characterized from various lactic acid bacteria including *Lactobacillus delbrueckii* and *L. lactis* (Klein et al., 1996; E. Luesink, O.P. Kuipers & W.M. de Vos, unpublished data).

Engineering lactose catabolism in L. lactis: overproduction of intermediates

The pathway by which lactose is transported and degraded in *L. lactis* is fully known and many of the genes involved in this catabolism have also been characterized (Figure 1). The *lac* operon genes involved in the lactose PTS and the tagatose–6P pathway have already been described (Figure 2). In addition, several glycolytic genes have been characterized, including a novel glycolytic operon with the gene order *pfk-pyk-ldh*, the monocistronic *gap* gene, and the *tpi* gene (Llanos et al., 1994; de Vos & Simons, 1994). However, the use of classical mutagenesis has shown the first examples of successful engineering of the lactose catabolism in *L. lactis*. In elegant work described by Thompson et al., (1985), a double mutant of *L. lactis* was constructed that was deficient in glucokinase and the mannose PTS, which also transports glucose. When mutant cells are grown on lactose, there is no phosphorylation of the glucose originating from the hydrolyzed lactose–6P by glucokinase or the PTS, and only the galactose–6P moiety is utilized (see Figure 1). Hence, lactose is quantitatively converted into glucose which is secreted and can not be transported into the cells in a phosphorylated form because of the absence of a glucose PTS. During growth on lactose the mutant *L. lactis* strain maintains a high intracellularly glucose concentration of approximately 100 mM without apparent toxic or bacteriostatic effects. This study

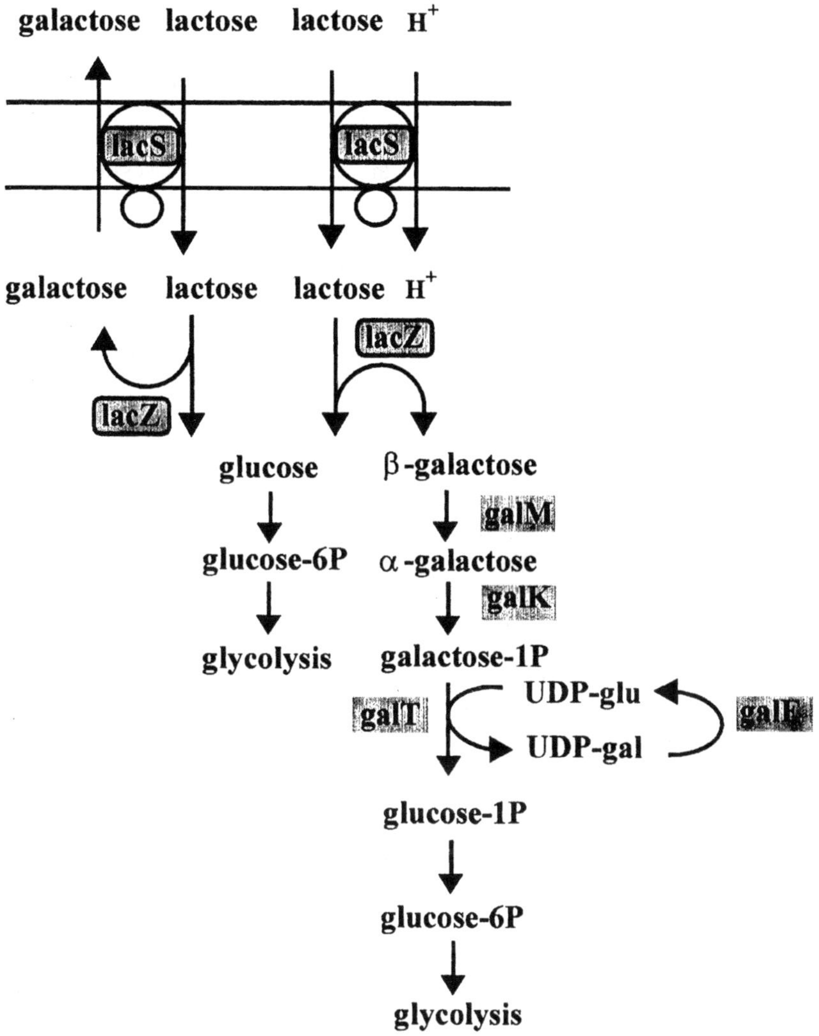

Figure 3. Schematic representation of the lactose transport and catabolism in *S. thermophilus*. Lactose-galactose exchange or lactose-proton symport systems catalyzed by LacS are indicated at the left and right, respectively. A similar system operates probably in *Lactobacillus bulgaricus*. Enzymes for which the genes have been identified and characterized in *S. thermophilus* are shaded; enzymes for which the genes have been subject to site-directed or classical mutagenesis are boxed. The conversion of ß-galactose into α- galactose is assumed to be catalyzed by the *galM* gene product in analogy with *E. coli* (Bouffard et al., 1994).

shows that it is possible to uncouple galactose and glucose metabolism in *L. lactis* cells, offering the possibility to exploit the glucose moiety from lactose for other purposes than cell growth (see below). In addition, these and similar strains could be used to generate sweet-tasting fermented milk products, such as buttermilk or quark where *L. lactis* strains are traditionally used.

Most dairy strains of *L. lactis* harbor multi-copy plasmids carrying the *lac* genes (de Vos & Vaughan, 1994). It is likely that the plasmid location reflects the recent acquisition of the *lac* genes rather than an adaptation to increase their expression, since *lac* gene expression is controlled by LacR and the intracellular tagatose–6- P concentration (van Rooijen et al., 1993; Figure 1). This inability to increase *lac* gene expression has been experimentally verified in a number of cases. First, no significant difference in *lacG* expression and growth rate on lactose was observed between strains harboring the *lac* genes either as a single copy in the chromosome or on the multicopy mini-lactose plasmid (van Rooijen et al., 1992). In addition, when the copy number of the *lac* genes was modulated no effect was found on the growth rate on lactose (Limsowtin et al., 1986). Finally, no effect on the rate of lactose degradation was found in strains where a two-fold increase in *lacG* gene expression was observed following an *in vivo* cloning experiment (Anderson & McKay, 1984). The increased LacG activity observed in the latter study could well be a result of a rearrangement of the *lac* genes or a specific effect on the *lacG* gene that is preceded by an intergenic promoter (Figure 2; de Vos & Gasson, 1989).

Some atypical *L. lactis* strains do not ferment lactose efficiently in spite of the presence of an active lactose PTS. In one of these, strain ATCC 7962, this is caused by a defect in the phospho- ß-galactosidase (LacG) activity resulting in the intracellular accumulation of high levels (more than 100 mM) of lactose–6P (Crow & Thomas, 1984). It would be of interest to determine to what level the *lac* genes are expressed in this strain since the real inducer, tagatose–6P is not formed. In addition, the question arises as to how much of the phosphorylated sugars, which usually are toxic, can be tolerated by lactococcal cells that could be used as production hosts for these valuable compounds.

Engineering lactose catabolism in Lactobacillus bulgaricus to reduce post-fermentation acidification

A well-known undesired property encountered in yoghurt production is the uncontrolled acidification by *Lactobacillus bulgaricus* at the end of the fermentation phase, resulting in a product which contains too much acid. Although a simple solution would be to reduce the number of lactobacilli, this may affect the fermentation and the protocooperation with *S. thermophilus* and, at the same time, jeopardize labeling since product laws dictate the presence in yoghurt of a certain number of viable, rod-shaped cells. In an attempt to reduce this post-fermentation acidification, the *lacZ* gene from *Lactobacillus bulgaricus* has been targeted for engineering studies (Schmidt et al., 1989). Using an *E. coli* expression system and random mutagenesis, a series of cold-sensitive mutations were detected in the *lacZ* gene (Adams et al., 1994). Further characterization showed these to affect two residues, Leu–316 and Pro–429, that are conserved in various ß-galactosidases. Strategies to use the cold- labile *lacZ* genes in *Lactobacillus bulgaricus* in order to prevent acidification during cold storage have been proposed (Mainzer et al., 1990). However, these have so far not been implemented in *Lactobacillus bulgaricus,* probably since the first reports on successful transformation of this industrial hosts were only recently described (Sasaki et al., 1993a). A way out from this bottleneck has been the development of spontaneous deletions that affect the *lacZ* gene of *Lactobacillus bulgaricus* (Mollet & Delley, 1990). These deletions have been analyzed in detail and comprise small and large deletions that partly involve the presence of a new IS element ISL3 (Mollet & Delley, 1991; Germond et al., 1995). *Lactobacillus bulgaricus* strains carrying these deletions do not utilize lactose and in combination with other components of a yoghurt starter these were effective in preventing post-fermentation acidification (Mollet & Hottinger, 1990).

Engineering galactose utilization in S. thermophilus

It has been known for long times that most dairy strains of *S. thermophilus* can not ferment galactose (Hutkins & Morris, 1987). As a result, stochiometric amounts of the galactose moiety are accumulating in the medium during growth on lactose due to the lactose/galactose exchange activity of the LacS transporter (Poolman 1993) (Figure 3). Various authors have reported on the characterization of galactose-fermenting mutants that were isolated under appropriate selective conditions (Thomas & Crow, 1984; Hutkins et al., 1985). Their comparison with the non-galactose fermenting parental strains indicated that the latter ones suffered from a defect in the induction mechanism for the rate-limiting key enzyme galactokinase. Recently, it was discovered that non-galactose fermenting strains do contain intact genes for the Leloir pathway but that these genes (specifically these of the *galKTE* operon) are not transcribed (Catzeddu et al., 1996) (Figure 2). Remarkably, galactose-fermenting mutants could be isolated that express the *gal* genes as a result of mutations that were mapped in the *gal* promoter region. The expression of these activated *gal* genes is under control of the apoinducer GalR, probably with galactose or a derivative thereof as inducer (E.E. Vaughan & W.M. de Vos, unpublished data). As a consequence, various strategies can now be devised to obtain stable mutants of *S. thermophilus* that can utilize galactose efficiently pNZ4000 located EPS gene cluster of *Lactococcus lactis*

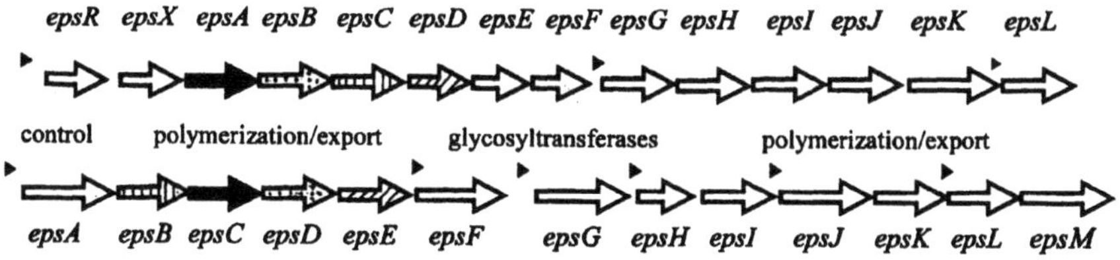

chromosomal EPS gene cluster of *Streptococcus thermophilus*

Figure 4. Comparison of the chromosomal *S. thermophilus* and pNZ4000-located *Lactococcus lactis eps* gene clusters (Stingele et al., 1996; van Kranenburg et al., 1996). Four genes that encode enzymes with significant identity have an identical shading. Putative functions based on homologies are indicated. Arrows indicate mapped (pNZ4000) or proposed (*S. thermophilus*) promoters.

and may have applications in dairy fermentations that require complete utilization of galactose.

Engineering exopolysaccharide production

The capacity of lactic acid bacteria to produce extracelluar polysaccharides has been well established (Cerning, 1990). Various homopolysaccharides are produced extracellularly, such as dextrans or glucans from sucrose, by secreted glycosyl transferases. In contrast, catabolized sugars are activated intracellularly, generating precursors that are used in the biosynthesis of extracellular heteropolysaccharides, here designated exopolysaccharides. Several of those exopolysaccharides have important functional properties and determine the viscosity of dairy products including yoghurt and specific Scandinavian fermented milk products, such as viili and longfil, made by lactococci. Significant progress has been made with the genetic analysis of production of the exopolysaccharides present in these products and *eps* genes have been characterized from *S. thermophilus* and *L. lactis* strains (Stingele et al., 1996; van Kranenburg et al., 1996). Gene cloning, expression and inactivation studies showed that the *eps* genes are located in large clusters comprising about a dozen genes (Figure 4). In *S. thermophilus* the *eps* gene cluster has been targeted by transposon mutagenesis and homologous recombination, and found to be one of the multiple loci involved in exopolysaccharide biosynthesis (Stingele & Mollet, 1995). In *L. lactis* the *eps* genes are flanked by insertion sequences and located on the 40-kb plasmid pNZ4000 which an be conjugally transferred to other lactococcal strains. Insertional inactivation by homologous recombination and antisense approaches were used to inactivate these plasmid-located *eps* genes. The structures of the exopolysaccharides produced by both lactic acid bacteria differ considerably. The repeating unit in the exopolysaccharide from *S. thermophilus* is a tetrasaccharide consisting of galactose and glucose, while in the *L. lactis* exopolysaccharide this is a pentasaccharide containing glucose, rhamnose, galactose and galactose–1P (Doco et al., 1990; Nakajima et al., 1992; van Kranenburg et al., 1996). However, four of the *eps* genes code for products that share considerable similarity with each other and, moreover, the organization, transcriptional direction, and deduced functions of the genes in the different *eps* clusters appear to be highly conserved (Figure 4).

An important bottleneck in the application of exopolysaccharides is the low level of production in lactic acid bacteria. In order to engineer exopolysaccharide production in a rational way, it is essential to know its biosynthetic pathway. So far, the biosynthesis of none of the exopolysaccharides has been elucidated in lactic acid bacteria. However, based on the structure of the *L. lactis* exopolysaccharide it can be assumed to occur via activation of the sugars glucose, mannose and galactose by their coupling to nucleotide sugars following reactions established in other bacteria (Sutherland, 1985; Reeves, 1994). Various specific enzymatic conversions will be encoded by the *eps* genes, including the transfer of the activated sugars resulting in a repeating unit most likely bound to an undecaprenyl phosphate lipid carrier, the export and polymerization, and possibly the regulation of production (Figure 1). However, other steps will be encoded by the household genes of the host, including glucose-P interconversions via phosphoglucomutase, which could be a key enzyme linking the lactose degradation pathway to the exopolysaccharide biosynthesis. Assuming that this linkage occurs at this branching point, it is tempting to speculate that it is possible to engineer overproduction of exopolysaccharides since the galactose moiety could be catabolized completely via the glycolysis, while the glucose moiety could be used for exopolysaccharide production (Figure 1). Uncoupling the galactose and glucose metabolism has previously been established in a *L. lactis* mutant impaired in glucokinase and glucose PTS activity that accumulated glucose (Thompson et al., 1985; see above). However, in the present case exopolysaccharide should start from a pool of phosphorylated sugars, for instance generated by affecting the expression of the *pfk* gene (Figure 1). Although accumulation of large amounts of lactose-P were tolerated by *L. lactis* deficient in LacG activity, overproduction of sugar-P could be toxic to the cells (Crow & Thomas, 1994; see above). Therefore, the flux via the phosphoglucomutase should be made sufficiently high and the question is whether this can be realized. Moreover, the bioenergetics and redox balance need also to be addressed. The catabolism of each galactose–6P molecule generates 2 ATP since one PEP will be used for transport and phosphorylation. This metabolic energy could be used to convert glucose into glucose–6-P and subsequently in the nucleotide-sugars but additional ATP should be gained to fuel other metabolic reactions and allow for growth. This could be realized by the production of acetate via acetate kinase rather than lactate from pyruvate, as could be realized in *L. lactis* strains deficient in lactate dehydrogenase by the activity of the pyruvate formate lyase under anoxic or pyruvate dehydrogenase complex under aerobic conditions (Hugenholtz, 1993; see also below). In the latter case, aeration may become a problem under conditions that the exopolysaccharide is highly overproduced. Finally, it has been suggested that the availability of the isoprenoid carrier is the most important factor affecting exopolysaccharide biosynthesis (Sutherland 1985). It should be investigated whether this also holds for lactic acid bacteria and the engineering strategy described above may allow to address this and other potential blockades that could obstruct exopolysaccharide overproduction.

With the availability of the *eps* genes various strategies can be selected to further improve the yield of exopolysaccharide production, to alter the structure of the exopolysaccharides, or to realize simultaneous production of different exopolysaccharides. An obvious approach is to change production by cloning the *eps* genes in suitable heterologous hosts. This was realized with the *S. thermophilus eps* gene cluster that could be expressed in a non- producing *L. lactis* strain (Stingele et al., 1996). However, only low amounts of exopolysaccharide were produced in this heterologous host and it was not established whether this had the expected structure. Another approach is to engineer the exopolysaccharide production in the production host itself or in related strains. This was realized with the plasmid-encoded lactococcal *eps* genes that could be conjugally transferred to the genetically well-characterized and well- transformable *L. lactis* strain MG1363 (van Kranenburg et al., 1996). In these and other production hosts, inactivation by site-directed or random mutagenesis of *eps* genes or relevant household genes coding for specific sugar activations, such as the *galE* gene, may affect the incorporation of activated nucleotide sugars or other steps in the polymerization process (Figure 1). Alternatively, existing or new *eps* genes may be overexpressed. This could result in mutant exopolysaccharides that have altered structures and properties. Such exopolysaccharide engineering offers not only the possibility to study structure-function relations of exopolysaccharides but also to create exopolysaccharides with novel properties that can be used as industrial biothickners.

Engineering the pyruvate metabolism

Pyruvate has a pivotal role in the sugar catabolism of lactic acid bacteria as the main molecule generated from all metabolic pathways (see Figure 1). It is the endproduct of the main substrate-level phosphorylation reactions as well as the phosphorylation cascade involved in the PTS and the precursor of the distinctive end product lactic acid. (Figure 4). The different reactions that may occur with pyruvate as substrate have recently been reviewed in detail (Hugenholtz 1993). Evidently, the main one is the conversion into the different isomers of lactic acid by L- or D-lactate dehydrogenases. This is the major electron sink in lactic acid bacteria under anaerobic conditions and limits the flexibility in batch culture of the metabolic conversions from pyruvate in absence of cosubstrates. However, in conjunction with the metabolic engineering of alternative redox pathways as described below, the conversions from pyruvate may be engineered by new approaches that are based on the accumulating

Table 3. Reactions involving oxygen or its toxic derivatives that are catalyzed by genetically characterized enzymes of lactic acid bacteria.

Enzyme	Reaction	Gene	Host	Reference
NADH:H_2O_2 oxidase	$NADH + H^+ + O_2 \rightarrow NAD^+ + H_2O_2$	*nox-1*	*Streptococcus mutans*	Higuchi et al., 1995
NADH:H_2O oxidase	$2\,NADH + 2\,H^+ + O_2 \rightarrow 2\,NAD^+ + H_2O$	*nox*	*Enterococcus faecalis*	Ross & Claiborne, 1992
		nox-2	*Streptococcus mutans*	Higuchi et al., 1994
NADH:peroxidase	$NADH + H^+ + H_2O_2 \rightarrow NAD^+ + 2\,H_2O$	*npr*	*Enterococcus faecalis*	Ross & Claiborne, 1991
superoxide dismutase	$O_2^- + 2H^+ \rightarrow H_2O_2$	*sod*	*Lactococcus lactis*	Sanders et al., 1995
haem-dependent catalase	$2\,H_2O_2 \rightarrow 2\,H_2O + O_2$	*katA*	*Lactobacillus sake*	Knauf et al., 1992
glutathione reductase	$NADPH^+ + H^+ + GSSG \rightarrow NADP^+ + 2\,GSH$	*gor*	*Streptococcus*	Pebay et al., 1995
	$2\,GSH + H_2O_2 \rightarrow GSSG + 2\,H_2O$		*thermophilus*	

knowledge of the pathways and characterization of the relevant genes. An important discovery that formed the basis for some of these approaches was the observation in the early 70s that a genetic defect in the production of lactate dehydrogenase resulted in an altered metabolism in a mutant from the homolactic *L. lactis* C2, which was obtained from a spontaneously occurring large colony (McKay & Baldwin, 1974; L.L. McKay & K.A. Baldwin, personal communication).

Engineering lactic acid production

The *ldh* genes coding for lactate dehydrogenases specific for either one of the two different lactic acid isomers have now been characterized in a variety of lactic acid bacteria (Table 3). While the L-lactate dehydrogenases belong to the well-known enzyme family comprising bacterial and eukaryal representatives, the D-lactate dehydrogenase constitutes a subfamily of the D-isomer-specific 2- hydroxyacid dehydrogenases (Griffin et al., 1992; Taguchi & Ohta 1991). Some lactic acid bacteria produce both L- and D-isomers of lactic acid and these were found to contain both types of *ldh* genes, designated *ldhL* and *ldhD*, respectively. Based on the availability of the *ldh* genes various inactivation strategies were designed and used to inactivate the coding regions for the various lactate dehydrogenases. This was described first for *Lactobacillus plantarum*, where the *ldhL* gene was inactivated successfully. However since this organism contained a functional *ldhD* gene, the resulting mutant still produced wild-type levels of D-lactate (Ferain et al, 1994).

In lactic acid bacteria that produced only one lactate isomer, *ldh* gene inactivation appeared to be more problematic. Inactivation of the single *ldh* in *S. mutans* could not be achieved suggesting it was a lethal event (Hillman et al., 1994). Moreover, a strain carrying a mutated copy coding for a thermolabile L-lactate dehydrogenase was generated by gene conversion and found to stop growing at a non- permissive temperature (Hillman et al., 1994; Chen et al., 1994). It has not been established what exactly causes this growth arrest and it can not be excluded that survival is dependent on specific physiological conditions. In contrast, inactivation of the single *ldh* gene in *L. lactis* was found to be possible since a spontaneous lactate dehydrogenase deficient mutant was obtained (McKay & Baldwin 1974). Using chromosomal integration strategies defined *ldh* mutants were realized in both *L. lactis* strains that could ferment lactose and in those that were lactose-deficient (C. Platteeuw & W.M. de Vos unpublished observations; Platteeuw et al., 1995; Gasson et al., 1996). Only single cross-over recombinations were obtained that did not show polar effects since the *ldh* is the last gene of the lactococcal glycolytic operon (Llanos et al., 1994). Finally, it has been reported that the *ldhD* gene from *Lactobacillus bulgaricus* has been replaced by gene conversion with the *ldhL* gene from *S. thermophilus* resulting in a yogurt starter that exclusively produced L-lactic acid (Sasaki et al., 1993b). However, so far these studies have not been further documented.

Recently, it was found to be possible to inactivate both the *ldhL* and *ldhD* genes in *Lactobacillus plantarum* (Ferain et al., 1996a). These elegant studies were used to demonstrate that D-lactate is incorporated with high efficiency instead of the normal terminal D- alanine in the muramyl pentapeptide of the peptidoglycan. Since the D- alanine containing pentapeptide is the target for the vancomycin, it follows that D-lactate producers have a high natural resistance to this and other glycopeptide antibiotics (Ferain et al., 1996a). Remarkably, vancomycin resistance was also observed in a *Lactobacillus plantarum* carrying a *ldhD* deletion since the formed L-lactate was converted into D-lactate via a lactate racemase activity and subsequently incorporated. No lactate dehydrogenase was detectable in the thus constructed strains of *L. lactis* and *Lactobacillus plantarum* and the metabolic end products were changed drastically (Platteeuw et al., 1995; Ferain et al., 1996b). Under anaerobic conditions the lactate dehydrogenase deficient *L. lactis* strain showed a mixed acid fermentation of lactose with high amounts of formate and ethanol (Platteeuw et al., 1995). However, aeration of this strain resulted in the increased production of acetate via the activity of pyruvate dehydrogenase complex while acetoin became the major end-product as was also observed with the spontaneous lactate dehydrogenase-deficient mutant (Hugenholtz, 1993) (Figure 4). Acetoin was also the main end product in the *ldhL* and *ldhD* double mutant of *Lactobacillus plantarum* grown under aerobic conditions on glucose, indicating that in both organisms the flux from pyruvate was rerouted from lactate in the direction of α-acetolactate (Figure 4). Culture supernatants of fermentations performed with the lactate dehydrogenase- deficient strains unexpectedly contained small amounts of lactic acid (Platteeuw et al., 1995; Ferain et al., 1996b). Although this lactic acid could be generated from medium components, such as by malolactic fermentation, it has also been suggested that it is a result of a D- hydroxyisocaproate dehydrogenase activity (Ferain et al., 1996b). These enzymes that are related to D-lactate dehydrogenases are able to reduce α-ketoacids to the corresponding α-hydroxyacid and several of their genes have now been detected in lactic acid bacteria (Table 3).

Engineering ethanol production

To show the feasibility of using pyruvate as a source for new products, a novel pathway was introduced in *L. lactis* by overexpressing the *pdc* and *adh* genes for pyruvate decarboxylase and alcohol dehydrogenase, respectively, from the efficient ethanol producer *Zymomonas mobilis* (Ingram et al., 1989). This was realized by generating a multicopy plasmid carrying a gene fusion of the heterologous genes with the promoter and coding region of the well- expressed lactococcal *prtP* gene (Marugg et al., 1996). The resulting *L. lactis* strains produced both the *Z.mobilis* pyruvate decarboxylase and alcohol dehydrogenase in addition to an endogenous alcohol dehydrogenase (Hugenholtz et al., 1996). In a lactate dehydrogenase proficient strain only traces of ethanol were detected. However, when the gene fusion was introduced in a strain carrying an inactivated *ldh* gene, large amounts of ethanol were obtained, up to 300 mM, in cells grown at low pH and with high glucose concentration. It remains to be established whether an even higher level of alcohol production may be obtained or whether this represents the maximum concentration of ethanol that can be tolerated by *L. lactis* cells. No aceetaldehyde was detected in the ethanol overproducing cultures. However, it is feasible that in *L. lactis* lacate dehydrogenase deficient strains in which the endogenous alcohol dehydrogenase is inactivated, overexpression of the *Z.mobilis pdc* gene results in the overproduction of acetaldehyde, another natural flavor compound.

Engineering diacetyl production

One of the products that can be generated from pyruvate is diacetyl, a desirable flavor component in dairy and other food products. This compound is naturally produced by lactic acid bacteria, notably *L. lactis* subsp. *lactis* var. *diacetylactis*, from citrate in a cometabolic fermentation of lactose (Hugenholtz, 1993). This fermentation yields mainly α-acetolactate that can be converted via an oxydative decarboxylation into diacetyl (Figure 5). There has been considerable interest in constructing *L. lactis* strains that overproduce diacetyl from lactose rather than from citrate. In order to do so, three different enzymatic conversions have been targetted by genetic approaches and the obtained results described below illustrate very well the problems, pitfalls and potential of metabolic engineering.

It has been suggested that more diacetyl may be produced by increasing the amount of the enzyme involved in the branching reaction from pyruvate to α-acetolactate. Two enzymes are known that catalyze this reaction. One is the α-acetolactate synthase encoded by the *als* gene, which is a catabolic synthase with a low affinity (K_m of 50 mM) for pyruvate and is widely distributed in *L. lactis* strains (Snoep et al., 1992; Marugg et al., 1994). The other is an anabolic acetohydroxy acid synthase encoded by the *ilvBN* gene, which is involved in branched amino acid biosynthesis and shows moderate affinity for pyruvate (K_m of 8 mM) but is inhibited by feed back control by the produced amino acids (Godon et al., 1993; Benson et al., 1996). The *ilvN* gene from a wild-type *L. lactis* strain was overexpressed in *L. lactis* MG1363 and when grown under aerated conditions on glucose this strain showed a three-fold higher production of acetoin than control cells (Benson et al., 1996). The IlvBN

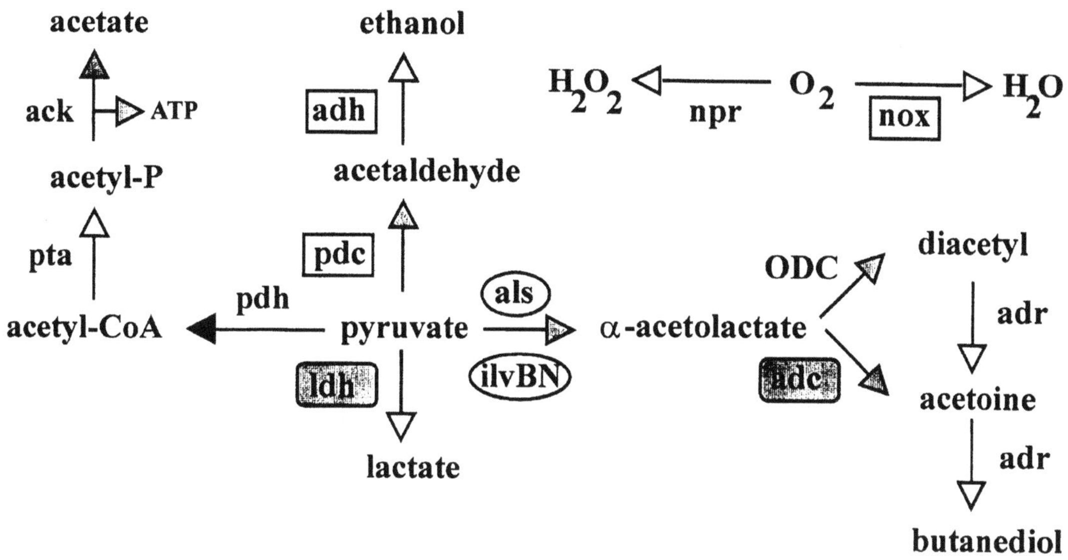

Figure 5. Schematic representation of the metabolic engineering strategies used in *L. lactis* to reroute pyruvate into other products such as diacetyl, ethanol and acetate. Enzymes for which the genes have been identified and characterized are in boxes or ellipses; enzymes for which the genes have been subject to site-directed or classical mutagenesis are shaded; enzymes for which the genes have been overexpressed are in ellipses; overproduced enzymes encoded by the *pdc* and *adh* genes from *Z.mobilis* and the *nox* gene from *S. mutans* are boxed. Reactions that are NADH producing have black and those that are NADH consuming have white arrowheads. ODC stands for oxydative decarboxylation.

overproduction level was not determined but probably was low since only 3 mM acetoin was obtained. The *als* gene from *L. lactis* MG1363 was overexpressed in a lactose-utilizing derivative of the same strain and this resulted in a more than 100-fold overproduction of α-acetolactate activity (Platteeuw et al., 1995). Under aerated conditions the α-acetolactate-overproducing strain produced high amounts of acetoin (more than 16 mM) accounting for about 40% of the flux from pyruvate. No α-acetolactate or diacetyl was produced in these strains due to the presence of an active α- acetolactate decarboxylase. Therefore, another approach has been directed to generate strains that are defective in α-acetolactate decarboxylase. One such strain, NIZO Ru4, has been well-studied and is a naturally occurring mutant of a citrate-utilizing *L. lactis* strain that has been widely used in industrial fermentations (Hugenholtz 1993). Other α-acetolactate decarboxylase deficient *L. lactis* mutants have now also been isolated by using a metabolic screening procedure and, like NIZO Ru4, produced high amounts of α- acetolactate and diacetyl but no acetoin during aerobic growth (Goupil N, Corthier G, Ehrlich SD & Renault, Abstract pD21, Lactic Acid Bacteria Conference, Cork, 1995). Recently, the *aldB* gene encoding α-acetolactate decarboxylase has been characterized from several *L. lactis* strains and used to generate deletion strains that show the same phenotype (Gasson et al., 1996; Benson et al., 1996).

In addition to the single gene based alterations, various combinations of the genetic modifications have been realized in *L. lactis.* The combination of IlvB production in an *aldB* deletion strain has been reported and this resulted in two- or three-fold increased amounts of diacetyl or α-acetolactate (Gasson et al., 1996). However, only low absolute amounts of these desired compounds were obtained under aerated conditions (up to a few mM). Another combination is the construction of a defined lactate dehydrogenase deficient strain which overproduced α-acetolactate synthase (Platteeuw et al., 1995). With this strain also low amounts of α- acetolactate were found but very high amounts of acetoin and even butanediol were obtained, accounting for 85% of the converted pyruvate.

The results of the different approaches indicate that significant changes in metabolic endproducts may be obtained. However, presently no *L. lactis* strains have been described that combine the three genetic alterations, i.e. inactivation of lactate dehydrogenase, overproduction of a α-acetolactate synthetase, and inactivation of the α-acetolactate decarboxylase. When these strains are constructed it is likely that the next bottleneck will be the stability of diacetyl, which may be degraded by the enzyme diacetyl reductase that also is involved in the further reduction of acetoin into butanediol (Figure 4). Although the gene for this enzyme has not been detected yet, various strains are known with low activity of diacetyl reductase, providing opportunities to obtain the desired diacetyl overproduction (Hugenholtz, 1993).

Engineering safe redox pathways

The metabolic inflexibility dictated by fermentation may be partly prevented when alternative electron acceptors can be used that alleviate the need to produce acids or other reduced compounds. Physiological studies have shown that lactic acid bacteria show a greater metabolic potential when the reduced cofactors, from which NADH is the most important, are regenerated by exogenous electron acceptors (Smart & Thomas, 1987; Condon, 1987; Hugenholtz, 1993). In most cases this also results in an increased fermentation yield since pyruvate dehydrogenase complex is activated and acetate may be produced via acetylphosphate, yielding an additional site for substrate phosphorylation (Figure 5).

Mixed-substrate fermentations

A limited number of lactic acid bacteria is capable of mixed substrate fermentations of carbohydrates in the presence of alternative organic electron acceptors, such as citrate, malate or fumarate (Poolman 1993; Hugenholtz, 1993). In addition, carbohydrates may be also used as an electron sink and the cometabolic fermentation of fructose by *Lactobacillus sanfransisco* results in mannitol as major end product (Bocker et al., 1994).

NADH: Oxidases

The simplest way to oxidize NADH, however, is by the reduction of molecular oxygen via the activity of NADH oxidases that may result in the formation of water or hydrogen peroxide. Activities of NADH:H_2O oxidase and NADH:H_2O_2 oxidase have been described in many lactic acid bacteria and several species may contain both oxidases (Condon, 1987; Higuchi et al., 1995). The last few years the genes for representatives of both classes of oxidases, which are biochemically well-studied flavoproteins, have been cloned and characterized (Table 4). Ross and Claiborne were the first to isolate one of these genes and characterized the *nox* gene for the NADH:H_2O oxidase from *E. faecalis* 10C1 coding for a 446-residue enzyme with homology with flavoprotein disulfide reductases (Ross & Claiborne, 1992). This was followed by the isolation of the gene for the NADH:H_2O oxidase from *S. mutans* NCIB 11723 (Matsumoto et al., 1996). The deduced sequence of the NADH:H_2O oxidases of *E. faecalis* and *S. mutans* showed considerable similarity (41% identity) with each other. Specifically, the FAD-binding and NADH binding sites are well-conserved, as is the region around the previously identified redox-active cysteine residue at position 42 in the *E. faecalis* enzyme (Ross & Claiborne, 1992). Remarkably, antibodies against the *S. mutans* NADH:H_2O oxidases showed no cross-reactivity against the *E. faecalis* oxidase but did react with enzymes from different lactic acid bacteria, including *L. lactis* and *S. thermophilus* (Higuchi et al., 1995). In conjunction with PCR-based methods based on conserved and functional regions in the NADH:H_2O oxidases this could allow for the isolation of *nox* genes from these and other lactic acid bacteria.

To date, only a single gene coding for a NADH:H_2O_2 oxidase has been characterized. This gene was designated *nox–1* since it was isolated from the same *S. mutans* strain (NCIB 11723) from which the gene for NADH:H_2O oxidase was characterized (later designated *nox–2*; Table 4) (Higuchi et al., 1994). Antibodies against the *S. mutans* NADH:H_2O_2 oxidase cross-reacted with cell-extracts of *L. lactis* and *S. thermophilus* (Higuchi et al., 1995). This property could be used in the further genetic characterization of the NADH:H_2O_2 oxidases of lactic acid bacteria, followed by the development of defined mutants that do not accumulate the toxic H_2O_2 during aeration.

Oxygen scavenger systems

Although aeration as a way to steer metabolic fluxes could have advantages because of its simple application and efficiency, there are many undesired side-effects in lactic acid bacteria. As a consequence of chemical and enzymatic conversions of oxygen, toxic derivatives may be generated, from which superoxides and hydrogen peroxide are the best described (Condon,

1987; see above). These may in fact be so deleterious to lactic acid bacteria that some species can not grow at all in the presence of oxygen (Archibald & Fridovich, 1981). However, various detoxification systems have found to operate in lactic acid bacteria and several of these have now been genetically characterized (Table 4). This analysis permits to increase the efficiency of these systems or to introduce them into strains where they may lack completely.

Hydrogen peroxide is usually detoxified by peroxidases and catalases. The *npr* gene coding for NADH peroxidase has been characterized from *E. faecalis* 10C1 and found to encode a 447- residue flavoprotein with similarities to the disulfide reductases, including 28% identity to the *E. coli* thioredoxin reductase (Ross & Claiborne, 1991). The enzyme has been well-characterized at the enzymatic and structural level and active site residues were identified by site-directed mutagenesis (Stahle et al., 1991; Parsonage et al., 1994). For the latter purpose the *npr* gene was functionally overexpressed in *E. coli* but overproduction the *E. faecalis* NADH peroxidase in lactic acid bacteria has yet not been described.

Catalases that detoxify H_2O_2 are only found in a limited number of meat lactic acid bacteria, including lactobacilli and pediococci (Hammes et al., 1990). Most of these catalases are heme- dependent and the *katA* gene for such an enzyme has been characterized from *Lactobacillus sake* LTH677 (Knauf et al., 1992). The *katA* gene was functionally expressed in *Lactobacillus casei*, which is naturally deficient in catalase, illustrating the possibility to increase the tolerance for H_2O_2 in lactic acid bacteria. The potential of the *katA* gene, however, is limited since the activity of the encoded catalase is dependent on exogenously added heme.

Free oxygen radicals can be detoxified in lactic acid bacteria by high internal levels of Mn^{2+} which are present in various species or the activity of superoxide dismutase which may have different metal cofactors (Archibald & Fridovich, 1981; Parker & Blake, 1988). The *sod* gene for superoxide dismutase has been cloned and characterized from *S. mutans* and *L. lactis* MG1363 and found to share significant similarity with each other and the group of manganese-dependent enzymes (Nakyama, 1994; Sanders et al., 1995). Mutations in the *sod* genes in either of these two lactic acid bacteria appeared to have no effect on the anaerobic metabolism but reduced the growth rate under aerated conditions. However, an intact *sod* gene is not essential for growth under oxic conditions indicating that there are alternative scavengers for superoxide radicals. A role for glutathione that also can detoxify free radicals or H_2O_2 has been suggested since this tripeptide with a reactive cysteine is present in high concentrations in some lactic acid bacteria, including *L. lactis* (Fahey et al., 1987). The resulting glutathione disulfide should be regenerated by a glutathione disulfide reductase (Table 4). Although this enzymatic activity had previously not been described in lactic acid and other gram-positive bacteria, recently the *gor* gene for an oxygen-inducible glutathione reductase with more than 60 % similarity to other bacterial and eukaryal enzymes has been characterized in *S. thermophilus* CNRZ 368 (Pebay et al., 1995). In view of the presence of glutathione concentrations in *L. lactis* that are comparable to those of gram-negative bacteria, it could be of interest to investigate the presence and expression of the *gor* gene in lactococci. Overexpression of *sod* or *gor* genes from lactic acid bacteria has not been reported but the functional overproduction of an *E. coli* superoxide dismutase as a fusion product in *L. lactis* and *Lactobacillus gasseri* has been described (Roy et al., 1993). The latter host does not contain an endogenous superoxide dismutase and hence this study illustrates the potential to increase the adaptation of lactic acid bacteria to oxydative stress.

In aerated cultures of lactic acid bacteria the activities of oxidases, peroxidases and superoxide dismutase are usually increased as compared to anaerobically grown cultures (Condon, 1987). This increase was also found for *L. lactis* where the NADH oxidase activity is induced 5–10 fold upon aeration (J. Hugenholtz, personal communication). However, the molecular details of this oxydative stress response have not yet been worked out. Gene fusion studies showed that transcription of the *L. lactis sodA* gene appears to be induced to a limited extent under aerobic growth conditions (Sanders et al., 1995). In addition, the promoter of the *npr* gene was preceded by a consensus OxyR binding site which appeared to be protected by *E. coli* OxyR in foot-print experiments (Ross & Claiborne, 1996). Since *E. faecalis* contains an enzyme which cross-reacted with the *E. coli* OxyR, it is tempting to assume that the oxydative stress response in lactic acid bacteria has features of that present in gram- negative bacteria.

Engineering oxidases

Since the activity of many intracellular metabolic and regulatory enzymes may be affected by the intracellular NADH/NAD ratio, it would be of great interest to modulate this ratio and determine its effect on the metabolic fluxes in lactic acid bacteria (Snoep et al., 1991; Snoep, 1992). Although a different NADH/NAD ratio could be partly realized by control of the aeration level, this could result in high amounts of the toxic H_2O_2, specifically in cells that also contain a NADH:H_2O_2 oxidase activity. This is the case in *L. lactis* which appears to convert half of the oxidized NADH into H_2O_2 (Higuchi et al., 1995). In the absence of mutants deficient in NADH:H_2O_2 oxidase activity, the *S. mutans nox*–2 gene was therefore overexpressed in *L. lactis* using the inducible nisin promoter (Kuipers et al., 1995). This resulted in a significant overproduction of the *S. mutans* NADH:H_2O oxidase amounting to a few percent of the total protein, which now can be used to further redirect the metabolism of *L. lactis* under oxic conditions (W. van de Zande, J. Hugenholtz & W.M. de Vos, unpublished results). Specifically, an effect of the NADH/NAD ratio on the activity of the pyruvate dehydrogenase complex would be expected since this enzyme is highly sensitive to the intracellular redox balance (Snoep et al., 1992). Metabolic simulations based on the kinetic model of the pyruvate metabolism have indicated that under aerobic conditions the overproduction of NADH oxidase in lactate dehydrogenase deficient strains would result in a drastic increase in acetoin production rates providing an additional and rational approach for the diacetyl engineering described above (Figure 5) (J.Snoep, J. Hugenholtz & H.V. Westerhoff, Abstract pD28, Lactic Acid Bacteria Conference, Cork, 1995).

Concluding remarks

It is evident that lactic acid bacteria have a much more flexible metabolism than previously assumed and that metabolic engineering allows for the further expansion of the potential of lactic acid bacteria as starter cultures in products or as cell factories in fermentors. However, even in a relatively simple lactic acid bacterium, e.g. the homolactic *L. lactis*, there are many branching reactions that may affect the final flux in the desired pathway. This was illustrated for the engineering of pyruvate metabolism, which is at the end of the sugar catabolic pipe line. Engineering at the onset of the glycolytic pathway for exopolysaccharide production presents an even greater challenge but will definitely allow for a better understanding of the production of these important compounds. Moreover, the consequences of global metabolic engineering should be carefully studied. Model studies should be performed in which the production can be modulated of global regulators, such as CcpA or HPr(Ser-P), or the concentration of important cofactors or substrates, such as the [NADH]/[NAD^+] and [ATP]/[ADP] ratios. This is now feasible in *L. lactis* by using well-controlled promoters such as the nisine-responding promoters in combination with the *ccpA*, *ptsH*, *nox* or *atpC* genes that recently have been identified (Kuipers et al., 1995; de Ruyter et al., 1996; O.P. Kuipers, E.J.Luesink, M.M. Beerthuyzen & W.M. de Vos, unpublished observations).

The rapid developments in the generation of partial and possibly whole genome sequences are expected to produce a wealth of information on selected lactic acid bacteria that should be followed up by further genetic, biochemical and physiological analysis. It is essential to integrate these studies with the metabolic control analysis in order to develop metabolic engineering concepts for the further improvement of lactic acid bacteria for present day and future products.

Acknowledgements

I am grateful to Al Claiborne, Francesca Stingele, Masako Higuchi and Jean Delcour for providing information prior to publication and my colleagues at NIZO, specifically Jeroen Hugenholtz, Richard van Kranenburg, Evert Vesink & Oscar Kuipers, for their contributions, critics, and collaborations on the metabolic and genetic engineering of lactic acid bacteria that was partly financed by contracts BIOT-CT91–0263 and BIOT-CT94–3055.

References

Adams RM, Youas S, Mainzer SE, Moon K, Palombella AL, Estell DA, Power SD & Schmidt BF (1994) Characterization of two cold- sensitive mutants of the ß-galactosidase from *Lactobacillus delbrueckii* susp. *bulgaricus*. J. Biol. Chem. 269: 5666–5672

Alpert CA & Chassy B (1990) Molecular cloning and DNA sequence of *lacE*, the gene encoding the lactose-specific Enzyme II of the phosphotransferase system of *Lactobacillus casei*: Evidence that a cysteine residue is essential for sugar phosphorylation. J. Biol. Chem. 265: 22561–22570

Anderson DG & McKay (1984) In vivo cloning of *lac* genes in *Streptocococus lactis* ML3. Appl. Environ. Microbiol. 47: 245–249

Archibald FS & Fridovich I (1981) Manganese, superoxide dismutase, and oxygen tolerance in some lactic acid bacteria. J. Bacteriol. 146: 928–936

Bernard N, Johnson K, Ferain T, Garmyn D, Hols P, Holbrook JJ & Delcour J (1994) NAD^{+}-dependent D–2-hydroxyisocaproate dehydrogenase of *Lactobacillus delbrueckii* subsp. *bulgaricus*. Gene cloning and enzyme characterization. Eur. J. Biochem. 224: 439–446

Benson KK, Godon JJ, renault P, Griffin HG & Gasson MJ (1996) Effect of *ilvBN*-encoded α-acetolactate synthase expression on diacetyl production in *Lactococcus lactis*. Appl. Microbiol. Biotechnol. 45: 107–111

Boyd DA, Cvitkovitch DG & Hamilton IR (1994) Sequence and expression of the genes for HPr (*ptsH*) and enzyme I (*ptsI*) of the phosphoenolpyruvate-dependent phosphotransferase transport system from *Streptocococus mutans*. Infect. Immunol. 62: 1156–1165

Bouffard GG, Rudd KK & Adhya SL (1994) Dependence of lactose metabolism upon mutarotase encoded in the *gal* operon in *Escherichia coli*. J. Mol. Biol. 244: 269–278

Bocker G, Stolz P & Hammes (1994) Progress in sourdough fermentation. In: Lactic 94 (Novel G & Le Querler) pp. 133–143. Adria Normandie and University of Caen, France

Catzeddu P, Vaughan EE, Deiana P & de Vos WM (1996) Transcriptional regulation and mutations that activate expression of the galactose operon in *Streptococcus thermophilus*. (Submitted)

Condon S (1987) Responses of lactic acid bacteria to oxygen. FEMS Microbiol. Rev. 46: 269–280

Cerning J (1990) Extracellular polysaccharides produced by lactic acid bacteria. FEMS Microbiol. Rev. 87: 113–130

Chen A, Hillman JD & Duncan M (1994) L-(+)-Lactate dehydrogenase deficiency is lethal in *Streptococcus mutans*. J. Bacteriol. 76: 1542–1545

Crow VL & Thomas (1984) Properties of a *Streptococcus lactis* strain that ferments lactose slowly. J. Bacteriol. 157: 28–34

de Ruyter PGGA, Kuipers OP, Beerthuyzen MM, van Alen-Boerrigter IJ & de Vos WM (1996) Functional analysis of promoters in the nisin gene cluster of *Lactococcus lactis*. J. Bacteriol. 178 (12) (in press)

Deutscher J & Sauerwald H (1986) Stimulation of dihydroxyacetone and glycerol kinase activity in *Streptococcus faecalis* by phosphoenolpyruvate-dependent phosphotransferase-dependent phosphorylation catalyzed by enzyme I and HPR of the phosphotransferase system. J. Bacteriol. 166: 829–836

Deutscher J, Kuster E, Bergstedt U, Charrier V & Hillen W (1995) Protein-kinase dependent HPr/CcpA interaction links glycolytic activity to carbon catabolite repression in Gram-positive bacteria. Mol. Microbiol. 15: 1049–1053

De Vos WM & Gasson MJ (1989) Structure and expression of the *Lactococcus lactis* gene for phospho-ß-galactosidase. J. Gen. Microbiol. 132: 331–340

De Vos WM, Boerrigter I, van Rooijen RJ, Reiche B & Hengstenberg W (1990) Characterization of the lactose-specific enzymes of the phosphotransferase system in *Lactococcus lactis*. J. Biol. Chem. 265: 22554–22560

De Vos WM & Simons G (1994) Gene cloning and expression systems in lactococci. pp. 52–105, in: Genetics and Biotechnology of Lactic Acid Bacteria (Gasson MJ & de Vos WM, Eds.) Chapman & Hall, London, UK

De Vos WM & Vaughan EE (1994) Genetics of lactose utilization in lactic acid bacteria. FEMS Microbiol. Rev. 15: 217–237

De Vos WM, Beerthuyzen MM, Luesink EL & Kuipers OP (1995) Genetics of the nisin operon and the sucrose-nisin conjugative transposon Tn*5276*. Dev. Biol.Stand. 85: 617–627

Doco T, Wieruzeski J.-M & Fournet B (1990) Structure of an exocellular polysaccharide produced by *Streptococcus thermophilus*. Carbohydr. Res., 198: 313–321

Driessen AJM, Poolman B, Kiewiet R & Konings WN (1987) Arginine transport in *Streptococcus lactis* is driven by a cationic exchanger. Proc. Natl. Acad. Sci. USA 84: 6093–6097

Duncan MJ & Hillman JD (1991) DNA sequence and in vitro mutagenesis of the gene encoding the fructose 1,6 diphosphate- dependent L-(+)lactate dehydrogenase of *Streptococcus mutans*. Infect. Immunol. 59: 3930–3934

Fahey RC, Brwon WC, Adams WB & Worsham MB (1978) Occurence of glutathione in bacteria. J. Bacteriol. 133: 1126–1129

Fath MJ & Kolter R (1993) ABC exporters. Microbiol. Rev. 57: 995–1017

Ferain T, Garmyn D, Bernard N, Hols P & Delcour J (1994) *Lactobacillus plantarum ldhL* gene: overexpression and deletion. J. Bacteriol. 176: 596–601

Ferain T, Hobbs Jr JN, Richardson J, Bernard N, Garmyn D, Hols P, Allen NE & Delcour J C (1996a) Genetic analysis of vancomycin resistance in *Lactobacillus plantarum*: disruption of *ldhD* and *ldhL* genes. (Submitted)

Ferain T, Schanck AN, Veiga da Cunha M & Delcour J (1996b) Distribution of end products from glucose and citrate metabolism in a *Lactobacillis plantarum* strain deficient for lactate dehydrogenase. (Submitted)

Gagnon G, Vandeboncoeur C Levesque RC & Frenette M (1992) Cloning, sequencing and expression in *Escherichia coli* of the *ptsI* gene encoding enzyme I of the phosphoenolpyruvate:sugar phosphotransferase transport system from *Streptococcus salivarius*. Gene 121: 71–78

Gagnon G, Vandeboncoeur C & Frenette M (1993) Phosphotransferase system of *Streptococcus salivarius*: characterization of the *pstH* gene and its product. Gene 136: 27–34

Gasson MJ & de Vos WM (1994) Genetics and Biotechnology of Lactic Acid Bacteria. Chapmann & Hall, Glasgow, UK

Gasson MJ, Benson K, Swindell S, Griffin H (1996) Metabolic engineering of the *Lactococcus lactis* diacetyl pathway. Le Lait 75: 33–40

Garmyn D, Ferain T, Bernard N, Hols P, Holbrook J & Delcour (1995) Cloning, nucleotide sequence and transcriptional analysis of the L- lactate dehydrogenase gene from *Pediococcus acidilactici*. Appl. Environ. Microbiol. 61: 266–272

Germond JE, Lapierre L, Delley M & Mollet B (1995) A new mobile genetic element in *Lactobacillus delbrueckii* subsp. *bulgaricus*. Mol. Gen. Genet. 248: 407–416

Godon JJ, Delorme C, Bardowski J, Chopin MC, Ehrlich SD & Renault P (1993) Gene inactivation in *Lactococus lactis*: branched chain animo acid biosynthesis. J. Bacteriol. 175: 4383–4390

Griffin HG, Swindell SR & Gasson (1992) Cloning, and sequence analysis of the gene encoding L-lactate dehydrogenase from *Lactococcus lactis*: evolutionary relationships between 21 diffrent LDH enzymes. Gene 122:, 193–197

Hammes WP, Bantleon A, Min S (1990) Lactic acid bacteria in meat fermentation. FEMS Microbiol. Rev. 87: 165–174

Higuchi M, Shimada M, Matsumoto J, Yamamoto Y, Rhaman A & Kamio (1994) Molecular cloning and sequence analysis of the gene encoding the H_2O_2-forming NADH oxidase from *Streptococcus mutans*. Biosci. Biotech. Biochem. 58: 1603–1607

Higuchi M, Matsumoto, Shimada M, Yamamoto Y & Kamio Y (1995) Occurrence, of the NADH oxidases corresponding to H_2O_2-forming oxidase and H_2O-forming oxidase among species of oral and non-oral streptococci. Oral. Microbial. Immunol. (in press)

Hillman JD, Chen A, Duncan M & Lee SW (1994) Evidence that L-(+)-lactate dehydrogenase deficiency is lethal in *Streptococcus mutans*. Infect Immunol. 62: 60–64

Hueck C & Hillen W (1995) Catabolite repression in *Bacillus subtilis*: a global regulatory mechanism for the gram-positive bacteria?. Mol. Microbiol. 15: 395–401

Hugenholtz J, Perdon L & Abee T (1993) growth and energy generation by *Lactococcus lactis* subsp. *lactis* biovar *diacetylactis* during citrate metabolism. Appl. Environ. Microbiol. 59: 4216–4222

Hugenholtz J (1993) Citrate metabolism in lactic acid bacteria. FEMS Microbiol. Rev. 12: 165–178

Hugenholtz J, Decates R, Simons G, Starrenburg MJC & de Vos WM (1996) Increased ethanol production by metabolic engineering of *Lactococcus lactis*. (Submitted)

Hutkins RW & Morris HA (1987) Carbohydrate metabolism by *Streptococcus thermophilis*: A review. J. Food Protect. 50 : 876–884

Hutkins RW, Morris HA & McKay LL (1985) Galactokinase activity in *Streptococcus thermophilus*. Appl. Environ. Microbiol. 50: 777–780

Ingram LO, Eddy CK, MacKenzie KF, Conway T, Altherthum (1989) Genetics of *Zymomonas mobilis* and ethanol production. Dev. Ind. Microbiol. 30: 53–69

Jensen PR, Michelsen O & Westerhoff HV (1993) Control analysis of the depedence of *Escherichia coli* physiology on the H+-ATPase. Proc. Natl. Acad. Sci. 90: 8068–8072

Jensen PR, van der Gugten AA, van Heeswijk WC, Rohwer J, Molenaar D, van Workum M, Richard P, Teusink B, Bakker BM, Kholodenko BN & Westerhoff HV (1995) Hierarchies in control. J. Biol. Sys. 3: 139–144

Kell DB & Westerhoff HV (1986) Metabolic control theory: its role in microbiology and biotechnology. FEMS Microbiol. Rev. 39: 305–320

Kim SF, Baek SJ & Pack MY (1991) Cloning and nucleotide sequence of the *Lactobacillus casei* lactate dehydrogenase gene. Appl. Environ. Microbiol. 57: 2431–2417

Konings WN, Lolkema JS & Poolman B (1995) The generation of metabolic energy by solute transport. Arch. Microbiol. 164: 235–242

Knauf HJ, Vogel RF & Hammes WP (1992) Cloning, sequencing, and phenotypic expression of *katA*, which encodes the catalase of *Lactobacillus sake* LTH677. Appl. Environ. Microbiol. 58: 832–829

Kochhar S, Chuar N & Hottinger H (1992) Cloning and overexpression of the the *Lactobacillus bulgaricus* NAD-dependent D-lactate dehydrogenase gene in *Escherichia coli:* purification and characterization of the recombinant protein. Biophys. Biochem. Res. Comm. 185: 705–712

Kuster E, Luesink EJ, de Vos WM & Hillen W (1996) Immunoloical cross-reactivity to catabolite control protein CcpA from *B. megaterium* is found in many Gram-positive bacteria. FEMS Microbiol Lett, (in press)

Kuipers OP, Beerthuyzen MM, de Ruyter PGGA, Luesink EJ & de Vos WM (1995) Autoregulation of nisin biosynthesis in *Lactococcus lactis* by signal transduction. J. Biol. Chem. 270: 27229–27304

Law J, Buist G, Haandrikman A, Kok J, Venema G & Leenhouts K (1995) A system to generate chromosomal mutations in *Lactococcus lactis* which allows fast analysis of targeted genes. J. Bacteriol. 177: 7011–7018

Llanos RM, Hillier AJ & Davidson BE (1992) Cloning, nucleotide sequence, expression and chromosomal location of *ldh*, the gene encoding L-(+)-lactate dehydrogenase from *Lactococcus lactis*. J. Bacteriol. 174: 6956–6964

Llanos RM, Marin CJ, Hillier AJ & Davidson BE (1993) Identification of a novel operon in *Lactococcus lactis* encoding enzymes for lactic acid synthesis: phosphofructokinase, pyruvate kinase and lactate dehydrogenase. J. Bacteriol. 175: 254–255

Limsowtin GKY, Davey GP & Crow VL (1986) Effect of gene dosage on expression of lactose enzymes in *Streptococcus lactis*. N.Z. Dairy Sci. Technol. 21: 151–156

Leong-Morgenthaler P, Zwahlen MC and Hottinger H (1991) Lactose metabolism in *Lactobacillus bulgaricus*: analysis of the primary structure and expression of the genes involved. J. Bacteriol. 173: 1951–1957

Lerch, H-P, Blocker H, Kallwas H, Hoppe J, Tsai H & Collins J (1989a) Cloning, sequencing and expression in *Escherichia coli* of the 2-D-hydroxycaproate dehydrogenase gene of *Lactobacillus casei*. Gene 78: 47–57

Lerch, H-P, Frank R H & Collins J (1989b) Cloning, sequencing and expression of the 2-D-hydrocxycaproate dehydrogenase-encoding gene of *Lactobacillus confusus* in *Escherichia coli*. Gene 83: 263–270

Lokman BC, Leer RJ, van Sorge R & Pouwels (1994) Promoter analysis and transcriptional regulation of *Lactobacillus pentosus* genes involved in xylose catabolism. Mol. Gen. Genet. 245: 117–125

London J (1990) Uncommon pathways of metabolsim among lactic acid bacteria. FEMS Microbiol. Rev. 87: 103–112

Mainzer SE, Yoast S, Palombella A, Adams Silva R, Pooman B, Chassy BM, Biozet B & Schmidt BF (1990) Pathway engineering of *Lactobacillus bulgaricus* for improved yoghurt. pp. 41–55. In: R.C. Chandan (ed.) Yoghurt: Nutritional and Health Properties. National Yoghurt Association, Virginia, US

Marugg JD, Goelling D, Stahl U, Ledeboer AM, Toonen MY, Verhue WM & Verrips CT (1994) Identification and characterization of the α-acetolactate synthase gene from *Lactococcus lactis* subsp. *lactis* biovar. *diacetylactis*. Appl. Environ. Microbiol. 60: 1390–1394

Marugg JD, van Kranenburg R, Laverman P, Rutten GA & de Vos WM (1996) Identical transcriptional control of the divergently transcribed *prtP* and *prtM* genes that are required for proteinase production in *Lactococcus lactis*. J. Bacteriol. 178: 1525–1531

Marty-Teysset C, Posthuma C, Lolkema JS, Schmitt P, Divies C & Konings WN (1996) Proton motive force generation by citrolactic fermentation in *Leuconostoc mesenteroides*. J. Bacteriol. 178: 2178–2185

Matsumoto J, Higichi M, Shimada M, Yamamoto Y & Y Kamio (1996) Molecular cloning and sequence analysis of the gene encoding the H_2O-forming NADH oxidase gene from *Streptococcus mutans*. Biosci. Biotech. Biochem. 60: 39–43

McKay LL & Baldwin KA (1974) Altered metabolism of *Streptococcus lactis* C2 deficient in lactate dehydrogenase. J. Dairy Sci. 57: 181–186

Minowa T, Iwata S, Sakai H, Masaki H & Ohta T (1989) Sequence and characteristics of the *Bifidobacterium longum* gene encoding L-lactate dehydrogenase and the primary structure of the enzyme: a new feature of the allosteric site. Gene 85: 161–168

Mollet B & Delley M (1990) Spontaneous deletion formation within the ß-galactosidase gene of *Lactobacillus bulgaricus*. J. Bacteriol. 172: 5670–5676

Mollet B & Delley M (1991) A ß-galactosidase deletion mutant of *Lactobacillus bulgaricus* reverts to an active enzyme by internal DNA sequence duplication. Mol. Gen. Genet. 227, 17–21

Mollet B & Hottinger H (1992) Yoghurt contenant de microorganismes vivants. European Patent Application 0 518 096

Nakyama K (1994) Nucelotide sequence of *Streptococcus mutans* superoxide dismutase gene and isolation of insertion mutants. J. Bacteriol. 174: 4928–4934

Nakajima H, Hirota T, Toba T & Adachi S (1992) Structure of the extracellular polysaccharide from slime-forming *Lactococcus lactis* subsp. *cremoris* SBT 0495. Carbohydr. Res. 224: 245–253

Otto R, Lageveen RG, Veldkamp H & Konings (1980) Generation of an electrochemical proton gradient in membrane vesicles of *Streptococcus cremoris*. Proc. Natl. Acad. Sci. USA 77: 5502–5506

Parker MW & Blake CCF (1988) Iron- and manganese-containnig superoxide dismutates can be distinguished by analysis of their primary structures. FEBS Lett. 229: 377–382

Parsonage D, Miller H, Ross RP & Claiborne A (1994) Purification and analysis of streptococcal NADH peroxidase expressed in *Escherichia coli*. J. Biol. Chem. 268: 3161–3167

Pebay M, Holl A-C, Simonet J-M, Decaris B (1995) Characterization of the *gor* gene of the lactic acid bacterium *Streptococcus thermophilus* CNRZ368. Res. Microbiol. 146: 317–383

Platteeuw C, Hugenholtz J, Starrenburg M, van Alen-Boerrigter IJ. & de Vos WM (1995) Metabolic engineering of *Lactococcus lactis*: Influence of the overproduction of α-acetolactate synthase in strains deficient in lactate dehyrogenase as a fuction of culture conditions. Appl. Environ. Microbiol. 61: 3967–3971

Poolman B (1993) Energy transduction in lactic acid bacteria. FEMS Microbiol. Rev. 12: 125–148

Poolman B, Royer TJ, Mainzer SE & Schmidt B.F. (1989) Lactose transport system of *Streptococcus thermophilus*: a hybrid protein with homology to the melibiose carrier and enzyme III of phosphoenolpyruvate-dependent phophotransferase systems. J. Bacteriol. 171: 244–253

Poolman B, Royer TJ, Mainzer SE, & Schmidt BF (1990) Carbohydrate utilization in *Streptococcus thermophilus*: characterization of the genes for aldose 1-epimerase (mutarotase) and UDPglucose 4-epimerase. J. Bacteriol. 172: 4037–4047

Poolman B, Knol J, Mollet B, Nieuwenhuis B & Sulter G (1995) Regulation of bacterial sugar-H^+ symport by phosphoenolpyruvate-dependent enzyme I/HPr-mediated phosphorylation. Proc. Natl. Acad. Sci. USA 92: 778–782

Poolman B, Knol J, van der Does C, Henderson PJF, Liang W-J, Leblanc G, Potcher T & Mus-Veteau I (1996) Cation and sugar selectivity determinants in a novel family of transport proteins. Mol. Microbiol., 19: 911–922

Postma PW, Lengeler JW & Jacobson GR (1993) Phosphoenolpyruvate- dependent carbohydate phosphotransferase systems of bacteria. Microbiol. Rev. 57: 543–594

Rauch PJG & de Vos WM (1992) Characterization of the novel nisin- sucrose conjugative transposon Tn*5276* and its insertion in *Lactococcus lactis*. J. Bacteriol. 174: 1280–1287

Reizer J (1989) Regulation of sugar uptake ands efflux in gram-positive bacteria. FEMS Microbiol. Rev. 63: 149–157

Reizer J, Sutrina SL, Saier MH, Stewart GC, Peterkofsy A & Reddy P (1989) Mechanistic and physiological consequences of HPr(Ser) phosphorylation on the activities of the phosphoenolpyruvate: sugar phosphotransferase system in gram-positive bacteria: studies with site- specific mutants of HPr. EMBO J. 8: 2111–2120

Reeves PR (1994) Biosynthesis and asembly of lipopolysaccharide. New. Compr. Biochem. 27: 281–314

Rosey EL and Stewart G (1993) Nucelotide and deduced amino acid sequences of the *lacR*, *lacABCD*, and *lacFE* genes encoding the repressor, tagatose–6-phosphate gene cluster, and sugar- specific phosphotransferase system components of the lactose operon of *Streptococcus mutans*. J. Bacteriol. 174: 6159–6170

Ross RP & Claiborne A (1991) Cloning, sequence and overexpression of the NADH peroxidase from *Streptococcus faecalis* 10C1. Structural relationship with the flavoprotein disulfide reductases. J. Mol. Biol. 221: 857–871

Ross RP & Claiborne A (1992) Molecular cloning and analysis of the gene encoding the NADH oxidase from *Streptococcus faecalis* 10C1. Comparison with NADH peroxidase and the flavoprotein disulfide reductases. J. Mol. Biol. 227: 658–671

Ross RP & Claiborne A (1996) Analysis of the OxyR-binding site associated with the NADH peroxidase gene in *Enterococcus faecalis* 10C1. (Submitted)

Roy DG, Klaenhammer TR & Hassan HM (1993) Cloning and expression of the manganese superoxide dismutase gene of *Escherichia coli* in *Lactococcus lactis* and *Lactobacillus gasseri* (1994) Mol. Gen. Genet. 239: 33–40

Russell RRB, Adus-Opoku J, Sutcliffe IC, Tao L & Ferretti JJ (1992) A binding-protein dependent transport system in *Streptococcus mutans* responsible for multiple sugar metabolism. J. Biol. Chem. 267: 4631–4637

Saier Jr MH, Ye JJ, Klinke S & Nino E (1996) Identification of an anaerobically induced phosphoenolpyruvate dependent fructose-specific phosphotransferase system and evidence for the Embden-Meyerhof glycolytic pathway in the heterofermentative bacterium *Lactobacillus brevis*. J. Bacteriol. 178: 314–316

Sato Y, Poy F, Jacobson GR & Kuramitsu (1989) Characterization and sequence analysis of the *scrA* gene encoding enzyme II^{scr} of the *Streptococcus mutans* phosphoenolpyruvate-dependent sucrose phosphotransferase system. J. Bacteriol. 171: 263–271

Sanders JW, Leenhouts K, Haandrikman AJ, Venema G & Kok J (1995) Stress response in *Lactococcus lactis*: Cloning, expression analysis and mutation of the lactococcal superoxide dismutase gene. J. Bacteriol. 177: 5254–5260

Sasaki T, Ito Y & Sasaki Y (1993a) Electrotransformation of *Lactoibacillus delbrieckii* subsp. *bulgaricus*. In: W.M. de Vos, J. Huis in 't Veld & B. Poolman (Eds.) FEMS Microbiol. Rev. 12: P8

Sasaki Y, Ito Y & Sasaki T (1993b) Gene conversion in transconjugants of *Lactobacillus delbruecki* subsp. *bulgaricus* using pAMß1 as an integration vector. In: W.M. de Vos, J. Huis in 't Veld & B. Poolman (Eds.) FEMS Microbiol. Rev. 12: P9

Schroeder CJ, Robert C, Lenzen G, McKay LL, and Mercienier A (1991) Analysis of the *lacZ* sequences from two *Streptococcus thermophilus* strains: comparison with the *Escherichia coli* and *Lactobacillus bulgaricus* ß-galactosidase sequences. J. Gen. Microbiol. 137, 369–380

Smart JB & Thomas TD (1987) Effect of oxygen on lactose metabolism in lactic streptococci. Appl. Environ. Microbiol. 53: 533–541

Schmidt BF, Adams RM, Requadt C, Power S & Mainzer SE (1989) Expression and nucleotide sequence of the *Lactobacillus bulgaricus* ß-galactosidase gene cloned in *Escherichia coli*. J. Bacteriol. 171, 625–635

Snoep JL (1992) Regulation of pyruvate catabolism in *Enterococcus faecalis*. A molecular aproach to physiology. Academic Thesis, University of Amsterdam, Amsterdam

Snoep JL, Teixeira de Mattos MJ & Neijssel OM (1991) Effect of the energy source on the NADH/NAD ratio and on pyruvate catabolism in anaerobic chemostrat cultures of *Enterococcus faecalis* NTC 775. FEMS Microbiol. Lett. 81: 63–66

Snoep MJ, Teieira de Mattos MJ, Starrenburg MJC & Hugenholtz J (1992) Isolation, characterization and physiological role of the pyruvate dehydrogenase complex and α-acetolactate synthase of *Lactococcus lactis* subsp. *lactis* var. *diacetylactis*. J. Bacteriol. 174: 4838–4841

Stahle T, Ahmed SA, Claiborne A & Schulz GE (1991) The structure of NADH peroxidase from *Streptococcus faecalis* 10C1 refined at 2.16 A resolution. J. Mol. Biol. 221: 1325–1344

Stucky K, Schich J, Klein JR, Heinrich B & Plapp R (1996) Characterization of *pepRI*, a gene coding for a potential transcriptional regulator of *Lactobacillus delbrueckii* subsp. *lactis* DSM729. FEMS Microbiol. Lett. 136: 63–69

Stingele F, Neeser J-R & Mollet B (1996) Identification and characterization of the *eps* (exopolysaccharide) gene cluster from *Streptococcus thermophilus* Sfi6. J. Bacteriol. 178: 1680–1690

Stingele F & Mollet B (1995) Homologous integration and transposition to identify genes involved in the production of exopolysaccharides in *Streptococcus thermophilus*. Dev. Biol. Stand. 85: 487–493

Sutherland IW (1972) Bacterial exopolysaccharides. Adv. Microbiol. Physiol. 8: 143–212

Taguchi H & Ohta T (1991) D-lactate dehydrogenase is a member of the D-isomer-specific 2-hydroxyacid dehydrogenase family. J. Biol. Chem. 266: 12588–12594

Tao L, Sutcliffe IC, Russell RRB & Ferretti JJ (1995) Regulation of the multiple sugar metabolism operon in *Streptococcus mutans*. Dev. Biol. Stand. 85: 434–350

Thomas TD & Crow VF (1984) Selection of galactose-fermenting *Streptococcus thermophilus* in lactose-limited chemostat cultures. Appl. Environ. Microbiol. 48: 186–191

Thompson J (1987) Regulation of sugar uptake and metabolism in lactic acid bacteria. FEMS Microbiol. Rev. 46: 221–231

Thompson J, Chassy BM & Egan W (1985) Lactose metabolism in *Streptococcus lactis*: studies with a mutant lacking glucokinase and mannose-phosphotransferase activities. J. Bacteriol. 162: 217–223

Thompson J & Gentry-Weeks CR (1994) Metabolism des sucres par les bacteries lactiques. In: Bacteries Lactiques (de Roissart H & Luquest FM, Eds) pp. 239–290. Lorica, Uriage, France

Van Dam K, van der Vlag J, Kholodenko BN & Westerhof HV (1993) The sum of the control coefficients of all enzymes on the flux control through a group-tranfer pathway can be as high as two. Eur. J. Biochem. 212: 791–799

Van Kranenburg R, Marugg JD, van Swam II, Willem NJ & de Vos WM (1996) Molecular characterization of the plasmid-located *eps* gene cluster coding for exopolysaccharide biosynthesis in *Lactococcus lactis*. (Submitted)

van Rooijen RJ, van Schalkwijk S & de Vos WM (1991) Molecular cloning, characterization, and nucleotide sequence of the tagatose 6- phosphate pathway gene cluster of the lactose operon of *Lactococcus lactis*. J. Biol. Chem. 266: 7176–7181

Van Rooijen RJ, Gasson MJ & de Vos WM (1992) Characterization of the promoter of the *Lactococcus lactis* lactose operon: Contribution of flanking sequences and LacR repressor to its activity. J. Bacteriol. 174: 2273–2280

Van Rooijen RJ & de Vos WM (1990) Molecular cloning, transcriptional analysis and nucleotide sequence of *lacR*, a gene encoding the repressor of the lactose phosphotransferase system of *Lactococcus lactis*. J. Biol. Chem. 265: 18499–18503

Van Rooijen RJ, Dechering KJ, Wilmink CNJ & de Vos WM (1993) Lysines 72, 80, 213, and aspartic acid 210 of the *Lactococcus lactis* LacR repressor are involved in the response to the inducer tagatose-6-phosphate leading to induction of *lac* operon expression. Protein Eng. 6: 208–215

Vaughan EE, David S & de Vos WM (1996) The lactose transporter in *Leuconostoc lactis* is a new member of the LacS subfamily of galactoside-pentose-hexuronide translocators. Appl. Environ. Microbiol. 62: 1547–1582

Wagner E, Gotz F & Bruckner R (1993) Cloning and characterization of the *scrA* gene encoding the sucrose-specific enzyme II of the phosphotransferase systen of *Staphylococcus carnosus*

Ye JJ, Reizer J, Cui X & Saier Jr. MJ (1994) Inhibition of the phosphoeneolpyruvate:lactose phosphotransferase system and activation of a cytoplasmic sugar-phosphate phosphatase in *Lactococcus lactis* by ATP-dependent metabolite-activated phosphorylation of serine 46 in the phosphocarrier protein HPr. J. Biol. Chem. 269: 11837–11844

Ye JJ & Saier Jr. MJ (1995) Cooperative binding of lactose and the phosphorylated phosphocarrier protein HPr(Ser-P) to the lactose/H^+ symport permease of *Lactobacillus brevis*. Proc. Natl. Acad. Sci. USA 92: 417–421

Antonie van Leeuwenhoek **70:** 243–251, 1996.

Lactococcus lactis and stress

Fabien Rallu, Alexandra Gruss & Emmanuelle Maguin
Institut National de la Recherche Agronomique Laboratoire de Génétique Microbienne, 78352 Jouy en Josas, France

Key words: stress, acid pH, adaptation, acid-resistant mutants

Abstract

It is now generally recognized that cell growth conditions in nature are often suboptimal compared to controlled conditions provided in the laboratory. Natural stresses like starvation and acidity are generated by cell growth itself. Other stresses like temperature or osmotic shock, or oxygen, are imposed by the environment. It is now clear that defense mechanisms to withstand different stresses must be present in all organisms. The exploration of stress responses in lactic acid bacteria has just begun. Several stress response genes have been revealed through homologies with known genes in other organisms. While stress response genes appear to be highly conserved, however, their regulation may not be. Thus, search of the regulation of stress response in lactic acid bacteria may reveal new regulatory circuits. The first part of this report addresses the available information on stress response in *Lactococcus lactis.*

Acid stress response may be particularly important in lactic acid bacteria, whose growth and transition to stationary phase is accompanied by the production of lactic acid, which results in acidification of the media, arrest of cell multiplication, and possible cell death. The second part of this report will focus on progress made in acid stress response, particularly in *L. lactis* and on factors which may affect its regulation. Acid tolerance is presently under study in *L. lactis*. Our results with strain MG1363 show that it survives a lethal challenge at pH 4.0 if adapted briefly (5 to 15 minutes) at a pH between 4.5 and 6.5. Adaptation requires protein synthesis, indicating that acid conditions induce expression of newly synthesized genes. These results show that *L. lactis* possesses an inducible response to acid stress in exponential phase.

To identify possible regulatory genes involved in acid stress response, we determined low pH conditions in which MG1363 is unable to grow, and selected at 37°C for transposition insertional mutants which were able to survive. About thirty mutants resistant to low pH conditions were characterized. The interrupted genes were identified by sequence homology with known genes. One insertion interrupts *ahrC*, the putative regulator of arginine metabolism; possibly, increased arginine catabolism in the mutant produces metabolites which increase the pH. Several other mutations putatively map at some step in the pathway of (p)ppGpp synthesis. Our results suggest that the stringent response pathway, which is involved in starvation and stationary phase survival, may also be implicated in acid pH tolerance.

Introduction

In the first part of this report, we will review *L. lactis* genes which demonstrate stress-inducible expression. Stress imposed by the environment (temperature, high salt, oxygen, and DNA damage) will be discussed. It should be noted that expression under a given condition may be due to either transcriptional, post-transcriptional, translational, or post-translational regulation, or a combination of these effects. Thus, increased expression under a given condition is not a proof that expression of a given gene is induced. Stress response can be affected by cell physiology and prior exposure to other stresses. It should be noted that the field of stress response is vast, and this report is restricted to progress made in *L. lactis*. Several reviews are recommended for further reading (Kellenberg; Georgopoulos & Welch, 1993; Hecker et al., 1996; Kolter, 1993; Hall et al., 1995).

The second part of this report is devoted more specifically to present information on acid stress, which is imposed during natural growth of the lactic acid bacteria. We report our recent results which suggest a regulation mechanism of acid resistance in *L. lactis.*

L. lactis *responses to stresses imposed by the environment*

Temperature stress. Most present information on stress response in *L. lactis* concerns heat shock. Genes involved in heat shock response are highly conserved, and several have been identified in *L. lactis*, including *dnaJ* (van Asseldonk et al., 1993), *grpE-dnaK* (Eaton et al., 1993), and *groELS* (Kim & Batt, 1993). The identified genes encode chaperones, which in *E. coli*, are necessary for the proper refolding of proteins, and/or for targetting denatured proteins for degradation by cytoplasmic proteases (which are also heat shock proteins; Herman et al., 1995; Parsell & Lindquist, 1993). These functions may be particularly important during heat shock, when protein denaturation is increased. The above genes respond to a heat shock by an induction of 2 to 100 - fold of mRNA (Arnau et al., 1996) or 2 to 3 fold of protein levels (Duwat et al., 1995). Induction is transient, with an mRNA peak reported at about 15 minutes (Arnau et al., 1996). It has also been shown that pre-adaptation of cells to a non-lethal elevated temperature improves survival in face of a lethal temperature challenge, indicating that these cells can undergo an adaptive heat shock response (Boutibonnes et al., 1991).

A question which has not yet been resolved in gram-positive bacteria, which contain no heat shock-specific sigma factor (see section on cross-protection), concerns the regulation of heat shock gene expression. Several, but not all heat shock promoters in *L. lactis*, other gram-positive bacteria, and even some gram-negative bacteria are adjacent to a palindromic sequence, called the CIRCE element (Zuber & Schuman, 1994, see Hecker et al., 1996 for review). A repressor protein was recently identified in several microorganisms which inhibits expression of the heat shock genes, probably by stabilizing CIRCE (Yuan & Wong, 1995). This repressor is encoded by a gene just upstream of *grpE-dnaK* in *L. lactis* (Eaton et al., 1993), and is thus a heat shock protein itself. Presumably, this repressor is active at low temperature, but somehow inactivated upon heat shock. Regulation of the repressor and identification of other possibly positive regulatory proteins remains to be determined.

Stress induced by salt, oxygen, or DNA damaging agents

Promoters which are sensitive to specific environmental conditions can be identified by screening random chromosomal integrated fusions with a reporter gene (Israelson et al., 1995; Law et al., 1995). This type of approach has been successfully used to identify several stress-sensitive genes. A fusion resulting in salt- dependent expression was thus isolated (J. Sanders, personal communication). A library of acid-responsive genes was also isolated in this way (Israelson et al., 1995; see section on acid stress).

Stress-responsive genes may also be identified by activity tests. Zymograms were performed to visualize activity of the manganese- dependent superoxide dismutase (SodA) of *L. lactis* (Sanders et al., 1995). Reverse genetics starting from the N-terminal sequence of the purified protein resulted in the cloning of the *sodA* gene. SodA activity was about two-fold greater in the presence of oxygen or of acid, than in buffered standing cultures, suggesting that *sodA* is a stress-responsive gene. A *sodA* mutant grows slowly in the presence of oxygen, suggesting that this enzyme is important in oxygen stress survival.

Another gene necessary for stress response is *recA* (Duwat et al., 1995; Duwat et al., 1995). This gene is known to be important for resistance to DNA damage stress, both by its recombination properties and its induction of a set of DNA repair genes known as SOS. In *L. lactis*, *recA* has a pleiotropic role in stress response (see section on cross-protection). After cloning and inactivating the gene, we observed that *recA* is important in the response to several environmental stresses, including DNA damage, oxygen, and heat stress, as the *recA* mutant is UV sensitive, oxygen sensitive, and temperature sensitive for growth (Duwat et al. 1995). RecA also appears to regulate a newly described heat shock protease named HflB (Herman et al., 1995, also called FtsH; Nilsson et al., 1994; Duwat et al., 1995). HflB/FtsH has a pleiotropic phenotype: In *E. coli*, HflB degrades σ^{32}, a heat shock sigma factor (Herman et al., 1995), lambda phage repressors (Herman et al., 1993), and uncomplexed SecY (Kihara et al. 1995). In *L. lactis*, an *ftsH* mutant is salt, heat and cold sensitive (Nilsson et al., 1994). The pleiotropic phenotype of *recA* has not been previously reported, and may indicate that stress responses in *L. lactis* may be more closely overlapping than in *E. coli* (Duwat et al., 1995). Interestingly, while *recA* is necessary for DNA repair, RecA protein levels do not appear to be induced by DNA damage (S. Sourice & P. Duwat, personal communication).

Cross protection by stress and starvation

The response of a cell to a particular stress seems to be very complex, and may differ greatly with the physiological state of the cell. The complexity seems to indicate the existence of several, overlapping levels of controls (see Kellenberg, Hecker et al., 1996 for review). While some of the genes induced by stress seem to be genuinely specific, others are induced by a wide variety of stresses, and are thus thought to be general stress response genes (Hecker et al., 1996). In *E. coli*, the sigma factor involved in general stress response is σ^S, the 'stationary phase' sigma factor (Loewen & Hengge-Aronis, 1994); in *B. subtilis*, the general stress response sigma factor appears to be σ^B (Hecker et al., 1996). In *L. lactis*, no stress-specific sigma factors have been thus far isolated. However, there is physiological evidence of cross-protection by exposing cells to a given stress: for example, subjecting *L. lactis* cells to irradiation resulted in improved survival against acid, ethanol, H_2O_2, or heat stress (Hartke et al., 1996). Our results showing that the *recA* gene is needed for full oxygen and heat resistance (Duwat et al., 1995) are consistent with these findings. Furthermore, in keeping with observations in *E. coli*, *L. lactis* non-adapted stationary phase cells were found to be more resistant to stress than exponentially growing cells (Hartke et al., 1995; Hartke et al., 1994; K. Heller, personal communication). These results suggest that a general stress response is induced both during starvation and when cells undergo a particular stress.

Protein synthesis has been implicated in the induction of stress response in *L. lactis*. Two-dimensional gels performed in independent studies confirm that heat (M. Kilstryp, personal communication), starvation (Kunji et al., 1993), and UV treatment (Hartke et al., 1996), each result in the appearance of additional proteins which are presumably involved in stress resistance. Surprisingly, it is reported that chloramphenicol or rifamycin treatment during acid adaptation (see below) did not seem to abolish acid stress resistance in *L. lactis* strain IL1403 (Hartke et al., 1995) (but was needed for cross-protection to H_2O_2 treatment Hartke et al., 1994). However, protein synthesis *is* required in *L. lactis* strain MG1363 (S. Condon, personal communication; this report). These differences remain to be explained.

L. lactis *responses to stresses resulting from cell growth*

Acid pH and stationary phase. Bacterial growth is a self-limiting process. When the medium can no longer provide the necessary nutrients, or if the environment produces a substance inhibitory for growth, bacteria stop multiplying and 'reset' internal conditions to adapt to the new conditions (Kolter, 1993). Milk fermentations by lactic acid bacteria (LAB) are extensively used to produce cheeses and other dairy products. In milk fermentations, LAB degrade lactose, resulting in lactate accumulation and consequent acidification of the media to as low as pH 4.0 (Piard & Mesmazeaud, 1991). Either acidification or lactate, or both, may inhibit further growth and metabolism, even if nutrients are still available (Piard & Mesmazeaud, 1991). Acidification by lactic acid is a desired property in cheesemaking, which contributes to cheese maturation and prevents spoilage by contaminating bacteria. However, too low a pH may give an undesirable taste, and may also result in death of the starter culture.

Until now, acid tolerance has been examined predominantly in enteric bacteria, where it may be an important pathogenicity factor (Hall et al. 1995; Slonczewski, 1992; Fang et al., 1992). Survival in acid conditions is affected by factors such as growth phase (Hall et al., 1995; Rowbury et al., 1992) and expression of specific sigma factors (Lee et al., 1995), or adaptation to low pH (Foster & Hall, 1991). Three overlapping systems of acid tolerance have been described in *Salmonella typhimurium* (Hall et al., 1995): In exponential phase cells, a shift to a slightly acidic pH (pH 5.8) induces enzymes which reestablish neutral internal pH (Foster & Hall, 1991). Upon acid shock (pH 4.5 or below), synthesis of 43 proteins is induced (Foster & Hall, 1991). The stationary phase sigma factor, σ^s, is also induced and is responsible for the induction of seven additional proteins (Lee et al., 1995). All these proteins are predicted to be involved in repair of acid pH damage and survival (Parsell & Lindquist, 1993; Foster, 1993; Visick & Clarke, 1995). Cells adapted in this way survive a challenge at a pH of 3.3. The second system of adaptation occur in stationary phase, but is σ^S independent: An acid shock in stationary phase *S. typhimurium* cells triggers the synthesis of some 15 proteins, of which only four are in common with the exponential phase acid- induced cells (Hall et al., 1994; Lee et al., 1994). The third system of acid resistance is σ^s dependent, but pH independent: it is the consequence of a global stress response which is induced in stationary phase (Hall et al., 1995; Hengge-Aronis, 1993). This system of overall increased stress response in stationary phase (induced by carbon starvation) has also been reported for *L. lactis* (Hartke et al., 1995; K. Heller, personal communication). Interestingly, acid-responsive promoters identified by a random fusion mutagenesis strategy (Israelson et al., 1995) revealed one acid-induced promoter which was active principally in stationary phase, and was also induced at low temperature. In this case, acid is necessary, but not sufficient for increased promoter activity.

The factors known to be involved in stress response are highly conserved (Parsell & Lindquist, 1993), suggesting that different organisms respond to stress in similar ways. Nevertheless, there are suggestions that *regulation* of stress responses may differ among organisms as exemplified above by the role of *recA* in different stress responses (Duwat et al., 1995). Another reason to suspect regulatory differences is that the bacteria themselves have rather different environments, and thus the stress response demands may be different. For LAB, a consequence of growth is the acidification of the media; acidity is a main growth-limiting factor. For enteric bacteria, aerobic growth does not result in significant decrease in pH; transition to stationary phase may precede acidification. However, acidic conditions can be generated *by* the environment, e.g., by passage in the intestines, during phagocytosis, or by coexistence with other fermenting microorganisms; acid resistance may thus be an important factor in pathogenicity.

We are analyzing the response of *L. lactis* strain MG1363 to acid stress. In these studies, we use HCl, rather than lactic acid to lower the pH, in order to eliminate the potentially lethal effects of lactate in the medium (Kashket, 1987). Our results indicate that adaptation of exponential phase cells at an intermediate acid pH induces tolerance at low pH. Induction of acid tolerance is more rapid than that reported for enteric bacteria. In addition, we have characterized about thirty transpositional mutants which have increased survival in low pH conditions at 37 °C. Sequence identification suggests that a homeostatic response may exist, and also implicates genes in the stringent response pathway in the regulation of acid stress response in *L. lactis*.

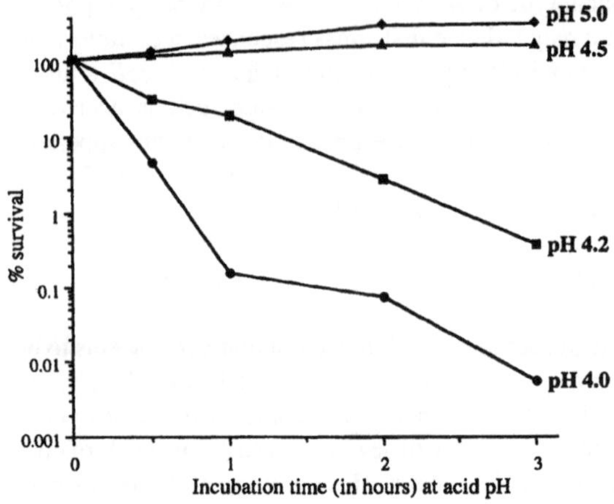

Figure 1. Spontaneous resistance of *L. lactis* to an acid stress. Exponential phase cultures of MG1363, grown in defined SA media (Jensen & Hammer, 1993) buffered at pH 7.0, were centrifuged and resuspended in media adjusted to pH 5.0 (diamonds), 4.5 (triangles), 4.2 (squares) or 4.0 (circles). After the given time intervals, samples were plated on M17 solid medium plus glucose. Survival is determined as the plating efficiency after the given incubation time divided by the plating efficiency at time zero (no acid shift).

Results and discussion of acid resistance studies

Acid sensitivity of L. lactis. An exponential phase culture of MG1363 (Gasson, 1983) was maintained in defined SA medium (Jensen & Hammer, 1993) at pH7.0, then centrifuged, and transferred to media with a pH of 5.0, 4.5, 4.2, or 4.0. Samples were plated after being incubated for given time intervals in different pH media (Figure 1). Good survival was observed for samples incubated at pH 4.5 or above. However, survival was markedly decreased with increasing incubation times at pH below 4.5, indicating that the cells are sensitive to acid pH conditions. These results show that *L. lactis* does not express constitutive resistance to acid pH. Using other *L. lactis* strains and different acidifying agents, similar results have been reported by our colleagues (Hartke et al., 1995; S. Condon, personal communication; K. Heller, personal communication).

Adaptive response to acid pH by L. lactis. An exponential phase culture of MG1363 was grown on defined buffered media (at neutral pH), centrifuged, and then adapted by transfer to pH 4.5 media for between five minutes and two hours. Following adaptation, cells were challenged at pH 4.0 and survival was determined (Figure 2). The unadapted control showed poor sur-

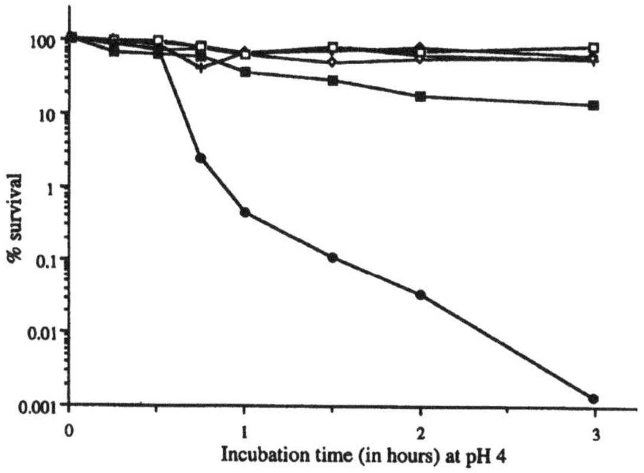

Figure 2. Adaptive response in *L. lactis*.
Exponential phase cultures of MG1363 were grown in defined SA media buffered at pH 7.0, and then centrifuged. Cell aliquots were tested for adaptive response. For the adaptation step, cells were first resuspended in pH 4.5 media for periods of 5 (black squares), 15 (black triangles), 30 (crosses), 60 (white diamonds) or 120 (white squares) minutes. For the challenge, cells were then centrifuged and transferred to pH 4.0 media. One aliquot of cells was resuspended directly in pH 4.0 media (black circles). Survival is determined as described in Figure 1, legend.

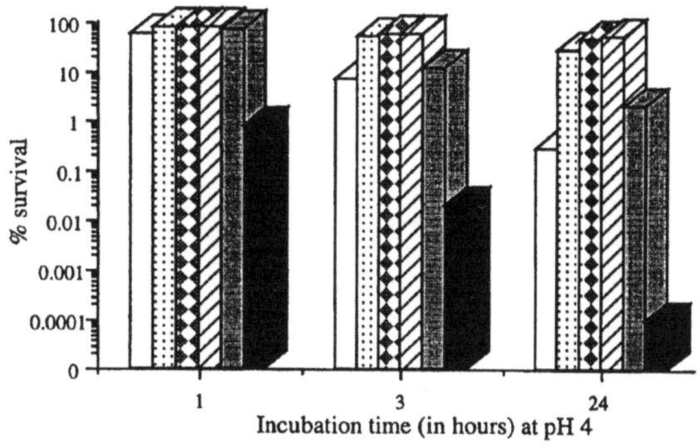

Figure 3. Effect on acid resistance of the pH used for adaptation. Adaptive response was examined using 5 minutes of adaptation (see Figure 2, legend), and varying the pH of the media in the adaptive step. Cell aliquots of exponential phase cultures of MG1363 (grown in defined SA media buffered to pH 7.0) were centrifuged and then pre- incubated 5 minutes at pH 6.5 (white bars), 6.0 (dotted bars), 5.5 (diamond bars), 5.0 (diagonally striped bars) or 4.5 (grey bars), then centrifuged and transferred to pH 4.0 media. One aliquot was transferred to pH 4.0 without adaptation (black bars). Cells were maintained in the pH 4.0 media for the times indicated. Survival is determined as described in Figure 1, legend.

vival after treatment at pH 4.0. In contrast, all adapted cultures survived well at pH 4.0. Fifteen minutes of adaptation was necessary to obtain complete survival. These experiments demonstrate that acid tolerance is induced in *L. lactis* after an adaptive step at a non-lethal acid pH.

The adaptive step was explored further. We examined the effects of adaptation in media having a pH between 4.5 and 6.5, with a duration of either five (Figure 3) or fifteen minutes (not shown) prior to challenge at pH 4.0. All adaptive conditions permitted good survival for up to 24 hours at pH 4.O, compared to loss of viability of non-adapted cultures. For five minute adaptation, incubations at pH 6.5 and pH 4.5 were less effective than at pH 5.0, 5.5, or 6.0 in inducing acid tolerance. Good acid tolerance was induced by a fifteen minute adaptation at every tested pH (data not shown). Taken together, these results confirm that *L. lactis* exhibits an inducible acid tolerance. Induction is best at adaptive pH of 5.0 to 5.5. Tolerance is long-lasting, as cells are fully acid resistant for at least 24 hours at pH 4.0 after the adaptive step.

To determine whether acid tolerance requires *de novo* protein synthesis, cultures were submitted to adaptive conditions, but in the presence of chloramphenicol (75 μg/ml), which blocks protein synthesis and is bacteriostatic (Figure 4). It was added to samples 10 minutes prior to the adaptive shift to pH 5.0; the acid challenge was subsequently carried out without chloramphenicol. A control in which cultures were treated with chloramphenicol and maintained at pH 5.0 retained full cell viability. However, cells that underwent adaptation in the presence of chloramphenicol were unable to survive the acid challenge. We thus conclude that MG1363 adaptation to acid stress (as perhaps opposed to IL1403 (Hartke et al., 1995)) requires protein synthesis.

The results reported here indicate that *L. lactis* adapts readily to acid stress, as it is fully induced after short (5 to 15 minutes) incubations, at pH levels varying between 4.5 and 6.5. The induced acid tolerance allows good survival for at least 24 hours at pH 4.0. These observations, and similar results reported by our colleagues (Hartke et al., 1995; S. Condon, personal communication; K. Heller, personal communication) suggest that acid protection of LAB, which are naturally acidifying bacteria, may confer a survival advantage for LAB in fermented products.

Isolation of transposition insertional mutants resistant to low pH conditions. To understand the genetic basis for acid tolerance, we isolated mutants that were able to survive under normally lethal acid pH con-

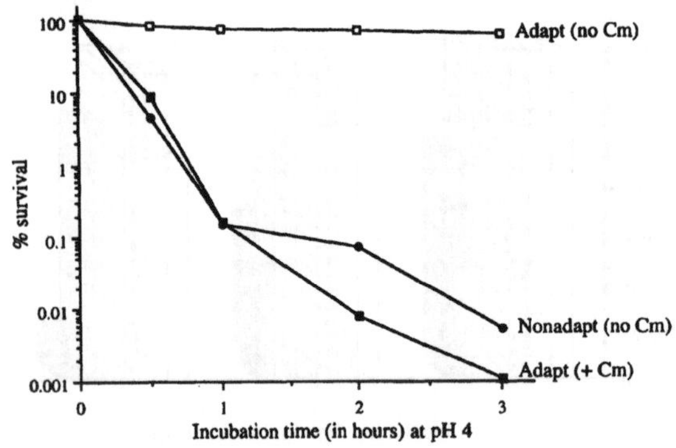

Figure 4. Chloramphenicol addition inhibits the adaptive response to acid pH.
Adaptive response was examined as described in Figure 2, using 15 minutes of adaptation at pH 5.0. One aliquot was shifted directly from pH 7.0 to pH 4.0 without prior adaptation (black circles; Nonadapt (no Cm)). One aliquot underwent adaptation with no treatment (white squares; Adapt (no Cm)). To another aliquot, chloramphenicol was added 10 minutes prior to, and maintained during the adaptation only; cells were then challenged at pH 4.0 (black squares; Adapt (+Cm)). A control treated with chloramphenicol at pH 5.0 but not submitted to acid challenge was totally viable (not shown).

ditions. We observed that growth on solid media at pH below 6.0 is thermosensitive; the plating efficiency is about 10^{-4} at a pH below 6.0 at 37 °C compared to 100% at 30 °C. We made use of this interesting phenomenon to isolate acid-resistant mutants at 37 °C, using a pG$^+$host::IS*S1*-based transposon mutagenesis system we recently developed (Maguin et al., 1996). pG$^+$host::IS*S1* is comprised of a thermosensitive replicon, pG$^+$host (Maguin et al., 1992), and a transposable element, IS*S1* (Huang et al., 1992; LeBourgeois et al., 1992). Transposition results in integration of the pG$^+$host plasmid flanked by copies of the IS*S1* element and is detected by plating at 37 °C (non-permissive temperature for pG$^+$host replication), selecting for an antibiotic resistance marker present on pG$^+$host (Maguin et al., 1996). The exact site of insertion can be determined by re-cloning pG$^+$host and flanking sequences directly from a chromosomal DNA preparation. Acid-resistant mutant colonies were selected. They arose at a frequency of about 0.2% of the total number of mutants isolated. For each of the mutants, either one or both junctions was cloned and sequenced. Sequence analyses revealed more than 30 independent mutations which conferred acid stress resistance at 37 °C. Several of the mutated genes were redundant, indicating that the mutational analysis was saturated. Genetic assignments were made on the basis of homology with known genes.

Identification of ahrC, *an acid-resistant mutant.* One of the identified genes shows significant homology with *ahrC*, a regulator of arginine synthesis and catabolism in *Bacillus subtilis* (Baumberg & Klingel, 1993). We propose that its inactivation results in constitutive expression of arginine metabolic enzymes in the arginine deiminase pathway (Abdelal, 1979). It was previously shown that arginine deiminase pathway is active at low pH and results in an increased environmental pH in *L. lactis* (Marquis et al., 1987; Casiano-Colon & Marquis, 1988). It is possible that acid tolerance of the *ahrC* mutant strain is due to increased arginine deiminase activity which restores environmental pH. Mutant strains with the potential to restore pH of the medium may have great importance in dairy fermentations, where overacidification of the product can cause taste deterioration.

Stringent response pathway involvement in acid tolerance. Several other mutants were sequenced, and showed homologies to genes already identified in other organisms (not shown). Interestingly, four of the mutants, *hpt* (Nilsson & Lauridsen 1992), *relA* or *spoT* (Cashel & Rudd, 1987; Mechold et al., 1993), *guaA* (Mäntsälä & Zalkin, 1992), and *deoB* (Burland et al., 1993), showed significant homologies with genes which may be implicated in nucleotide metabolism. More specifically, in *E. coli*, each of these gene products can be placed in the biosynthesis pathway of (p)ppGpp, the stringent response regulator (Figure 5). Some of these genes were found more than once in the mutagenesis. What is the potential involvement of stringent response in acid tolerance? The answer may come in part from what is already known about (p)ppGpp. In *E. coli*, (p)ppGpp synthesis is induced by starvation of carbon, nitrogen (Cashel & Rudd, 1987), or phosphate (Spira et al., 1995), and by other stresses (Cashel & Rudd, 1987). The consequences of (p)ppGpp synthesis are pleiotropic (Cashel & Rudd, 1987). By affecting specificity of RNA polymerase, (p)ppGpp directly alters the expression profile of the cell. Among its many effects, (p)ppGpp was also found to affect expression of σ^S, the stationary phase and general stress response sigma factor (Loewen & Hengge-Aronis, 1994; Gentry et al., 1993; Lange et al., 1995). Induction of σ^S also results in the expression of a specific set of genes. Thus, (p)ppGpp modifies the pattern of expressed genes both directly, by interacting with

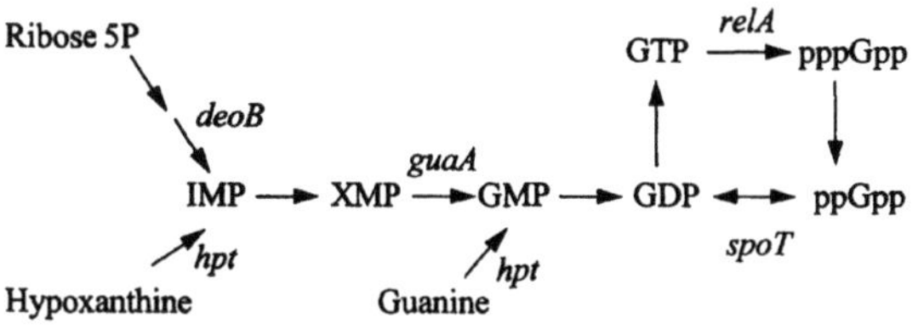

Figure 5. Acid resistant mutants which may affect stringent response pathway.
Four acid resistant transposon insertional mutants of MG1363 have been characterized, and the interrupted genes have been partially sequenced. Homology searches indicate that these genes potentially affect biosynthesis of (p)ppGpp, the regulator of stringent response. The biosynthesis pathway of (p)ppGpp and its precursors is shown above; genes for which we obtained mutants are indicated. Note that these genes also participate in other pathways. One of the genes we isolated has homology with both *relA* and *spoT*, and has not yet been characterized further.

RNA polymerase, and indirectly, by its role in the induction of σ^S.

We can put this information together with what is known on acid stress response: σ^S was shown in *S. typhimurium*, *E. coli*, and *Shigella flexneri* to be needed for acid tolerance (Lee et al., 1995; Small et al., 1994). Since (p)ppGpp is required for full induction of σ^S, and σ^S is required for acid tolerance, we would thus predict that (p)ppGpp also modulates acid tolerance. The mutants we obtained seem to confirm this prediction. We will examine the adaptive response of the *L. lactis* mutant strains in both exponential and stationary phases. Further characterization of these mutants, as well as a direct demonstration of (p)ppGpp effects on acid tolerance are in progress. To our knowledge, this is the first report of a potentially direct link between stringent response and acid tolerance.

What are the effects of these acid resistant mutants on other types of stress? Several lines of evidence suggest that acid resistance may affect other stress responses: First, it has been reported that acid tolerance also enhances tolerance to other stress, although the converse seems to be less sure (Lee et al., 1995). We have evidence that *L. lactis* stress responses are overlapping, as heat and oxygen stress responses both require *recA* (Duwat et al., 1995). HflB, which is implicated in heat shock response (Herman et al., 1995; Duwat et al., 1995; Nilsson et al., 1994; Herman et al., 1993), is also induced by acid pH in *L. lactis* (Ullerup et al., 1995). Second, it is known that (p)ppGpp plays a pleiotropic role in gene expression and survival in stationary phase (Cashel & Rudd, 1987; Gentry et al., 1993; Lange et al., 1995; Nystrom, 1993). Therefore, acid resistant mutants affecting (p)ppGpp expression are expected to selection procedure to isolate thermoresistant mutants of the *L. lactis recA* strain. Remarkably, this selection gave rise to many of the same mutants as those reported here, and which we tentatively place in the (p)ppGpp biosynthesis pathway (P. Duwat, personal communication). This strongly suggests that acid-resistance in *L. lactis* can also affect survival after other types of stress. This may have particular importance in stationary phase cells, where cells have increased resistance to other stresses, such as heat shock and ethanol shock.

Another unanswered question in *L. lactis* concerns the existence of a stationary phase sigma factor. In enteric bacteria, σ^S is necessary for sustained acid tolerance. Yet, despite searches to find a σ^S homologue in *L. lactis*, only the vegetative sigma factor of *L. lactis* was found (Gansel et al., 1993; Araya et al., 1993). Mutational analyses may lead us to the identification of either stress- specific sigma factors, or factors which modulate activity of the vegetative sigma factor. Such studies may clarify whether acidification of the medium is the real trigger for entry of *L. lactis* into stationary phase.

Applications of acid-resistant mutants. Acidification during milk fermentation has the double role of developing the taste of the final product, and of preventing growth of contaminating bacteria. Nevertheless, it could be desirable to stop over-acidification which may occur, for example, upon prolonged storage. Furthermore, survival of LAB in acidified cultures is also desirable, particularly from the industrial point of view where probiotic effects of viable LAB may give additional value to dairy products. The types of mutants characterized here may fit the criteria of a good industrial strain, and may thus have considerable interest in dairy fermentation.

Acknowledgements

The authors are particularly grateful to Patrick Duwat for constant discussion of this work with relation to his own unpublished results, and to our colleagues for communicating their results prior to publication. We are also grateful to all the members of our team who have listened and contributed to the advancement of this project. Part of this work has been supported by the CEE BRIDGE program, contract #BIO2-CT94–3055.

References

Abdelal AT (1979) Arginine catabolism by microorganisms. Ann. Rev. Microbiol. 33: 139–168

Araya T, Ishibashi N, Shimamura S, Tanaka K & H Takahashi (1993) Genetic and molecular analysis of the *rpoD* gene from *Lactococcus lactis*. Biosci. Biotech. Biochem. 57: 88–92

Arnau J, Sorensen KI, Appel KF, Vogensen FK & K Hammer (1996) Analysis of heat shock gene expression in Lactococcus lactis MG1363. Microbiology, 142: 1685–1691

van Asseldonk M, Simons A, Vissier H, de Vos WM & G Simons (1993) Cloning, nucleotide sequence, and regulatory analysis of the Lactococcus dnaJ gene. J. Bacteriol. 175: 1637–1644

Baumberg S & U Klingel (1993) Biosynthesis of arginine, proline, and related compounds. In: Bacillus subtilis and Other Gram-Positive Bacteria. pp. 299–306 In: AL Sonenshein, R Losick, JA Hoch (Eds.), ASM press

Boutibonnes P, Gillot B, Auffray Y & B Thammavongs (1991) Heat shock induces thermotolerance and inhibition of lysis in a lysogenic strain of *Lactococcus lactis*. Int. J. Food Microbiol. 14: 1–9

Burland V, G Plunkett III, DL Daniels & FR Blattner (1993) DNA sequence and analysis of 136 kilobases of the *Escherichia coli* genome : organizational symmetry around the origin of replication. Genomics 16: 551–561

Cashel M & K Rudd (1987) The stringent response, pp. 1410–1438 In: FC Neidhardt, JL Ingraham, KB Low, B Maganasik, M Schaechter and HE Umbarger (Ed.), *Escherichia coli* and *Salmonella typhimurium*: Cellular and Molecular Biology, Vol. 2. American Society for Microbiology, Washington, D.C

Casiano-Colon A & RE Marquis (1988) Role of the arginine deiminase system in protecting oral bacteria and an enzymatic basis for acid tolerance. Appl. Environ. Microbiol. 54: 1318–1324

Duwat P, SD Ehrlich & A Gruss (1995) The *recA* gene of *Lactococcus lactis*: Characterization and involvement in oxidative and thermal stress. Mol. Microbiol. 17: 1121–1131

Duwat P, S Sourice, SD Ehrlich & A Gruss (1995) The *recA* gene involvement in oxidative and thermal stress in *Lactococcus lactis*. In 'Genetics of Streptococci, Enterococci and Lactococci'. Ferretti JJ, Gilmore MS, Klaenhammer TR & Brown F (Eds.), Dev. Biol. Stand. Basel Karger, publ. Vol 85. pp 455–467

Eaton, T., C. Shearman & M. Gasson. 1993. Cloning and sequence analysis of the *dna*K region of *L. lactis* subsp. *lactis*. J. Gen. Microbiol. 139: 3253–3264

Fang FC, Libby SJ, Buchmeir NA, Loewen PC, Switala J, Harwood J & D.G. Guiney. (1992) The alternative σ factor KatF (RpoS) regulates *Salmonella* virulence. Proc. Natl. Acad. Sci. USA. 89: 11978–11982

Foster JW & HK Hall (1991) Inducible pH homeostasis and the acid tolerance response of Salmonella typhimurium. J. Bacteriol. 173: 5129–5135

Foster JW (1993) The acid tolerance response of *Salmonella typhimurium* involves transient synthesis of key acid shock proteins. J. Bacteriol. 175: 1981–1987

Gansel X, Hartke A, Boutibonnes P & Y Auffray (1993) Nucleotide sequence of the *Lactococcus lactis*NCDO 763 (ML3) *rpoD* gene. Biochem. et Biophysica Acta 1216: 115–118

Gasson M (1983) Plasmid complements of *Streptococcus lactis* NCDO712 and other lactic Streptococci after protoplast-induced curing. J. Bacteriol. 154: 1–9

Gentry DR, Hernandez VJ, Nguyen LH, Jensen DB & M Cashel (1993) Synthesis of the stationary-phase sigma factor ss is positively regulated by ppGpp. J. Bacteriol. 175: 7982–7989

Georgopoulos C & WJ Welch (1993) Role of the major heat shock proteins as molecular chaperones. Ann. Rev. Cell. Biol. 9: 601–634

Hall HK, Karem KL & JW Foster (1995) Molecular responses of microbes to environmental pH stress. Adv. Microbial Physiol. 37: 229–272

Hartke A, S Bouche, X Gansel, P Boutibonnes & Y Auffray (1994) Starvation-induced stress resistance in *Lactococcus lactis* subsp. *lactis* IL1403. Appl. Environ. Microbiol. 60: 3474–3478

Hartke A, Giard JC, Benachour A & Y Auffray (1995) The acid-stress response of Lactococcus lactis subsp. lactis IL1403: Changes in protein synthesis and cross-protection. 7eme Colloque du Club des Bactéries Lactiques Paris 13–15 Septembre 1995. Aff. S5

Hartke A, Bouché S, Boutibonnes P & Y Auffray (1996) UV-inducible proteins and UV-induced Cross-protection against acid, ethanol, H_2O_2 or heat treatments in *Lactococcus lactis* subsp. *lactis* (submitted).

Hecker M, Schumann W & U Völker (1996) Heat- shock and general stress response in *Bacillus subtilis*. Mol. Microbiol. 19:. 417–428

Hengge-Aronis R (1993) The role of *rpoS* in early stationary-phase gene regulation in *Escherichia coli* K12. pp. 171–200 In: "Starvation in Bacteria" S Kellenberg (Ed.), Plenum Press

Herman C, Ogura T, Tomoyasu T, Hirage S, Akiyama Y, Ito K, Thomas R, D'Ari R & P Bouloc (1993) Cell growth and λ phage development controlled by the same essential *Escherichia coli* gene *ftsH*/*hflB*. Proc. Natl. Acad. Sci. USA 90: 10861–10865

Herman C, D Thévenet, R D'Ari & P Bouloc (1995) Degradation of σ^{32}, the heat shock regulator in *Escherichia coli*, is governed by HflB. Proc. Natl. Sci. USA. 92: 3516–3520

Huang DC, Novel M, Huang XF & G Novel (1992) Nonidentity between plasmid and chromosomal copies of ISS1-like sequences in Lactococcus lactis subsp. lactis CNRZ270 and their possible role in chromosomal integration of plasmid genes. Gene 118, 39–46

Israelson H, Madsen S, Vrang A, Hansen E & E Johansen (1995) Cloning and partial characterization of regulated promoters from *Lactococcus lactis* Tn*917-lacZ* integrants with the new promoter probe vector pAK80. Appl. and Environ. Microbiol. 61: 2540–2547

Jensen PR & K Hammer. (1993) Minimal requirements for exponential growth of *Lactococcus lactis*. Appl. Environ. Microbiol. 59: 4363–4366

Kashket ER (1987) Bioenergetics of lactic acid bacteria: cytoplasmic pH and osmotolerance. FEMS Microbiol. Rev. 46: 233–244

Kellenberg S. Starvation in Bacteria. Plenum Press. The reader is referred particularly to chapters 6 through 10

Kihara A, Akiyama Y & K Ito (1995) FtsH is required for proteolytic elimination of uncomplexed forma of SecY, an essential protein translocase subunit. Proc. Natl. Acad. Sci. USA 92: 4532–4536

Kim SG & CA Batt (1993) Cloning and sequencing of the *Lactococcus lactis* subsp. *lactis groESL* operon. Gene 127: 121–126

Kolter R (1993) The stationary phase of the bacterial life cycle. Annu. Rev. Biochem. 47: 855–874

Kunji ERS, Ubbink T, Matin A, Poolman B & WN Konings (1993) Physiological responses to Lactococcus lactis ML3 to altering conditions of growth and starvation. Arch. Microbiol. 159: 372–379

Lange R, Fischer D & R Hengge-Aronis (1995) Identification of transcriptional start sites and the role of ppGpp in the expression of rpoS, the structural gene for the ss subunit of RNA polymerase in Escherichia coli. J. Bacteriol. 177: 4676–4680

Law J, Buist G, Haandrikman A, Kok J, Venema G & K Leenhouts (1995) A system to generate chromosomal mutations in

Lactococcus lactis which allows fast analysis of targeted genes. J. Bacteriol. 177: 7011–7018

LeBourgeois P, Lautier M, Mata M & P Ritzenthaler (1992) New tools for the physical mapping of *Lactococcus lactis* strains. Gene 11: 109–114

Lee IS, Slonczewski JL & JW Foster (1994) A low- pH-inducible, stationary-phase acid tolerance response in *Salmonella typhimurium*. J. Bacteriol. 176: 1422–1426

Lee IS, J Lin, HK Hall, B Bearson & JW Foster (1995) The stationary-phase sigma factor s^s (RpoS) is required for a sustained acid tolerance response in virulent *Salmonella typhimurium*. Mol. Microbiol. 17: 155–167

Loewen PC & R Hengge-Aronis (1994) The role of sima factor σ^s (KatF) in bacterial global regulation. Ann. Rev. Microbiol. 48: 53–80

Maguin E, P Duwat, T Hege, D Ehrlich & A Gruss (1992) New thermosensitive plasmid for Gram-positive bacteria. J. Bacteriol. 174: 5633–5638

Maguin E, H Prévost, SD Ehrlich & A Gruss (1996) Efficient insertional mutagenesis in Lactococci and other Gram-positive bacteria. J. Bacteriol. 178: 931–935

Mäntsälä P & H Zalkin (1992) Cloning and sequence of *Bacillus subtilis purA* and *guaA*, involved in the conversion of IMP to AMP and GMP. J. Bacteriol 174: 1883–1890

Marquis RE, GR Bender, DR Murray & A Wong (1987) Arginine deiminase system and bacterial adaptation to acid environments. Appl. Environ. Microbiol. 53: 198–200

Mechold V, Steiner K, Vettermann S & H Malke (1993) Genetic organization of the streptokinase region of the Streptococcus equisimilis H46A chromosome. Mol. Gen. Genet. 241: 129–140

Nilsson D & AA Lauridsen (1992) Isolation of purine auxotrophic mutants of *Lactococcus lactis* and characterization of the gene *hpt*encoding hypoxanthine guanine phosphoribosyltransferase. Mol. Gen. Genet. 235: 359–364

Nilsson D, AA Lauridsen, T Tomoyasu & T Ogura (1994) A *Lactococcus lactis* gene encodes a membrane protein with putative ATPase activity that is homologous to the essential *Escherichia coli ftsH* gene product. Microbiol. 140: 2601–2610

Nystrom T (1993) Global systems approach to the physiology of the starved cell. pp. 129–150. In: "Starvation in Bacteria" S Kellenberg (Ed.), Plenum Press

Parsell DA & S Lindquist (1993) The function of heat- shock proteins in stress tolerance: degradation and reactivation of damaged proteins. Ann. Rev. Genet. 27: 437–496

Piard JC & M Desmazeaud (1991) Inhibiting factors produced by lactic acid bacteria. 1- Oxygen metabolites and catabolism end-products. Lait. 71: 525–541

Rowbury RJ, M Goodson & AD Wallace (1992) The PhoE porin and transmission of the chemical stimulus for induction of acid resistance (acid habituation) in Escherichia coli. J. Appl. Bacteriol. 72: 233–243

Sanders JW, Leenhouts K, Haandrikman AJ, Venema G & J Kok (1995) Stress response in *Lactococcus lactis*: Cloning, expression analysis, and mutation of the superoxide dismutase gene. J. Bacteriol. 177: 5254–5260

Slonczewski JL (1992) pH-regulated genes in enteric bacteria. ASM News. 58: 140–144

Small P, D Blankenhorn, D Welty, E Zinser & J Slonczewski (1994) Acid and base resistance in *Escherichia coli* and *Shigella flexneri*: Role of *rpoS* and growth pH. J. Bacteriol. 176: 1729–1737

Spira B, Silberstein N & E Yagil (1995) Guanosine 3',5'-Bispyrophosphate (ppGpp) synthesis in cells of *Escherichia coli* starved for P_i. J. Bacteriol. 177: 4053–4058

Ullerup A, Saxild HH & D Nilsson (1995) Regulation of *Lactococcus lactis ftsH* expression. 7eme Colloque du Club des Bactéries Lactiques Paris 13–15 Septembre 1995. Aff. G8

Visick JE & S Clarke (1995) Repair, refold, recycle: how bacteria can deal with spontaneous and environmental damage to proteins. Mol. Microbiol. 16: 835–845

Yuan G & SL Wong (1995) Isolation and characterization of *Bacillus subtilis groE* regulatory mutants: Evidence for *orf39* in the *dnaK* operon as a repressor gene in regulating the expression of both *groE* and *dnaK*. J. Bacteriol. 177: 6462–6468

Zuber U & W Schuman (1994) CIRCE, a novel heat shock element involved in regulation of heat shock operon *dnaK* of *Bacillus subtilis*. 176: 1359–1363

Antonie van Leeuwenhoek **70:** 253–267, 1996.

Physiology of pyruvate metabolism in *Lactococcus lactis*

Muriel Cocaign-Bousquet, Christel Garrigues, Pascal Loubiere & Nicolas D. Lindley*
Centre de Bioingnierie Gilbert Durand, UMR CNRS & Lab Ass INRA, Institut National des Sciences Appliques, Complexe Scientifique de Rangueil, 31077 Toulouse cedex, France. (author for correspondence)*

Key words: pyruvate metabolism, mixed acid fermentation, glycolysis

Abstract

Lactococcus lactis, a homofermentative lactic acid bacterium, has been studied extensively over several decades to obtain sometimes conflicting concepts relating to the growth behaviour. In this review some of the data will be examined with respect to pyruvate metabolism. It will be demonstrated that the metabolic transformation of pyruvate can be predicted if the growth-limiting constraints are adequately established. In general lactate remains the major product under conditions in which sugar metabolism via a homolactic fermentation can satisfy the energy requirements necessary to assimilate anabolic substrates from the medium. In contrast, alternative pathways are involved when this energy supply becomes limiting or when the normal pathways can no longer maintain balanced carbon flux. Pyruvate occupies an important position within the metabolic network of *L. lactis* and the control of pyruvate distribution within the various pathways is subject to co-ordinated regulation by both gene expression mechanisms and allosteric modulation of enzyme activity.

Introduction

The description of pyruvate metabolism in homofermentative lactic acid bacteria appears to be simple. More than 90% of pyruvate is converted to lactate during industrial sugar fermentations. However, in an increasing number of cases diversion of this simple conversion has been observed, leading to production of a number of other metabolites. In fact, pyruvate is the key metabolic intermediate in lactic acid bacteria and metabolism is strongly regulated in the process of pyruvate production and conversion. In this overview, a variety of nutitional factors are mentioned that effect the activity of enzymes involved in pyruvate metabolism. These changes in activity can lead to important variations in end product formation.

Lactic acid bacteria (LAB), unlike many of the other widely studied bacteria of industrial importance, metabolise sugars predominantly to generate biochemical energy. Anabolic precursor metabolites are obtained from other components of the medium. It is, therefore, the energy generating aspect of the catabolic pathways which needs to be considered when examining metabolic regulation. The 'growth supporting' substrates (a variety of sugars) can best be defined as those compounds able to be fermented at rates and by pathways which can provide the necessary flux of biochemical energy to facilitate the inter-conversion and synthesis of cell material from preformed nitrogen-containing organic matter. Other carbon substrates may be metabolised and partially catabolised in so much as their consumption has an influence on the energy status of the cell.

Metabolic pathways of sugar fermentation

Before examining the factors specifically regulating pyruvate metabolism it is important to briefly review the catabolic pathways involved in sugar metabolism. Most of the sugars are taken up by the cells via either PEP-dependent phosphotransferase systems (PTS) involving coupled transport and phosphorylation of the sugar (Thompson, 1978, 1979; Yamada, 1987; Benthin et al., 1993), or via permease systems (Thompson et al., 1985; Crow & Thomas, 1984; Romano et al., 1987) in which sugar transport is followed by kinase-mediated phosphorylation of the free sugar within the cytosol (Bisset & Anderson,1974). The contribution of each transport mechanism to overall sugar uptake has never been adequately assessed and speculation tends to predominate as to which mechanism is functional. Most probably a variety of uptake mechanisms contribute to the global transport and relative proportions of each system will depend upon the prevailing nutritional environment.

Once phosphorylated, sugars are catabolised by relatively simple linear pathways (Figure 1) whose composition depends upon the nature of the sugar. Many monosaccharides enter central metabolism via glucose-6-phosphate (G6P) and their catabolism follows a classical glycolysis to pyruvate with a net gain of 2 ATP per sugar once transport requirements have been taken into consideration. The vast majority of sugars enter the central pathways as G6P though fructose enters as FDP (due to the synthesis of F1P by the PTS^{fru}) and the Gal6P moiety of lactose transported via the PTS^{lac} enters glycolysis at the level of triose-phosphates after transformation via the tagatose pathway. Galactose entering the cytoplasm via permease transport or via the hydrolysis of lactose by β-galactosidase (Farrow, 1980) enters metabolism as G6P but requires the operation of the Leloir pathway involving galactose kinase, UDP-glucose epimerase/transferase and the P-glucose isomerase.

Regulation of homolactic metabolism

Glycolysis generates the ATP necessary for the biosynthesis of cell material but also NADH, which in the absence of respiratory activity needs to be recycled via reactions involving the reduction of metabolic intermediates to liberate the fermentation end-products. This reducing equivalent wastage occurs primarily at the level of pyruvate and under non-limiting growth conditions involves the production of lactic acid. The allosteric control of many enzymes of glycolysis by FDP has been reported (see below) and hence phosphofructokinase and fructose diphosphate aldolase might be expected to play an important role in glycolytic regulation. Indeed, in *L. lactis*, FDP aldolase has been studied in some detail (Crow & Thomas, 1982) and found to have an affinity for FDP of 1.1 mM. Measured intracellular pools of FDP are an order of magnitude higher than this value under conditions of homolactic fermentation and hence the enzyme operating under substrate saturated conditions might contribute significantly to

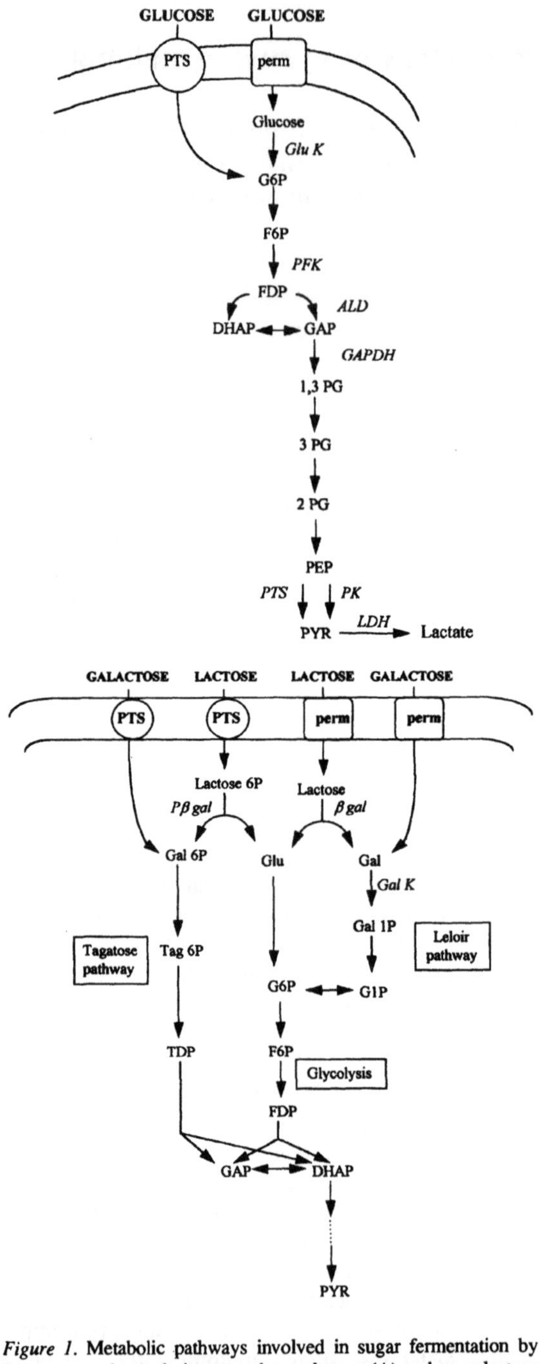

Figure 1. Metabolic pathways involved in sugar fermentation by *Lactococcus lactis* during growth on glucose (A) and on galactose or lactose (B). PTS: phosphotransferase system; perm: permease; (P)β gal: (Phospho) β galactosidase; Gal K: galactokinase; Glu K: glucokinase; PFK: phosphofructokinase; PK: pyruvate kinase; LDH: lactate dehydrogenase.

the control of pathway flux. The controlling influence of this key enzyme, however, has never been studied and it is generally believed that the glyceraldehyde-phosphate dehydrogenase (GAPDH) exerts most influence over the flux through glycolysis. Poolman et al. (1987a) studied this enzyme and estimated a flux control coefficient of 0.9 when applying metabolic control analysis. It should be remembered that the sum of the flux control coefficients of each enzyme in a linear pathway as defined by metabolic control theory (Kell & Westerhoff, 1986) should be equal to 1. This indicates an extremely important role for this enzyme but the results were obtained with non-proliferating cells and under conditions (use of enzyme inhibitors) which might have led to over-estimation of the control coefficient of GAPDH. Metabolic control theory gives a precise quantification of the contribution of each enzyme to control of carbon flux but the values obtained cannot be extrapolated to growth conditions other than those used experimentally. Furthermore, it is often seen that diminishing the activity of any enzyme has a significant effect on carbon flux while increasing the activity of that same enzyme has a more constrained effect. This can be interpreted as evidence that many enzymes share the control, i.e., protein conservation has evolved a metabolic structure in which no single enzyme is produced in significant excess under conditions in which high carbon throughput is favoured. Despite such arguements, it is clear that GAPDH has a major role in regulating carbon flux through glycolysis in *L. lactis*. Many related homolactic bacteria implicated in oral pathology posses an NADPH-dependent GAPDH activity (Yamada, 1987; Crow & Wittenberg, 1979) though this enzyme has not been found in *L. lactis*. The genetic organisation of the glycolytic enzymes is different to that frequently encountered among aerobic Gram-positive bacteria (Cancilla et al., 1995) though the implications for this from a regulation viewpoint are not yet clear.

The enzymes more closely associated with pyruvate metabolism have been examined and shown to be subject to considerable biochemical regulation. The pyruvate kinase which, together with the PTS mechanism, generates the pyruvate, has been shown to be subject to activation effects by various glycolytic metabolites: FDP, G6P, F6P, Tag6P, Gal6P, TagDP but also by GAP and DHAP. This mechanism modifies the affinity of the enzyme (Thomas, 1976a) for PEP from a low affinity (Km = 5 mM) to a high affinity status (Km = 0.14 mM) leading to a low intracellular PEP concentration (Thompson, 1978; Thompson & Torchia, 1984; Lohmeier-Vogel et al., 1986). In other words, when the initial stages of glycolysis, or parallel pathways, are operating rapidly, high concentrations of C_6 compounds and both GAP and DHAP are observed and the pyruvate kinase is activated. To understand why such a mechanism has developed, it is necessary also to take into account that the PTS provides the second reaction generating pyruvate and that these same activators are implicated in the control of PTS sugar transport. Thus, in order to assure an efficient transformation of PEP to pyruvate under substrate excess conditions in which permease transporters are probably dominant, an antagonistic control is exerted leading to a shift in reactions favouring the pyruvate kinase and hence regenerating the ATP necessary for direct intracellular phosphorylation of the permeated sugar. This antagonistic control is also seen in the manner in which inorganic phosphate and ATP inhibit the pyruvate kinase activity (Collins & Thomas, 1974; Thompson & Torchia, 1984) but activate the PTS.

The catabolism of sugars leads to pyruvate, a major branch point metabolite whose further metabolism determines the nature of the fermentation. Since *L. lactis* strains are generally considered to be homolactic, the normal pathway accounting for the vast majority of the generated pyruvate involves the reductive transformation of pyruvate to lactate via lactate dehydrogenase (LDH). This reaction enables the recycling of the reduced coenzymes produced in glycolysis thus maintaining energetic equilibrium. As for other key enzymes, phosphorylated metabolites play an important role in regulating this activity: both FDP and TagDP are known to activate the enzyme (Thomas, 1976b) while PEP and inorganic phosphate have a negative effect on LDH activity (Yamada, 1987; Konings et al., 1989). Indeed, LDH is a tetrameric protein stabilised by a phosphate with an affinity for pyruvate of approximately 2 mM (Thomas et al., 1980). The various subspecies of *L. lactis* used in the dairy industry seem to be extremely similar as regards the LDH activity and recent molecular characterisation of the *ldh* gene shows no significant difference in the putative amino acid structure (Swindell et al., 1994). Interestingly, the *ldh* gene is part of an operon containing the genes for phosphofructokinase and pyruvate kinase enzymes expressed to high levels but also subject to strict control by the phosphate potential of the cell (Llanos et al., 1992, 1993).

From these data a model has been developed to explain the homolactic behaviour of the strain which is coherent with the majority of the experimental observa-

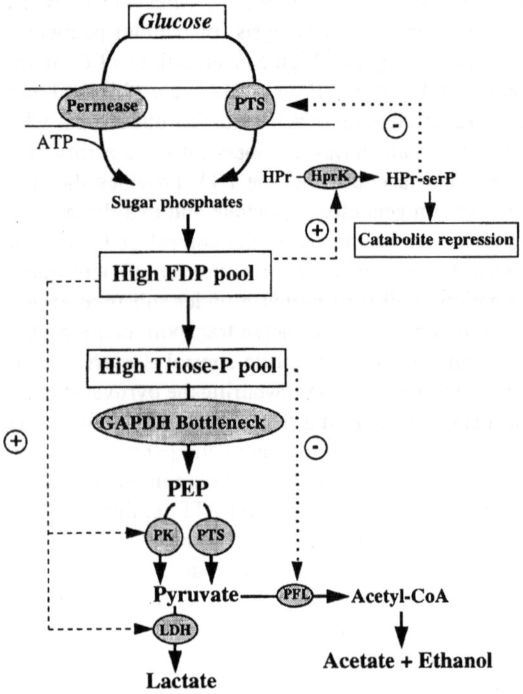

Figure 2. Central role of FDP in co-ordinating glycolytic flux both upstream and downstream of the rate-limiting GAPDH during homolactic fermentation of glucose by *L.lactis*. HPrK: HPr kinase; GAPDH, glyceraldehyde dehydrogenase; PK, pyruvate kinase; PFL, pyruvate-formate lyase; LDH, lactate dehydrogenase. ⊕, activation of an enzyme activity; ⊖, inhibition of an enzyme activity.

tions. When substrate is in excess, the flux through the catabolic pathways is important and adequate to supply the ATP required for cell growth. Under such conditions the synthesis of cell biomass from other organic compounds present in the medium limits growth. The regulation of glycolysis has been correlated with the high level of FDP which activates both pyruvate kinase and lactate dehydrogenase, and a low level of PEP, presumably resulting directly from the activation of these enzymes. The result is a linear metabolism of sugar conversion to lactate of constant thermodynamic efficiency. The glycolytic pathways articulate around GAP dehydrogenase whose controlling influence (Poolman et al., 1987) would lead to a situation in which all metabolites downstream of the triose phosphates would tend to be present at relatively low concentrations, phenomena accentuated by the activated PK and LDH activities. Thus the upper portion of the catabolic pathways can be envisaged as being substrate saturated with correspondingly high pools of phosphorylated sugars, while the lower pathway, common to all pathways of sugar catabolism, is substrate limited due to the GAP dehydrogenase bottleneck. Within such a model, FDP clearly plays a key role modulating the metabolic potential of the cell by a cascade of allosteric, post-translational and catabolite repression phenomena (Figure 2), but other phosphorylated metabolites are also implicated in the modulation of enzyme activity. Indeed FDP has less effect on PK than other metabolites (Thomas, 1976a): G6P is the principal activator (more significant effect on Vm at lower concentrations) in a hierarchic control cascade which can be expressed as G6P > F6P > DHAP > GAP > FDP. The confusion which has developed in this respect is linked to the fact that FDP is the metabolite whose intracellular concentration is highest and therefore easiest to measure leading to an over-simplification of the phenomena involved. Recent progress as regards the role of FDP as an alarmone controlling the carbon catabolite repression cascade via the phosphorylation of HPr in Gram-positive bacteria (Ye et al., 1994; Veyrat et al., 1994; Saier et al., 1996) needs also to be taken into account, though further research is necessary to identify the extent to which modulation of gene expression reinforces the biochemical control of enzyme activity within *L. lactis*.

Mixed acid fermentation

Under certain conditions, *L. lactis* deviates significantly from a homolactic fermentation and minor products (<10% of total carbon consumption) derived from pyruvate become far more important. Such a metabolism is quite distinct from a heterolactic metabolism in which glycolysis is replaced by a variant of the pentose pathway involving a phosphoclastic enzyme generating directly a C_3 and a C_2 residue from pentose-P (Figure 3). In the case of mixed acid fermentation (Figure 4) glycolysis continues to generate pyruvate, but further metabolism involves a considerably diminished flux through LDH. This response was first observed in carbon-limited chemostat cultures in which presumably other organic nutrients are in excess. Under such conditions, pyruvate is metabolised via either pyruvate formate lyase (PFL) or pyruvate dehydrogenase (PDH) to give either formate or CO_2 respectively, and acetate + ethanol mixtures. Thus, the mixed acid fermentation and the extent of the deviation of pyruvate away from lactate will result

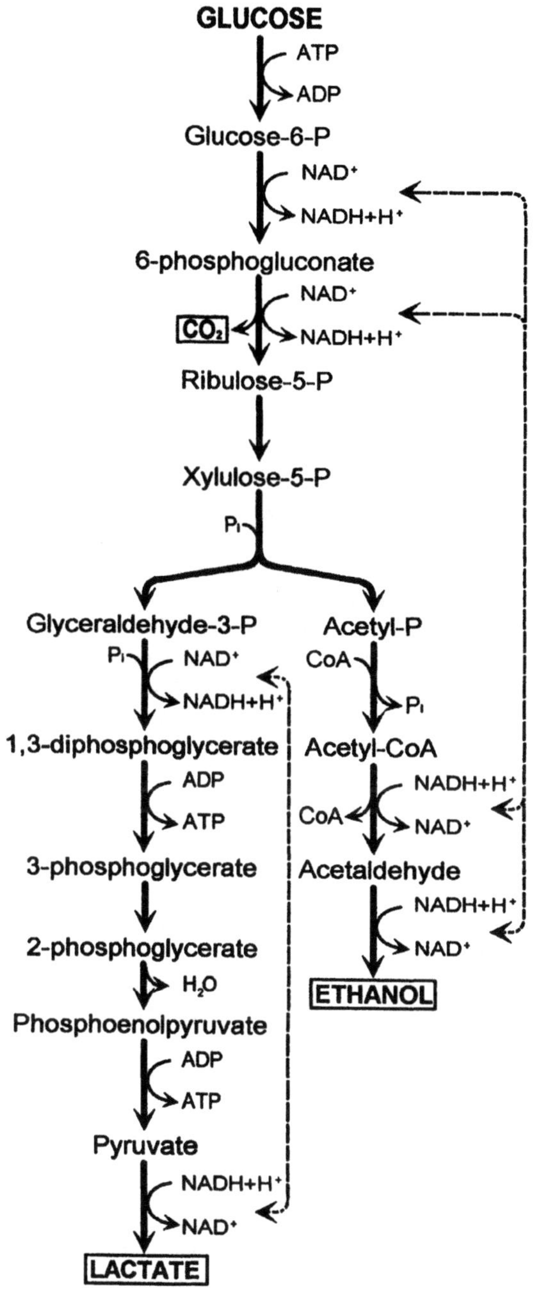

Figure 3. Heterofermentative metabolism of sugars by lactic acid bacteria.

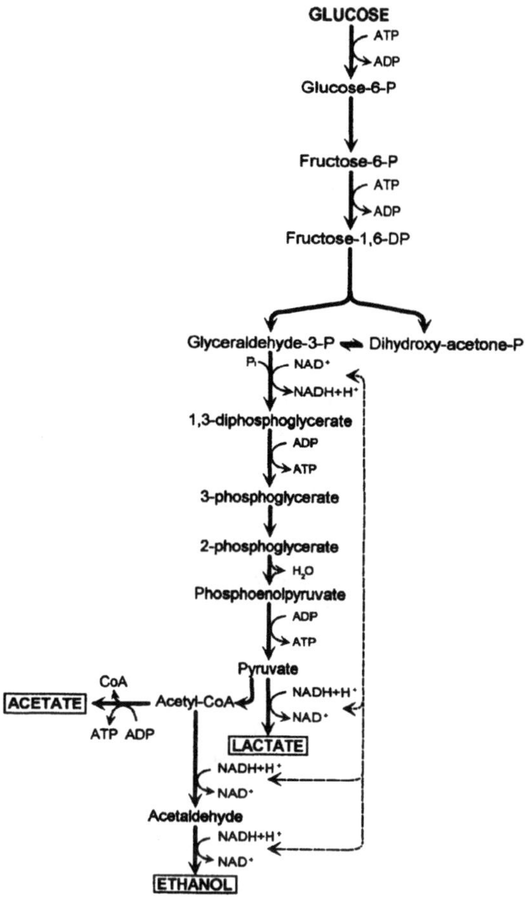

Figure 4. Mixed acid fermentation of sugars by *L. lactis*.

from the competition which will occur between LDH and PFL or PDH.

In strictly anaerobic conditions, pyruvate formate lyase activity is usually detected though the extreme sensitivity of this enzyme to oxygen (Takahashi et al., 1982; Abbe et al., 1982) requires extensive attention to analytical techniques. The enzyme transforms pyruvate to formate and acetyl-CoA and is inhibited by both GAP and DHAP (Thomas et al., 1980; Takahashi et al., 1982). This inhibition of PFL activity by triose phosphates, present at high intracellular concentrations during homolactic fermentations, is probably sufficient to explain why the majority of pyruvate is transformed to lactate under conditions in which the LDH is activated. This orientation was thought to be reinforced by a difference in affinity for pyruvate: the PFL has been shown to have a lower affinity (Km = 7 mM, Thomas et al., 1980) than LDH (Km = 2 mM), though recent work in our laboratory indicates that this value may be falsely high. Under strict anaerobic conditions, we have found that the PFL activity of a *L. lactis* strain isolated from vegetal matter has a Km for pyruvate of 1 mM. The biochemical control of PFL has been most extensively studied in *E. coli* (Knappe & Sawers, 1990) in which activity is controlled by enzymatic activase/deactivase activities which modify the redox state of the enzyme. As opposed to the radical form of the enzyme that is irreversibly deactivated by oxygen, the non radical form of PFL is insensitive to oxygen, this constituting a protection mechanism against oxygen. However, this modulation of the active form of the protein appears not to exist in the lactic bacteria (Yamada, 1987) making the enzyme irreversibly deactivated by oxygen. For the moment little is known of the factors governing the expression of this activity in *L. lactis* though subject to complex transcriptional control in *E. coli* (Knappe & Sawers, 1990).

The alternative reaction involves a reductive decarboxylation of pyruvate via the pyruvate dehydrogenase complex yielding acetyl-CoA and CO_2. The significant difference between this reaction and that of the pyruvate formate lyase is that a CO_2 rather than formate is produced and hence an additional reduced coenzyme is generated. In gram-negative bacteria, the PDH activity is subject to a variety of allosteric regulations but no evidence for this exists in gram- positive species for which the only known inhibition is related to the biochemical redox potential. The sensitivity to this factor, as represented by the NADH:NAD ratio, is linked to the inactivation of Enzyme 3 (Figure 5) of the complex (Snoep et al., 1992a) and the variation between species depends upon the relative amount of this protein within the enzyme complex. In the case of *Enterococcus faecalis*, this component is strongly expressed and hence the PDH remains active under anaerobic conditions while Enzyme 3 is only weakly expressed in *L. lactis* and is hence rapidly inactivated under anaerobic conditions (Snoep et al., 1993a).

Irrespective of the enzyme involved in acetyl-CoA formation, further metabolism of this intermediate to give either acetate or ethanol involves either the phosphotransacetylase / acetate kinase, or the aldehyde and alcohol dehydrogenases respectively. Acetate formation generates a supplementary ATP while ethanol production enables two reduced co-enzymes to be recycled. When pyruvate formate lyase is employed, reducing equivalent equilibrium can be maintained by an equimolar partition of acetyl-CoA between acetate and

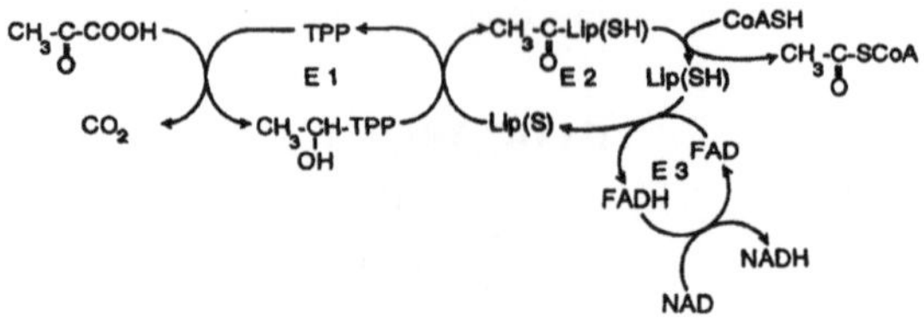

Figure 5. Biochemical structure of the multienzyme pyruvate dehydrogenase complex. E 1: pyruvate decarboxylase; E 2: dihydrolipoamide succinyltransferase; E 3: dihydrolipoamide dehydrogenase.

ethanol with a net increase in ATP gain of 50% relative to a homolactic fermentation for that part of the pyruvate deviated towards the mixed acid fermentation. No obvious gain can be expected from using the PDH since the additional NADH synthesised requires that all acetyl-CoA is converted to ethanol to maintain coenzyme equilibrium. However, since PDH activity is only observed under aerobic conditions in which oxygen is present, additional pathways of reduced coenzyme wastage may be involved (see below). The mixed acid fermentation is likely to predominate when anaerobic conditions are encountered if sugar consumption is rate limiting, i.e., when the rate of formation of new cell material is limited by the availability of ATP rather than the availability of anabolic carbon metabolites.

Influence of carbon source limitation on pyruvate metabolism

In conditions of carbon excess, i.e. batch fermentations, sugars are generally metabolised to yield a homolactic fermentation. However, under conditions in which glucose availability is limited, e.g., carbon-limited chemostat conditions, a diminished production of lactate is observed (particularly at low dilution rates in which true carbon- limitation occurs) with a corresponding increase in the quantity of fermentation products derived from the action of pyruvate formate lyase: formate plus acetate/ethanol mixtures (Thomas et al., 1979). This shift from a homolactic fermentation towards a mixed acid fermentation can be correlated with a decrease in lactate dehydrogenase specific activity, but also with a significantly lower intracellular concentration of this enzyme's principle activator, FDP (Thomas et al., 1979). Work with non-proliferating cells (Thompson, 1978; Thompson & Torchia, 1984; Lohmeier-Vogel et al., 1986) also demonstrated that the intracellular pools of both GAP and DHAP were also significantly diminished. Since these metabolites are responsible for an inhibitory deactivation of PFL in many organisms, it would appear that a concerted control of pyruvate metabolism operates under carbon limitation, so as to redirect carbon away from the energetically less favourable pathway of lactate formation. This would effectively increase ATP production under conditions in which sugar metabolism is inadequate to supply the anabolic pathways with the required biochemical energy.

Despite the importance of this shift, little work has been published as regards the expression of PFL activity in *L. lactis*. In *Streptococcus mutans*, specific activity of PFL was seen to increase by a factor of five under carbon limitation, further consolidating the cells potential to modify its homolactic fermentation (Thompson & Gentry- Weeks, 1994). Thus it would seem unlikely that GAPDH has the same controlling constraint on carbon flux through glycolysis when growth is carbon-limited. Indeed, the modified metabolite pools (low FDP, DHAP, GAP, high PEP...) are more coherent with a pathway bottleneck at the level of pyruvate and it would be logical to assume that sugar transport may become the major controlling influence, though this reflects the lack of available substrate rather than the specific activity of the transporter itself. The low FDP pool and high PEP potential would in fact lead to maximal activity of the PTS mechanism, most probably the major transporter at low substrate concentrations.

Mixed acid fermentation under carbon excess conditions

Galactose metabolism results in a fermentation end product profile in which significant amounts of C_2 compounds are produced, though lactate remains the major product (Thomas et al., 1980). As was the case for glucose-limited chemostat cultures, this shift was correlated to a diminished flux through lactate dehydrogenase and an increase in pyruvate formate lyase activity. Again, the diminished metabolite pools upstream of GAPDH (FDP, GAP, DHAP) would consolidate the observed changes in enzyme activity.

The metabolism of galactose is believed to be directly related to the pathway employed and the corresponding transport system. A homolactic fermentation has been attributed to PTS transport and the tagatose pathway while a mixed acid fermentation involves permease activity coupled to the Leloir pathway. However, this hypothesis is incomplete since *L. lactis* subsp *cremoris* retains a mixed acid fermentation in the absence of Leloir pathway, i.e. all galactose is metabolised via the tagatose pathway (Thompson, 1980; Thomas et al., 1980). Furthermore, lactose in strains harbouring the lactose plasmid is metabolised via both pathways but retains a homolactic fermentation. The phenomena really provoking this shift are almost certainly the flux through the pathways and the effects of the corresponding metabolite pool concentrations on enzyme activities rather than the use of specific pathways.

Maltose metabolism deviates significantly from the homolactic fermentation and constitutes an interesting metabolic model since the general requirements outlined above are not entirely satisfied. Maltose is transported via a permease and most probably phosphorylated by an inorganic phosphate dependent phosphorylase to yield glucose–1-P and glucose, both of which will be further metabolised via glycolysis after transformation to glucose–6-P (Qian et al., 1994; Sjoberg & Hahn-Hagerdal, 1989). While FDP pools are low (and inorganic phosphate high) explaining the diminished flux through lactate dehydrogenase, the triose-phosphate pools remain high provoking a strong inhibition of the PFL. The result of such a metabolism is that the shift towards a mixed acid fermentation is less pronounced than during carbon-limited cultures or galactose-grown cells (Lohmeier-Vogel et al., 1986; Sjoberg & Hahn-Hagerdal, 1989). Interestingly, such a situation provokes some accumulation of exopolysaccharides, due principally to the low activity of the phosphoglucomutase activity (Qian et al., 1994).

The metabolism of pentose sugars presents a somewhat different situation and is often considered to involve a truely heterolactic fermentation, i.e. phosphoketolase activity leading to C3- and C2-unit formation directly from the pentose-phosphate (Kandler, 1983). The capacity to metabolise pentose sugars appears to be strain dependent and the extent to which mixed acid fermentation occurs perhaps reflects the different capacities of the strains used (Kandler, 1983; Ishizaki et al., 1992; Westby et al., 1993). In our laboratory, growth was found to be possible for *L. lactis* on ribose, xylose and gluconic acid, all of which lead to pentose-phosphate formation though ribose and xylose are taken up by permeases while gluconate is transported via a PTS mechanism (Thompson & Gentry-Weeks, 1994). The metabolism via phosphoclastic pathway has not been examined at the enzyme level but end-product profiles were not those to be expected from a classical heterolactic fermentation. Moreover, the close agreement between formate concentration and the amount of acetate and ethanol produced indicate

Table 1. Batch growth of Lactococcus lactis NCDO 2118 on defined media containing all essential amino acids. Influence of sugar uptake rate on growth and end-product profiles as expressed as the % of pyruvate recovered as lactate or as products of pyruvate-formate lyase activity (formate, acetate and ethanol)

Substrate	Specific rates		% Pyruvate distribution	
	Sugar uptake (mmol C_6/g.h)	Growth (h–1)	LDH	PFL
Glucose	15.0	0.55	93	7
Galactose	9.3	0.20	67	33
Lactose	5.1	0.17	3	97

that a mixed acid fermentation occured and hence pentose phosphates must presumably be directed back into glycolysis by transketolase/transaldolase reactions of the pentose pathway.

It is often remarked that the contribution of the mixed acid fermentation only exceeds 10% of carbon flux under conditions in which the growth rate is considerably affected. This, at first sight, suggests that this shift might be growth-rate related. In our hands, the correlation is not so straight forward since a number of substrates showing roughly the same growth rate show a significant variation in the extent to which the mixed acid fermentation operates (Figure 6). Preliminary results suggest that the causative phenomenon is the rate at which sugars can be taken up by the cells. Thus, under growth conditions in which sugar metabolism via the homolactic pathway is inadequate to yield the ATP necessary, the mixed acid fermentation is progressively activated. The mechanisms governing gene expression remain to be determined, but it should be remembered that the gene for lactate dehydrogenase is carried on the same operon as the genes coding for key glycolytic enzymes, phosphofructokinase and pyruvate kinase (Llanos et al., 1992, 1993).

The hypothesis that the flux through the central pathways determines the cells capacity to metabolise pyruvate provides a unified theory that explains equally well the behaviour of *L. lactis* on certain sugars and chemostat behaviour. In both cases, sugar uptake plays a key role though one can be viewed as a biological limitation, while in the chemostat the fermentation protocol determines the behaviour, i.e., in batch culture the sugar transport activity may impose a limitation on the resulting carbon flux, while in the chemostat the substrate availability (residual concentration) is limiting. To examine further this concept, we have investigated the manner in which the metabolism of lactose is perturbed in *L. lactis* strains lacking the lactose-PTS and associated tagatose pathways enzymes. Certain strains of *L. lactis* isolated from vegetable matter do not possess the plasmid encoding such proteins (Crow et al., 1983; McKay et al, 1972). For a long time these strains were believed to be unable to grow at the expense of lactose. However, the use of minimal medium (Cocaign-Bousquet et al., 1995) thus overcoming certain problems associated with the use of complex media, has enabled us to demonstrate clearly that such strains are able to grow on lactose, albeit at considerably diminished growth rates. If the biochemical model proposed above is correct, it would be expected that such strains, in which lactose permease is the only manner to take up lactose, would have slow rates of sugar consumption and show a mixed acid fermentation. Indeed, such was the case with virtually no lactate at all being produced. Interestingly, the growth rate of the strain on lactose is similar to that on galactose (Table 1) but much lower than that obtained on glucose, for which a homolactic fermentation was seen. Galactose metabolism yields mixtures of lactate and formate/acetate/ethanol. Examination of the sugar uptake rate shows that lactose is taken up at considerably slower rates than galactose and one might postulate that the extent to which the metabolism is limited by uptake capacity is greater in the case of lactose. The additional energy derived from the increased production of acetate is sufficient however to sustain similar growth rates (Table 1) and the estimated Y_{ATP} remains constant. How then does the cell control this shift? Intracellular metabolite measurements have shown a good correlation between the concentrations of FDP and triose-P and pyruvate flux through LDH and PFL, phenomenon accentuated by the modified expression of each enzyme. Similar shifts have been seen in strains having undergone mutation selection to inactivate P-β-galactosidase activity and/or the lactose PTS transport (Demko et al, 1972; Crow & Thomas, 1984). Mixed acid fermentation was seen in these strains, as was the case for *S. mutans* lacking a PTS^{gluc} activity (Thompson & Gentry-Weeks, 1994), however no details concerning the effect on sugar consumption or the degree of the diversification of the fermentation are available. Leblanc et al. (1979) observed that the loss of the plasmid encoding the PTS^{lac}, P-β-Gal and the tagatose pathway in dairy lactococcal strains grown on galactose diminished lactate production from 90% of products recovered to 75% with a corresponding increase in acetate/ethanol production. No data was given for lactose. Of course, from a practical point of

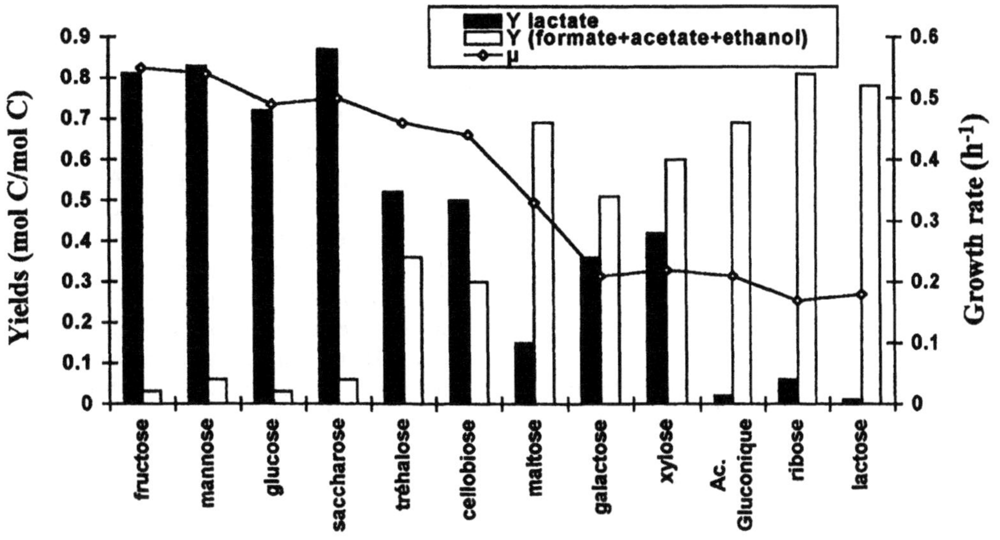

Figure 6. Effect of substrate on growth rate and fermentation profile during batch cultivation of *Lactococcus lactis* NCDO 2118 in a defined medium.

view, the strains thus produced would remain minority within a population due to the greatly diminished growth potential and be unlikely to result in significant spoilage of product. One might however consider whether such strains have any potential role to play in metabolic engineering strategies since they possess phenotypes virtually identical to those of ldh^- strains which now form the basis of many genetic engineering strategies for volatile compound overproduction (Hugenholtz et al., 1994). Such strategies aim at directing pyruvate through the acetolactate synthase (ALS) reaction towards diacetyl (Platteeuw et al., 1995; Benson et al., 1996) by deletion or attenuation of the usual pyruvate consuming activities but grow slowly. Growth of the vegetable strains on lactose may facilitate just such a modified metabolism, with the additional advantage that such strains express all the enzymes necessary for the synthesis of branched chain amino acids and hence possess the high affinity ALS (Godon et al., 1992) as well as the low affinity enzyme habitually associated with acetolactate synthesis in *L. lactis* (Marrug et al., 1994). It is too early to assess the feasibility of such a nutritional approach to flavour compound production but the use of 'natural' strains may avoid the public's unfavourable reception to genetic engineering for food compounds.

If the concept is accepted that sugar (and hence energy) limitation leads to loss of homolactic fermentation, then the unspoken but implicit correlative is that under such conditions, the organic matter necessary for cell synthesis is in excess. This was examined using progressively simplified media in our laboratory with the vegetal strain. It was seen that removing amino acids from the media progressively diminished growth rate and hence, might be expected to shift the extent to which energy availability limited cell proliferation. Indeed, the observed end- product profile was seen to favour lactate production as the media was simplified though the homolactic behaviour was never completely restored. This adds further evidence to the model but suggests that a delicate equilibrium between carbon and energy flux towards biomass is involved rather than a simple relationship between sugar catabolism and pyruvate fermentation. Of course, the logical extrapolation of such a hypothesis is that whenever a homolactic fermentation is observed, growth is being limited either by the availability of other organic components of the growth medium, or the cells capacity to assimilate such compounds.

The physicochemical environment

Effect of aeration

In the mixed acid fermentation, only the competition between LDH and PFL have been discussed. Since PFL is inactivated by oxygen one might ask how *L.*

$$NADH + H^+ + O_2 \xrightarrow{\text{NADH oxidase}} NAD + H_2O_2$$

$$2\,NADH + 2H^+ + O_2 \xrightarrow{\text{NADH oxidase}} 2NAD + 2H_2O$$

$$NADH + H^+ + H_2O_2 \xrightarrow{\text{NADH peroxidase}} NAD + 2H_2O$$

$$4\,O^{\cdot} + 2H^+ \xrightarrow{\text{superoxide dismutase}} O_2 + H_2O_2$$

$$\text{pyruvate} + Pi + O_2 \xrightarrow{\text{pyruvate oxidase}} \text{Acetyl-P} + CO_2 + H_2O_2$$

Figure 7. Enzymes involved in oxygen metabolism in lactic acid bacteria.

lactis responds to the presence of air. In conditions of excess substrate, metabolism of sugars leading to either a homolactic fermentation (glucose, lactose) or mixed acid fermentation (galactose), is sometimes deviated towards an increased production of acetate at the expense of lactate and/or ethanol (Smart & Thomas, 1987). Certain strains are reported to accumulate small amounts of pyruvate. This shift towards acetate production is related to the induction of various enzymes able to oxidise NADH (NADH oxidase, NADH peroxidase) in aerated cultures (Smart & Thomas, 1987; Hanson and Haggstrom, 1984; Cogan et al., 1989). These enzymes are able to recycle the reduced coenzymes directly without recourse to carbon compound oxido-reduction reactions (Figure 7). The requirement to produce reduced end-products is diminished and acetate production can be favoured, thus yielding more ATP (Condon, 1987). This metabolic response should not be confused with respiration since no electron transport phosphorylation is involved though an effective but indirect energetic gain is achieved due to the increased production of acetate. Competition between lactate and/or aldehyde and alcohol dehydrogenases and the O_2 consuming NADH oxidising enzymes is responsible for any shift towards acetate production though this may also involve a modified level of *ldh* gene expression since lactate dehydrogenase has been shown to diminish to approximately half the anaerobic activity in some strains (Smart & Thomas, 1987). The consequence of this is an increased availability of pyruvate. The PFL activity is inactivated by O_2 and hence an alternative pathway of pyruvate disposal is necessary. One might be tempted to ask whether the oxygen consumption is to provoke an increased ATP yield or to remove the oxygen from the medium thus enabling an active PFL to be synthesised. However, in the absence of an active PFL, other enzymes are able to metabolise pyruvate, notably the pyruvate dehydrogenase complex, but also in certain LAB a pyruvate oxidase (Sedewitz et al., 1984a, b; Zitzelberger *et al.*, 1984). However, no evidence has yet been presented to suggest that this enzyme exists in *L. lactis*. The action of NADH oxidase has been demonstrated in many micro-organisms and this enzyme is widespread but not ubiquitous within LAB. Two forms of this enzyme exist, one of which leads to H_2O production while the other yields H_2O_2. The first type is difficult to distinguish from the H_2O_2-producer since frequently an NADH peroxidase exists, though separate enzyme activities have been demonstrated in *S. mutans* (Higuchi et al., 1993). Superoxide dismutase activity is widespread in LAB and also induced in aerated cultures. This enzyme, like Mn^{2+} cations, effectively removes the toxic radical oxygen (Archibald & Fridovich, 1981a, b).

As regards the oxygen effect, the enormous variety of media used to cultivate *L. lactis* most probably account for the somewhat variable results obtained. As yet, it cannot really be decided whether such variation is due to the strain variability as regards either the oxidising enzymes themselves or their capacity to take up and assimilate amino acids, peptides, bases, etc., or to the media composition used. While some authors claim an increase in acetate production, other have observed no significant influence of aeration (Smart & Thomas, 1987, Starrenburg & Hugenholtz, 1991). Once again it is necessary to take into account the extent to which each culture is energy-limited or carbon-limited and to recall that unlike many bacteria, these two fluxes essential for cell growth are quite separate and independently controlled in *L. lactis*.

Aeration might be expected to improve cell growth in certain media but it must be remembered that both NADH oxidase and superoxide dismutase lead to H_2O_2 production which may lead to autoinhibition in certain strains (Grufferty & Condon, 1983). The homofermentative LAB in general, and *L. lactis* in particular, do not possess a catalase activity (Zitzelberger et al., 1984) and only NADH peroxidase can avoid H_2O_2 accumulation. It is therefore essential to maintain a correct balance and though details are not available, one might expect that the enzyme possessing the highest affinity for NADH would be the peroxidase, thereby avoiding an excessive accumulation of H_2O_2 in those strains possessing both NADH oxidase and NADH peroxidase activity.

Effect of pH

Little is known of the effect of pH on the metabolism of *L. lactis* but in *Enterococcus faecalis* the partition between pyruvate flux passing via the pyruvate dehydrogenase and the PFL is strongly influenced by the broth pH. It must be remembered that many acid- producing bacteria do not maintain pH homeostasis but rather a constant pH gradient across the cell membrane. This is certainly the case for *L. lactis* (Cook & Russell, 1994) in which intracellular ATP was also seen to decrease as pH diminished. However, this effect appears to be dependent upon the manner in which the experiments are undertaken since intracellular pH can be maintained in washed cells when the external medium is acidified with mineral acids (Poolman et al, 1987b). The PDH is dominant at pH values of 5.5–6.5 while the PFL becomes dominant at pH > 7. This shift was attributed to an effect of redox potential rather than pH itself and involved changes in the specific activity of each enzyme, i.e., the concentration of enzyme (Snoep et al., 1990, 1991). Such effects were no doubt further enhanced by the effect a modified redox potential would have on the various dehydrogenase enzymes, all of whose activity will depend upon the ratio of oxidised to reduced coenzymes. Furthermore, the nutritional requirements were also shown to influence this distribution since addition of lipoic acid under acidic conditions further accentuated the flux through the PDH at the expense of PFL (Snoep et al., 1993b). As mentioned above, the PDH has a somewhat different structure in *L. lactis* and hence coexistence of significant activity of these two enzymes will be restricted to a limited range of nutritional environments as regards both oxygenation and acidity.

Minor fermentation products

If the major pathways described above account for the vast majority of the carbon flux from sugars, much interest is being stimulated by the minor by-products which take on considerable importance as flavour compounds in natural food products (Figure 8). When lactose is consumed in complex media (Kaneko et al., 1990, 1991) or in lipoate-limited synthetic media (Cogan et al., 1989), some strains produce, in the presence of oxygen, alternative products such as diacetyl or acetoin. These compounds are synthesised from pyruvate and are only produced in significant quantities under conditions in which the normal metabolism of pyruvate is perturbed. Such by-products are synthesised via a specific pathway involving acetolactate synthase (ALS), enzyme containing a similar first subunit to the PDH complex, which is responsible for pyruvate decarboxylation into hydroxy-ethyl-TPP. The product of the ALS reaction involving two moles of pyruvate is acetolactate; this compound can accumulate in the medium in certain conditions (Jordan & Cogan, 1988) but is unstable and subject to spontaneous transformation to diacetyl via a purely chemical reaction in the presence of O_2. The acetolactate synthase normally implicated in diacetyl production is an isoenzyme possessing a low affinity for pyruvate (Km = 40–50 mM) and hence unlikely to be operative under normal growth conditions in which the pyruvate pool is low (Snoep et al., 1992b). It is believed that this enzyme may play a role in detoxifying pyruvate and the gene coding for this enzyme has been cloned and sequenced (Marugg et al., 1994).

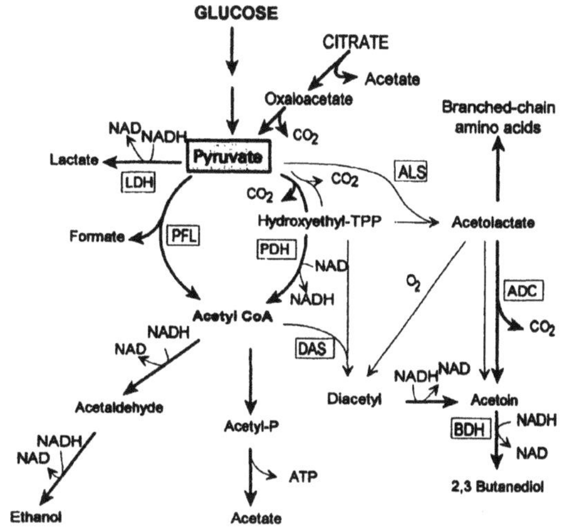

Figure 8. Pyruvate metabolism in *Lactococcus lactis*. LDH: lactate dehydrogenase; PDH: pyruvate dehydrogenase; PFL: pyruvate formate lyase; ALS: acetolactate synthase; ADC: acetolactate decarboxylase; BDH: butanediol dehydrogenase; DAS: diacetyl synthase.

If this enzyme is normally associated with diacetyl production, *L. lactis* also possess a second acetolactate synthase activity coded by the *ilv BN* genes (Godon, 1992) which is involved in the initial reaction of the pathway of branched-chain amino acid synthesis and repressed by the presence of isoleucine. The possible role of this enzyme in diacetyl production has never been investigated in detail since most strains of *L. lactis* require branched-chain amino acids for growth. The strain *L. lactis* NCDO 2118 has no such auxotrophy (Loubiere et al., 1996) and expresses this enzyme under certain growth conditions, but lacks many of the other characteristics considered essential for obtaining detectable diacetyl accumulation i.e. citrate consumption capacity. Though *L. lactis* NCDO 2118 is apparently prototroph for branched-chain amino acids by the single omission technique, minimal medium (containing only five amino acids) requires either a suitable supplement of these amino acids (Cocaign-Bousquet et al., 1995) or threonine (Loubiere et al., 1996), whose deamination by the *ilv A* gene product yields ketobutyrate. This compound is a substrate of acetolactate synthase, but has also been implicated in the regulation of *ilv B* expression, inducing higher levels of expression in *Corynebacterium glutamicum* (Eggeling et al., 1987) and *Salmonella tiphymurium* (Shaw & Berg, 1980) when added to the growth medium. This type of regulation may provide a useful strategy for diacetyl production in strategies employing the acetolactate synthase of the amino acids pathway (Benson et al., 1996), thus avoiding the necessity to provoke a tenfold increase in the pyruvate concentration.

Acetolactate may also be biologically transformed to less financially interesting compounds than diacetyl. As mentioned above, acetolactate is an intermediary compound in branched-chain amino acid synthetic pathways, but may also be decarboxylated to acetoin by the acetolactate decarboxylase coded by *ald B*. This enzyme is activated by valine, leucine and isoleucine (Monnet et al., 1994; Phalip et al., 1994) in *L. lactis diacetylactis*. Furthermore, the expression of the gene is activated by the presence of leucine, thus providing the basis of an interesting metabolic phenomena in which the pyruvate flux into amino acid synthetic pathway can be oriented towards carbon metabolites when the amino acids are present (Godon, 1992). The result is acetoin production, which in turn may be reduced to butanediol by a dehydrogenase activity (Crow, 1990). In *L. lactis* two butanediol dehydrogenase activities exist leading to two isomers of butanediol. This same enzyme is able to reduce acetoin to butanediol but also diacetyl into acetoin and is most probably the enzyme which previously known as acetoin reductase or diacetyl reductase. A further enzyme activity, the diacetyl synthase, has been proposed to explain the production of diacetyl from pyruvate and acetyl CoA (Kaneko et al., 1989, 1990, 1991) but this pathway is no longer considered to play any significant role in diacetyl production in *L. lactis*. This conclusion is supported by C^{13}-NMR analysis (Verhue & Tjan, 1991; Ramos et al., 1994).

In addition to this group of products derived from pyruvate, research in our laboratory has recently shown that *L. lactis* produces trace amounts of succinate, presumably via a reductive TCA cycle and though few details of TCA cycle activities are available as yet the anaplerotic reaction PEP carboxylase has been demonstrated. Indeed when precise carbon balances are established (feasible on the defined media used) it is clear that a variety of non-identified minor products accumulate in the media whose physiological role and possible commercial value remain to be determined.

Effects of auxiliary substrates on pyruvate metabolism

Citrate metabolism

Citrate is present in milk at concentration of 8–9 mM and is cometabolised together with the sugars by many strains of LAB including certain *L. lactis*. The strain specific nature of citrate metabolism is due to the presence in certain strains of a plasmid encoding for citrate permease (Sesma et al., 1990; Smith et al., 1992). This metabolism of citrate is important since it generates an additional source of pyruvate without the production of reduced co- enzymes. This occurs since citrate is metabolised via citrate lyase yielding acetate and oxaloacetate, whose further metabolism via oxaloacetate decarboxylase leads to the production of pyruvate (Figure 8). The result of such a metabolism is that under conditions of aerobiosis and in acidic environments, the fermentation of a reconstituted milk medium leads to a significant increase in flavour compound production in the presence of citrate (Drinan et al., 1976; Hugenholtz & Starrenburg, 1992). Indeed, it is now known that citrate can be metabolised in the absence of a sugar, producing acetate, formate and flavour compounds, though such a metabolism occurs only in a narrow range of pH values (Starrenburg & Hugenholtz, 1991).

The increase in diacetyl production during citrate metabolism can be correlated to an increase in acetolactate synthase activity and a partial repression of butanediol dehydrogenase (Cogan, 1981). While this latter activity is derepressed immediately after complete depletion of citrate and is hence directly related to the presence of citrate, the effect on acetolactate synthase activity is less clear. While it is possible that citrate has a direct effect on enzyme expression the increased pyruvate pool resulting from citrate metabolism may also be involved. Certainly such an increase will favour the flux through acetolactate synthase.

Amino acid metabolism

For many years, amino acids were considered uniquely as a source of assimilable carbon for biomass formation but recent work in our laboratory has shown that certain essential amino acids may also contribute to the fermentation pattern. This was most obvious for serine in minimal medium for which ^{14}C-serine was predominantly recovered as lactate (Loubiere et al., 1996) and as such may play a role as an auxiliary substrate somewhat akin to citrate. Furthermore, the depletion of serine in such media led to a shift in the acetate/ethanol ratio with ethanol appearing as a product only after total removal of serine. Other amino acids are known to be catabolised but this generally is restricted to decarboxylation reactions, e.g. arginine conversion to ornithine with a corresponding gain in ATP (Poolman, 1993).

Conclusion

Pyruvate occupies a key position within the metabolic network of *L. lactis* and as such is subject to considerable regulation. Current understanding of this metabolic node is adequate to envisage some degree of exploitation within a biotechnological context, but at a scientific level we are only able to see the consequences of the regulation mechanisms involved. This absence of mechanistic understanding is most obvious at the level of gene expression for which little is known. Biochemical knowledge is more advanced, though certain areas remain rather speculative, and a more structured approach to metabolic control within a modelisation framework may promote a more rigorous interpretation of the current data. Since pyruvate metabolism represents the convergence of catabolic and fermentation pathways, while also being generated directly via the various PTS sugar transport mechanisms, it would seem likely that we will need to incorporate extensive energy-signalling mechanisms into such models before the details become apparent. This information is essential for the next generation of lactic acid bacteria, constructed pragmatically via genetic engineering strategies, to satisfy market demand. The fact that such strains do not yet meet with public approval should not block the basis research necessary to construct customised cell factories.

Acknowledgement

The authors would like to thank Mr Henri FEAU for preparing the artwork included in this paper.

References

Abbe K, Takahashi S & Yamada T (1982) Involvement of oxygen-sensitive pyruvate formate-lyase in mixed acid fermentation by *Streptococcus mutans* under strictly anaerobic conditions. J. Bacteriol. 152: 175–182

Archibald FS & Fridovich I (1981a) Manganese and defenses against oxygen toxicity in *Lactobacillus plantarum*. J. Bacteriol. 145: 442–451

— (1981b) Manganese, superoxide dismutase, and oxygen tolerance in some lactic acid bacteria. J. Bacteriol. 146: 928–936

Benson KH, Godon JJ, Renault P, Griffin HG & Gasson MJ (1996) Effect of *ilvBN*-encoded a-acetolactate synthase expression on diacetyl production in *Lactococcus lactis*. Appl. Microbiology Biotechnol. 45: 107–111

Benthin S, Nielsen J & Villadsen J (1993) Two uptake systems for fructose in *Lactococcus lactis* subsp *cremoris* FD1 produce glycolytic and gluconeogenic fructose phosphates and induce oscillations in growth and lactic acid formation. Appl. Env. Microbiol. 59: 3206–3211

Bisset DL & Anderson RL (1974) Lactose and D-galactose metabolism in group N Streptococci: presence of enzymes for both the D-galactose 1-phosphate and D-tagatose 6-phosphate pathways hydrolysing enzymes in *Streptococcus lactis* and *Streptococcus cremoris* and also in some species of Streptococci. J. Bacteriol. 117: 318–320

Cancilla MR, Davidson BE, Hillier AJ, Nguyen NY & Thompson J (1995) The *Lactococcus lactis* triosephosphate isomerase gene, *tpi*, is monocistronic. Microbiology 141: 229–238

Cocaign-Bousquet M, Garrigues C, Novak L, Lindley ND & Loubiere P (1995) Rational development of a simple synthetic medium for the sustained growth of *Lactococcus lactis*. J. Appl. Bacteriol. 79: 108–116

Cogan TM (1981) Constitutive nature of the enzymes of citrate metabolism in *Streptococcus lactis* subsp *diacetylactis*. J. Dairy Res. 48: 489–495

Cogan JF, Walsh D & Condon S (1989) Impact of aeration on the metabolic end-products formed from glucose and galactose by *Streptococcus lactis*. J. Appl. Bacteriol. 66: 77–84

Collins LB & Thomas TD (1974) Pyruvate kinase of *Streptococcus lactis*. J. Bacteriol. 120: 52–58

Condon S (1987) Responses of lactic acid bacteria to oxygen. FEMS Microbiol. Rev. 46: 269–280

Cook GM & Russell JB (1994) The effect of extracellular pH and lactic acid on pH homeostasis in *Lactococcus lactis* and *Streptococcus bovis*. Curr. Microbiol. 28: 165–168

Crow VL (1990) Properties of 2,3-butanediol dehydrogenases from *Lactococcus lactis* subsp *lactis* in relation to citrate fermentation. Appl. Environ. Microbiol. 56: 1656–1665

Crow VL & Wittenberger CL (1979) Separation and properties of NAD^+ and $NADP^+$-dependent glyceraldehyde-3-phosphate dehydrogenases from *Streptococcus mutans*. J. Biol. Chem. 254: 1134–1142

Crow VL & Thomas TD (1982) D-tagatose 1,6-diphosphate aldolase from lactic Streptococci: purification, properties, and use in measuring intracellular tagatose 1,6-diphosphate. J. Bacteriol. 151: 600–608

— (1984) Properties of a *Streptococcus lactis* strain that ferments lactose slowly. J. Bacteriol. 157: 28–34

Crow VL, Davey GP, Pearce LE & Thomas TD (1983) Plasmid linkage of the D-tagatose–6 phosphate pathway in *Streptococcus lactis*: Effect on lactose and galactose metabolism. J. Bacteriol. 153: 76–83

Demko GM, Blanton SJB & Benoit RE (1972) Heterofermentative carbohydrate metabolism of lactose-impaired mutants of *Streptococcus lactis*. J. Bacteriol. 112: 1335–1345

Drinan DF, Tobin S & Cogan TM (1976) Citric acid metabolism in hetero- and homofermentative lactic acid bacteria. Appl. Environ. Microbiol. 31: 481–486

Eggeling I, Cordes C, Eggeling L & Sahm H (1987) Regulation of acetohydroxy-acid synthase in *Corynebacterium glutamicum* during fermentation of a-ketobutyrate to L-isoleucine. Appl. Microbiol. Biotechnol. 25: 346–351

Farrow JAE (1980) Lactose hydrolysing enzymes in *Streptococcus lactis* and *Streptococcus cremoris* and also in some other species of Streptococci. J. Appl. Bacteriol. 49: 493–503

Godon JJ (1992) Régulation génétique de la synthèse des acides aminés branchés chez *Lactococcus lactis*. Thèse, Université Paris XI

Godon JJ, Chopin MC & Ehrlich SD (1992) Branched-chain amino acid biosynthesis genes in Lactococcus lactis subsp lactis. J. Bacteriol. 174: 6580–6589

Grufferty RC & Condon S (1983) Effect of fermentation sugar on hydrogen peroxide accumulation by *Streptococcus lactis* C10. J. Dairy Res. 50: 481–489

Hanson L & Haggstrom HM (1984) Effects of growth conditions on the activities of superoxide dismutase and NADH-oxidase / NADH-peroxidase in *Streptococcus lactis*. Curr. Microbiol. 10: 345–352

Higuchi M, Shimada M, Yamamoto Y, Hayashi T, Koga T & Kamio Y (1993) Identification of two distinct NADH oxidases corresponding to H_2O_2-forming oxidase induced in *Streptococcus mutans*. J Gen. Microbiol. 139: 2343–2351

Hugenholtz J & Starrenburg MJC (1992) Diacetyl production by different strains of *Lactococcus lactis* subsp *lactis* var *diacetylactis* and *Leuconostoc* spp. Appl. Microbiol. Biotechnol. 38: 17–22

Hugenholtz J, Starrenburg MJC & Weerkamp AH (1994) Diacetyl production by Lactococcus lactis: optimilisation and metabolic engineering. In: ECB6 Proceedings of the 6th European Congress on Biotechnology (Eds.), L Alberghina, L Frontali & P Sensi. Elsevier Science B.V., Amsterdam, Netherlands

Ishizaki A, Ueda T, Tanaka K & Stanbury P F (1992) L-Lactate production from xylose employing *Lactococcus lactis* IO–1. Biotech. Lett. 14: 599–604

Jordan KN & Cogan TM (1988) Production of acetolactate by *Streptococcus diacetylactis* and *Leuconostoc* ssp. J. Dairy Res. 55: 227–238

Kandler O (1983) Carbohydrate metabolism in lactic acid bacteria. Antonie van Leeuwenhoek 49: 209–224

Kaneko T, Watanabe Y & Suzuki H (1989) Enhancement of diacetyl production by a diacetyl-resistant mutant of citrate-positive *Lactococcus lactis* subsp. *lactis* 3022 and by aerobic conditions of growth. J. Dairy Sci. 73: 291–298

Kaneko T, Takahashi M & Suzuki H (1990) Acetoin fermentation by citrate-positive *Lactococcus lactis* subsp. *lactis* 3022 grown aerobically in the presence of hemin or Cu^{2+}. Appl. Environ. Microbiol. 56: 2644–2649

Kaneko T, Watanabe Y & Suzuki H (1991) Differences between *Lactobacillus casei* 2206 and citrate-positive *Lactococcus lactis* subsp. *lactis* 3022 in the characteristics of diacetyl production. Appl. Env. Microbiol. 57: 3040–3042

Kell DB & Westerhoff HV (1986) Metabolic control theory: its role in microbiology and biotechnology. FEMS Microbiol. Rev. 39: 305–320

Knappe J & Sawers G (1990) A radical-chemical route to acetyl-CoA: the anaerobically induced pyruvate formate-lyase system of *Escherichia coli*. FEMS Microbiol. Rev. 75: 383–398

Konings WN, Poolman B & Driessen AJM (1989) Bioenergetics and solute transport in Lactococci. CRC Crit. Rev. Microbiol. 16: 419–475

LeBlanc DJ, Crow VL, Lee LN & Garon CF (1979) Influence of the lactose plasmid on the metabolism of galactose by *Streptococcus lactis*. J. Bacteriol. 137: 878–884

Llanos RM, Hillier A & Davidson BE (1992) Cloning, nucleotide sequence, expression and chromosomal location of *ldh*, the gene encoding L-Lactate dehydrogenase, from *Lactococcus lactis*. J. Bacteriol. 174: 6956–6964

Llanos RM, Harris CJ, Hillier AJ & Davidson BE (1993) Identification of a novel operon in *Lactococcus lactis* encoding three enzymes for lactic acid synthesis: phosphofructokinase, pyruvate kinase, and lactate dehydrogenase. J. Bacteriol. 175: 2541–2551

Lohmeier-Vogel EM, Hahn-Hagerdahl B, & Vogel HJ (1986) Phosphorus–31 NMR studies of maltose and glucose metabolism in *Streptococcus lactis*. Appl. Microbiol. Biotechnol. 25: 43–51

Loubiere P, Novak L, Cocaign-Bousquet M & Lindley ND (1996) Besoins nutritionnels des bacteries lactiques: interactions entre flux de carbone et d'azote. Lait 76, (in press)

Marugg JD, Goelling D, Stahl U, Ledeboer AM, Toonen MY, Verhue WM & Verrips CT (1994) Identification and caracterization of the a-acetolactate synthase gene from *Lactococcus lactis* subsp *lactis* biovar diacetylactis. Appl. Env. Microbiol. 60: 1390–1394

McKay LL, Baldwin KA & Zottola EA (1972) Loss of lactose metabolism in lactic Streptococci. Appl. Microbiol. 23: 1090–1096

Monnet C, Phalip V, Schmitt P & Divies C (1994) Comparison of a-acetolactate decarboxylase in *Lactococcus* spp and *Leuconostoc* spp. Biotech. Lett. 16: 257–262

Phalip V, Monnet C, Schmitt P, Renault P, Godon JJ & Divies C (1994) Purification and properties of the a-acetolactate decarboxylase from *Lactococcus lactis* subsp *lactis* NCDO 2118. FEBS Lett. 351: 95–99

Platteeuw C, Hugenholtz J, Starrenburg M, van Alen-Boerrigter I & de Vos WM (1995) Metabolic engineering of *Lactococcus lactis*: influence of the overproduction of a-acetolactate synthase in strains deficient in lactate dehydrogenase as a function of culture conditions. Appl. Environ. Microbiol. 61: 3967–3971

Poolman B (1993) Energy transduction in lactic acid bacteria. FEMS Microbiol. Rev. 12: 125–148

Poolman B, Bosman B, Kiers J & Konings WN (1987a) Control of glycolysis by glyceraldehyde-3-phosphate dehydrogenase in *Streptococcus cremoris* and *Streptococcus lactis*. J. Bacteriol. 169: 5887–5890

Poolman B, Driessen AJM & Konings WN (1987b) Regulation of solute transport in streptococci by external and internal pH values. Microbiol. Rev. 51: 498–508

Qian N, Stanley GA, Hahn-Hagerdal B & Radstrom P (1994) Purification and characterisation of two phosphoglucomutases from *Lactococcus lactis* subsp *lactis* and their regulation in maltose and glucose utilizing cells. J. Bacteriol. 176: 5304–5311

Ramos A, Jordan KN, Cogan TM & Santos H (1994) ^{13}C nuclear magnetic resonance studies of citrate and glucose cometabolism by *Lactococcus lactis*. Appl. Environ. Microbiol. 60: 1739–1748

Romano AH, Brino G, Peterkofsky A & Reizer J (1987) Regulation of b-galactoside transport and accumulation in heterofermentative lactic acid bacteria. J. Bacteriol. 169: 5589–5596

Saier MH, Chaivaix S, Cook GM, Deutscher J, Reizer J & Ye GJJ (1996) Catabolite repression and inducer control in Gram-positive bacteria. Microbiology 142: 217–230

Sedewotz B, Schleifer KH & Gotz F (1984a) Purification and biochemical characterization of pyruvate oxidase from *Lactobacillus plantarum*. J. Bacteriol. 160: 273–278

Sedewitz B, Schleifer KH & Gotz F (1984b) Physiological role of pyruvate oxidase in aerobic metabolism of *Lactobacillus plantarum*. J. Bacteriol. 160: 462–465

Sesma F, Gardiol D, De Ruiz Holgado AP & De Mendoza D (1990) Cloning of the citrate permease gene of *Lactococcus lactis* subsp. *lactis* biovar *diacetylactis* and expression in *Escherichia coli*. Appl. Environ. Microbiol. 56: 2099–2103

Shaw K & Berg CM (1980) Substrate channeling: a-ketobutyrate inhibition of acetohydroxy acid synthase in *Salmonella typhimurium*. J. Bacteriol. 143: 1509–1512

Sjoberg A. & Hahn-Hagerdal B (1989) b-glucose-1-phosphate, a possible mediator for polysaccharide formation in maltose-assimilating *Lactococcus lactis*. Appl. Env. Microbiol. 55: 1549–1554

Smart JB & Thomas TD (1987) Effect of oxygen on lactose metabolism in *Lactic Streptococci*. Appl. Env. Microbiol. 53: 533–541

Smith MR, Hugenholtz J, Mikoczi P, Ree E, Bunch AW & Bont JAM (1992) The stability of the lactose and citrate plasmids in *Lactococcus lactis* biovar. *diacetylactis*. FEMS Microbiol. Lett. 96: 7–12

Snoep JL, Teixeira de Mattos MJ, Postma PW & Neijssel OM (1990) Involvement of pyruvate dehydrogenase in product formation in pyruvate-limited anaerobic chemostat cultures of *Enterococcus faecalis* NCTC 775. Arch. Microbiol. 154: 50–55

— (1991) Effect of the energy source on the NADH / NAD ratio and on pyruvate catabolism in anaerobic chemostat cultures of *Enterococcus faecalis* NCTC 775. FEMS Microbiol. Lett. 81: 63–66

Snoep JL, Teixeira de Mattos MJ, Starrenburg MJC & Hugenholtz J (1992a) Isolation, characterization, and physiological role of the pyruvate dehydrogenase complex and a-acetolactate synthase of *Lactococcus lactis* subsp. *lactis* bv *diacetylactis*. J. Bacteriol. 174: 4838–4841

Snoep JL, De Graef MR, Westphal AH, De Kok A, Teixeira de Mattos MJ & Neijssel OM (1992b) Pyruvate catabolism during transient state conditions in chemostat cultures of *Enterococcus faecalis* NCTC 775: importance of internal pyruvate concentrations and NADH / NAD^+ ratios. J Gen. Microbiol. 138: 2015–2020

— (1993a) Differences in sensitivity to NADH of purified pyruvate dehydrogenase complexes of *Enterococcus faecalis*, *Lactococcus lactis*, *Azotobacter vinelandii* and *Escherichia coli*: implications for their activity in vivo. FEMS Microbiol. Lett. 114: 279–284

Snoep JL, Van Bommel M, Lubbers F, Teixeira de Mattos MJ & Neijssel OM (1993b) The role of lipoic acid in product formation by *Enterococcus faecalis* NCTC 775 and reconstitution *in vivo* and *in vitro* of the pyruvate dehydrogenase complex. J Gen. Microbiol. 139: 1325–1329

Starrenburg MJC & Hugenholtz J (1991) Citrate fermentation by *Lactococcus* and *Leuconostoc* spp. Appl. Env. Microbiol. 57: 3535–3540

Swomdell SR, Griffin HG & Gasson J (1994) Cloning, sequencing and comparison of three lactococcal L-lactate dehydrogenase genes. Microbiol. 140: 1301–1305

Takahashi S, Abbe K & Yamada T (1982) Purification of pyruvate formate-lyase from *Streptococcus mutans* and its regulatory properties. J. Bacteriol. 149: 1034–1040

Thomas TD (1976a) Activator specificity of pyruvate kinase from lactic Streptococci J. Bacteriol. 125: 1240–1242

— (1976b) Regulation of lactose fermentation in group N Streptococci. Appl. Env. Microbiol. 32: 474–478

Thomas TD, Ellwood DC & Longyear MC (1979) Change from homo- to heterolactic fermentation by *Streptococcus lactis* resulting from glucose limitation in anaerobic chemostat cultures. J. Bacteriol. 138: 109–117

Thomas TD, Turner KW & Crow VL (1980) Galactose fermentation by *Streptococcus lactis* and *Streptococcus cremoris*: pathways, products, and regulation. J. Bacteriol. 144: 672–682

Thompson J (1978) In vivo regulation of glycolysis and characterisation of sugar: phosphotransferase systems in *Streptococcus lactis*. J. Bacteriol. 136: 465–476

— (1979) Lactose metabolism in *Streptococcus lactis*: phosphorylation of galactose and glucose moieties in vivo. J. Bacteriol. 140: 774–785

— (1980) Galactose transport systems in *Streptococcus lactis*. J. Bacteriol. 144: 683–691

Thompson J & Torchia DA (1984) Use of ^{31}P nuclear magnetic resonance spectroscopy and ^{14}C fluorography in studies of glycolysis and regulation of pyruvate kinase in *Streptococcus lactis*. J. Bacteriol. 158: 791–800

Thompson J, Chassy BM & Egan W (1985) Lactose metabolism in *Streptococcus lactis*: studies with a mutant lacking glucokinase and mannose-phosphotransferase activities. J. Bacteriol. 162: 217–223

Thompson J & Gentry-Weeks CR (1994) Mtabolisme des sucres par les bactries Lactiques. In: Bactries lactiques, Vol 1. (Eds.), De Roissart H et Luquet, FM Lorica, Uriage. pp 239–290

Verhue WM & Tjan FSB (1991) Study of the citrate metabolism of *Lactococcus lactis* subsp *lactis* biovar *diacetylactis* by means of ^{13}C nuclear magnetic resonance. Appl. Environ. Microbiol. 57: 3371–3377

Veyrat A, Monedero V & Perez-Martinez G (1994) Glucose transport by the phosphoenolpyruvate : mannose phosphotransferase system in *Lactobacillus casei* ATCC 393 and its role in carbon catabolite repression. Microbiology 140: 1141–1149

Westby A, Nuraida L, Owens JD & Gibbs PA (1993) Inability of *Lactobacillus plantarum* and other lactic acid bacteria to grow on D-ribose as sole source of fermentable carbohydrate. J. Appl. Bacteriol. 75: 168–175

Yamada T (1987) Regulation of glycolysis in Streptococci. In: Sugar Transport and Metabolism in Gram-Positive Bacteria. (Eds.), J. Reizer and A. Peterkofsky. Ellis Howood Series in Biochemistry and Biotechnology. pp 69–93

Ye JJ, Reizer J & Saier MH (1994) Regulation of deoxyglucose phosphate accumulation in *Lactococcus lactis* vesicles by metabolite activated , ATP-dependent phosphorylation of serine-46 in HPr of the phosphotransferase system. Microbiology 140: 3421–3429

Zitzelberger W, Gotz F & Schleifer KH (1984) Distribution of superoxide dismutases, oxidases, and NADH peroxidases in various Streptococci. FEMS Microbiol. Lett. 21: 243–246

APPLICATIONS

Antonie van Leeuwenhoek **70:** 271–297, 1996.

Acceleration of cheese ripening

P.F. Fox, J.M. Wallace, S. Morgan, C.M. Lynch, E.J. Niland & J. Tobin
Department of Food Chemistry, National Food Biotechnology Centre, University College, Cork, Ireland

Abstract

The characteristic aroma, flavour and texture of cheese develop during ripening of the cheese curd through the action of numerous enzymes derived from the cheese milk, the coagulant, starter and non- starter bacteria. Ripening is a slow and consequently an expensive process that is not fully predictable or controllable. Consequently, there are economic and possibly technological incentives to accelerate ripening. The principal methods by which this may be achieved are: an elevated ripening temperature, modified starters, exogenous enzymes and cheese slurries. The advantages, limitations, technical feasibility and commercial potential of these methods are discussed and compared.

Introduction

The original objective of cheese manufacture was to conserve the principal nutrients in milk, i.e., lipids and proteins. This was achieved by a combination of acidification, dehydration, low redox potential and salting. Although a few minor cheese varieties are dehydrated sufficiently, or contain a sufficiently high level of NaCl, to prevent microbiological and/or enzymatic changes during storage, the composition of most varieties permits biological and enzymatic activity, i.e., ripening (maturation), during storage. The characteristics of the individual cheese varieties develop as a result of the biochemical changes that occur during ripening, as determined by curd composition, microflora, residual coagulant and residual milk enzymes.

Although cheese ripening is a very complex biochemical process (for reviews, see Fox et al., 1993, 1995, 1996), it primarily involves glycolysis, lipolysis and proteolysis, together with numerous secondary changes that are responsible for the characteristic flavour and texture of each cheese variety. These changes are catalysed by: (1) residual rennet, (2) starter bacteria and their enzymes, (3) secondary cultures and their enzymes, (4) non-starter adventitious microflora and their enzymes, and (5) indigenous milk enzymes.

Most ($\sim$ 98%) of the lactose in milk is removed in the whey but fresh cheese curd contains 0.7 to 1.5% lactose which is fermented, mainly to L-lactic acid, in all cheese varieties to give a pH of $\sim$ 5.0. Lactose is usually completely fermented within, at most, a few weeks; however, a high level of salt may cause its incomplete fermentation (Thomas & Pearce, 1981). Lactic acid may be metabolized to propionic acid, acetic acid and CO_2, as in Swiss- type cheeses, to H_2O and CO_2, eg, in Camembert, or to D-lactate and some acetate as in Cheddar, Dutch and Italian varieties.

Only limited lipolysis occurs in most cheese varieties, notable exceptions being Blue cheeses and some Italian varieties. Lipases secreted by *P. roqueforti* are the principal lipolytic agents in blue cheeses, the characteristic peppery flavour of which is due to methyl ketones produced by partial β-oxidation of free fatty acids. An exogenous lipase, pregastric esterase, is the principal lipolytic agent in Italian cheeses.

Proteolysis occurs in all cheese varieties, ranging from limited, e.g., Mozzarella, to very extensive, e.g., Blue, Parmesan and extra-mature Cheddar. Proteolysis is largely responsible for the textural changes in most varieties, makes a direct contribution to flavour, e.g., peptides and amino acids (and perhaps off-flavour, e.g., bitterness), produces substrates (amino acids) for the generation of sapid compounds, e.g., amines, acids, thiols and thioesters, and facilitates the release of sapid compounds from the cheese mass during mastication. Proteolysis is perhaps the most important reaction during cheese ripening, with the exception of blue and Italian varieties, in which lipolysis and fatty acid oxidation dominate, although proteolysis is also very important.

When the objective of cheese production was primarily the conservation of milk constituents, then the more stable the product, i.e., the less change, the better. While storage stability is still important, it is no longer the primary objective of cheese manufacture, a consistently high quality being the target. Since ripening is expensive, acceleration of ripening, especially of low-moisture, slow-ripening varieties, is desirable, provided that the proper balance can be maintained.

Objective of accelerating ripening

Proteolysis appears to be rate-limiting in the maturation of most cheese varieties and hence has been the focus of most research on the acceleration of ripening. Acceleration of ripening is most pertinent for low-moisture, slow-ripening varieties and most published work has been on Cheddar. Techniques for the acceleration of ripening are also applicable to low-fat cheeses which ripen more slowly than their full-fat counterparts.

The extensive literature on the acceleration of cheese ripening has been reviewed by Law (1978, 1980, 1982, 1984, 1987), Moskowitz & Noelck (1987), Fox (1988/89), El-Soda and Pandian (1991), El-Soda (1993) and Wilkinson (1993). This article will concentrate on recent work by our group rather than attempt to review again the whole subject.

Proteolysis in naturally ripened cheese

Since accelerating proteolysis is the usual objective of accelerated ripening of cheese, elucidation of the extent and type of proteolysis in naturally-ripened cheese would appear to be a desirable prerequisite. Considerable progress has been made on this subject during the past 10 years or so.

The development and standardization of methods for the quantitation and characterization of proteolysis in cheese is essential for studies on cheese ripening. The subject has been reviewed by Grappin et al. (1985), Rank et al. (1985), Fox (1989), IDF (1991), McSweeney & Fox (1993) and Fox et al. (1995) and will not be discussed further here; suffice it to say that the methods used fall into 3 principal categories:

1. Quantitation of nitrogen soluble in various extractants/precipitants [water, pH 4.6 buffers, 2–12% TCA, 30–70% ethanol, phosphotungstic acid (PTA)], usually by Kjeldahl but less frequently by the method of Lowry or dye-binding methods, absorbance of 280 nm or amino group-reactive agents, eg, TNBS, ninhydrin, fluorescamine or o-phataldialdehyde. The suitability of these methods has been compared by Wallace and Fox (1994); the Lowry modification of the biuret method appears to give best results.
2. Release of amino groups, as quantified by reaction with TNBS, ninhydrin, fluorescamine or o-phataldialdehyde.
3. Electrophoresis, usually in urea-containing polyacrylamide gels, or chromatography, usually RP-HPLC of small water-soluble peptides or IE-HPLC of larger, water-insoluble peptides.

Proteolysis in mature cheeses, especially in Cheddar, Blue and Parmesan, is so complex that fractionation of the cheese is necessary to fully appreciate its extent. Various fractionation methods were compared by Kuchroo & Fox (1983); some of these have been developed further and combined into a protocol, Figure 1 (O'Sullivan & Fox, 1990; Singh et al., 1994; Fox et al., 1994) which is used by many other investigators, sometimes in modified form.

Contribution of individual agents to proteolysis

The proteolytic enzymes involved in the ripening of cheese originate from 4, and in some varieties 5, sources: (1) Milk, (2) Coagulant, (3) Starter bacteria, (4) Non-starter lactic acid bacteria (NSLAB), (5) Secondary/adjunct microorganisms, e.g., *P. roqueforti* (Blue cheeses), *P. camemberti* (Camembert and Brie), *Br. linens*, yeasts and *Micrococcus* (surface smear cheeses), *P. freudenreichii* subsp *shermanii* (Swiss varieties); these organisms dominate the ripening of cheeses in which they are used through their proteolytic and/or lipolytic activity and secondary metabolism, e.g., β-oxidation of fatty acids (Blue cheeses), amino acid catabolism (smear ripened cheeses) and/or lactate metabolism (Swiss varieties and Camembert). *L. lactis* ssp *lactis* var. *diacetylactis* and *Leuconostoc* spp, components of the starter for Dutch-type cheeses, metabolise citrate to diacetyl and CO_2, which are important for flavour and eye development, respectively; traditionally, secondary/adjunct starters are not used in Cheddar-type cheeses but, as discussed in Section 4.4, the development and application of such cultures are among the promising approaches toward accelerating ripening.

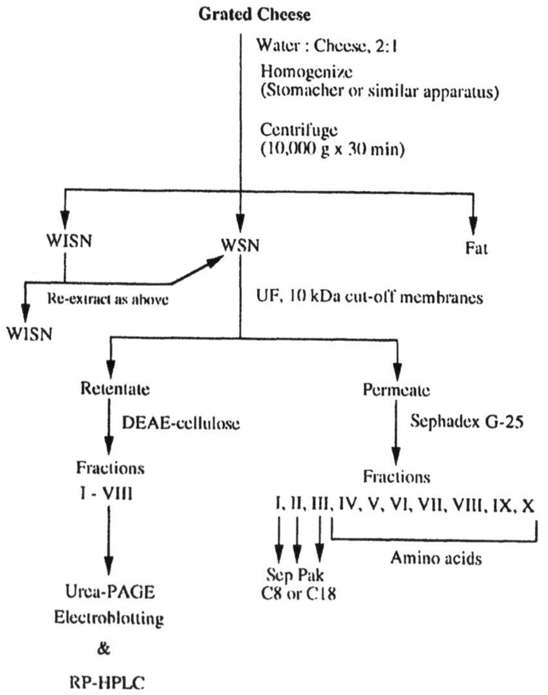

Figure 1. Scheme for the fractionation of cheese nitrogen (Fox et al., 1994).

Starting with the pioneering work of Mabbitt et al. (1955), techniques have been developed which permit the elimination of one or more of the 5 ripening agents, permitting assessment of their contribution to ripening. These methods have been reviewed by Fox et al. (1993). Essentially, they involve:

1. Selection of milk with a very low microbiological count, followed by pasteurization (or perhaps microfiltration) to give essentially sterile milk.
2. Manufacture of cheese in a sterile environment (enclosed vats, sterile room or laminar airflow unit) to eliminate adventitious contaminants (NSLAB).
3. Use of an acidogen, usually glucono-δ-lactone, to replace the acidifying function of the starter.
4. Inactivation of the coagulant, after it has hydrolysed κ- casein, by heating and/or pH adjustment.
5. Inhibition of plasmin (the principal milk proteinase) by 6- aminohexanoic acid.

Studies on such model cheeses have shown (Mabbitt et al., 1959; Visser, 1977a, b, c; Visser & de Groot-Mostert, 1977; O'Keeffe et al., 1976a, b, 1978; Le Bars et al., 1975; Lynch et al., 1996a, b) that:

1. The coagulant is principally responsible for the formation of water- or pH 4.6-soluble N and the changes indicated by PAGE but makes little contribution to the formation of small peptides or free amino acids (TCA- or PTA-soluble N).
2. Although plasmin contributes relatively little to the formation of water- or pH 4.6-soluble N, and even less to the formation of TCA- or PTA-soluble N, it does hydrolyse some β-casein with the formation of γ- caseins and proteose peptones. The contribution of plasmin is more important in high-cooked cheeses, e.g., Swiss and Parmesan, than in Cheddar or Dutch types due to the extensive inactivation of the coagulant and to more extensive activation of plasminogen, possibly due to the inactivation of inhibitors of plasminogen activators at high cook temperatures (Farkye & Fox, 1991).
The contribution of the indigenous acid proteinase (cathepsin D) has not been established; its specificity is similar to that of chymosin (McSweeney et al., 1995) and is probably overshadowed by chymosin in normal cheese.
3. Although lactic acid bacteria are weakly proteolytic, they possess a very comprehensive proteolytic system which is necessary for their extensive growth in milk which contains very little free amino acids and small peptides. The proteolytic system of *Lactococcus* and, to a lesser extent, *Lactobacillus* has been studied extensively at the molecular, biochemical and genetic levels; the extensive literature has been reviewed by Thomas & Pritchard (1987), Tan et al. (1993), Visser (1993), Monnet et al. (1993), Pritchard and Coolbear (1993), Law & Haandrikman (1996) and summarized by Fox & McSweeney (1996a, b).
The proteolytic system consists of a cell wall-associated proteinase, 3 or 4 intracellular proteinases (Stepaniak et al., 1995), 2 intracellular endopeptidases (PepO, PepF), at least 3 aminopeptidases (PepN, PepA and PepC), X-prolyl dipeptidyl aminopeptidase (PepX), iminopeptidase, at least one tripeptidase, at least one general dipeptidase and a number of proline-specific peptidases; a carboxypeptidase has not been reported in *Lactococcus* and in only a few species of *Lactobacillus*. Through their concerted action, the proteolytic system of LAB can hydrolyse casein to amino acids (Figure 2).
Studies on controlled microflora cheeses have shown that the proteolytic system of starter LAB contributes little to primary proteolysis in cheese,

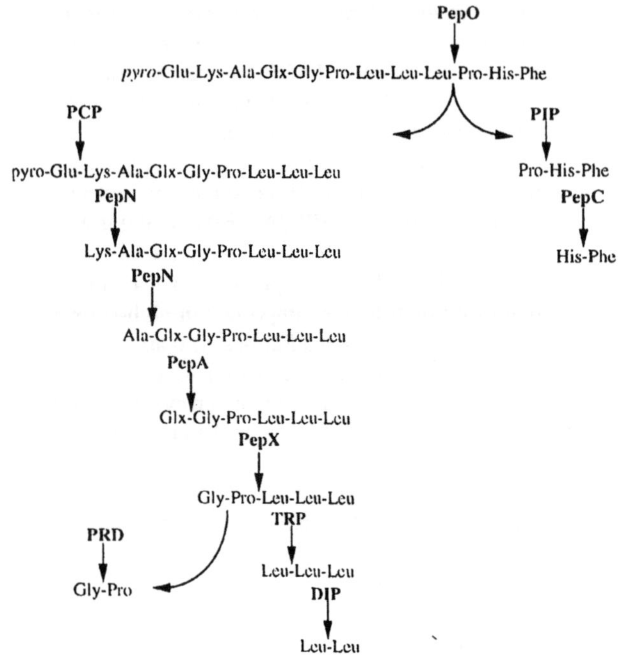

Figure 2. Schematic representation of the hydrolysis of a hypothetical dodecapeptide by the combined action of endo- and exopeptidases of *Lactococcus* spp.

as detected by PAGE or the formation of water- or pH 4.6-soluble peptides, but is mainly responsible for the formation of small peptides and amino acids (i.e., TCA- or PTA-soluble N).

4. Although cheese made from pasteurized milk in modern mechanized plants contains very few NSLAB immediately after manufacture (typically < 50 cfu/g), these multiply rapidly to reach, typically, 10^7 cfu/g within about 2 months. During this period, the starter cells which are present initially at, typically, 10^9 cfu/g, decrease to 10^7 cfu/g; hence, in long-ripened cheeses, e.g., Cheddar, NSLAB dominate the viable microflora throughout most of the ripening period.

The cheeses investigated in all the above-mentioned studies on controlled microflora cheeses were intended to be free of NSLAB and hence their contribution to proteolysis, or ripening in general, was not studied. Recent studies in our laboratory (Lane & Fox, 1996; Lynch et al., 1996a, b) using chemically acidified (GDL) or biologically acidified (starter) cheese, with or without an adjunct culture of non-starter lactobacilli, showed that lactobacilli appeared to produce many of the same peptides as the starter, although at a slower rate, accelerated the production of free amino acids and certain volatile compounds and influenced the flavour of the cheeses.

The specificity of chymosin, plasmin, cathepsin D and lactococcal cell wall-associated proteinases on individual caseins in solution are known and have been summarized by Fox et al. (1994, 1995, 1996) and Fox & McSweeney (1996a, b).

Proteolysis in cheese

The principal water-insoluble peptides are produced either from α_{s1}-casein by chymosin or from β-casein by plasmin; several minor water-insoluble peptides remain to be identified (McSweeney et al., 1994a; Mooney & Fox, unpublished). In mature (> 6 mo) Cheddar, all α_{s1}-CN is hydrolyzed at Phe$_{23}$-Phe$_{24}$ to yield α_{s1}-CN f1–23 and α_{s1}-CN f24–199 (α_{s1}-I). About 50% of the latter is hydrolysed at Leu$_{101}$-Tyr$_{102}$ to yield α_{s1}-CN f24–101 and α_{s1}-CN f102–199. The bonds Phe$_{32}$-Gly$_{33}$ and Leu$_{109}$-Glu$_{110}$ are also hydrolysed, probably in α_{s1}-CN f24–199. The bond Leu$_{109}$-Glu$_{110}$ was not found by McSweeney et al. (1993b) to be hydrolysed in α_{s1}-CN in solution. Although the bond Trp$_{164}$-Tyr$_{165}$ in α_{s1}-CN in solution is hydrolysed rapidly (McSweeney et al., 1993b; Exterkate et al., 1995), it does not appear to be hydrolysed in cheese - at least no peptide commencing with Tyr$_{165}$ has yet been identified. Although α_{s1}-CN is readily hydrolysed by plasmin (McSweeney et al., 1993c; Le Bars & Gripon, 1993), it does not appear to be hydrolysed to a significant extent by plasmin in cheese.

β-Casein in solution is readily hydrolysed by chymosin, especially at Leu$_{192}$-Tyr$_{193}$ and also at Ala$_{189}$-Phe$_{190}$, Leu$_{165}$-Ser$_{166}$, Gln$_{167}$-Ser$_{168}$, Leu$_{163}$-Ser$_{164}$, Leu$_{139}$-Leu$_{140}$, Leu$_{127}$-Thr$_{128}$. However, very little of the primary product, β-CN f1–192, is normally produced in cheese. Instead, ~ 50% of the β-casein is hydrolysed by plasmin at Lys$_{28}$-Lys$_{29}$, Lys$_{105}$-His$_{106}$ and Lys$_{107}$- Glu$_{108}$, yielding β-CN f1–28 (PP8f), f1–105/107 (PP5), f28–105/107 (PP8s), f29–209 (γ^1-CN), f106–209 (γ^2-CN) and f108–209 (γ^3-CN).

The principal water-insoluble peptides are common to several internal-bacterially ripened cheeses, i.e., Cheddar, Cheddar types (English territorials), Edam, Gouda, Maasdammer, Emmental and Parmesan. At least the bonds Phe$_{23}$-Phe$_{24}$ in α_{s1}-casein and Lys$_{28}$-Lys$_{29}$, Lys$_{105}$- His$_{106}$ Lys$_{107}$-Glu$_{108}$ in β-casein are cleaved in all these cheeses. The bond Leu$_{101}$-Glu$_{102}$ of α_{s1}-CN is hydrolysed in all except Emmental and Parmesan, possibly because chymosin is extensively inactivated in these cheeses; the bond Phe_{23}-Phe_{24} in these cheeses may be hydrolysed by cathepsin D rather than by chymosin.

Many of the water-soluble peptides in Cheddar have been isolated and identified (Singh et al., 1994, 1995, 1996; Fox et al., 1994); these are summarized in Figures 3 and 4. The N-terminal of many of these peptides corresponds to a chymosin (α_{s1}-CN) or a plasmin (β-CN) cleavage site while that of many others corresponds to a known lactococcal CEP cleavage site. Few of the isolated peptides contain a primary chymosin (α_{s1}- CN) or plasmin (β-CN) cleavage site, suggesting that lactococcal CEP cleaves polypeptides produced from α_{s1}-CN by chymosin or from β-CN by plasmin rather than the intact proteins. The C-terminal of many of the peptides does not correspond to a known chymosin, plasmin or lactococcal CEP cleavage site, suggesting: (1) carboxypeptidase activity, which has not been identified in *Lactococcus* and in only a few *Lactobacillus* strains, (2) unreported lactococcal CEP cleavage sites, or (3) activity of other proteinases, eg, from NSLAB or intracellular proteinases from *Lactococcus* or NSLAB or endopeptidases (PepO, PepF).

Significantly, the vast majority of water-soluble peptides are produced from the N-terminal half of α_{s1}- or β-casein; only 4 peptides from α_{s2}-CN and none from κ- CN have been identified so far.

The small peptides in other varieties have not been studied as extensively as those in Cheddar; although some peptides are common to several varieties, RP-HPLC indicates varietal-specific profiles (see Fox & McSweeney, 1996a, b).

There appears to be general agreement that the intensity of cheese flavour correlates with the concentration of free amino acids (Aston et al., 1983). In mature Cheddar and Parmesan, ~ 3 and 20% of total amino acids are free. The concentrations of individual amino acids in a selection of cheeses are summarized in Table 1. Free amino acids contribute directly to cheese flavour and serve as substrates for other flavour-generating reactions, eg, deamination, decarboxylation, desulfuration, Strecker reaction, Maillard reaction.

It is clear from the discussion in this section that considerable information is now available on the extent and nature of proteolysis in Cheddar cheese and to a lesser extent in a number of other varieties. Attempts to accelerate ripening must aim to accelerate in a balanced way the key flavour-generating proteolytic reactions. Unfortunately, the key reactions have not yet been identified - a detailed comparison of the peptide profiles in Cheddar cheeses (or other varieties) varying in quality from poor to excellent would appear to be a desirable objective.

Table 1. Free Amino Acid Composition of Selected Cheeses

Amino Acid	Cheddar mmol/g[a]	Swiss mmol/g[b]	Gouda mg/g[c]	Blue[d]
Ala	4.04	5.66	0.50	1729
Arg	6.37	nd	1.18	445
Asn	-	-	-	-
Asp	11.65	5.40	0.45	860
Cys	0.37	nd	-	-
Gln	-	-	-	-
Glu	21.62	9.79	3.38	4775
Gly	4.14	1.74	0.37	466
His	-	-	0.22	449
Ile	3.58	-	0.65	1929
Leu	21.19	15.72	3.34	3015
Lys	7.80	1.12	2.02	2231
Met	2.95	0.42	0.66	1111
Phe	9.02	6.40	1.67	1785
Pro	2.95	2.41	-	3076
Ser	4.00	2.65	0.54	1045
Thr	3.78	2.80	1.55	1123
Trp	-	-	-	552
Tyr	3.75	1.03	1.00	1154
Val	9.47	8.36	1.13	2408
Orn	nd	nd	-	77
Asn,Gln	-	-	-	876
Citrulline	-	-	-	423
γ-ABA	-	-	-	221

nd = not detected
- = not analysed
γ-ABA = g-amino butyric acid
[a] Wood et al. (1985). 8 m old cheese analyzed by capillary GC.
[b] Wood et al. (1985). 2 m old Swiss-type cheese analyzed by capillary GC.
[c] Visser, F.M.W. (1977c). 6 m old cheese made with starter *Lactococcus lactis* ssp *cremoris*.
[d] Ismail & Hansen (1972). 248 d old Danablue cheese analyzed by amino acid analyzer, results expressed as mg amino acid residue/15.7 g total N.

In the following sections, published studies on accelerated ripening of cheese will be discussed in the context of the preceding discussion on proteolysis during normal ripening.

Methods for accelerating cheese ripening

The methods used to accelerate ripening can be categorised into 6 groups: (1) elevated ripening tempera-

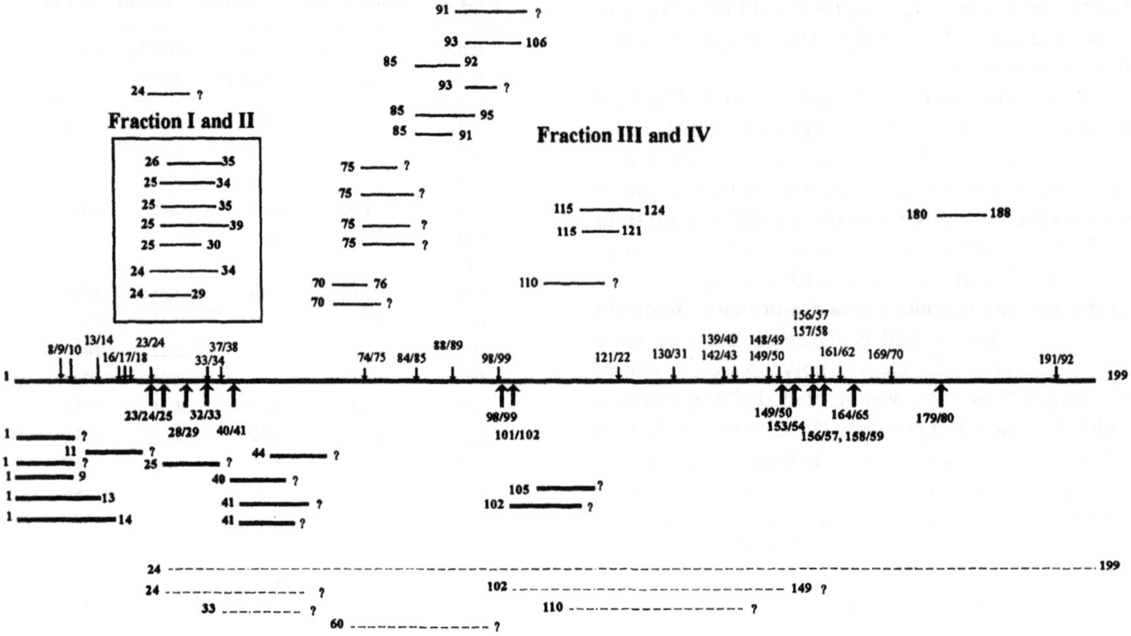

Figure 3. α_{s1}-Casein-derived peptides isolated from the water-insoluble (- - -) or from the diafiltration permeate (____) or retentate (▬) of the water-soluble fraction of Cheddar cheese. ? Incomplete sequence (from Singh et al., 1996).

tures, (2) exogenous enzymes, (3) chemically or physically modified cells, (4) genetically modified starters, (5) adjunct cultures, (6) cheese slurries (Table 2). These methods aim to accelerate cheese ripening either by increasing the level(s) of putative key enzymes or by making the conditions under which the 'endigenous' enzymes in cheese operate more favourably for their activity.

Elevated temperature

Traditionally, cheese was ripened in caves or cellars, probably at 15–20 °C for much of the year. Since the introduction of mechanical refrigeration for cheese ripening rooms in the 1940s, the use of controlled ripening temperatures has become normal practice in modern factories. These range from 22–24 °C for Parmesan and Emmental, 12–20 °C for mould and smear-ripened cheeses, 12–14 °C for Dutch varieties to 6–8 °C for Cheddar; thus, the ripening temperature for Cheddar is exceptionally low. The ripening temperature for most varieties is profiled - the above temperatures are the 'maximum' in the profiles and are usually maintained for 4–6 weeks, usually to induce the growth of a desired microflora, after which the cheese is transferred to a much lower temperature, e.g., 4 °C, e.g., Emmental or mould-ripened cheeses. Again, Cheddar is an exception since it is normally ripened at 6–8 °C throughout.

The scope for accelerating the ripening of most cheese varieties by increasing the ripening temperature is quite limited since fat will exude from the cheese > 20 °C. However, this approach has potential for Cheddar and offers the simplest method for accelerating ripening: no additional costs are involved (indeed savings may accrue from reduced refrigeration costs) and there are no legal barriers. However, considering the numerous complex biochemical reactions that occur during ripening, it is unlikely that all reactions will be accelerated equally at elevated temperatures and unbalanced flavour or off-flavours may result.

Based on the results of a study on the influence of starter type, number of NSLAB and ripening temperature (6 or 13 °C) on the flavour of Cheddar cheese, Law et al. (1979) concluded that ripening temperature was the most important single factor in determining flavour intensity, irrespective of the type of starter and number of NSLAB. The time required to reach maturity was at least 50% less at 13 than at 6 °C. Bitterness was more marked at the lower temperature, possibly due to the

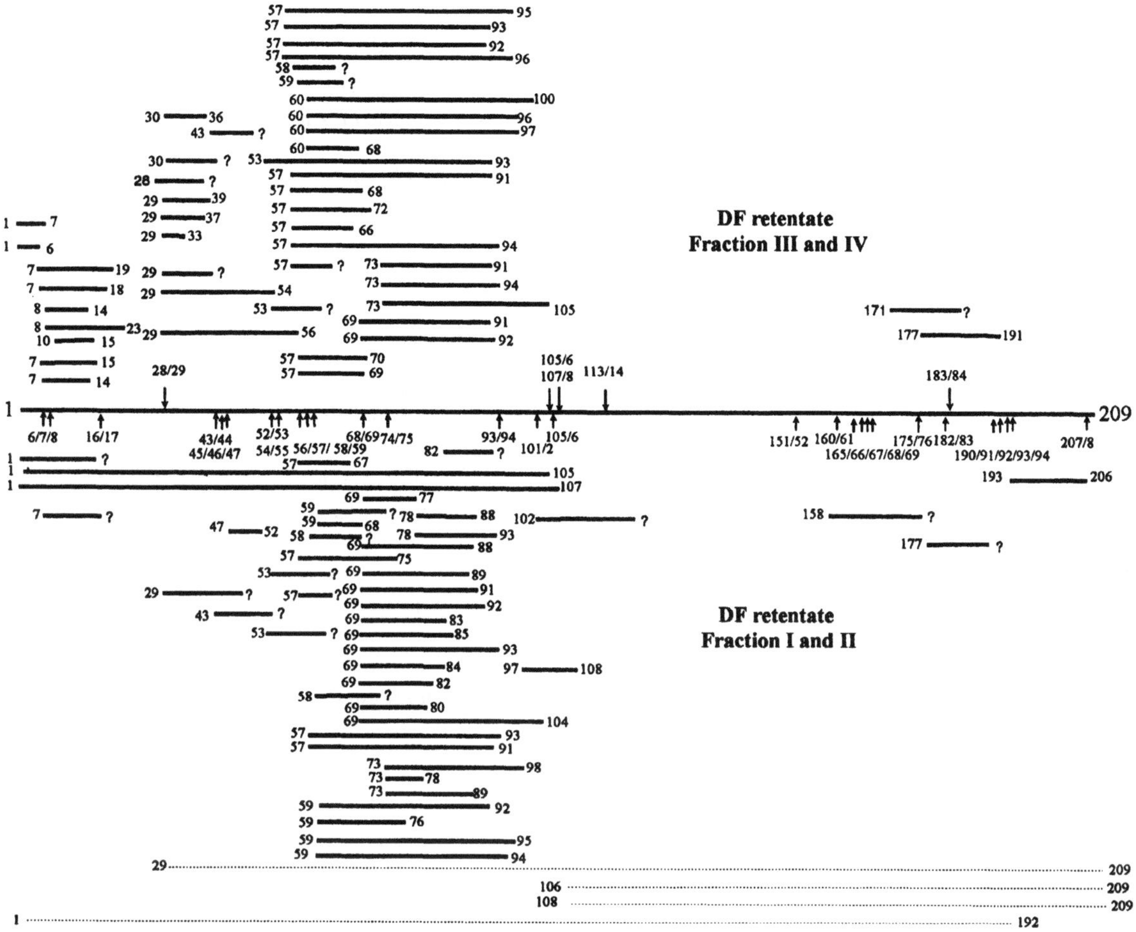

Figure 4. β-Casein-derived peptides isolated from the water-insoluble (- - -) or from the diafiltration retentate (▬) of the water-soluble fraction of Cheddar cheese. ? Incomplete sequence (from Singh et al., 1996).

lower intensity of Cheddar flavour, or perhaps because the peptidases needed to hydrolyse the bitter peptides required the higher temperature for adequate activity.

A very comprehensive study on the effect of temperature (8–20 °C, applied consistently or profiled), alone or in combination with lactose-negative starter or exogenous proteinase (Neutrase), was conducted in Australia in the 1980s by Aston, Dulley, Fedrick and collaborators (Aston et al., 1983; Fedrick et al., 1983; Fedrick & Dulley, 1984; Fedrick, 1987). Ripening was monitored by various chemical and rheological criteria and compositional parameters specified. The essential conclusions of this study, which was reviewed in detail by Fox (1988/89), were that it is possible to reduce the maturation time by ~ 50% by ripening at 13–15 °C but it was emphasised that cheese should be of good compositional and microbiological quality.

High populations of NSLAB present a risk of off-flavour development in cheese ripened at elevated temperatures. Fryer (1982) recommended that Cheddar cheese should contain < 10^3 NSLAB/g at hooping and that their numbers should be controlled by rapid cooling to < 10°C. If the cooled cheeses are held for 14 days, NSLAB grow slowly, producing only lactic acid, and should not exceed 10^6/g at 14 days. After this initial period at a low temperature, balanced ripening could be accelerated at a relatively high temperature without a risk of off-flavour development (Fryer, 1982).

The potential of elevated temperatures to accelerate the ripening of Cheddar was also studied by

Table 2. Methods for accelerating cheese ripening

Method	Advantage	Limitations/Problems
Elevated temperature	Effective; no legal barriers; technically simple; no cost, perhaps saving	Non-specific; increased risk of microbial spoilage; applicable to relatively few varieties, e.g. Cheddar
Exogenous enzymes		
rennet	Natural additive; cheap	not effective
plasmin	Indigenous milk enzyme; effective	expensive
Other proteinases/ peptidases	Low cost; specific action; choice of flavour options	Limited choice of useful enzymes; possible legal barriers; difficult to incorporate uniformly; risk of over-ripening; limited commercial use to date
Lipases	Traditional for certain cheese varieties	Risk of rancidity; very limited general use
Selected/activated/modified starters		
Selected starters Enzyme profile Rapid lysis	Normal additives;	None
Attenuated starters	Easily incorporated; natural enzyme profile	May be expensive
Lysozyme treatment Heat or freeze shocked Solvent treated Neutralized starters Mutant cultures, e.g. Lac^-		
Other types of bacterial cells	Easily incorporated; range of enzyme options?	Perhaps legal problems in some cases
Genetically engineered starters	Easily incorporated; desirable enzyme profiles	Possible legal barriers; key enzymes not yet identified
Adjunct starters	Natural microflora; appear to be effective; flavour options; commercially available	Careful selection required
Cheese slurries/ high moisture cheese	Very rapid flavour development; commercially used	High risk of microbial spoilage; suitable only as a food ingredient
Addition of free amino acids to cheese curd	choice of flavour	May be too expensive; limited work to date

Folkertsma et al. (1996). Commercially-made Cheddar cheeses were cooled either rapidly or slowly and ripened for various time/temperature combinations at 8, 12 or 16 °C. NSLAB grew very slowly in the cheese which was cooled rapidly and ripened at 8 °C, although ripening temperature had little influence on the final numbers of NSLAB (10^7–10^8 cfu/g). Proteolysis (monitored by the formation of water- and PTA-soluble N and total free amino acids, urea-PAGE and RP-HPLC) and lipolysis were accelerated by increasing the ripening temperature. Cheeses ripened at 16 °C received good flavour scores early during ripening but their texture deteriorated after prolonged storage at 16 °C. Ripening at 12 °C was considered to be optimal and ripening could be accelerated or decelerated by increasing or lowering the temperature at any stage of the process.

According to El Soda & Pandian (1991), the use of an elevated temperature to accelerate the ripening of Cheddar cheese is likely to be limited to those made under very hygienic conditions in commercial factories. However, since most Cheddar is now made in highly automated plants from pasteurized milk with initial low counts, elevated ripening temperatures appear to be feasible; at least, ripening at temperatures as low as 6°C is unnecessary unless a very slow rate of ripening is desired, for whatever reasons.

The ripening of Manchego cheese can be accelerated and flavour intensified, especially in cheese made from pasteurized milk, by ripening at an elevated temperature (16 °C); an earlier study had shown that ripening at 20 °C had a negative effect on cheese quality although proteolysis and lipolysis were accelerated compared with cheese ripened at 10 °C (Gaya et al., 1990).

Exogenous enzymes

A number of options are available, ranging from the quite conservative to the more exotic.

Coagulant

Since the coagulant is principally responsible for primary proteolysis in most cheese varieties (see Section 3.1), it might be expected that ripening could be accelerated by increasing the level or activity of rennet in the cheese curd. Although, Exterkate and Alting (1995) suggested that chymosin is the limiting proteolytic agent in the initial production of amino N in cheese, several studies (Stadhouders, 1960; Creamer et al., 1987; Guinee et al., 1991; Johnston et al., 1994) have shown that increasing the level of rennet in cheese curd (achieved by various means) does not accelerate ripening and in fact probably causes bitterness. However, as far as we are aware, the combined effect of increasing rennet level and starter and/or NSLAB population has not been investigated.

The natural function of chymosin is to coagulate milk in the stomach, thereby increasing the efficiency of digestion. It is fortuitous that chymosin is not only the most efficient milk coagulant but also gives best results in cheese ripening. However, it seems reasonable to suggest that the efficiency of chymosin in cheese ripening could be improved by protein engineering. The chymosin gene has been cloned and expressed in several microorganisms (*Kluyveromyces marxianus* var. *lactis*, *E. coli* and *Aspergillus niger* var. *awamori*) and chymosin from such sources is now used widely, but not universally, in commercial cheese manufacture, with excellent results (see Teuber, 1990; IDF, 1992). The gene for the acid proteinase of *R. miehei* has also been cloned and expressed in *A. oryzae*, and the product is commercially available (Marzyme GM; Texel, Cheshire, UK). In all these cases, the parent gene has not been modified but a number of studies on the genetic engineering of chymosin have been published (see Fox & McSweeney, 1996b). As far as we know, the cheesemaking properties of such mutants have not been assessed.

As discussed in Section 3.1, chymosin has very little activity on β-casein in cheese, probably because the principal chymosin- susceptible bond in β-casein, Leu_{192}-Tyr_{193}, is in the hydrophobic C-terminal region of the molecule which appears to interact hydrophobically in cheese, rendering this bond inaccessible. However, *C. parasitica* proteinase preferentially hydrolyses β- casein in cheese (possibly because its preferred cleavage sites are in the hydrophilic N-terminal region) without causing flavour defects (Rea & Fox, unpublished). A rennet containing chymosin and *C. parasitica* proteinase might be useful for accelerating ripening.

Plasmin

Plasmin contributes to proteolysis in cheese, especially of high- cooked varieties in which chymosin is extensively or totally inactivated (see Section 3.1). Plasmin is associated with the casein micelles in milk, which can bind at least 10 times the amount of plasmin normally present (Farkye & Fox, 1992) and is totally and uniformly incorporated into cheese curd, thus overcoming one of the major problems encountered with the use of exogenous proteinase to accelerate cheese ripening.

Addition of exogenous plasmin to cheesemilk accelerated the ripening of cheese made from that of milk without off-flavour development (Farkye & Fox, 1992; Farkye & Landkammer, 1992; Kelly, 1995). At present, plasmin is too expensive for use in cheese on a commercial scale. Perhaps the gene for plasmin can be cloned in a suitable bacterial host which could be engineered to excrete the enzyme. Since milk normally contains 4 times as much plasminogen as plasmin, an alternative strategy might be to activate indigenous plasminogen by adding a plasminogen activator, eg, urokinase, which also associates with the casein micelles. However, the cost of this approach may also be excessive.

Since plasmin is a trypsin-like enzyme, trypsin, which is relatively cheap and readily available commercially, may also be suitable for accelerating ripening. Careful use of trypsin has been reported (Madkor & Fox, 1994) to accelerate ripening but these findings must be confirmed. Since trypsin is more proteolytic than plasmin, greater care is required in its use.

Exogenous proteinases

The possibility of accelerating ripening through the use of exogenous (non-rennet) proteinases has attracted considerable attention over the past 20 years. The principal problems associated with this approach, which has been reviewed by Law (1984, 1987) and Fox (1988/89), are ensuring uniform distribution of the enzyme in the curd and the prohibition of exogenous enzymes in many countries.

The earliest reports on the use of exogenous enzymes to accelerate the ripening of Cheddar cheese appear to be those of Kosikowski and collaborators who investigated various combinations of commercially available acid and neutral proteinases, lipases, decarboxylases and lactases (see Fox, 1988/89 for references). Acid proteinases produced pronounced bitterness but the addition of certain neutral proteinases and peptidases with the salt gave a marked increase in flavour after 1 month at 20 °C but an overripe, burnt flavour and free fluid were evident after 1 month at 32 °C. Incorporation of the enzyme- treated cheese in processed cheese gave a marked increase in Cheddar flavour at 10% addition and a very sharp flavour at 20%. Good quality medium-sharp Cheddar could be produced in 3 months at 10 °C through the addition of combinations of selected proteinases and lipases. Up to 60% enzyme-treated (fungal lipases and proteinases) UF retentate could be successfully incorporated into processed cheese.

On the assumption that a mixture of enzymes is likely to be more effective at accelerating ripening than a single enzyme, Law (1980) described the results of Cheddar cheesemaking trials in which a proteinase-peptidase preparation from a *Pseudomonas* culture was incorporated into the curd at salting (the organism secreted an extracellular proteinase and released intracellular peptidases when grown in media containing surfactants). A low level of enzyme addition accelerated flavour development, especially during the early stages of ripening, but larger amounts of enzyme caused bitterness and other off- flavours.

Law and Wigmore (1982a, b, 1983) compared the influence of acid, neutral and alkaline proteinases on proteolysis and flavour development in Cheddar cheese. Neutrase (*B. subtilis*), which enhanced flavour development at a low level of enzyme addition but caused bitterness at higher levels, was considered to be the most promising of the enzymes tested, possible because it is unstable in cheese (and hence its activity is somewhat limited), whereas acid proteinases are more stable. Use of an optimum level of Neutrase reduced the ripening time by ~ 50% but enzyme-treated cheese had a softer body and was more brittle than control cheeses of the same age. A combination of Neutrase and streptococcal cell-free extract (CFE) gave better results than Neutrase alone. Although increasing the level of CFE progressively increased proteolysis, flavour intensity did not increase *pro rata*, suggesting that subsequent amino acid transformations to sapid compounds were rate-limiting and were not catalysed by the enzymes in the CFE. This combined enzyme preparation was commercialized by Imperial Biotechnology, London, and marketed as 'Accelase'. Its use in several large-scale commercial cheesemaking trials was described by Fullbrook (1987). However, in spite of the claimed success of the Accelase in pilot-scale and commercial-scale studies, it has not been commercially successful and, as far as we are aware, is not currently available; its commercial failure may be due to the prohibition on the use of exogenous enzymes (other than rennet) in cheesemaking in the UK.

Frick et al. (1984) reported that proteinase 11 (a neutral proteinase from *A. oryzae*; Miles Marshall) added to Colby cheese curd at salting accelerated ripening without bitter flavour development. However, Fedrick et al. (1986a) could not confirm this; the lowest level of this enzyme that gave detectable flavour enhancement also resulted in bitterness that intensified with increasing level of added enzyme; proteinase P11 produced a slightly higher level of bitterness for comparable levels of proteolysis than Neutrase. [Law & Wigmore (1982a, b) had found that the acid proteinase of *A. oryzae* was unsuitable for cheese ripening.]

The results of a comparative study on proteolysis and textural changes in granular Cheddar cheeses supplemented with Neutrase, calf lipase, Neutrase plus calf lipase or NaturAge (a culture-enzyme mixture; Miles Marshall) was reported by Lin et al. (1987). TCA- soluble N increased rapidly in all proteinase-supplemented cheeses but free amino acid levels increased more slowly. Textural changes reflected gross proteolysis better than the formation of free amino acids. Unfortunately, the flavours of the cheeses were not reported.

The combined influence of Neutrase, a lac$^-$ prt$^-$ starter and ripening temperature (8 or 15 °C) was studied by Fedrick et al. (1986b). All treatments accelerated ripening compared to the control at 8 °C. Storage at 15 °C was the most effective single treatment, reducing ripening time by > 50%. Neutrase alone gave ~ 25% reduction. A slightly bitter flavour was noted in the Neutrase-treated cheeses but did not significantly affect panel preferences or grade until late in ripening.

A proteinase (*P. candidum*) - peptidase (*Lc. lactis* or *Lb. casei*) preparation for accelerating the ripening of Dutch, Tilsit or Lowicki-type cheese was described by Kalinowski et al. (1979, 1982). The enzyme, added to the cheesemilk, accelerated proteolysis and approximately halved the ripening time. Addition of a CFE from *Lb. casei, Lb. helveticus* or *Lb. bulgaricus* to the curd accelerated proteolysis and lipolysis in Cheddar cheese but the cheeses were bitter after 2 months (El Soda et al., 1981, 1982).

Guinee et al. (1991) reported that Neutrase, FlavourAge FR (a lipase-proteinase preparation from *A. oryzae*) or extra rennet added to Cheddar curd at salting accelerated flavour development when the cheese was ripened at 5°C for a relatively short period (4–5 months) but excessive proteolysis and associated flavour and body defects occurred on further storage, especially at a higher temperature. According to Wilkinson et al. (1992), neither FlavourAge FR nor DCA 50 (a proteinase-peptidase blend; Imperial Biotechnology, London) caused substantial acceleration of flavour development and in some cases led to off-flavours and textural defects.

Addition of exogenous proteinases to curd. With the exception of rennet and plasmin (which adsorbs on casein micelles), the incorporation and uniform distribution of exogenous proteinases throughout the cheese matrix poses several problems: (1) proteinases are usually water-soluble, and hence when added to cheesemilk, most of the added enzyme is lost in the whey, which increases cost, (2) enzyme-contaminated whey must be heat-treated if the whey proteins are recovered for use as functional proteins; the choice of enzyme is limited to those that are inactivated at temperatures below those that cause thermal denaturation of whey proteins, (3) according to Law & King (1985), the amount of Neutrase that should be added to milk to ensure a sufficient level of enzyme in the curd (Law & Wigmore, 1982a) caused a 25–80% decrease in rennet coagulation time, yielded a soft curd and at least 20% of the β- casein was hydrolysed at pressing; presumably, this would reduce cheese yield, which was not measured.

Consequently, most investigators have added enzyme, usually diluted with salt to facilitate mixing, to the curd at salting. Since the diffusion coefficient of large molecules, like proteinases and lipases, is very low, this method is applicable only to Cheddar-type cheeses, which are salted as chips at the end of manufacture, and not to surface- salted (brine or dry) cheeses which include most varieties. Even with Cheddar-type cheeses, the enzyme will be concentrated at the surface of chips, which may be quite large. Uneven mixing of the salt-enzyme mixture with the curd may lead to 'hot spots' where excessive proteolysis and lipolysis, with concomitant off-flavours, may occur.

Enzyme encapsulation offers the possibility of overcoming the above problems. The microcapsules, being sufficiently large, are occluded in the curd; the main problem is to achieve the release of the enzymes after curd formation. Several studies on the microencapsulation of enzymes for incorporation into cheese have been reported (see Fox, 1988/89; Pandian & El-Soda, 1991; Wilkinson, 1993; Skeie, 1994). Although microcapsules added to milk are incorporated efficiently into cheese curd, the efficiency of enzyme encapsulation is low, thus increasing cost. As far as we know, encapsulated enzymes are not being used commercially in cheese production.

Exogenous lipases

Lipolysis is a major contributor, directly or indirectly, in flavour development in strong-flavoured cheeses, eg, hard Italian, Blue varieties, Feta. Rennet paste or crude preparations of pre-gastric esterase (PGE) are normally used in the production of Italian cheeses (see Nelson et al., 1977; Fox, 1988/89; Kilara, 1985; Fox

& Stepaniak, 1993). *M. miehei* lipase may also be used for Italian cheeses, although it is less effective than PGE; lipases from *P. roqueforti* and *P. candidum* may also be satisfactory.

The ripening of blue cheese may be accelerated and quality improved by added lipases (see Fox, 1988/89; Kilara, 1985; Fox & Stepaniak, 1993). A Blue cheese substitute for use as an ingredient for salad dressings and cheese dips can be produced from fat-curd blends by treatment with fungal lipases and *P. roqueforti* spores (see Fox, 1988/89; Kilara, 1985; Fox & Stepaniak, 1993, for references).

Although Cheddar-type and Dutch-type cheeses undergo little lipolysis during ripening, it has been claimed that addition of rennet paste or gastric lipase improves the flavour of Cheddar cheese, especially that made from pasteurized milk; several patents have been issued for the use of lipases to improve the flavour of 'American' or 'processed American' cheeses (see Nelson et al., 1977; Kilara, 1985). The enzyme mixtures used by Kosikowski and collaborators (see Fox, 1988/89) to accelerate Cheddar cheese ripening contained lipases. Law & Wigmore (1985) reported that the addition of PGE or *M. miehei* lipase, with or without Neutrase, to Cheddar cheese curd had a negative effect on flavour quality.

FlavorAge contains a unique lipase from a strain of *A. oryzae* which has an exceptionally high specificity for C_6-C_8 acids and forms micelles, $\sim 0.2\ \mu m$ in diameter, in aqueous media as a result of which $\sim$ 94% of the enzyme added to milk is recovered in the cheese curds (Arbige et al., 1986). According to these authors, FlavourAge accelerated the ripening of Cheddar cheese; the formation of short-chain fatty acids paralleled flavour intensity in Cheddar cheese. In contrast to the FFA profile caused by PGE, which liberated high concentrations of butanoic acid, the FFA profile in cheese treated with FlavourAge was similar to that in the control cheese except that the level of FFA was much higher (Arbige et al., 1986).

Frick et al. (1984) compared the fatty acid profiles in Colby cheese to which FlavourAge or Miles 600 lipase plus proteinase was added. The latter produced a Romano-type flavour while FlavourAge produced a flavour more typical of an aged Cheddar at similar enzyme activities. Addition of an unspecified lipase to Samsoe yielded a cheese with a flavour closely resembling that of Greek Kasseri cheese (Jensen, 1970). Feta cheese produced from cow's milk with a blend of *Lc. lactis* and *Lb. casei* as starter and a blend of kid and lamb PGEs developed the body, flavour and texture of authentic Feta cheese (Efthymiou & Mattick, 1964). The flavour of Egyptian Ras cheese was improved by addition of PGE or lipases from *M. miehei* or *M. pusillus* (El Shibiny et al., 1978). Low levels of PGE improved and accelerated flavour development in Domiati cheese but prolonged ripening led to rancid off-flavours in enzyme-treated cheeses (El Neshawy et al., 1982). The flavour of Latin America White cheese was improved by low levels of pre-gastric esterase (Torres & Chandan, 1981).

Selected, activated or modified starters

Since the proteolytic system of the starter bacteria is responsible for the formation of small peptides and amino acids and probably for flavour development in cheese (Section 3.1), it seems obvious to exploit these enzymes to accelerate ripening; at least 4 approaches to do so have been employed.

Selected starters

The primary function of starters is to produce acid at a reliable and predictable rate. Traditionally, cheesemakers relied on the indigenous microflora of milk or on 'slop-back' natural starters for acid production. Such methods are still used for artisinal cheeses and even for such famous varieties as Parmesan. However, selected, undefined starters have been used for Cheddar, Dutch and Swiss cheeses since the beginning of this century and have been refined and improved progressively over the years. In the case of Cheddar, cocktails of phage-unrelated, single-strain starters were introduced in New Zealand by Whitehead in the 1930s, and are now widely used in New Zealand, Australia, Ireland, USA and probably elsewhere.

The principal criterion applied in the selection of single-strain starters is phage-unrelatedness; other important criteria include the ability to grow well and produce acid at the temperature profile used in cheesemaking and inter-strain compatibility (Martley & Lawrence, 1972; Crow et al., 1993); selection is usually made by the protocol of Heap and Lawrence (1976). Bitterness is a common problem with fast acid-producing strains, apparently because these strains have high heat tolerance and usually reach high numbers in the cheese curd (Lemieux & Simard, 1991, 1992). Fast acid-producing strains are usually *Lc. lactis* ssp *lactis*; consequently, strains of *Lc. lactis* ssp *cremoris* are now usually used as starters for Cheddar cheese.

Although the selection protocol of Heap and Lawrence (1976) does not include specific criteria for the selection of starter strains with the ability to produce high quality cheese, commercial experience has provided evidence for the exclusion of strains with undesirable cheesemaking properties, eg, bitterness, and the use of strains that more or less consistently produce high quality cheeses. The scientific selection of starter strains with desirable cheesemaking properties is hampered by the lack of precise knowledge as to which enzymes are most important.

Selection based on enzyme profiles. Lactococcal strains differ considerably with respect to total and cell wall-associated proteinase activity (Coolbear et al., 1994; Crow et al., 1994); however, no information is available on the comparative cheesemaking properties of these strains. Breen and Fox (unpublished) studied the cheesemaking properties of 19 single-strain starters in Cheddar cheese manufactured on a small (20 L) scale; results indicated considerable inter-strain variations in proteolysis, lipolysis and sensory quality (Figure 5). Unfortunately, information is not available at present on the enzyme complement of these strains. The influence of starter strain on the sensory properties of Cheddar cheese was also demonstrated by Muir et al. (1996). Further studies on the cheesemaking properties, preferably on a large scale, and the enzyme complement of single-strain *Lactococcus* starters is warranted.

The only extracellular enzyme in *Lactococcus* is the cell wall- associated proteinase. The peptidases are intracellular, although some may have a peripheral location (Tan et al., 1992). The esterase(s) and phosphatase(s) are also intracellular. The significance of lactococcal exopeptidases in cheese quality is unclear but they are responsible for the production of free amino acids and probably thus influence flavour development (see Section 3.1). Dephosphorylation of casein-derived peptides occurs during ripening (Singh et al., 1995, 1996). The significance of dephosphorylation is not known although Martley & Lawrence (1972) suggested that phosphatase activity was an important attribute of starters.

Selection based on starter cell lysis. Since the growth of lactococci ceases at or shortly after the end of curd manufacture (Martley & Lawrence, 1972; Visser, 1977b), their intracellular enzymes are ineffective until the cells die and lyse. Generally, *Lc. cremoris* cultures die faster than *Lc. lactis* ssp *lactis* strains although there is considerable interstrain variation within each subspecies (Martley & Lawrence, 1972; Visser, 1977b; Chapot-Chartier et al., 1994; Wilkinson et al., 1994b; O'Donovan, 1994). Information on the rate of lysis of *Lactococcus* species in cheese is rather limited but available evidence indicates substantial inter-strain differences (Wilkinson et al., 1994b; Chapot-Chartier et al., 1994).

It would be expected that the sooner starter peptidases are released through lysis, the sooner they can participate in proteolysis and hence the faster the rate of ripening. However, the stability of lactococcal exopeptidases in cheese is unknown. If they are unstable, it is possible that enzymes released early during ripening through accelerated lysis may contribute little to flavour development since the concentration of suitable peptides is low at this time. The stability of some intracellular marker enzymes was studied by Wilkinson et al. (1994a) who found that PepX activity was quite unstable (15% of initial activity remained after 24 h) in a cheese slurry system (pH 5.17). The other enzyme activities studied (glucose–6-phosphate hydrogenase and lactate dehydrogenase) were also relatively unstable. In contrast, Chapot- Chartier et al. (1994) found that PepX and PepC/N activities were stable in an extract of St. Paulin cheese (pH 5.8). Further research in this area appears warranted.

The release of intracellular peptidases into the matrix of St. Paulin cheese as a consequence of lysis was confirmed by Chapot-Chartier et al. (1994). Cheese made with fast-lysing *Lc. lactis* subsp *cremoris* AM2 developed higher levels of amino nitrogen than that made with slow-lysing *Lc. lactis* subsp *lactis* NCDO 763; lower levels of bitterness were reported in the cheese made with the fast-lysing starter. Wilkinson et al. (1994b) reported that the production of free amino acids was 5 times faster in Cheddar cheese made using a fast-lysing strain (AM2) than in cheese made using a slow-lysing strain (HP); the latter cheese was bitter.

Considering the presumed importance of cell lysis, a number of authors have attempted to accelerate ripening by increasing the rate of starter lysis. Four principal approaches have been investigated:

Selection of naturally fast-lysing strains. There have been few systematic studies on rate of lysis of *Lactococcus* but many known fast-lysing strains have undesirable cheesemaking properties, eg, slow acid production or phage sensitivity. Further studies in this area are warranted.

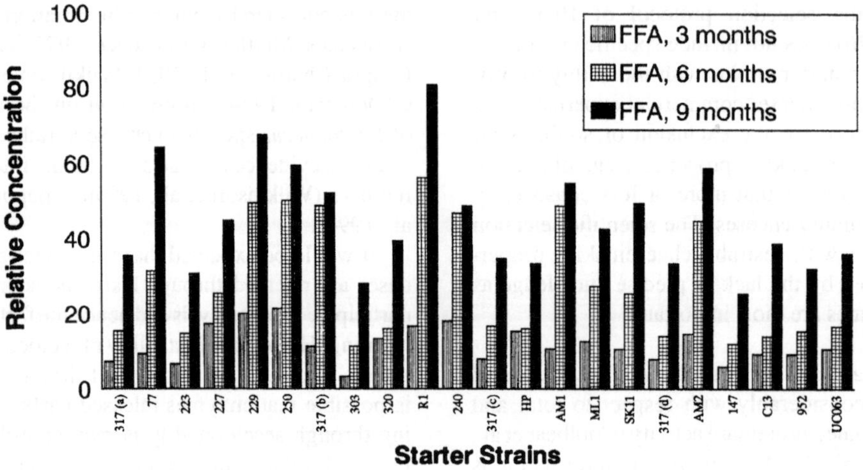

Figure 5. Relative concentration of free amino acids in Cheddar cheeses manufactured using single strain *Lactococcus* starters (Breen D & Fox PF, unpublished).

Thermoinducible lysis. Feirtag & McKay (1987a, b) isolated a *Lactococcus* mutant which underwent lysis during the cooking of Cheddar cheese (38 to 40 °C) because it harboured a thermoinducible prophage. The authors speculated that such strains would be useful for accelerating cheese ripening through the early release of intracellular enzymes; however, extensive loss of these enzymes in the whey may occur. No further reports on this or similar strains appear to have been published. Should this approach prove effective in practice, it should be possible to construct thermoinducible mutants of any desired starter by genetic techniques.

Bacteriophage-assisted lysis. Crow et al. (1996) accelerated the lysis of *Lc. lactis* subsp *lactis* ML8 using its homologous phage, ml_8. Although the phage was added to the milk at the start of cheese- making, it did not adversely affect acid production during manufacture. Phage treatment accelerated the decline in viable starter numbers during ripening, accelerated the release of free amino acids and ammonia and reduced bitterness. Phage-induced lysis may have potential for accelerating ripening but the technique may be unacceptable to cheese manufacturers due to fears of unpredictable acid production.

Bacteriocin-induced lysis. In recent years, there has been a surge of interest in broad- spectrum bacteriocins, mostly due to their potential to preserve foods against spoilage and pathogenic organisms. The classical example of a commercially successful bacteriocin is nisin (Daeschel, 1993). The potential applications of narrow-spectrum bacteriocins have not been investigated but they have been studied extensively at the molecular level.

Morgan et al. (1995) identified a narrow-spectrum bacteriocin producer, *L. lactis* subsp *lactis* biovar. *diacetylactis* DPC3286, which differed from other lactococcal bacteriocin producers in that it exhibited a bacteriolytic effect on sensitive lactococci; generally, lactococcal bacteriocins exhibit either a bacteriostatic or bactericidal effect. Analysis of strain DPC3286 revealed that it is both proteinase- and lactose-negative and that bacteriocin production is encoded on a 78-kb plasmid, pSM78. DPC3286 produces three bacteriocins, lactococcins A, B and M, all of which have been studied in detail (van Belkum et al., 1989, 1991a; Venema et al., 1993, 1994). The genetic organization of the genes encoding the lactococcins was found to be highly conserved between DPC3286 and *Lc. lactis* subsp. *cremoris* 9B4 (the strain investigated by van Belkum et al., 1989). The mechanism of action of lactococcins A and B has been identified (van Belkum et al., 1991b; Venema et al., 1993). Since neither A nor B is capable of lysing cells and although the mechanism of action of lactococcin M has not been reported, it is thought that a combination of the three lactococcins may initiate lysis.

All Cheddar cheese starter cultures tested were found to be sensitive to the bacteriolytic activity of DPC3286, although to different extents. The potential of DPC3286, used as an adjunct culture to accelerate the lysis of starter lactococci and consequently the maturation of Cheddar cheese made using *Lc. lactis* subsp. *cremoris* HP, a strain which exhibits a low level of autolysis and produces bitter cheese (O'Donovan, 1994; Wilkinson et al., 1994b), was investigated.

Laboratory-scale (3 L) cheesemaking trials were carried out to determine a suitable ratio of adjunct to starter culture. Since DPC3286 is Lac^- and Prt^-, it does not contribute to acid production during manufacture but its lytic effect on the starter reduced the rate of acid production by the starter culture. Since acid production is critical in cheese manufacture, a level of adjunct that increased the manufacturing time by not more than 30 min was established.

Two pilot-scale trials (500 L) were then conducted using HP with differing levels (0.0–0.225%) of bacteriocin-producing adjunct. In the second trial, a bacteriocin-negative adjunct was included as a control (this strain differs from the bacteriocin-positive strain only in that it lacks the plasmid pSM78, responsible bacteriocin production). The cheeses, which were within the compositional range for Cheddar, were ripened at 8 °C. Lysis was monitored by the release of intracellular enzymes assayed in 'cheese-juice' expressed from cheese under hydraulic pressure. Greater release of intracellular LDH was observed in cheeses containing the bacteriocin-producing adjunct than in the control cheeses, indicating that the adjunct promoted lysis of the HP starter. In trial 1, 0.03 or 0.125% adjunct resulted in average increases of 26 and 66%, respectively, in LDH activity over a 6-month ripening period, relative to the control; corresponding values for intracellular PepX were 60 and 180%. In trial 2, cheese with a 0.225% inoculum of the bacteriocin-producing adjunct exhibited average increases of 62 and 33% LDH and PepX activity over the control while cheeses containing the bacteriocin-negative adjunct had similar LDH activity to control cheese but the level of PepX was increased.

In trial 1, the total concentration of free amino acids in the cheeses containing 0.03 or 0.125% bacteriocin-producing adjunct was 26 and 47% higher, respectively, than in the control. In trial 2, the bacteriocin-producing strain increased the level of free amino acids by 22%. RP- HPLC of cheese juice showed distinctly different peptide profiles for the experimental and control cheeses; the former contained an increased level of hydrophilic peptides, which may indicate reduced bitterness. The experimental cheeses received higher grades than controls for both flavour/aroma and body/texture.

This study revealed that Cheddar manufactured with the bacteriocin-producing strain, DPC3286, as a starter adjunct exhibited increased levels of starter cell lysis, higher concentrations of free amino acids, a reduction in bitterness and higher grading scores; the adjunct did not inhibit the growth of non-starter lactic acid bacteria (since it is a narrow spectrum bacteriocin producer). This novel method for increasing starter cell lysis in Cheddar cheese has many advantages over more conventional methods for accelerating cheese ripening: it requires no special legal approval, avoids the occurrence of hot spots since the bacteriocin-producing cells are distributed throughout the cheese curd and involves no extra costs for specialized equipment. However, extension of the cheese make-time may be a cause for concern.

Attenuated starters

Since the starter plays a key role in cheese ripening it might be expected that increasing cell numbers would accelerate ripening. However, Lowrie & Lawrence (1972) reported that, at least in the case of Cheddar, high numbers of starter cells are associated with bitterness. Not all authors (e.g., Stadhouders et al., 1983) agree that bitterness is related simply to starter cell numbers and suggest that too much or the wrong type of proteolytic activity is responsible, e.g., too little peptidase activity relative to proteinase activity. In fact, a number of authors (see Fox, 1988/89 for references) reported that stimulating starter growth, eg, by adding starter autolysate, protein hydrolysate or trace metals accelerated ripening; this approach appears to run contra to the view that high starter cell numbers cause bitterness. Perhaps the significance of starter cell numbers on cheese ripening should be reinvestigated.

An alternative to the use of high starter cell numbers is the addition of attenuated starter cells to the cheese milk, the rational being to destroy the acid-producing ability of the starter (since excessively rapid acid development is undesirable), but causing as little denaturation of the cell's enzymes as possible. The discussion in the preceding paragraph suggests that adding attenuated cells might cause bitterness but this has not been reported to be a problem, the opposite usually being reported. However, most or all of the studies on the use of attenuated starters have been on varieties other than Cheddar.

Five alternative treatments/approaches have been investigated for the production of attenuated starters.

Lysozyme treatment. Law et al. (1976) report that the addition of lysozyme-treated cells to a level equivalent to 10^{10} cells/g cheese had little influence on the rate of flavour development in Cheddar cheese although the level of free amino acids was increased up to 3 fold compared with controls. Law (1980) considered that while the procedure is suitable for laboratory-scale studies, lysozyme is too expensive for commercial, large-scale cheesemaking; a cheaper supply of lysozyme may render this approach viable.

Heat- or freeze-shocked cells. The lactic acid-producing ability of lactic acid bacteria can be markedly reduced by a sub-lethal heat treatment while only slightly reducing proteinase and peptidase activities; heating at 59 or 69 °C for 15 sec was optimal for mixed mesophilic and lactobacilli cultures, respectively (Pettersson & Sjostrom, 1975). When concentrates of heat-shocked cultures were added to cheese milk at a level of 2% (v/v), ~ 90% of the added cells were entrapped in the curd but entrapment efficiency decreased at higher levels of addition. Proteolysis in Swedish household cheese was increased and quality improved by addition of the heat-shocked cells to the cheesemilk, *Lb. helveticus* being the most effective. The extent of proteolysis increased *pro rata* with the level of heat-shocked *Lb. helveticus* culture added but not for a mesophilic culture, suggesting some limiting factor in the latter. Bitterness was not observed in any of the cheeses.

Essentially similar results were reported by Bartels et al. (1987a) for Gouda cheese. Heat shocking at 70 °C for 18 sec was found to be optimal and 2% addition was almost as effective as 4%. Of several thermophiles investigated, *Lb. helveticus* gave best results; *Lb. bulgaricus* and one strain of *Str. thermophilus* had negative effects on flavour quality due mainly to bitterness. An acetaldehyde-like or yoghurt flavour was noted in most of the cheeses containing heat-shocked lactobacilli.

Heat-shocked (67 °C × 10 s) *Lb. helveticus* cells accelerated amino nitrogen formation and enhanced flavour development in Swedish hard cheese; although Neutrase when added alone accelerated proteolysis, it caused bitterness which was eliminated when both heat-shocked *Lb. helveticus* cells and Neutrase were added to the curd (Ardö & Pettersson, 1988). The effect of heat treatment on the proteolytic system of *Lb. delbrueckii* ssp *bulgaricus* was studied by López-Fandiño & Ardö (1991).

Freeze-thawing also kills bacteria without inactivating their enzymes. Addition of freeze-shocked *Lb. helveticus* CNRZ 32 cells to cheesemilk markedly accelerated proteolysis and flavour development in Gouda cheese without adverse effects (Bartels et al., 1987b). The greatest flavour difference between the control and experimental cheeses was observed after 5 weeks of ripening. Addition of untreated *Lb. helveticus* cells also accelerated proteolysis but caused off-flavours. *Lb. helveticus* peptidases appeared to be capable of degrading and debitterizing bitter peptides.

Solvent-treated cells. Exterkate (1979, 1984) and Exterkate & de Veer (1987) reported that treatment of starter cells with n-butanol activated some membrane-bound proteinases and peptidases, presumably by increasing accessibility for substrate. Addition of a suspension of butanol-treated cells to cheese milk accelerated ripening slightly and, perhaps more importantly, reduced the intensity of bitter flavour compared to control cheeses (Stadhouders et al., 1983). This approach is probably impractical for use in cheesemaking at present because of its complexity and possible legal barriers.

Neutralized inactivated cultures. Shchedushnov and D'Yachenko (1974) described a method for the preparation of inactive starter (*Lactobacillus* spp) by continuous neutralization of the growth medium (whey or skim milk) using marble chips. After 3 days, most of the cells had died but their proteolytic enzymes remained active. Addition of the inactivated starter (1–1.5%), together with the regular starter (1–1.5%), to milk for Cheddar cheese intensified proteolysis and accelerated ripening.

Mutant starters. Because the rate of acid development is a critical factor in cheese manufacture, the amount of normal starter cannot be increased without producing an atypical cheese. This has led to consideration of the use of Lac^- mutants, incorporation of which does not affect the rate of acid development but provides additional proteinases and peptidases.

The use of a Lac^-, Prt^- mutant, *Lc. lactis* C2, to accelerate cheese ripening was described by Grieve & Dulley (1983). Mutant concentrates containing ~ 10^{11} cfu/ml were added to the cheese milk to give levels of starter cells in the curd at milling 10–60 times higher than in the control cheese. Proteolysis was accelerated in the experimental cheeses, flavour quality was improved and flavour development was advanced by up to 12 weeks over controls. Exposure of some control and experimental cheeses to elevated ripening temperatures (20 °C) for one month further increased proteolysis and advanced flavour development. This work was extended by Aston et al. (1983) and Fedrick et al. (1986b) who studied the combined effects of Lac$^-$ starter, exogenous proteinase and elevated temperatures on cheese ripening; regardless of the other treatments employed, supplementation with Lac$^-$ starter accelerated ripening.

Richardson et al. (1983) recommended the use of Prt$^-$ starters to reduce bitterness in cheese. It was claimed (Oberg et al., 1986) that the rate of proteolysis in Cheddar cheese made using Prt$^-$ starters was similar to that in control cheese, but this was not confirmed by Law et al. (1992) who found considerably higher levels of small peptides and free amino acids in cheese made using Prt$^+$ starter than in those made with a Prt$^-$ mutant.

Lac$^-$ *Lactococcus* strains with high exopeptidase activity are commercially available as cheese additives. A selection of such cultures obtained from Chr. Hansen's Laboratories (Reading, UK) was assessed by Tobin and Fox (unpublished) in Cheddar cheese with a controlled microflora. The cheeses containing individual Lac$^-$ *Lactococcus* mutants consistently received higher scores for flavour and body than the controls. Proteolysis and lipolysis in these cheeses are being studied.

The current active programmes on the genetics of lactic acid bacteria will probably lead to the development of Lac$^-$ starters with superior cheese ripening properties, e.g., with increased proteinase and/or peptidase or perhaps other activities that may be important in the rate of cheese ripening and/or quality.

Other types of bacterial cells as additives. *Pseudomonas* spp are extremely proteolytic bacteria. They produce very active, heat-stable extracellular proteinases and lipases which have been studied extensively (see McKellar, 1987), owing to their spoilage potential in dairy products, meat and fish. *Pseudomonas* spp also possess a range of intracellular peptidases which have been the subject of relatively little research: an aminopeptidase (Shamsuzzaman & McKellar, 1987; Gobbetti et al., 1995) and a dipeptidase (Gobbetti & Fox, 1996) from *Ps. fluorescens* and an iminopeptidase and a dipeptidase from *Ps. tolaasii* (Baral, 1995) have been studied. *Ps. tolaasii* also possesses a carboxypeptidase, which has not been isolated (Baral, 1995). Since *Pseudomonas* spp are strict aerobes, they will not grow in or on vacuum-packed cheese. Hence, washed *Pseudomonas* cells (i.e., washed free of extracellular proteinase and lipase) should serve as a useful source of peptidases. Niland & Fox (1996) reported a preliminary study on the use of washed *Ps. tolaasii* cells to accelerate the ripening of Cheddar. Washed cells added to cheesemilk at 10^5–10^8 cfu/ml were entrapped in the curd to give approx 10^6–10^9 cfu/g of fresh curd. The cells died very quickly (to $\sim 10^4$ cfu/g after 4 weeks). Even at 10^9 cfu/g, the *Pseudomonas* enzymes did not affect proteolysis as detected by PAGE but did increase the concentration of WSN and amino acids and accelerated textural and flavour development without the occurrence of off-flavours. Inoculation of cheesemilk with 10^7–10^8 cfu/ml was necessary to have a significant effect. Such a large inoculum may be uneconomic although the ability of *Pseudomonas* to grow on cheap minimal media would reduce production costs. It may also be possible to select or genetically engineer strains with very high peptidase activity; unfortunately, as for genetically engineering starter strains, the key peptidase(s), or other key enzymes, required for accelerated ripening are not yet known.

Genetically engineered starters

The considerable knowledge now available on the genetics of cell wall-associated proteinase and many of the intracellular peptidases makes it possible to specifically modify the proteolytic system of starter *Lactococcus*.

The gene for the neutral proteinase (Neutrase) of *B. subtilis* was cloned in *L. lactis* UC317 by McGarry et al. (1995). Cheddar cheese manufactured with this engineered culture as the sole starter underwent very extensive proteolysis and the texture became very soft within 2 weeks at 8 °C. The cheese was not tasted but its aroma was satisfactory. By using a blend of unmodified and Neutrase-producing cells as starter, a more controlled rate of proteolysis was obtained and ripening was accelerated (McGarry et al., 1994). An 80:20 blend of unmodified:modified cells gave best results. Since the genetically-modified cells were not food grade, the cheese was not tasted but the results appear sufficiently interesting to warrant further investigation when a food-grade modified mutant becomes available.

Since free amino acids are widely believed to make a major contribution, directly or indirectly, to flavour development in cheese, the use of a starter with increased aminopeptidase activity would appear to be attractive. Two studies have been reported (McGarry et al., 1995; Christensen et al., 1995) on the use of a starter genetically engineered to super-produce aminopeptidase N; although the release of amino acids was accelerated, the rate of flavour development and its intensity were not, suggesting that the release of amino acids is not rate limiting. The availability of *Lactococcus* mutants lacking up to 5 peptidases (Mierau et al., 1996) should facilitate identification of key peptidases and hence the engineering of mutants that superproduce these peptidases.

Adjunct starters

The fourth group of contributors to the ripening of cheese are non- starter lactic acid bacteria (NSLAB) which may originate in the milk, especially if raw milk is used, or the cheesemaking environment (equipment, air, personnel). Cheese is quite a hostile environment (low pH, low E_h, lack of fermentable carbohydrate, probiotics produced by the starter) and consequently very few genera of bacteria can grow or even survive in properly made cheese. Apart from *Clostridium* spp, which can grow in the interior of most cheeses (Cheddar types are the major exceptions) unless adequate precautions are taken, NSLAB are the principal bacteria capable of growth in the interior of cheese.

Although NSLAB have been reported to include *Micrococcus, Pediococcus* and *Enterococcus* (in special cases), the predominant species are mesophilic lactobacilli, which may be the only non-starter bacteria present (Jordan & Cogan, 1993). In Cheddar and Dutch-type cheeses made from high-quality pasteurized milk in modern enclosed automated plants, the number of NSLAB is < 50 cfu/g in 1 day-old cheese. These grow at a temperature-dependent rate to $\sim 10^7$, typically within about 2 months in the case of Cheddar. As discussed in Section 3.1, the significance of NSLAB to cheese ripening and quality is unclear; experiments on cheese with a controlled microflora suggest that they perform a similar proteolytic function to starter *Lactococcus* but are less effective.

There is a widely held view that cheese made from raw milk ripens faster and develops a more intense flavour than cheese made from pasteurized milk, suggesting that the indigenous microflora may be responsible. However, pasteurization causes other changes in addition to killing the indigenous microorganisms, e.g., inactivation of indigenous enzymes, denaturation of whey proteins, minor shifts in milk salts. The development of microfiltration permits the removal of indigenous microorganisms without other concomitant changes.

The ripening of Cheddar cheese made from raw, pasteurized or microfiltered milk was compared by McSweeney et al. (1993a). The cheeses made from pasteurized or microfiltered milk were essentially similar with respect to proteolysis, lipolysis, microflora and quality but were considerably different from the cheeses made from raw milk. Flavour developed faster and more intensely in the raw milk cheeses than in the other two cheeses, although it was considered atypical of modern Cheddar. The number of NSLAB was about 10-fold higher in the raw milk cheese than in the others (10^8 compared with 10^7 cfu/g) and were more heterogeneous. Essentially similar results were reported by Bouton & Grappin (1995) for Gruyère Comte cheese made from raw or microfiltered milk. It is concluded from these studies that the microflora of raw milk cheese makes a significant and perhaps a positive contribution to cheese quality.

The development of a more intense flavour in raw milk cheese has stimulated interest in *Lactobacillus* cultures for addition to pasteurized milk to simulate the quality of raw milk cheese. Such cultures are now available from commercial starter suppliers but little scientific information is available on their performance. Published studies include: Puchades et al. (1989), Broome et al. (1990), Lee et al. (1990) and McSweeney et al. (1994b). In all of these studies, low numbers of selected mesophilic lactobacilli were added to the cheesemilk; there is general agreement that the lactobacilli modified proteolysis: in particular, they resulted in the formation of a higher concentration of free amino acids than in the control cheese and improved sensoric quality.

Two further studies on the use of mesophilic lactobacilli as adjunct starters have been completed in our laboratory using cheese made with a controlled microflora (Lynch et al., 1996a, b). In one study, adjunct cultures of 4 species of mesophilic lactobacilli (*Lb. plantarum, Lb. casei* ssp *pseudoplantarum, Lb. casei* ssp *casei* and *Lb. curvatus*) were added individually to the cheese milks at a level of $\sim 10^3$ cfu/ml; a fifth uninoculated vat served as control. Numbers of lactobacilli in the experimental cheeses ex-press were 10^4 to 10^5 cfu/g and increased rapidly during the first month of ripening, reaching a maximum in all cases of 10^7 to 10^8 cfu/g after ~ 3 months. Numbers of *Lb. casei* ssp *casei* and *Lb. curvatus* showed no decline to the end of ripening but numbers of *Lb. plantarum* and *Lb. casei* ssp *pseudoplantarum* decreased by ~ 1 log cycle between 3 and 6 months. The control cheeses remained free of 'wild' NSLAB for 34 (trial 1) and 97 (trial 2) days and their numbers were always at least 2 log cycles lower than the number of lactobacilli in the experimental cheeses.

The 4 month-old control and experimental cheeses received similar scores for flavour intensity and flavour acceptability. After 6 months, the control cheeses received the highest scores for flavour intensity but the lowest scores for flavour acceptability; slight bitterness in the control cheeses detected by some graders may have accounted for this. *Lb. plantarum* and *Lb. casei* ssp *pseudoplantarum* improved flavour acceptability to a greater extent than *Lb. casei* or *casei* and *Lb. curvatus*.

Urea-PAGE showed essentially no differences in proteolysis between the cheeses and only minor quantitative differences between the water-soluble extracts. However, the level of total free amino acids (FAA) was higher in the experimental cheeses than in the controls towards the end of ripening, in agreement with earlier studies. The effectiveness of *Lb. plantarum* and *Lb. casei* ssp *pseudoplantarum* as adjuncts has not been reported previously; they appear to warrant further investigation.

The second study was designed to assess the influence of a mixed *Lactobacillus* adjunct culture (comprising of strains of *Lb. casei* ssp *casei*, *Lb. casei* ssp *pseudoplantarum*, *Lb. plantarum* and *Lb. curvatus*) on the ripening of Cheddar cheese acidified by starter or with lactic acid and glucono-δ-lactone (GDL) (O'Keeffe et al., 1976a). Numbers of lactobacilli in the adjunct-containing cheeses were 10^6 to 10^7 cfu/g of curd ex-press and $\sim 10^8$ cfu/g after 1 month; numbers decreased slightly thereafter. The control starter cheese remained free of 'wild' NSLAB for $\sim$ 1 month while the chemically-acidified control cheeses remained free for only 1 or 2 weeks; in both control cheeses, the number of NSLAB remained at least 2 log cycles lower than in the experimental cheeses throughout ripening.

Adjunct lactobacilli considerably intensified the flavour of the GDL/NSLAB cheese in comparison to the GDL control; however, the flavour was considered unacceptable by many members of the panel and was downgraded for flavour acceptability. The starter-acidified cheeses (with or without adjunct lactobacilli) received considerably higher grades for flavour intensity and flavour acceptability than GDL cheeses. The starter/NSLAB cheese received slightly higher scores for flavour intensity but slightly lower scores for flavour acceptability than the starter control cheese. The sensory data suggest that the adjunct was capable of intensifying but not necessarily improving cheese flavour.

Urea-PAGE showed essentially no differences in primary proteolysis between the cheeses. Levels of WSN were higher in the starter-acidified than in the chemically-acidified cheeses and NSLAB influenced WSN development to only a very minor extent. Both starter-acidified cheeses had considerably higher levels of FAA than the chemically-acidified cheeses, highlighting the importance of the starter in FAA formation. Both the GDL/NSLAB and starter/NSLAB cheeses had higher levels of FAA than their corresponding controls throughout ripening, probably due to increased peptidase activity contributed by the adjunct lactobacilli.

RP-HPLC of 70% ethanol-soluble fractions of the cheeses showed few differences in peptide profiles between the starter and starter/NSLAB cheeses but there were major differences between the chemically and starter-acidified cheeses. Quantitative and qualitative differences were also apparent between the chromatograms of GDL and GDL/NSLAB cheeses.

The results of these studies show that adjunct cultures of mesophilic lactobacilli do influence proteolysis in Cheddar cheese during ripening (to a greater extent in the absence of a starter than in its presence), mainly at the level of FAA formation. While the adjunct used did not accelerate ripening to a significant extent, some modification of cheese flavour was achieved. The use of a single species as adjunct appears to be a more promising than a 'cocktail' of species in enhancing the sensory properties of Cheddar. Further research is being undertaken to confirm this.

Although the *Lactobacillus* strains used in the studies by McSweeney et al. (1994b) and Lynch et al. (1996a, b) were isolated from the highly flavoured raw milk cheese studied by McSweeney et al. (1993a), the impact of these strains when used either individually or as cocktails was very much less than the more heterogeneous indigenous microflora of raw milk. Although the flavour of the raw milk cheese studied by McSweeney et al. (1993a) was atypical of 'modern' Cheddar, it was very much more intense than that of the cheeses made from pasteurized or microfiltered milk. Lactobacilli clearly have the potential to modify cheese flavour and accelerate flavour development but further research is required to select the best strains.

Secondary cultures

The final agent involved in the ripening of many cheese varieties are the secondary starters, especially *Propionibacterium, Brevibacterium, Penicillium* and some yeasts. As discussed in Section 3.1, these cultures play key and characterizing roles in the ripening of cheeses in which they are used. With the exception of some Swiss varieties, cheeses in which secondary starters are used have relatively short ripening times, due to a relatively high moisture content and the very high activity of the secondary starter.

Apart from the production of blue cheese substitutes (as discussed in Section 4.2.4), we are not aware of work on the accelerated ripening of cheeses with a secondary starter and they will not be discussed in this review.

Cheese 'slurries'

The greatest acceleration of ripening has been achieved using the slurry system introduced by Kristoffersen et al. (1967) and refined by Singh & Kristoffersen (1970, 1971b, 1972). Cheddar cheese curd slurried in ~ 3% NaCl to ~ 40% cheese solids developed full flavour in 4–5 days at 30–35 °C when reduced glutathione (100 mg/kg) was included. A relationship was shown between flavour development and the formation of active SH groups and free fatty acids. Ripening of slurries made from chemically acidified curd showed the importance of rennet, lactic acid starter, glutathione and pH (~ 5.3, to retard the growth of undesirable bacteria). Addition of cheese slurries to cheese milk or cheese curd was reported (Abdel Baky et al., 1982a, b) to accelerate the ripening of Cephalotyre 'Ras' cheese. Inclusion of proteinases, lipases or trace elements in the slurries improved their effectiveness.

Ripened Cheddar cheese slurries have been successfully incorporated into processed cheese up to ~ 20% of the blend (Sutherland, 1975). Dulley (1976) reported that addition of a slurry of ripened cheese, which was considered to serve as a source of lactobacilli, to cheese milk reduced the ripening time of the resultant cheese by ~ 25%. A similar principle was described by von Bockelmann and Lodkin (1974): mature cheese was homogenized in Na_3 citrate and added to cheese milk before manufacture; ripening of the resultant cheese was accelerated, apparently due to an increased population of lactobacilli in the cheese.

'Cheese slurries' have been adopted for the rapid ripening of Brick (Kristoffersen et al., 1967), Feta (Zerfiridis & Kristoffersen, 1970) & Swiss (Singh and Kristoffersen, 1971a). A fast-ripening procedure for the preparation of particulate Blue cheese of normal composition was described by Harte & Stine (1972).

Cheese slurries have been used as models in which to study the biochemistry of cheese ripening, to screen the suitability of proteinases and lipases for use as cheese additives and of bacterial cultures as starter or adjuncts. The principal attractions of cheese slurries for such purposes are the short ripening time, the low cost and the possibility of including numerous parameters in a single study which is not possible with cheesemaking, even on a pilot scale. While acknowledging these important advantages, we believe that slurries do not approximate the composition of cheese sufficiently closely and are suitable only for screening cultures or enzymes.

A variant of the slurry system is used in the Novo process for the production of 'Enzyme Modified Cheese' (EMC). Medium-aged cheese is mixed with water, 'emulsified', homogenized and pasteurized to control the indigenous microflora. After cooling, exogenous enzymes, e.g., 'Palatase' (a *M. miehei* lipase) and/or proteinase, are added at the required level and the mixture incubated at 40 °C for 12–96 h. The mixture is then repasteurized (66–72 °C for 4–8 min) to yield an EMC paste that may be used in processed cheese, soups, dips, dressings, snack foods, etc. EMC has 5–20 times the flavour impact of the mild Cheddar from which it is made and may be used at 2–3% of a processed cheese blend.

A similar approach is used by the Miles Marshall Company in the preparation of Marstar enzyme-modified cheese products for use in processed cheeses or in recipes containing cheese (Talbott & McCord, 1981). A range of enzyme preparations is available for Cheddar, Romano and Swiss cheese. The enzyme preparations are added to cheese pastes 40–55% (of total solids) at a level of ~ 1.0%. An essentially similar protocol was described by Lee et al. (1986): mild cheese or fresh cheese curd was mixed with cream, treated with a combination of neutral proteinase (Neutrase) and a lipase (preferably a mixture of gastric and pregastric lipases or *M. miehei* lipase) and incubated at 35 °C for ~ 48 h. Several other methods have been described for the preparation of enzyme-modified cheese (see Lee et al., 1986; Kilara, 1985).

Enzyme-modified cheeses, produced by propriety technology, are used commercially by processed cheese manufacturers. While such products may be very bitter and do not resemble cheese flavour, they apparently do intensify the flavour of processed cheese products and cheese ingredients.

Effect of adding free amino acids to cheddar cheese curd in flavour development

The most abundant amino acids in Cheddar cheese are glutamic acid, leucine, valine, isoleucine, lysine and phenylalanine (Law et al., 1976; Hickey et al., 1983; Puchades et al., 1989; Wilkinson, 1993); histidine and alanine are also present at high concentrations (Broome et al., 1990). The concentration of total amino acids is considered not to be directly responsible for Cheddar flavour but the release of certain amino acids, particularly glutamic acid, methionine and leucine, coincides with flavour development (Broome et al., 1990). Leucine and methionine are considered to be the main contributors to cheesy flavour in the water-soluble extract of Cheddar (Kowalewska et al., 1985; Marsili, 1985; Aston & Creamer, 1986).

Amino acids undergo various catabolic reactions, such as deamination, decarboxylation, transamination and side chain modification, yielding α-keto acids, NH_3, amines, aldehydes, acids or alcohols (Gripon et al., 1991). These degradation products are thought to play a significant role in the formation of specific cheese flavours (Hemme et al., 1982; Fox et al., 1993, 1995). Amino acid catabolism is less intense in Cheddar than in varieties in which moulds or non-lactic bacteria are present. Sulphur compounds, e.g., methanethiol, are major flavour components in washed-rind or mould-ripened cheeses. Methanethiol can be produced from methionine by strains of *Brevibacterium linens* (Law & Sharpe, 1978; Hemme et al., 1982; Ferchichi et al., 1985). It has been reported (Manning, 1979; Law, 1981) that the concentration of methanethiol in Cheddar cheese correlates closely with flavour intensity and its absence from headspace volatiles coincides with the lack of typical Cheddar flavour and aroma (Manning & Price, 1977). Production of methanethiol in Cheddar, which does not contain *Br. linens*, is thought to arise from non-enzymatic decomposition of methionine (Wainwright et al., 1972; Law & Sharpe, 1978) or by the combination of H_2S, produced from cysteine, with methionine (Hemme et al., 1982). A cystathionine β-lyase, capable of producing methanethiol, dimethyldisulphide dimethyltrisulphide from methionine, has been isolated by Alting et al. (1995) from *Lc. lactis* ssp *cremoris* B78. The enzyme was active under conditions equivalent to those in cheese and may be responsible for the biosynthesis of sulfur- containing compounds in cheese, although methionine was a poor substrate for the isolated enzyme. Methanethiol is a precursor of other sulphur compounds, e.g., hydrogen sulphide, carbonyl sulphide and dimethyl sulphide, which can be produced either by the cheese microflora and their enzymes or non-enzymatically (Adda et al., 1982; Hemme et al., 1982). Hydrogen sulphide is desirable in Cheddar but it is not essential for balanced flavour development and may even cause a sulphur flavour defect in ripened cheese if its concentration is too high (Law, 1981; Hemme et al., 1982).

Since free amino acids are released rather slowly during cheese ripening, a study was undertaken by Wallace & Fox (1996) to assess the possibility of accelerating flavour development in Cheddar cheese by adding free amino acids to the curd at salting.

According to Wood et al. (1985), the total concentration of free amino acids in Cheddar is $\sim$ 84 mmol/kg ($\sim$ 11 g/kg). In this experiment, cas-amino acids were added to milled Cheddar curd with the salt at concentrations of 0, 1.4, 2.8, 5.7 and 8.5 g/kg (cheeses A, B, C, D and E, respectively). Proteolysis was monitored by measuring water and PTA-soluble N, RP-HPLC and amino acid analysis on the water-soluble extract (WSE), and urea-PAGE of cheese and WSE. Cheeses were graded after 1, 3 and 6 months on the basis of flavour and texture.

The composition of the experimental and control cheeses were similar and within the expected ranges for Cheddar. Very small differences in the concentration of WSN were observed between the control and experimental cheeses throughout ripening. Increases in PTA-SN were more pronounced in the experimental cheeses and were directly proportional to the level of cas-amino acids added to the curd. Urea-PAGE showed no differences, either quantitative or qualitative, between the control and experimental cheeses or their WSE's at any stage of ripening. RP-HPLC indicated that the control and experimental cheese E contained considerably lower levels of all major peptides than cheeses B, C and D, suggesting that low concentrations of amino acids activated proteolysis but a very high concentration appeared to be inhibitory.

Free amino acid levels, which were proportional to the amount added, remained static or decreased slightly during the first 5 weeks of ripening; however, concentrations increased substantially (1–2 g/kg cheese) in all cheeses between 1 and 3 months and especially between 3 and 6 months, particularly in those supplemented with intermediate levels of cas-amino acids. Cheeses C and D showed increases of 2.5 and 3 g of total amino acids per kg cheese between 3 and 6 months of ripening. Although cheese E had the highest concentration of total amino acids throughout ripening, substantially greater increases in the concentration of amino acids, particularly serine, isoleucine, leucine and phenylalanine, occurred in all other experimental cheeses and the control during ripening (e.g., the increase in free amino acids was 0.8 g/kg higher in cheese C than in cheese E), suggesting that while intermediate levels of free amino acids enhanced peptidolytic activity, higher levels tended to be inhibitory.

The principal amino acids in all 1 day old cheeses were glutamic acid, proline, arginine and leucine; high concentrations of NH_3 were also present in all cheeses. The concentrations of NH_3, lysine and proline decreased during the first 5 weeks of ripening (particularly in the experimental cheeses, the lost being proportional to the level of amino acids added), suggesting that these amino acids were catabolised by the cheese microflora. In agreement with previous workers (Law et al., 1976; Hickey et al., 1983; Puchades et al., 1989; Wilkinson, 1993), there was a substantial increase in the concentrations of leucine, glutamic acid and phenylalanine during ripening, with leucine being the dominant amino acid after 6 months. There was also a substantial increase in the concentration of ammonia in all cheeses, which has not been reported previously. Arginine appeared to be catabolised rapidly in all cheeses during the latter half of ripening, perhaps by intracellular lactococcal enzymes or by NSLAB which are capable of utilising Arg, perhaps producing ornithine (which was not monitored in this study).

Off-flavours were detected in all experimental cheeses (but not in the control) when they were first graded after 5 weeks. Cheese E was described as having a very advanced flavour at 3 months but was 'overripe', with an 'unclean' flavour and a weak, pasty texture, at 6 months. Cheese D had a slight burnt off-flavour up to 3 months but after 6 months it had the best flavour and texture. The quality of cheese C was also very high (although it received a lower flavour intensity score than cheese E at 6 months, its flavour was described as superior). Cheese B was downgraded due to a bitter, over-acid flavour; the control was also bitter at 6 months.

It is concluded that addition of intermediate levels of free amino acids to Cheddar cheese curd during manufacture has a beneficial effect on the development of cheese flavour. Amino acids appear to stimulate proteolysis, particularly secondary proteolysis involving the breakdown of small peptides to free amino acids, either due to the activation of peptidases, increased cell lysis or increased growth of NSLAB (which was not studied). The products of amino acid catabolism were not studied; perhaps this would merit further study as these products are thought to be major contributors to cheese flavour. While cas-amino acids are expensive and would not be practical for industrial-scale use, it may be possible to economically manufacture a protein hydrolysate by acid hydrolysis for use in accelerated cheese ripening.

Prospects for accelerated ripening

There is undoubtedly an economic incentive to accelerate the ripening of low-moisture, highly-flavoured, long-ripened cheeses. Although consumer preferences are tending towards more mild- flavoured cheeses, there is a considerable niche market for more highly flavoured products. While the ideal might be to have cheese ready for consumption within a few days, this is unlikely to be attained and in any case it would be necessary to stabilize the product after reaching optimum quality, eg, by heat treatment, as is used in the production of enzyme-modified cheeses.

Although the possibility of using exogenous (non-rennet) proteinases, and in some cases peptidases, attracted considerable attention for a period, this approach has not been commercially successful for which a number of factors may be responsible: (1) primary proteolysis is probably not the rate-limiting reaction in flavour development, (2) the use of exogenous enzymes in cheese is prohibited in several countries, and (3) uniform incorporation of enzymes is still problematic and the use of encapsulated enzymes is not viable at present. Because it can be easily incorporated into cheese curd, is an indigenous enzyme active in natural cheese and has narrow specificity, producing non-bitter peptides, plasmin may have potential as a cheese ripening aid; however, at present it is too expensive but its cost may be reduced via genetic engineering.

Attenuated cells appear to have given useful results in pilot-scale experiments but, considering the mass of cells required, the cost of such cells would appear to be prohibitive for commercial use, except perhaps in special circumstances. Selected peptidase-rich, Lac^-/Prt^- *Lactococcus* cells added as adjuncts have given promising results but further work is required and they may not be cost-effective.

We believe that the selection of starter strains according to scientific principles holds considerable potential. Such selection procedures are hampered by the lack of information on the key enzymes involved in ripening. Preliminary studies on the significance of early cell lysis have given promising results and further studies are warranted; bacteriocin-induced lysis appears to be particularly attractive.

The ability to genetically engineer starters holds enormous potential but results to date using genetically engineered starters have been disappointing. Again, identifying the key enzymes in ripening is essential for the success of this approach. It is hoped that current research on cheese ripening will identify the key sapid compounds in cheese and hence the critical, rate-limiting enzymes. Genetic manipulation of Lac^-/Prt^- adjunct *Lactococcus* will also be possible when key limiting enzymes have been identified. We believe that adjunct starters, especially lactobacilli, hold considerable potential. It appears to be possible to produce cheese of acceptable quality without lactobacilli but they appear to intensify (Cheddar) cheese flavour and offer flavour options. The volume of literature published on starter adjuncts has been rather limited to date; further work will almost certainly lead to the development of superior adjuncts. There is the obvious possibility of transferring desirable enzymes from lactobacilli to starter lactococci.

At present, elevated ripening temperatures (~ 15 °C) offer the most effective, and certainly the simplest and cheapest, method for the accelerating ripening of Cheddar, which is usually ripened at unnecessarily low temperatures; however, this approach is less applicable to many other varieties for which relatively high ripening temperatures are used at present.

The key to accelerating ripening ultimately depends on identifying the key sapid compounds in cheese. This has been a rather intractable problem; work on the subject commenced nearly 100 years ago and has been quite intense since about 1960, i.e. since the development of gas chromatography. Although as many as 400 compounds which might be expected to influence cheese taste and aroma have been identified, it is not possible to describe cheese flavour precisely. Until such information is available, attempts to accelerate ripening will be speculative and empirical.

References

Abdel Baky AA, El Fak AM, Rabia AM & El Neshewy AA (1982a) Cheese slurry in the acceleration of Cephalotyre 'Ras' cheese ripening. J. Food Prot. 45: 894–897

Abdel Baky AA, El Neshewy AA, Rabia AHM & Farahat SM (1982b) Ripening changes in Cephalotyre 'Ras' cheese slurries. J. Dairy Res. 49: 337–341

Adda J, Gripon JC & Vassal L (1982) The chemistry of flavour and texture generation in cheese. Food Chem. 9: 115–129

Alting AC, Engels WJM, van Schalkwijk S & Exterkate FA (1995) Purification and characterization of cystathionine β-lyase from *Lactococcus lactis* subsp. *cremoris* B78 and its possible role in flavor development in cheese. Appl. Environ. Microbiol. 61: 4037–4042

Arbige MV, Freund PR, Silver SC & Zelro JT (1986) Novel lipase for Cheddar cheese flavor development. Food Technol. 40(4): 91–98

Ardö Y & Pettersson H-E (1988) Accelerated cheese ripening with heat treated cells of *Lactobacillus helveticus* and a commercial proteolytic enzyme. J. Dairy Res. 55: 239–245

Aston JW & Creamer LK (1986) Contribution of the components of the water-soluble fraction to the flavour of Cheddar cheese. N.Z. J. Dairy Sci. Technol. 21: 229–248

Aston JW, Durward IG & Dulley JR (1983) Proteolysis and flavour development in Cheddar cheese. Aust. J. Dairy Technol. 38: 55–65

Aston JW, Grieve PA, Durward IG & Dulley JR (1983) Proteolysis and flavour development in Cheddar cheeses subjected to accelerated ripening treatments. Aust. J. Dairy Technol. 38: 59–65

Baral A (1995) Isolation and Characterization of Enzymes from *Pseudomonas tolaasii*. PhD Thesis, National University of Ireland, Cork

Bartels HJ, Johnson ME & Olson NF (1987a) Accelerated ripening of Gouda cheese. 1. Effect of heat-shocked thermophilic lactobacilli and streptococci on proteolysis and flavor development. Milchwissenschaft 42: 83–88

Bartels HJ, Johnson ME & Olson NF (1987b) Accelerated ripening of Gouda cheese. 1. Effect of freeze-shocked *Lactobacillus helveticus* on proteolysis and flavor development. Milchwissenschaft 42: 139–144

Bouton Y & Grappin R (1995) Comparison de la qualite de fromages a pate pressee cuite fabriques a partir de lait cru on microfiltre. Le Lait 75: 31–44

Broome MC, Krause DA & Hickey MW (1990) The use of non-starter lactobacilli in Cheddar cheese manufacture. Aust. J. Dairy Technol. 45: 67–73

Chapot-Chartier M-P, Deniel C, Rousseau M, Vassal L & Gripon J-C (1994) Autolysis of two strains of *Lactococcus lactis* during cheese ripening. Int. Dairy J. 4: 251–269

Christensen JE, Johnson ME & Steele JC (1995) Production of Cheddar cheese using a *Lactococcus lactis* ssp *cremoris* SK11 derivative with enhanced aminopeptidase activity. Int. Dairy J. 5: 367–379

Coolbear T, Pillidge CJ & Crow VL (1994) The diversity of potential cheese ripening characteristics of lactic acid starter bacteria. 1. Resistance to cell lysis and levels and cellular distribution of proteinase activities. Int. Dairy J. 4: 697–721

Creamer LK, Iyer M & Lelievre J (1987) Effect of various levels of rennet addition on characteristics of Cheddar cheese made from ultrafiltered milk. N.Z. J. Dairy Sci. Technol. 22: 205–214

Crow VL, Holland R, Pritchard GG & Coolbear T (1994) The diversity of potential cheese ripening characteristics of lactic acid starter bacteria: The levels of subcellular distributions of peptidase and esterase activities. Int. Dairy J. 4: 723–742

Crow VL, Martley FG, Coolbear T & Roundhill SJ (1996) The influence of phage-assisted lysis of *Lactococcus lactis* subsp *lactis* ML8 on Cheddar cheese ripening. Int. Dairy J. 6: in press

Crow VL, Coolbear T, Holland R, Pritchard GG & Martley FG (1993) Starters as finishers: Starter properties relevant to cheese ripening. Int. Dairy J. 3: 423–460

Daeschel MA (1993) Applications and interactions of bacteriocins from lactic acid bacteria in foods and beverages. In: Hoover DG & Steenson LR (Eds) Bacteriocins of Lactic Acid Bacteria (pp 63–91). Academic Press, New York

Dulley JR (1976) The utilization of cheese slurries to accelerate the ripening of Cheddar cheese. Aust. J. Dairy Technol. 31: 143–148

Efthymiou CC & Mattick JF (1964) Development of a domestic Feta cheese. J. Dairy Sci. 47: 593–598

El Neshawy AA, Abdel Baky AA & Farahat SM (1982) Enhancement of Domiati cheese flavour with animal lipase preparations. Dairy Ind. Intern. 47(2): 29, 31

El Shibiny S, Soliman MA, El-Bagoury E, Gad A & Abd El-Salam MH (1978) Development of volatile fatty acids in Ras cheese. J. Dairy Res. 45: 497–500

El-Soda M (1993) Accelerated maturation of cheese. Int. Dairy J. 3: 531–544

El-Soda M & Pandian S (1991) Recent developments in accelerated cheese ripening. J. Dairy Sci. 74: 2317–2335

El Soda M, Desmazeaud MJ, Abou Donia S & Badran A (1982) Acceleration of cheese ripening by the addition of extracts from *Lactobacillus helveticus, Lactobacillus bulgaricus* and *Lactobacillus lactis* to the cheese curd. Milchwissenschaft 37: 325–327

El Soda M, Desmazeaud MJ, Abou Donia S & Kamal N (1981) Acceleration of cheese ripening by the addition of whole cells or cell free extracts from *Lactobacillus casei* to the cheese curd. Milchwissenschaft 36: 140–142

Exterkate FA (1979) Effect of membrane perturbing treatments on the membrane-bound peptidases of *Streptococcus cremoris* HP. J. Dairy Res. 46: 473–484

Exterkate FA (1984) Location of peptidases outside and inside the membrane of *Streptococcus cremoris*. Appl. Environ. Microbiol. 47: 177–185

Exterkate FA & Alting AC (1995) The role of starter peptidases in the initial proteolytic events leading to amino acids in Gouda cheese. Int. Dairy J. 5: 15–28

Exterkate FA & de Veer GJCM (1987) Efficient implementation of consecutive reactions by peptidases at the periphery of the *Streptococcus cremoris* membrane. Appl. Environ. Microbiol. 53: 1482–1486

Exterkate FA, Alting AC & Slangen CJ (1995) Conversion of α_{s1}-casein-(24–199)-fragment and β-casein under cheese conditions by chymosin and starter peptidases. System. Appl. Microbiol. 18: 7–12

Farkye NY & Fox PF (1991) A preliminary study of the contribution of plasmin to proteolysis in Cheddar cheese: cheese containing plasmin inhibitor, 6aminohexanoic acid. J. Agric. Food Chem. 39: 786–788

Farkye NY & Fox PF (1992) Contribution of plasmin to Cheddar cheese ripening: effect of added plasmin. J. Dairy Res. 59: 209–216

Farkye NY & Landkammer CF (1992) Milk plasmin activity influence on Cheddar cheese quality during ripening. J. Food Sci. 57: 622–624, 639

Fedrick IA (1987) Technology and economics of the accelerated ripening of Cheddar cheese. Proc. Ann. Conf. Victorian Div. Dairy Ind. Assoc. Aust., Melbourne

Fedrick IA & Dulley JR (1984) The effect of elevated storage temperatures on the rheology of Cheddar cheese. N.Z. J. Dairy Sci. Technol. 19: 141–150

Fedrick IA, Aston JW, Durward IF & Dulley JR (1983) The effect of elevated ripening temperatures on proteolysis and flavour development in Cheddar cheese. I. High temperature storage midway during ripening. N.Z. J. Dairy Sci. Technol. 18: 253–260

Fedrick IA, Aston JW, Nottingham SM & Dulley JR (1986a) The effect of neutral fungal protease on Cheddar cheese ripening. N.Z. J. Dairy Sci. Technol. 21: 9–19

Fedrick IA, Cromie SJ, Dulley JR & Giles JE (1986b) The effects of increased starter populations, added neutral proteinase and elevated temperature storage on Cheddar cheese manufacture and maturation. N.Z. J. Dairy Sci. Technol. 21: 191–203

Feirtag JM & McKay LL (1987a) Isolation of *Streptococcus lactis* C2 mutants selected for temperature sensitivity and potential use in cheese manufacture. J. Dairy Sci. 70: 1773–1778

Feirtag JM & McKay LL (1987b) Thermoinducible lysis of temperature sensitive *Streptococcus cremoris* strains. J. Dairy Sci. 70: 1779–1784

Ferchichi M, Hemme D, Nardi M & Pamboukjian N (1985) Production of methanethiol from methionine by *Brevibacterium linens* CNRZ918. J. Gen. Microbiol. 131: 715–723

Folkertsma B, Fox PF & McSweeney PLH (1996) Acceleration of Cheddar cheese ripening at elevated temperatures. Int. Dairy J. 6: in press

Fox PF (1988/89) Acceleration of cheese ripening. Food Biotechnol. 2: 133–185

— (1989) Proteolysis during cheese manufacture and ripening. J. Dairy Sci. 72: 1379–1400

Fox PF & McSweeney PLH (1996a) Proteolysis in cheese during ripening. Food Rev. Int., in press

Fox PF & McSweeney PLH (1996b) Rennets: their role in milk coagulation and cheese ripening. In: Law BA (Ed) The Microbiology and Biochemistry of Cheese and Fermented Milk, 2nd edn., Chapman & Hall, London, in press

Fox PF & Stepaniak L (1993) Enzymes in cheese technology. Int. Dairy J. 3: 509–530

Fox PF, McSweeney PLH & Singh TK (1995) Methods for assessing proteolysis in cheese. In: Malin EL & Tunick MH (Eds) Chemistry of Structure Function Relationships in Cheese (pp 161–194). Plenum Press, New York

Fox PF, Singh TK & McSweeney PLH (1994) Proteolysis in cheese during ripening. In: Andrews AT & Varley J (Eds) Biochemistry of Milk Products (pp 1–31). Royal Society of Chemistry, Cambridge

Fox PF, Singh TK & McSweeney PLH (1995) Biogenesis of flavour compounds in cheese. In: Malin EL & Tunick MH (Eds) Chemistry of Structure/Function Relationships in Cheese (pp 59–98). Plenum Press, New York

Fox PF, Law J, McSweeney PLH & Wallace J (1993) Biochemistry of cheese ripening. In: Fox PF (Ed) Cheese: Chemistry, Physics and Microbiology, Vol 1 (pp 389–438). Chapman & Hall, London

Fox PF, O'Connor TP, McSweeney PLH, Guinee TP & O'Brien NM (1996) Cheese: Physical, chemical, biochemical and nutritional aspects. Adv. Food Nutr. Res. 39: 163–328

Frick CM, Hicks CL & O'Leary J (1984) Use of fungal enzymes to accelerate cheese ripening. J. Dairy Sci. 67: suppl. 1, 89

Fryer TF (1982) The controlled ripening of Cheddar cheese. Proc. XXI Intern. Dairy Congr. (Moscow) 1(1): 485

Fullbrook P (1987) Biotechnology and related developments as they apply to cheese and other dairy products. Soc. Dairy Technol. Spring Conf., Apr. 26–29

Gaya P, Medina M, Rodriguez-Marin MA & Nunez M (1990) Accelerated ripening of ewes' milk Manchego cheese: The effect of elevated ripening temperatures. J. Dairy Sci. 73: 26–32

Gobbetti M & Fox PF (1996) Isolation and characterization of a dipeptidase from *Pseudomonas fluorescens* ATCC 948. J. Dairy Sci., in press

Gobbetti M, Corsetti A & Fox PF (1995) Purification and characterization of an intracellular aminopeptidase from *Pseudomonas fluorescens* ATCC 948. J. Dairy Sci. 78: 44–54

Grappin R, Rank TC & Olson NF (1985) Primary proteolysis of cheese proteins during ripening. J. Dairy Sci. 68: 531–540

Grieve PA & Dulley JR (1983) Use of *Streptococcus lactis* lac$^-$ mutants for accelerating Cheddar cheese ripening. 2. Their effect on the rate of proteolysis and flavour development. Aust. J. Dairy Technol. 38: 49–54

Gripon J-C, Monnet V, Lamberet G & Desmazeaud MJ (1991) Microbial enzymes in cheese ripening. In: Fox PF (Ed) Food Enzymology, Vol 1 (pp 131–169). Elsevier Applied Science, London

Guinee T, Wilkinson M, Mulholland E & Fox PF (1991) Influence of ripening temperature, added commercial enzyme preparations and attenuated, mutant (lac$^-$) *Lactococcus lactis* starter on the proteolysis and maturation of Cheddar cheese. Ir. J. Food Sci. Technol. 15: 27–51

Harte BR & Stine CM (1977) Effects of process parameters on formation of volatile acids and free fatty acids in quick-ripened Blue cheese. J. Dairy Sci. 60: 1266–1272

Heap HA & Lawrence RC (1976) The selection of starter strains for cheesemaking. N.Z. J. Dairy Sci. Technol. 11: 16–20

Hemme D, Bouillanne C, Metro F & Desmazeaud M-J (1982) Microbial catabolism of amino acids during cheese ripening. Sci. des Aliments 2: 113–123

Hickey MW, van Leeuwen H, Hillier AJ & Jago GR (1983) Amino acid accumulation in Cheddar cheese manufactured from normal and ultrafiltered milk. Aust. J. Dairy Technol. 38: 110–113

IDF (1991) Chemical Methods for Evaluation of Proteolysis in Cheese Maturation. Bulletin 261. International Dairy Federation, Brussels

IDF (1992) Fermentation-produced Enzymes and Accelerated Ripening in Cheesemaking. Bulletin 269. International Dairy Federation, Brussels

Ismail AA & Hansen K (1972) Accumulation of free amino acids during cheese ripening of some types of Danish cheese. Milchwissenschaft 27: 556–559

Jensen F (1970) Free fatty acids in lipase activated cheese. Proc. XVIII Intern. Dairy Congr. (Sydney) IE: 365 (abstr.)

Johnston KA, Dunlop FP, Coker CJ & Wards SM (1994) Comparisons between the electrophoretic pattern and textural assessment of aged Cheddar cheese made using various levels of calf rennet or microbial coagulant (Rennilase 46L). Int. Dairy J. 4: 303–327

Jordan KN & Cogan TM (1993) Identification and growth of non-starter lactic acid bacteria in Irish Cheddar cheese. Ir. J. Agric. Food Res. 32: 47–55

Kilara A (1985) Enzyme-modified lipid food ingredients. Process. Biochem. 20(2): 35–45

Kalinowski L, Frackiewicz E & Janiszewska L (1982) Acceleration of cheese ripening by the use of complex enzyme preparation. Proc. XXI Intern. Dairy Congr (Moscow). I (Book 2): 500

Kalinowski L, Frackiewicz E, Janiszewska L, Pawlik A & Kikolska D (1979) Enzymic preparation for ripening milk protein products. US Patent 4, 158, 607

Kelly AL (1995) Variations in Total and Differential Milk Somatic Cell Counts and Plasmin Levels and their Role in Proteolysis and Quality of Milk and Cheese. PhD Thesis, National University of Ireland, Cork

Kowalewska J, Zelazowska H, Babuchowski A, Hammond EG, Glatz BA & Ross F (1985) Isolation of aroma-bearing material from *Lactobacillus helveticus* culture and cheese. J. Dairy Sci. 68: 2165–2171

Kristoffersen T, Mikolajcik EM & Gould IA (1967) Cheddar cheese flavor. IV. Directed and accelerated ripening process. J. Dairy Sci. 50: 292–297

Kuchroo CN & Fox PF (1983) A fractionation scheme for the water-soluble nitrogen in Cheddar cheese. Milchwissenschaft 38: 389–391

Lane CN & Fox PF (1996) Contribution of starter and added lactobacilli to proteolysis in Cheddar cheese during ripening. Int. Dairy J. (in press)

Law BA (1978) The accelerated ripening of cheese by the use of non-conventional starters and enzymes - a preliminary assessment. Doc. 108 (pp 40–48). International Dairy Federation, Brussels

— (1980) Accelerated ripening of cheese. Dairy Ind. Int. 45(5): 15, 17, 19, 20, 22, 48

— (1981) The formation of aroma and flavour compounds in fermented dairy products. Dairy Sci. Abstr. 43: 143–154

— (1982) Cheeses. In: Rose AH (Ed) Economic Microbiology, Vol 7, Fermented Foods (pp 147–198). Academic Press, London

— (1984) The accelerated ripening of cheese. In: Davies FL & Law BA (Eds) Advances in the Microbiology and Biochemistry of Cheese and Fermented Milk (pp 209–228). Elsevier Applied Science Publishers, London

— (1987) Proteolysis in relation to normal and accelerated cheese ripening. In: Fox PF (Ed) Cheese: Chemistry, Physics and Microbiology, Vol 1 (pp 365–392). Elsevier Applied Science, London

Law BA & King JS (1985) Use of liposomes for proteinase addition to Cheddar cheese. J. Dairy Res. 52: 183–188

Law BA & Sharpe ES (1978) Formation of methanethiol by bacteria isolated from raw milk and Cheddar cheese. J. Dairy Res. 45: 267–275

Law BA & Wigmore A (1982a) Accelerated cheese ripening with food grade proteinases. J. Dairy Res. 49: 137–146

— (1982b) Microbial proteinases as agents for accelerated cheese ripening. J. Soc. Dairy Technol. 35: 75–76

— (1983) Accelerated ripening of Cheddar cheese with a commercial proteinase and intracellular enzymes from starter streptococci. J. Dairy Res. 50: 519–525

— (1985) Effect of commercial lipolytic enzymes on flavour development in Cheddar cheese. J. Soc. Dairy Technol. 38: 86–88

Law BA, Castanon MJ & Sharpe ME (1976) The contribution of starter streptococci to flavour development in Cheddar cheese. J. Dairy Res. 43: 301–311

Law BA, Hosking ZD & Chapman HR (1979) The effect of some manufacturing conditions on the development of flavour in Cheddar cheese. J. Soc. Dairy Technol. 32: 87–90

Law J & Haandrikman A (1996) Proteolytic enzymes of lactic acid bacteria. Int. Dairy J. 6 (in press)

Law J, Fitzgerald GF, Daly C, Fox PF & Farkye NY (1992) Proteolysis and flavor development in Cheddar cheese made with the single starter strains *Lactococcus lactis* ssp *lactis* UC317 or *Lactococcus lactis* ssp *cremoris* HP. J. Dairy Sci. 75: 1173–1185

Le Bars D & Gripon J-C (1993) Hydrolysis of α_{s1}-casein by bovine plasmin. Le Lait 73: 337–344

Le Bars D, Desmazeaud MJ, Gripon J-C & Bergere JL (1975) Etude du role des micro-organismes et de leurs enzymes dans la maturation des fromages. I. -Fabrication aseptique d'un caille modelle. Le Lait 55: 377–389

Lee BH, Laleye LC, Simard RE, Munsch MH & Holley RA (1990) Influence of homofermentative lactobacilli on the microflora and soluble nitrogen components in Cheddar cheese. J. Food Sci. 55: 391–397

Lee CR, Lin CR & Melachouris N (1986) Process for preparing intensified cheese flavor product. US Patent 4, 594, 595

Lemieux L & Simard RE (1991) Bitter flavour in dairy products. I. A review of the factors likely to influence its development, mainly in cheese manufacture. Le Lait 71: 599–636

— (1992) Bitter flavour in dairy products. II. A review of bitter peptides from the caseins: their formation, isolation and identification, structure masking and inhibition. Le Lait 72: 335–382

Lin JCC, Jeon IJ, Roberts HA & Milliken GA (1987) Effects of commercial food grade enzymes on proteolysis and textural changes in granular Cheddar cheese. J. Food Sci. 52: 620–625

Lõpez-Fandiño R & Ardö Y (1991) Effect of heat treatment on the proteolytic/ peptidolytic enzyme system of a *Lactobacillus delbrueckii* subsp *bulgaricus* strain. J. Dairy Res. 58: 469–475

Lowrie RJ & Lawrence RC (1972) Cheddar cheese flavour. IV. A new hypothesis to account for the development of bitterness. N.Z. J. Dairy Sci. Technol. 7: 51–53

Lynch CM, McSweeney PLH, Fox PF, Cogan TM & Drinan FD (1996a) Contribution of starter and non-starter lactobacilli to proteolysis in Cheddar cheese with a controlled microflora. J. Dairy Res. (submitted)

Lynch CM, McSweeney PLH, Fox PF, Cogan TM & Drinan FD (1996b) Manufacture of Cheddar cheese under controlled microbiological conditions, with or without adjunct lactobacilli. Int. Dairy J. (in press)

Mabbitt LA, Chapman HR & Berridge NJ (1955) Experiments in cheesemaking without starter. J. Dairy Res. 22: 365–373

Mabbitt LA, Chapman HR & Sharpe ME (1959) Making Cheddar cheese on a small scale under controlled bacteriological conditions. J. Dairy Res. 26: 105–112

Madkor SA & Fox PF (1994) Ripening of Cheddar cheese containing added trypsin. Egypt. J. Dairy Sci. 22: 93–106

Manning DJ (1979) Cheddar cheese flavour studies: II. Relative flavour contributions of individual volatile components. J. Dairy Res. 46: 523–529

Manning DJ & Price JC (1977) Cheddar cheese aroma - the effect of selectively removing specific compounds from cheese headspace. J. Dairy Res. 44: 357–361

Marsili R (1985) Monitoring chemical changes in Cheddar cheese during aging by high performance liquid chromatography and gas chromatography techniques. J. Dairy Sci. 68: 3155–3161

Martley FG & Lawrence RC (1972) Cheddar cheese flavour. II. Characteristics of single strain starters associated with good or poor flavour development. N.Z. J. Dairy Sci. Technol. 7: 38–44

McGarry A, El-Kholi A, Law J, Coffey A, Daly C, Fox PF & Fitzgerald GF (1994) Impact of manipulating the lactococcal proteolytic system on ripening and flavour development in Cheddar cheese. Proc. 4th Meeting BRIDGE T-Project, Oviedo, Spain, p 32 (abstr.)

McGarry A, Law J, Coffey A, Daly C, Fox PF & Fitzgerald GF (1995) Effect of genetically modifying the lactococcal proteolytic system on ripening and flavor development in Cheddar cheese. Appl. Environ. Microbiol. 60: 4226–4233

McKellar RC (Ed) (1989) Enzymes of Psychrotrophs in Raw Food. CRC Press Inc., Boca Raton, FL

McSweeney PLH & Fox PF (1993) Cheese: Methods of chemical analysis. In: Fox PF (Ed) Cheese: Chemistry, Physics and Microbiology, Vol 1, 2nd edn. (pp 341–388). Chapman and Hall, London

McSweeney PLH, Fox PF & Olson NF (1995) Proteolysis of bovine caseins by cathepsin D: preliminary observations and comparison with chymosin. Int. Dairy J. 5: 321–336

McSweeney PLH, Fox PF, Lucey JA, Jordan KN & Cogan TM (1993a) Contribution of the indigenous microflora to the maturation of Cheddar cheese. Int. Dairy J. 3: 613–634

McSweeney PLH, Olson NF, Fox PF, Healy A & Hϕjrup P (1993b) Proteolytic specificity of chymosin on bovine α_{s1} casein. J. Dairy Res. 60: 401–412

— (1993c) Proteolytic specificity of plasmin on bovine α_{s1}casein. Food Biotechnol. 7: 143–158

McSweeney PLH, Pochet S, Fox PF & Healy A (1994a) Partial identification of peptides from the water-insoluble fraction of Cheddar cheese. J. Dairy Res. 61: 587–590

McSweeney PLH, Walsh EM, Fox PF, Cogan TM, Drinan FD & Castelo-Gonzalo M (1994b) A procedure for the manufacture of Cheddar cheese under controlled bacteriological conditions and the effect of adjunct lactobacilli on cheese quality. Ir. J. Agric. Food Res. 33: 183–192

Mierau I, Junji ERS, Venema G, Poolman B & Kok J. (1996). Peptidases and growth of *Lactococcus lactis* in milk. Le Lait 76: 25–32

Monnet V, Chapot-Chartier MP & Gripon J-C (1993) Les peptidases des lactocoques. Le Lait 73: 97–108

Morgan SM, Ross RP & Hill C (1995) Bacteriolytic activity due to the presence of novel lactococcal plasmid encoding lactococcins A, B and M. Appl. Environ. Microbiol. 61: 2995–3001

Moskowitz GJ & Noelck SS (1987) Enzyme-modified cheese technology. J. Dairy Sci. 70: 1761–1769

Muir DD, Banks JM & Hunter EA (1996) Sensory properties of Cheddar cheese: Effect of starter type and adjunct. Int. Dairy J. 6: 407–432

Nelson JH, Jensen RG & Pitas RE (1977) Pregastric esterase and other oral lipases - A review. J. Dairy Sci. 60: 327–362

Niland EJ & Fox PF (1996) Use of *Pseudomonas tolaasii* as an adjunct for Cheddar cheese. Int. Dairy J. (in press)

Oberg CJ, Davis LH, Richardson GH & Ernstrom CA (1986) Manufacture of Cheddar cheese using proteinase-negative mutants of *Streptococcus cremoris*. J. Dairy Sci. 69: 2875–2981

O'Donovan C (1994) An Investigation of the Autolytic Properties of Different Strains of Lactococci during Cheddar Cheese Ripening. MSc Thesis, National University of Ireland, Cork

O'Keeffe AM, Fox PF & Daly C (1978) Proteolysis in Cheddar cheese: role of coagulant and starter bacteria. J. Dairy Res. 45: 465–477

O'Keeffe RB, Fox PF & Daly C (1976a) Manufacture of Cheddar cheese under controlled bacteriological conditions. Ir. J. Agric. Res. 15: 151–155

— (1976b) Contribution of rennet and starter proteases to proteolysis in Cheddar cheese. J. Dairy Res. 43: 97–107

O'Sullivan M & Fox PF (1990) A scheme for the partial fractionation of cheese peptides. J. Dairy Res. 57: 135–139

Pettersson HE & Sjostrom G (1975) Accelerated cheese ripening: a method for increasing the number of lactic starter bacteria in cheese without detrimental effect on the cheesemaking process and its effect on the cheese ripening. J. Dairy Res. 42: 313–326

Pritchard G & Coolbear T (1993) The physiology and biochemistry of the proteolytic system in lactic acid bacteria. FEMS Microbiol. Rev. 12: 179–206

Puchades R, Lemieux L & Simard RD (1989) Evolution of free amino acids during the ripening of Cheddar cheese containing added lactobacilli strains. J. Food Sci. 54: 885–888

Rank TC, Grappin R & Olson NF (1985) Secondary proteolysis of cheese during ripening: A review. J. Dairy Sci. 68: 801–805

Richardson GH, Ernstrom CA, Kim JM & Daly C (1983) Proteinase negative variants of *Streptococcus cremoris* for cheese starters. J. Dairy Sci. 66: 2278–2286

Shamsuzzaman K & McKellar RC (1987) Peptidases of 2 strains of *Pseudomonas fluorescens*: partial purification, properties and action in milk. J. Dairy Res. 54: 283–293

Shchedushnov EV & D'Yachenko PF (1974) Activation of the enzymatic processes in the manufacture of cheese. Proc. XIX Intern. Dairy Congr. (New Delhi), IE: 696–697

Singh S & Kristoffersen T (1970) Factors affecting flavor development in Cheddar cheese slurries. J. Dairy Sci. 53: 533–536

— (1971a) Accelerated ripening of Swiss cheese curd. J. Dairy Sci. 54: 349–354

— (1971b) Influence of lactic culture and curd milling acidity on flavor of Cheddar curd slurries. J. Dairy Sci. 54: 1589–1594

— (1972) Cheese flavor development using direct acidified curd. J. Dairy Sci. 55: 744–749.

Singh TK, Fox PF & Healy A (1995) Water-soluble peptides in Cheddar cheese: Isolation and identification of peptides in the UF retentate of water-soluble fractions. J. Dairy Res. 62: 629–640

— (1996) Water-soluble peptides in Cheddar cheese: Isolation and identification of peptides in the UF retentate of water-soluble fractions. 2. J. Dairy Res. 63 (in press)

Singh TK, Fox PF, Hϕjrup P & Healy A (1994) A scheme for the fractionation of cheese nitrogen and identification of principal peptides. Int. Dairy J. 4: 111–122

Skeie S (1994) Developments in microencapsulation science applicable to cheese research and development. A review. Int. Dairy J. 4: 573–595

Stadhouders J (1960) De eiwithydrolyse tijdens de kaasrijping de enzymes die het eiwit in kaas hydrolyseren. Neth. Milk Dairy J. 14: 83–110

Stadhouders J, Hup G, Exterkate FA & Visser S (1983) Bitter flavor in cheese. I. Mechanism of the formation of the bitter flavour defect in cheese. Neth. Milk Dairy J. 37: 157–167

Stepaniak L, Gobbetti M & Fox PF (1995) Isolation and characterization of intracellular proteinases from *Lactococcus lactis* subsp *lactis* MG1363. Proc. Conf. on Lactic Acid Bacteria, 22–26 October 1995, p 111, Cork, Ireland

Sutherland BJ (1975) Rapidly ripened cheese curd slurries in processed cheese manufacture. Aust. J. Dairy Technol. 30: 138–142

Talbott LL & McCord C (1981) The use of enzyme modified cheeses for flavoring processed cheese products. Paper No. 1981–14, Proc. 2nd Biennial Marshall International Cheese Conference. Madison, WI, September 15–18

Tan PST, Poolman B & Konings WN (1993) The proteolytic enzymes of *Lactococcus lactis*. J. Dairy Res. 60: 269–286

Tan PST, Chapot-Chartier MP, Pos KM, Rosseau M, Boquien CY, Gripon J-C & Konings WN (1992) Localization of peptidases in lactococci. Appl. Environ. Microbiol. 58: 285–290

Teuber M (1990) Production of chymosin (EC 3.4.23.4) by microorganisms and its use in cheesemaking. Bulletin 251. International Dairy Federation, Brussels, pp 3–15

Thomas TD & Pearce K (1981) Influence of salt on lactose fermentation and proteolysis in Cheddar cheese. N.Z. J. Dairy Sci. Technol. 16: 253–259

Thomas TD & Pritchard G (1987) Proteolytic enzymes and dairy starter cultures. FEMS Microbiol. Rev. 46: 245–268

Torres N & Chandan RC (1981) Flavor and texture development in Latin American White cheese. J. Dairy Sci. 64: 2161–2169

van Belkum MJ, Hayema BJ, Geis A, Kok J & Venema G (1989) Cloning of two bacteriocin genes from a lactococcal bacteriocin plasmid. Appl. Environ. Microbiol. 55: 1187–1191

van Belkum MJ, Hayema BJ, Jeeninga RE, Kok J & Venema G (1991a) Organisation and nucleotide sequence of two lactococcal bacteriocin plasmid. Appl. Environ. Microbiol. 57: 492–498

van Belkum MJ, Kok J, Venema G, Holo H, Nes IF, Konings WN & Abee T (1991b) The bacteriocin lactococcin A specifically increases the permeability of lactococcal cytoplasmic membranes in a voltage- independent, protein mediated manner. J. Bacteriol. 173: 7934–7941

Venema K, Abee T, Haandrikman AJ, Leenhouts KJ, Kok J, Konings WN & Venema G (1993) Mode of action of lactococcin B, a thiol- activated bacteriocin from *Lactococcus lactis*. Appl. Environ. Microbiol. 59: 1041–1048

Venema K, Haverkart RE, Abee T, Haandrikman AJ, Leenhouts KJ, de Leij L, Venema G & Kok J (1994) Mode of action of LciA,yhe lactococcin A immunity protein. Mol. Microbiol. 14: 521–532

Visser FMW (1977a) Contribution of enzymes from rennet, starter bacteria and milk to proteolysis and flavour development in Gouda cheese. 1. Description of cheese and aseptic cheesemaking techniques. Neth. Milk Dairy J. 31: 120–133

— (1977b) Contribution of enzymes from rennet, starter bacteria and milk to proteolysis and flavour development in Gouda cheese. 2. Development of bitterness and cheese flavour. Neth. Milk Dairy J. 31: 188–209

— (1977c) Contribution of enzymes from rennet, starter bacteria and milk to proteolysis and flavour development in Gouda cheese. 3. Protein breakdown: analysis of the soluble nitrogen and amino acid fractions. Neth. Milk Dairy J. 31: 210–239

Visser FMW & de GrootMostert AEA (1977) Contribution of enzymes from rennet, starter bacteria and milk to proteolysis and flavour development in Gouda cheese. 4. Protein breakdown: a gel electrophoretical study. Neth. Milk Dairy J. 31: 247–264

Visser S (1993) Proteolytic enzymes and their relation to cheese ripening and flavor: an overview. J. Dairy Sci. 76: 329–350

von Bockelmann I & Lodin LO (1974) Use of a mixed microflora of ripened cheese as an additive to starter cultures for hard cheese. Proc. XIX Intern. Dairy Congr. (New Delhi). IE: 441 (abstr.)

Wainwright T, McMahon JF & McDowell J (1972) Formation of methional and methanethiol from methionine. J. Sci. Food Agric. 23: 911–914

Wallace JM & Fox PF (1994) Comparison of different methods for quantitation of peptides/amino acids in the water-soluble fraction of Cheddar cheese. Proc. FLAIR-SENS 3: 41–58

— (1996) Effect of adding free amino acids to Cheddar cheese curd on proteolysis, flavour and texture development. Int. Dairy J., submitted

Wilkinson MG (1993) Acceleration of cheese ripening. In: Fox PF (Ed) Cheese: Chemistry, Physics and Microbiology, Vol 1, 2nd edn., (pp 523–555). Chapman & Hall, London

Wilkinson MG, Guinee TP & Fox PF (1994a) Factors which may influence the determination of autolysis of starter bacteria during Cheddar cheese ripening. Int. Dairy J. 4: 141–160

Wilkinson MG, Guinee TP, O'Callaghan DM & Fox PF (1994b) Autolysis and proteolysis in different strains of starter bacteria. J. Dairy Res. 61: 249–262

— (1992) Effect of commercial enzymes on proteolysis and ripening in Cheddar cheese. Le Lait 72: 449–459

Wood AF, Aston JW & Douglas GK (1985) The determination of free amino acids in cheese by capillary column gas-liquid chromatography. Aust. J. Dairy Technol. 40: 166–169

Zerfiridis G & Kristoffersen T (1970) Accelerated ripening of Feta curd. Proc. 18th Intern. Dairy Congr. (Sydney). IE: 351 (abstr.)

Antonie van Leeuwenhoek **70**: 299–316, 1996.

Barriers to application of genetically modified lactic acid bacteria

C. T. Verrips* & D. J. C. van den Berg
*Unilever Research Laboratorium Vlaardingen, Oliver van Noortlaan 120, 3133 AT Vlaardingen The Netherlands (*author for correspondence)*

Key words: lactic acid bacteria, rDNA technology, risk assessment and consumer acceptance

Abstract

To increase the acceptability of food products containing genetically modified microorganisms it is necessary to provide in an early stage to the consumers that the product is safe and that the product provide a clear benefit to the consumer. To comply with the first requirement a systematic approach to analyze the probability that genetically modified lactic acid bacteria will transform other inhabitants of the gastro- intestinal (G/I) tract or that these lactic acid bacteria will pick up genetic information of these inhabitants has been proposed and worked out to some degree. From this analysis it is clear that reliable data are still missing to carry out complete risk assessment. However, on the basis of present knowledge, lactic acid bacteria containing conjugative plasmids should be avoided. Various studies show that consumers in developed countries will accept these products when they offer to them health or taste benefits or a better keepability. For the developing countries the biggest challenge for scientists is most likely to make indigenous fermented food products with strongly improved microbiological stability due to broad spectra bacteriocins produced by lactic acid bacteria. Moreover, these lactic acid bacteria may contribute to health.

Introduction

In the mid-seventies, in spite of the efforts of scientists to evaluate the consequences of the rDNA technology before starting to develop this technology, public opinion about this technology was quite negative. For a considerable part, this was due to poor communication. Instead of addressing the public in an understandable language, scientists discussed the complex aspects of this emerging technology in their own jargon. Moreover, new discoveries in science pose more questions than provide answers (Campbell, 1990). These questions and uncertainties were communicated by the media and sometimes even lead to fear by the public. This fear was further increased by books like 'Playing God' or 'The Boys of Brazil'. It is understandable that the majority of the public rejected this technology which resulted in unprecedented constraints to carry out research in biosciences.

However, the rejection of rDNA technology is not unique in the history of science and technology. Most scientific breakthroughs, like Copernicus theory of the universe, Darwin's evolution theory, or Mendel's laws and even Fleming's penicillin were received with disbelieve or fear. Even scientists are sometimes dogmatic and reject new discoveries as was shown with Temin's discovery of reverse transcriptase, one of the crucial elements in rDNA technology. Technological developments like printing, steam engines, bicycles, or airplanes were received also by the public with mistrust and often fear. Negative attitudes towards technology are much more pronounced in Northern European countries and parts of the US where immigrants from Northern Europe live, than in Southern European, Asian, or South American countries. A clear example of positive perception in Southern Europe of scientific discoveries were those made by Louis Pasteur, who was considered by the French people as a hero. However, we also have to remind Galileo Galilei, who was banned by the church of Rome because of his revolutionary ideas.

Beside the fact that rDNA technology is really a new technology, the public perception of this technology was negatively influenced by the lack of understanding of biology and genetics. However, it was understood that the rDNA technology changes the key molecule of living systems and this caused emotional reactions and concern by a considerable amount of people in Europe and the US. They were under the impression that some of these products may affect their health adversely. Scientists, representatives of governments, and the private industry have communicated this technology to the consumer in such an unstructured and complicated way, that it offered opponents of this technology plenty of opportunities to attack. Figure 1 shows a simple division of the various potential applications of rDNA technology which proved to be very helpful to keep the discussion focused, and free of emotional or ethical aspects.

The acceptance of rDNA food products by consumers depends on a number of factors. The main factors are:

1. Absence of any (perceived) risk of foods containing rDNA;
2. Benefits to the consumers;
3. Clarity and timing of communication to the consumer and environmental organizations on why and how the genetic modification was performed;
4. Positive effects on the environment.

The logic behind the order of these factors is that if, according to the best possible risk assessment, the manufacturer of the food product cannot provide information to the authorities that show that the risk is 'absent', there is no way that such a product will enter the market. However, when the consumer does not see a clear personal benefit, either in terms of quality, health, convenience, shelf life, or price, it is very unlikely that the he or she will buy this product. Assuming that the first two criteria are met, then it becomes the job of the manufacturer and the retailer to communicate why this new product has been developed, how it was done, why it is safe, and what the benefits are to the various organizations and subsequently to the consumer. This communication should take place in an early stage of product development. It is very important in discussions with authorities and organizations that when claims are made that the new process or product contributes to a better environment, such statements are supported by a life cycle analysis of the old process/product versus the new process/product. At least in one case, the addition of phytase to animal feed to reduce the manure production, the environmental benefit was so clear that acceptance by consumer and environmental organizations was not a problem.

Risk of foods containing rDNA products or organisms

It is essential to define the terms, hazard, and risk before starting any discussion on this subject. *Hazard* is defined as the potential (toxicological or ecotoxicological) harmful intrinsic property of the product encoded by the newly constructed genetic material or by the host carrying this genetic information. *Risk* is the probability of *hazard* occurring.

To prove the absence of *risk* is impossible, and therefore 'absence' has to be defined in clear figures. Fortunately, in the food industry a number of risk calculations have been made for various types of products, and it is therefore possible to quantify risks that have proven to be acceptable to authorities, consumer organizations, as well as the public. Even more important is, that based on these risk assessments, the food industry managed to obtain and maintain an extremely good record of safety, in spite of some incidents. For an *extremely hazardous* microorganism as *Clostridium botulinum*, heat treatments for non-microbiologically stable, non-chilled distributed foods have been developed, and are now described in various codes of practice. The prescribed heat treatment will result in a destruction of the most heat resistant *Clostridium botulinum* spores with a factor 10^{12}. Such a heat treatment normally guarantees that statistically one in 10^{12} cans will be contaminated with one spore of *Clostridium botulinum* (Smelt, 1980). For food products that may support growth of the less *hazardous Salmonella* species, heat treatments or other physical decontamination methods are applied resulting in the probability that less than one in 10^8 product units will be contaminated. Clear criteria for spoilage of microorganisms are not present. However, in the dairy industry it is generally accepted that for chilled distributed fresh dairy products, a probability of spoilage by moulds or yeast during the limited shelf life should be less than one in 10^4 product units. Although as such, the risk assessment for these microorganisms look straight forward, this is not the real situation. Lack of reliable data on the contamination of the product before heat treatment, or the probability of contamination during the filling procedure during manufacturing of fresh dairy products, result in uncertainties. Often this leads to an even more severe heat treatment of the product, or extreme decontamination processes for air and packaging material in filling procedures. However, the hazards of toxigenic and pathogenic microorganism are clear. A risk can be calculated and we guess that the calculated risk is between 1 and 1000 times the actual risk. The figures given above will be used as yardsticks for risk assessments of genetically modified microorganisms in food products, assuming that a hazardous situation can be created by transfer of genetic information from the original host to a recipient microorganism or vice versa.

The situation for rDNA modified lactic acid bacteria is quite different from toxigenic or pathogenic bacteria as discussed before. Lactic acid bacteria have been used for centuries in human food products, and consequently these bacteria are generally recognized as safe (so called GRAS status), so the *hazard* is zero. Tables 1A and 1B summarize the application of lactic acid bacteria in European and Indigenous Fermented Foods respectively. Recently, a workshop on the 'Safety of Lactic Acid Bacteria' has been organized by the Lactic Acid Bacteria Industrial Platform of the EU. The main point of discussion was whether certain members of lactic acid bacteria are involved in human infections, such as endocarditis. In particular *Lactobacillus rhamnosus* has been isolated from clinical cases (Klein et al., 1992; Gasser, 1994). As *L. rhamnosus* is a functional microorganism in cheese manufacturing, and a common inhabitant of the oral cavity, it is very difficult to trace the origin of the *L. rhamnosus* species involved in endocarditis. Although the participants of the meeting concluded that it is very unlikely that *L. rhamnosus* is the causative microorganism (Adams & Marteau, 1995), this microorganism has been placed by the 'Berufsgenossenschaft der Chemischen Industrie' in group II (small risk) and not in group I as the other lactic acid bacteria. Also, the involvement in human diseases of *Enterococcus faecalis* and *E. faecium* is under discussion (Jett et al., 1994) although during the workshop it has been concluded that foods containing *Enterococci* have a long history of use without established risk and that no cases of infection have been linked to the consumption of fermented foods or probiotics. Nevertheless the idea that microorganisms found or used in the production of food products are always safe should not be applied any more. However, for the vast majority of lactic acid bacteria there is no doubt about their safety for human consumption.

The consequence of the use of a GRAS organism is that such host cell can be considered as intrinsically free of any *hazard*. The next questions to be answered are:

(1) Does the newly introduced gene codes directly or indirectly for a hazardous property in the intrinsically safe host organism, if not,
(2) do the lactic acid bacteria serve as host cell for production of a single rDNA gene product, which gene product or metabolites produced with this gene product will be isolated and added to a food product or,
(3) do the recombinant organism encoding a new gene product and metabolites made by this gene product remain in the food product. Will the food product be pasteurized or otherwise treated in such a way that the host lactic acid bacterium is killed with an efficiency comparable to a pasteurization process of 15 seconds at 80 °C or,
(4) do the recombinant organism encoding a new gene product and metabolites made by this gene product remain in the food product.

(1) Newly introduced gene codes for a hazardous product

It is obvious, that when the answer on question (1) is positive, the development of food products based on this genetically modified microorganism (GMO) should be stopped immediately. Consequently question (1) will not be further discussed. What will be discussed in some detail are the spheres A 1–3, B 1–3 and C 1–3 of Figure 1 that represent products defined under well defined single food components, well defined food products that contain inactivated rDNA modified lactic acid bacteria and food products containing living lactic acid bacteria respectively.

(2) Well defined single food components

This type of rDNA products, like chymosin (Maat et al., 1981; Teuber, 1990; van den Berg et al., 1990), α-galactosidase (Overbeeke, 1989, Giuseppin et al., 1993) and, endoxylanase (Maat et al., 1992) are on the market. The host for these products are GRAS organisms, although not lactic acid bacteria. Many protocols for the approval of this type of products are applied in different countries. The scheme used in The Netherlands for single food components (Figure 2) proved to be suitable to obtain approval in other European countries as well. Although it is not absolutely required, it is useful to determine first whether the product belongs to sphere A1, A2 or A3 of Figure 1 before walking through the decision tree depicted in Figure 2. Especially whether the vector used to transform the host cell is a self replicating vector, or is integrated at a defined locus of the host chromosome, and/or if the vector is free of any resistance marker or non-essential

Table 1A. Main functional lactic acid bacteria in European Fermented Foods[1]

Product name	Substrate	Main lactic acid bacteria
Baked Goods	Wheat	*Lactobacillus plantarum*
		acidophilus
		delbrueckii
		brevis
		buchneri
		fermentum
		s.francisco
Wine & Brandy	Grapes	*Leuconostoc gracile*
		oenos
		Lactobacillus plantarum
		casei
		fructivorans
		hilgardii
		brevis
		Pediococcus cerevisiae
Cheese & Dairy Products	Milk	*Brevibacterium linens*
		Lactococcus lactis
		cremoris
		Lactobacillus casei
		helveticus
		bulgaricus
		plantarum
		Leuconostoc cremoris
		Pediococcus acidilactici
		pentosaceus
		Streptococcus thermophilus
		Enterococcus faecium
Fermented vegetables/fruits	Cabbage & Cucumbers	*Lactobacillus brevis*
		plantarum
		Leuconostoc mesenteroides
		Pediococcus cerevisiae
	Olives	*Lactobacillus plantarum*
		paracasei
		brevis
		delbrueckii
		Streptococcus sp.
		Pediococcus sp.
		Leuconostoc sp.
Sausages	Meat	*Lactobacillus curvatus*
		lactis
		plantarum
		sake
		Pediococcus acidilactici
		pentosaceus
		Micrococcus caseolyticus

[1] Deducted from Biotechnology Vol.5, Chapter 1–8.

Table 1B. Main functional lactic acid bacteria in Indigenous Fermented Foods[2]

Product name (Country)	Substrates	Main lactic acid bacteria
Banku (Ghana)	Maize, cassava	Lactic acid bacteria
Burukutu (Nigeria)	Sorghum, cassava	Lactic acid bacteria
Busa (Egypt)	Rice, millet	*Lactobacillus* sp.
Dawadawa (Nigeria)	Locust bean	Lactic acid bacteria
Dosai (India)	Black gram and rice	*Leuconostoc mesenteroides*
Hamanatto (Japan)	Whole soybean, wheat flour	*Streptococcus* sp.
		Pediococcus sp.
Idli (India)	Rice, black gram	*Leuconostoc mesenteroides*
Kecap (Indonesia)	Soybean, wheat	*Lactobacillus* sp.
Kimchi (Korea)	Vegetables (seafood, nuts)	Lactic acid bacteria
Kishk (Egypt)	Wheat, milk	Lactic acid bacteria
Mahewu (S. Africa)	Maize	*Lactobacillus delbrueckii*
Miso (China, Japan)	Rice and soybean	*Lactobacillus* sp.
	Rice and cereals	*Lactobacillus* sp.
Ogi (Nigeria)	Maize	Lactic acid bacteria
Puto (Philippines)	Rice	Lactic acid bacteria
Sorghum beer (S.Africa)	Sorghum, maize	Lactic acid bacteria
Soybean milk (Asia)	Soybean	Lactic acid bacteria
Soy sauce (Asia)	Soybean and wheat	*Lactobacillus* sp.
		Pediococcus sp.
Tarhana (Turkey)	Wheat and milk	Lactic acid bacteria

[2]Deducted from Biotechnology Vol.5, Chapter 1–8.

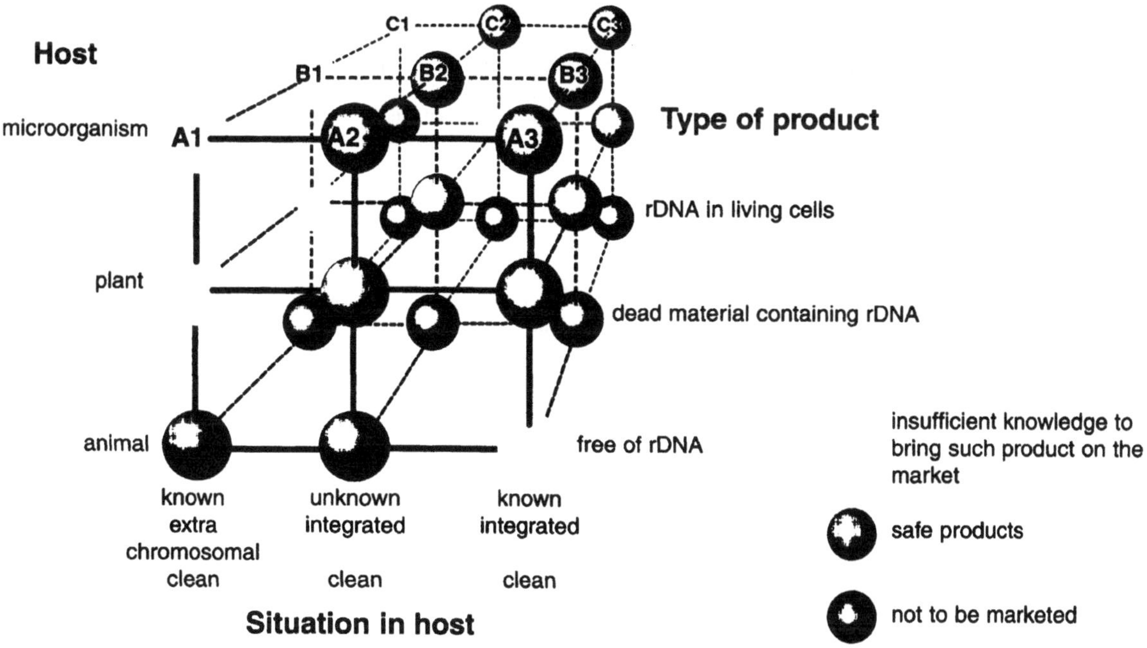

Figure 1. Matrix for the first evaluation of the risk of rDNA products on basis of three criteria: x-axis:Type of vector (epichromosomal, known and unknown integrated in chromosome) y-axis:Type of host (animal, plant or microorganism) z-axis:Type of end product (free of rDNA, contains inactivated rDNA or intact rDNA).

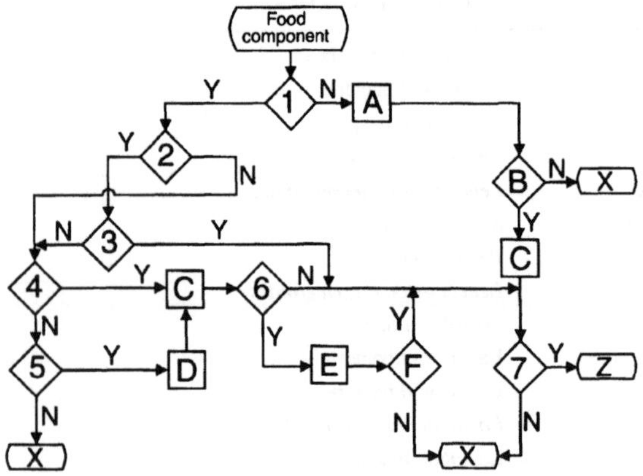

Figure 2. Decision scheme for single food components make by rDNA technology (see also text).

Actions and Results

C. This product should be analyzed as a novel food

X. It is not allowed to bring food products containing ingredients produced with this GMO on the market

Z. Food products containing ingredients produced with this GMO are approved

Questions

1. Does the unmodified microorganism have a record of safe use in food products?
2. Can on the basis of feeding and/or toxicological studies the unmodified microorganism be considered as safe in food products?
3. Is there sufficient knowledge and documentation that the new genetic material codes for (a) product(s) that is (are) acceptable in food products?
4. Does the GMO or an inherent part of it or the product(s) encoded by the new genetic material remain in the food product?
5. It is intended that the modified microorganism fulfills a functional role in the gastrointestinal tract of the consumer?
6. Has the intended functionality been demonstrated?
7. Is the modified microorganism free of genes encoding antibiotic resistance?
8. May the consumption of the food, in particular the GMO or an inherent part of it, or the product(s) encoded by the new genetic material in the intended or expected consumed quantities result in any negative aspect on the health of the consumer?
9. Is it possible to reduce the quantity of the GMO or an inherent part of it or the product(s) encoded by the new genetic material to an acceptable level?
10. Is the physical state or the integration into the chromosome of the host of the new genetic material fully known?
11. Does the integration of the new genetic material disturb the metabolism of the host in such a way that hazardous products may be formed?
12. Does a 90-day feeding trial with the food product containing the GMO or an inherent part of it show that the introduction of the new genetic material into the host does not have an effect on the metabolism of the host cell resulting in (a) hazardous compound(s)?

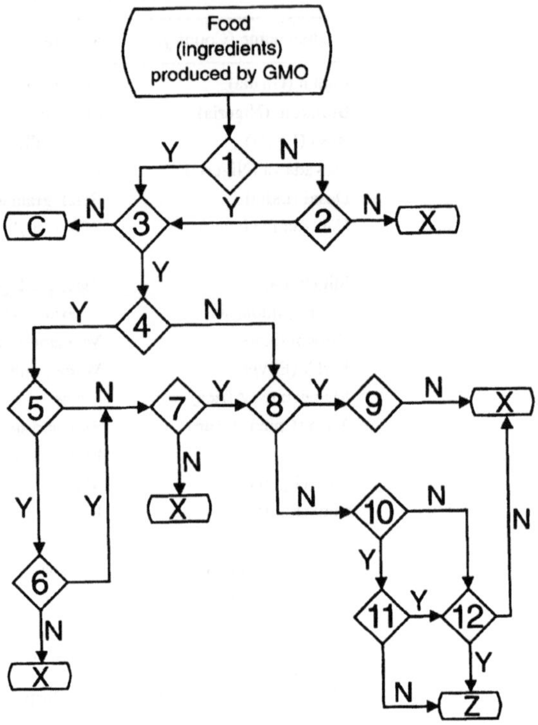

Figure 3. Decision scheme for food products made by genetically modified organisms (see also text).

Entry, Actions and Results

A. Carry out evaluation studies to determine the safety of the product and make specifications

C. Make new specifications

D. Apply a process to reduce the level of undesirable components

E. Carry out a 90-day feeding trial

X. It is not allowed to bring the component on the market

Z. The component is approved for use in foods

Questions

1. Is the use of the *component* in foods allowed at this moment?
2. Does the *component* comply with existing specifications on identity and purity?
3. Are the existing specifications sufficient to control the presence of undesirable site components or too high levels of the intended *component*?
4. Are the levels of known components within the safety specifications?
5. Is it possible to reduce the level of undesirable components during processing in order to comply with the existing specifications?
6. Is the possible that the product contains unknown components?
7. If the intended or assumed consumption of the *component* results in a change in eating habits will the new habit still be considered as safe?

B. Does the evaluation show that the *component* is safe?

F. Does the 90-day feeding trial show that the *component* is safe?

Note: Questions B and F are not (yet) included in the Dutch decision trees as separate questions but form part of *action A and question 5*, respectively.

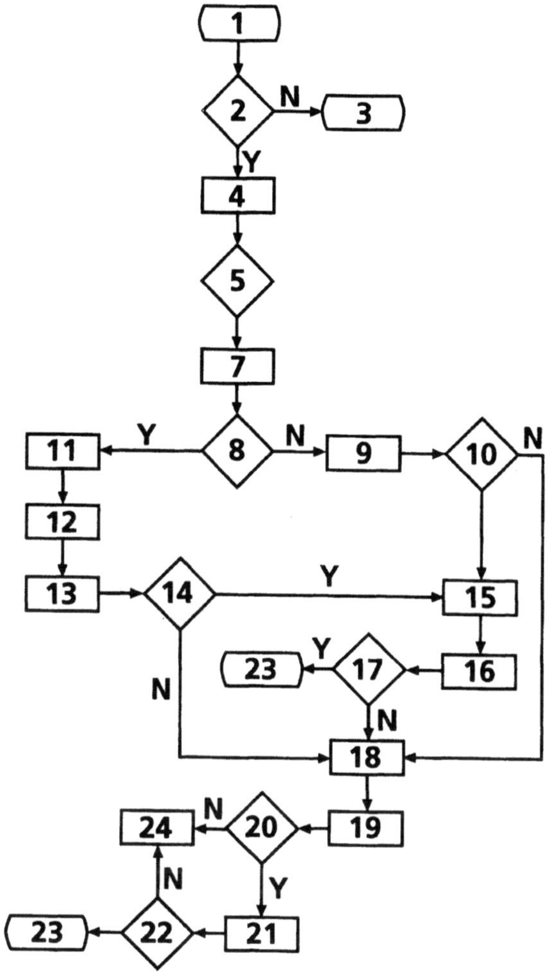

Figure 4. Decision scheme for the evaluation of the potential risk of rDNA lactic acid bacteria in food products. (I) Evaluation of events in the gastro-intestinal tract (see Table 2).

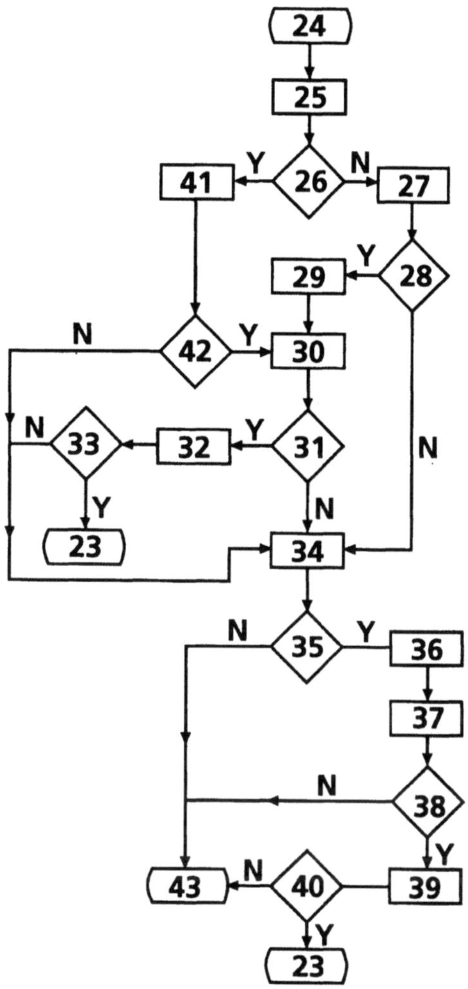

Figure 5. Decision scheme for the evaluation of the potential risk of rDNA lactic acid bacteria in food products. (II) Evaluation of events in the sewage and soil systems (see Table 2).

foreign DNA, are important issues in discussions with authorities and consumer organizations.

(3) Well defined rDNA killed microorganisms in food products

As far as known to the authors this type of products are not (yet) on the market, although Gist-brocades obtained approval in the UK for a Baker's yeast that was modified in such a way that two enzymes in the degradation of maltose (maltose permease and maltose hydrolase) were placed under constitutive promoters to shorten the proofing time of dough (Osinga et al., 1988). For this type of products no clear decision models are available. However, a modification of the model applied in The Netherlands for food ingredients produced by GMOs (Figure 3) can be used. In principle, also the model applied by the FDA for approval of 'Food derived from new plant varieties' (Verrips, 1995) can be applied, if some words are changed.

Before discussing the decision model presented in Figure 3 it is essential to determine whether the product belongs to sphere B1 or B2 or B3 of Figure 1 that represent respectively dead microorganisms containing extrachromosomal rDNA and integrated rDNA either

Table 2. A proposal for a structured assessment of the risk related to the introduction of genetically modified lactic acid bacteria in (or as) food products (Figs. 4 and 5)

Entry, Questions, Actions and Results:

FOOD PRODUCTS

1 (E). Genetically modified lactic acid bacterium (GMO);

2 (Q). Has the product containing the GMO formal clearance based on animal feeding trials ?

3 (R). The release of this GMO cannot be evaluated before it has this formal clearance.

4 (A). Determine the amount of product that comes in the environment (into the sewage system either from factories or households or directly in the soil) before consumption (= spilled product).

5 (Q). Is $V\ (spilled) > a\ V\ (produced)$?

6 (A). Go to *action 24* to evaluate the behaviour of g.m.o. (and if appropriate the s.t.o.) in spilled products and continue with *action 7* for consumed product.

7 (A). Determine the distribution of the residence times of GMO in the gastro/intestinal (g/i) tract of the consumers. Take the time corresponding with 95 % of this distribution curve as $t(r)$.

8 (Q). Will the GMO lyse with $P(b) > b$ in the G/I-tract ?

9 (A). Determine the probability $P(c)$ that intact cells of the GMO transfer genetic information to normal inhabitants of the G/I-tract. Use in these studies $t(r)$ as contact time and the conditions of the G/I-tract.

10 (Q). Is $P(c) > c$?

11 (A). Although the correct procedure will be the determination where lysis will occur and the determination of the distribution of the time of intact cells in the G/I-tract, a worst-case scenario is used assuming the concentration of intact cells is not changed by the lysis and that the intact cells can transfer their genetic information during $t(r)$ to other microorganisms in the G/I tract.

12 (A). As described in *action 11* a worst-case scenario is used to determine the probability that DNA of lysed GMO cells transform other microorganisms of the G/I tract. Use for these studies $t(r)$ as contact time and the G/I-tract conditions.

13 (A). Determine the probability $P(f)$ that DNA originating from lysed GMO transforms normal inhabitants of the G/I tract (resulting in transformed inhabitant).

14 (Q). Is $P(f) > f$?

15 (A). Determine whether the transformed inhabitants obtain an advantage over untransformed inhabitants in the G/I-tract: $A(i)$. Define $A(i)$ in either faster growth rates $t'(g)$; better adhesion h'; or higher production of certain metabolites $p(x)'$, x = 1, ...

16 (A). Determine the probability $P(e)$ that any of the events described under *action 14* will result in the formation of a hazardous (= transformed inhabitant produce toxins or that will replace beneficial microorganism in the G/I tract) microorganisms.

17 (Q). Is $\{P(c) + P(f)\} * P(e) > e$?

18 (A). Determine also the probability $P(d)$ that the GMO will be transformed by genetic material originating from the common G/I tract microorganisms (= modified GMO).

19 (A). Determine whether the modified GMO gains an advantage over untransformed GMO: $B(i)$.
Define $B(i)$ in either faster growth rates $t''(g)$; better adhesion h''; or higher production of certain metabolites $\{p(x)'', x = 1.....\}$.

20 (Q). Is $P(d) > d$?

21 (A). Determine the probability $P(g)$ that any of the events described under *action 21* will result in the formation of an hazardous GMO

22 (Q). Is $P(d) * P(g) > g$?

23 (A). This Risk is unacceptable and the GMO should not be released.

24 THIS IS THE END OF THE RISK ASSESSMENT IN THE G/I TRACT. THE NEXT PHASE WILL BE THE RISK ASSESSMENT OF GMOs IN THE ENVIRONMENT (Figure 5).

25 (A). Determine the average residence times of the GMO or modified GMO in sewage ($t(r2)$ *and* $t(r3)$)

26 (Q). Will the GMO lyse with $P(h) > h$ in the 'sewage and soil system' (S-system)?

27 (A). Determine the probability $P(i)$ that the GMO will transfer its genetic information to other inhabitants of the S-system. Use in these studies $t(r2)$ and $t(r3)$ respectively as contact time and the various S-system conditions to determine $P(i)$.

28 (Q). Is $P(i) > i$?

Table 2. continued

29 (A).	Determine whether the transformed inhabitants gain an advantage (or new property) over untransformed inhabitants of the s-system: *C(i)*. Define *C(i)* in either faster growth rates $t'''(g)$; better survival s''' or higher production of certain metabolites $\{p(\chi)''' , x = 1....\}$.
30 (A).	Determine the probability *P(j)* that any of the events described under *action 29* will result in the formation of a hazardous microorganism.
31 (Q).	Is $P(i) + P(k) * P(j) > j$?
32 (A).	Determine the probability *P(l)* that a transformed hazardous microorganism in the s-system ((secondary transformed organism (STO)) will enter the food chain.
33 (Q).	Is $P(i) + P(k) * P(j) * P(l) > l$?
34 (A).	Determine the probability *P(m)* that the GMO will be transformed by DNA of the normal inhabitants of the S-system.
35 (Q).	Is $P(m) > m$?
36 (A).	Determine whether the transformed GMO gains an advantage or new property over untransformed GMO in the s-system: *D(i)*. Define *D(i)* in either faster growth rates : $t''''(g)$; better survival: s'''': or higher production of certain metabolites $\{p(\chi)'''' , x = 1.....\}$.
37 (A).	Determine the probability *P(n)* that any of the events described under *action 37* will result in the formation of a hazardous modified GMO.
38 (Q).	Is $P(m) * P(n) > n$?
39 (A).	Determine the probability *P(q)* that the modified GMO will enter the food chain.
40 (Q).	Is $P(m) * P(n) * P(q) > q$?
41 (A).	Determine the probability *P(k)* that the DNA originating from lysed GMO transforms the normal inhabitants of the s-system.
42 (Q).	Is $P(k > k$?
43 (R).	Provided that the new genetic information of the (surviving) GMO still has its original configuration, the GMO can be released.

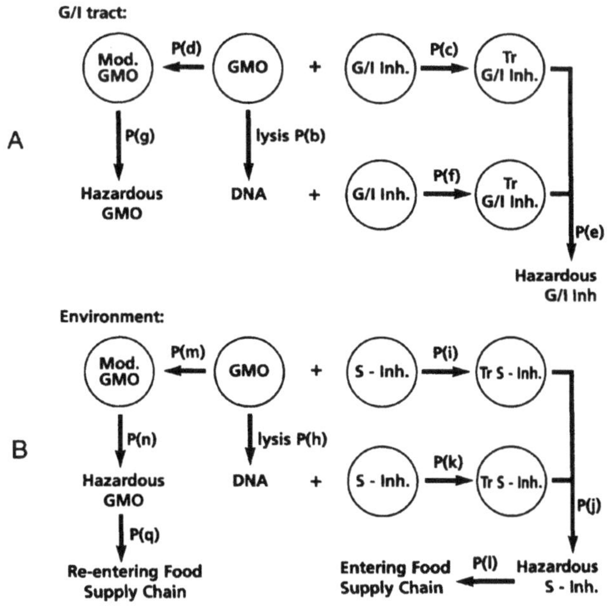

Figure 6. A. Schematic representation of the major events that may occur in the G/I tract and may result in modification of the GMO used or modification of the inhabitants of the G/I tract by conjugation and transformation processes. B. Schematic representation of the major events that may occur in the S-system and may result in modification of the GMO used or modification of the inhabitants of the S-system by conjugation and transformation processes.

Table 3. Estimates of probabilities of transfer of genetic information by/to lactic acid bacteria

Biological process	Probability	Probability function	References values
Lysis of GMO in G/I tract	P(b)	0.75 – 0.99	Marteau & Rambaud (1993)
Conjugative transfer DNA from GMO to lactic acid bacterium of G/I tract	$P(c_1)$	$10^{-3} - 10^{-10}$	El Alami et al. (1992)
Conjugative transfer DNA from GMO to non lactic acid bacterium of G/I tract	$P(c_2)$	$10^{-7} - 10^{-10}$	Langella et al. (1993)
Conjugative transfer DNA from lactic acid bacterium G/I tract to GMO	$P(d_1)$	$10^{-3} - 10^{-10}$	Langella et al. (1993), El Alami et al. (1992)
Conjugative transfer DNA from non lactic acid bacterium G/I tract to GMO	$P(d_2)$		Rood & Cole (1991)
Transformation of GMO by DNA of lactic acid bacteria of the G/I tract	$P(d_3)$	$10^{-8} - 10^{-10}$	El Alami et al. (1992)
Transformation of GMO by DNA of inhabitant G/I tract	$P(d_4)$	$10^{-8} - 10^{-10}$	
Formation hazardous inhabitant G/I tract	P(e)	$< 10^{-8}$	Guinee (1977), Israel et al. (1979) Isberg & Falkow (1985)
Transformation of lactic acid bacterium of G/I tract with DNA of GMO	$P(f_1)$	$<< 6.5\ 10^4/\mu g$	Derived from Tannock et al. (1994)
Transformation of lactic acid bacterium of G/I tract with DNA of GMO	$P(f_2)$	$< 10^{-8}$	Derived from Tannock et al. (1994)
Formation hazardous GMO	P(g)	$< 10^{-10}$	Guinee (1977), Israel et al. (1979)

on a known or unknown chromosomal locus. Bringing a food product on the market made by rDNA lactic acid bacteria that contain resistance markers is (even if the microorganism is killed) a difficult issue to discuss with consumer organizations and in fact they are right. Although the probability of the spread of genes encoding for antibiotic resistance is extremely low, one simply should not take any risk, just because the use of a resistance marker is convenient for scientists to select the transformed microorganism. So situation B1 and preferably also situation B2 of Figure 1 should be avoided. Moreover, in all cases the absence of antibiotic markers is strongly recommended.

(4) Food products containing living rDNA microorganisms

The authors are not aware that a product containing living rDNA microorganisms has been approved. This type of products is different from foods derived from new plant varieties, although the well known transgenic tomato Flavr Savr™, which is on the market, contains living cells. However, the probability that these living tomato cells will grow in the environment is extremely small. This is completely different for food products containing living lactic acid bacteria. A quite extensive decision scheme to cope with this type of products has been proposed, even taking into account the potential transfer of genetic information of the lactic acid bacteria to other microorganisms either in the gastro-intestinal tract (G/I tract) or in the environment (S-system) as described by Verrips (1995). There are two key questions: (i) what is the probability of transfer of rDNA encoded information to human cells; (ii) what is the probability of transfer of rDNA encoded information to normal inhabitants of the G/I tract or the probability that the G.M.O. will be transformed by DNA of inhabitants of the G/I tract.

(i). In these schemes it is assumed that the probability that genetic material encoding functional properties will be transferred from a genetically modified lactic acid bacteria to human cells is zero. This is based on the convincing experiments of Israel et al. (1979) and Chan et al. (1979). In these experiments the complete genomes of the Polyoma virus on either a plasmid or a lambda phage were transferred to *E. coli* K12 and 5×10^9 transformed cells were injected into mice. No transformed mice cells could be detected and the conclusion was that the probability of transfer of polyoma DNA present in *E. coli* was 10^{10} times less than the transfer of viral polyoma DNA. Also feeding trials with 100 mice were performed and these studies showed clearly that the probability of polyoma DNA transfer was less than 5×10^{-14}. Such a figure is even less than the probability of botulism in canned products and for that reason the probability is considered to be zero. Simplified forms of the original decision schemes are presented in Figures 4, 5 and Table 2.

(ii). These decision schemes also deal with the probability that living rDNA modified lactic acid bacteria (GMO) transform inhabitants of the intestine or that this GMO is transformed with genetic material from inhabitants of the G/I tract. The probability that the GMO enters the S- system and transforms microorganisms there (Klijn et al., 1995b), or that GMOs are transformed by soil and sewage microorganisms and re-enter the foods chain, are also part of these decision schemes. Figure 6 presents these events systematically. From this figure it can be concluded that a number of probabilities should be determined to assess in detail the risks involved with the introduction of living rDNA lactic acid bacteria in food products. Only for a few cases sufficient data and knowledge are available to carry out the risk assessment properly. Although extremely unlikely, here it is assumed that the intrinsically safe lactic acid bacteria, can by the uptake of genetic information from other inhabitants in the G/I tract, change into a hazardous one. On the other hand an inhabitant either of the G/I tract or S-system can pick up genetic information of lysed rDNA lactic acid bacteria and acquire new properties. A considerable research programme in order to answer quantitatively these questions is needed. Fortunately, some studies with lactic acid bacteria have been carried out and there are some data on genetic transfer of other microorganisms. In Table 3 a guestimate is presented of the probabilities of the events that may occur in the G/I tract. A number of comments on these figures have to be made:

During the workshop on 'Safety of Lactic Acid Bacteria' some calculations have been made that are helpful for quantification of the probabilities of Table 3. It is reasonable to assume that a person in Europe consumes daily 200 g of a naturally fermented food product. This means a daily intake of $1\text{–}10 \times 10^9$ living lactic acid bacteria and this number can be used as starting point for all the estimations of the probabilities. The most important assumption is that a probability that transformation of a GMO in the G/I tract occurs is less than 10^{-4} (compare this figure with the acceptability that a food product contains spoilage organism). Starting with $1\text{–}10 \times 10^9$ donor cells such event should be less than one donor cell 10^{13} to 10^{14} cells. It is important to define precisely the donor cell as this cell is not always the original GMO. El Alami et al. (1992) and Vogel et al. (1992) have demonstrated that plasmid transfer can take place in milk and fermenting sausages respectively. Very useful as model for events in the G/I tract are the batch fermentation studies of El Alami et al. (1992) on the transfer of plasmids between donor *Lc. lactis* subsp. *lactis* strains IL 2674, IL 2682 and IL 2683 and the recipient strain *Lc. lactis* subsp. lactis strain CNRZ 268M3. In stirred reactors non-self transmissible plasmids were not transferred (Probability $< 10^{-10}$) and self transmissible plasmids were transferred with frequencies between 10^{-7} to 10^{-8} in 12 hours. In static reactors this frequency is in the order of 2×10^{-4} to 2×10^{-6}. Vogel et al. (1992) studied the transfer of plasmids in sausage fermentations and concluded that in fermenting sausages the transfer rate of a conjugative plasmid from *Lactobacillus curvatus* (pAMβ1) to *L. curvatus* (pNZ12) was about 10^{-6}, a figure close to that in model systems. Therefore it is necessary to determine the probabilities of these events in more detail before introducing any GMO-containing food product (Figure 6 a,b).

P(b). It is assumed that lactic acid bacteria pass the stomach without any reduction in number, although more realistic is to take into account the lysis of lactic acid bacteria in the stomach (Marteau & Rambaud, 1993). In the G/I tract two scenarios have been worked out, one in which all lactic acid bacteria stay alive, and one in which all cells are lysed and that DNA is liberated in the G/I tract and that this DNA has a size that can transform other bacteria. The most realistic scenario should take into account both transformation of naked DNA and conjugation processes. From the studies of Marteau & Rambaud (1993) it can be concluded that about 1 % or less of *L. bulgaricus*, *L. acidophilus* and *S. thermophilus* survive these harsh conditions but about 25% of *Bifidobacterium* sp. survive. So the probability of lysis in or before the G/I tract, *P(b)*, is between 0.75 and 0.99. In the subsequent steps of the risk assessment the growth of the survivors as reviewed by Marteau & Rambaud (1993) should be taken into account. The volume of the G/I tract is about 7 litre. However, in the estimations of gene transfer events only the caecum (about 4 litre) is considered to be important as in the caecum rather high numbers of bacteria are present. The estimated number of inhabitants of the caecum are *Bacteroides* and *Eubacteria* 10^{11}/g; (Anaerobe) *Streptococci* and *Bifidobacterium* 10^{10}/g, *Escherichia* 10^7/g, *Lactobacilli* 10^6/g, *Veillonella* and *Clostridium* each

10^3/g (Tannock 1990). Accepting a probability of 10^{-4} that an inhabitant of the G/I tract will be transformed by a GMO the transfer frequency should be less than 10^{-4} : $(4 \times 10^6 \times 10^{11})$ for *Bacteroides* and less than 10^{-4} : $(4 \times 10^6 \times 10^3)$ for *Clostridia*. Important is that most bacteria in the G/I tract are not in a competent state. Taking that only one in 10^6 cells is in a competent state the estimated frequencies should be $< 10^{-15}$ for *Bacteroides* and $< 10^{-7}$ for *Clostridia*.

$P(c_1)$ and $P(d_1)$. As described above the exchange of genetic information between various lactic acid bacteria has been studied both in batch and continuous cultures by several groups. These data give also some indication on the probability of these events in the G/I tract and S-system. Moreover El Alami et al. (1992) determined that on solid media the transfer frequency is in the order of 10^{-2} to 10^{-3}. In the above described sausage fermentation studies a probability of transfer of a conjugative plasmid of 10^{-6} has been found. The most convincing data of transfer of plasmids in the G/I tract has been described by Brockmann et al. (1996). In germ-free rat feeding trials the probability of transfer from the conjugative plasmid pAMβ1 from a donor *Lactococcus lactis* (about 10^7 cells/g) to a recipient *L. lactis* (about 10^8 cells/g) is in the order of 10^{-3} to 10^{-4} taking the figures determined from samples of caecum, colon and faeces. From all these date it seems reasonable to assume a probability in the gut system of 10^{--5}. This is far beyond the acceptable probability and therefore food products containing lactic acid bacteria with conjugative plasmids should not be marketed before more evidence on their safety is provided. Brockmann et al. (1996) carried out similar studies with the non-conjugative plasmid (pLMP1). Using the same cell densities no transfer has been found, so the probability is $< 10^{-8}$. The probability in the actual G/I tract will be much less, as the fate of the donor cells is quite high. Most probably the transfer frequency will be below 10^{-12}, which is close to the acceptable limit. More studies as carried out by Brockmann will result in a more precise figure, either below or above the acceptable limit.

$P(c_2)$ Langella et al. (1993) describe a streptococcal conjugative plasmid pIP501 that encodes transfer functions which allow its transmission into a wide variety of gram positive bacteria. The frequency can be estimated between 10^{-7} to 10^{-10} transconjugants/recipient strain. Also these probabilities are too high. Brockmann et al. (1992) have found that in non-germ free rats the conjugative plasmid pAMβ1 can be transferred from *L. lactis* to *Enterococcus faecalis*. From this result in can be concluded that the probability of transfer of a conjugative plasmid from a Gram positive can occur with a probability of about 10^{-10} donor cells, a figure clearly above the proposed acceptable limit.

$P(d_2)$. Clostridium perfringens is commonly found in the G/I tract of humans as well in sewage and soil. In a number of different strains of *C. perfringens* conjugative plasmids have been found, often these plasmids contain tetracyclin or chloramphenicol resistance (Rood & Cole, 1990). Unfortunately, clear data on the transfer of these plasmids via conjugation to lactic acid bacteria are lacking.

$P(d_3)$. The importance of the stability of integrated rDNA on the probability of transfer of this rDNA to other cells is not known. The stability of integrated rDNA in lactic acid bacteria is very high. Without any selection pressure even a single copy integrant proved to be stable for more than 100 generations. This means a probability of loss of this gene is less than 10^{-13}. However, the integrated sequence should be free of a replicon (Leenhouts et al., 1990). From their work it can be estimated that the loss of erythromycin resistance was much less than one in 10^{13} cells. Consequently, the probability $P(d_3)$ that other inhabitants of the G/I tract are transformed by chromosomal DNA originating from the GMO is $<10^{-13}$.

$P(d_4)$. Rood & Cole (1991) constructed a *C. perfringens/E. coli* shuttle vector and this vector could electrotransform *C. perfringens*. On this basis we assume that the probability will be in the order of 10^{-8} to 10^{-10} transformed GMO/μg DNA. The frequency of electroporation is most likely 10^4 times higher than normal transformation, so it is unlikely that the proposed acceptable probability will be exceeded.

$P(e)$. An important issue is the probability that harmless inhabitants of the G/I tract or the sewage and soil system become pathogenic. Various groups have tried to reconstruct the pathogenicity of *E. coli* K12, a strain isolated from the gut, that lost four of the five essential properties of the pathogenicity of the original strain, being transferred on rich media for many generations. This concerns notably the property of adhesion in the gut, the production of enterotoxins, and resistance against phagocytosis. Studies of Guinee (1977) proved that the probability of reconstruction of the pathogenicity will be less than 10^{-10}. On the other hand, Isberg & Falkow (1985) were able to render *E. coli* K12 into a for cultured animal cells invasive microorganism by transferring a 3.2 Kb plasmid encoding *virulence* genes from *Yersinia pseudoturberculosis*. From these studies a probability fac-

tor for invasion between 9×10^{-2} to 5×10^{-5} can be deducted. However, these experiments were done with genetic material of which the gene products were directly involved in the pathogenicity (so hazard 100% certain), whereas in GMO discussed here, lactic acid bacteria, are intrinsically safe. Therefore, the probability of the creation of a hazardous inhabitant of the G/I tract by transfer of genetic material from the GMO to these inhabitants $P(e)$ can be considered as very small, certainly less than 10^{-10}. The probability that a hazardous GMO will be created can not be deduced from these experiments, but the guess that $P(g)$ will be considerable less than 10^{-10} seems reasonable, but studies are necessary to get realistic data.

$P(f_1)$. A broad host range plasmid isolated from *L. reuteri* (Tannock et al., 1994) could transform *Bacillus subtilis, Streptococcus sanguis, Staphylococcus aureus, Enterococcus faecalis* and several *Lactobacillus* spp. Remarkable was that the transformation into competent cells of *Streptococcus sanguis* was $6.5 \times 10^4/\mu g$ DNA. This high number demonstrates that transformation of plasmid DNA from GMOs to other (potential) inhabitants can not be ruled out. However, these studies were conducted with competent cells. It is unlikely that cells in the G/I tract will have that physiological state. The probability $P(f_1)$ can be estimated as less than 10^{-8}, which probability might be just acceptable or just too high.

$P(f_2)$. On the other hand Tannock et al. (1994) showed that even using electrotransformation, only 15 transformed *Enterococcus faecalis*/μg DNA could be obtained. Although it is not allowed to extrapolate data from electrotransformation to normal transformation processes, it is extremely likely that the probability of the latter is much lower. Taking into account the amount of free DNA in G/I tract that probability will be much less than 10^{-8}.

The survival of lactic acid bacteria in the environment is not very well studied. For the detection of *Lactococcus* species, Klijn et al. (1991) developed a sensitive method based on analysis of the hypervariable region of 16S rRNA using PCR and specific DNA probes. This method was used to study the survival of *Lactococcus* species in the waste flow of a cheese factory and in the environment of cattle (Klijn et al., 1995a, c). However, studies do not provide quantitative data to calculate $P(h)$. Therefore, the two extreme scenarios of 100% intact cells and 100% lysis have to be worked out in the second part of the risk assessment. Similar to the considerations for $P(e)$ the probability of $P(j)$ can be estimated as 10^{-10}. The probability $P(n)$ will just as $P(j)$ be also in the order of 10^{-10}. An important difference between the risk assessment in the G/I tract and in the S-system is that in the latter system the concentration of recipient cells is very low. However, the probability that the modified GMO's or the modified inhabitants of the S-systems re-enter the Food chain should be included. From our long experience in food factories we estimate the probability that (modified) lactic acid bacteria re-enter the food chain between 10^{-6} to 10^{-10} (Verrips, unpublished results). The probability of (modified) inhabitants of the S-system to re-enter the food chain is for canned foods in non chlorinated water in the order of 10^{-4}. If the cooling water is properly treated this probability is at least 10^4 times less.

To summarize the present situation, it is still not possible to carry out a complete and reliable risk assessment, and that is certainly a big problem to bring these products on the market. Especially the use of (conjugative) plasmids should be avoided. When a new gene (that does not code for a hazardous property) is integrated stably in the chromosome the data that are available show that with a very high probability products containing such GMO will get permission to enter the market.

Benefits to the consumer

The selection of food products by consumers depends on factors such as quality, functionality (including health and nutritional functionality), natural or 'Green', and convenience (Verrips, 1991). These factors should be improved to offer benefits to the consumers. At present it will not be easy to convince the consumer that rDNA modified lactic acid bacteria will provide any benefit to them. Most of the present targets are related to the improvement of phage resistance, controlled and/or accelerated proteolysis, stabilization of the proteolytic capability, introduction of genes encoding for hydrolytic enzymes to improve fermentation capacity or the usage of normally non-fermentable sugars, etc (McKay & Baldwin, 1990). Achievement of these targets will result in benefits to the manufacturer. The majority of studies, directed to fulfil the demands of the consumer, is related to improved processing and at the best this will result in a marginally lower price of the product. According to consumer studies carried out at Unilever Research Vlaardingen, this is not a major factor in consumer habits. However, in at least four areas significant ben-

Table 4. Potential health benefits of foods prepared with lactic acid bacteria

1.	Improved nutritional value, in particular higher levels of B vitamins and certain amino acids like methionine, lysine and tryptophan
2.	Antagonistic action towards enteric pathogens in the intestines, especially important for patents suffering from disorders such as functional diarrhoea, mucous colitis, ulcerated colitis, diverticulitis and antibiotic colitis
3.	Improved lactose utilization, which is of importance to negroes and some caucasians;
4.	Conversion of potential pre-carcinogens into less harmful compounds
5.	Inhibitory action towards some types of cancer, in particular cancers of the g/i tract by degradation of pre-carcinogens and simulation of the immune system
6.	Hypercholesterolemic action of lactic acid bacteria
7.	Naturally occurring or rDNA vaccinal epitopes

Table 5. Attitudes to applications of genetic manipulation (adapted from (A) Martin & Tait (1992) & (B) Heijs et al. (1993)

		Comfortable	Neutral	Uncomfortable
Microbial Production of bioplastics	(B)	91	6	3
Cell fusion to improve crops	(B)	81	10	10
Extension shelf life tomatoes	(B)	71	11	19
Anti blood clotting enzymes produced by rats	(B)	65	14	22
Medical research	(A)	59	23	15
Making medicines	(A)	57	26	13
Making crops to grow in the Third World	(A)	54	25	19
Mastitis resistant cows by genetic modification of cows	(B)	52	16	31
Producing disease resistant crops	(A)	46	29	23
Chymosin production by yeast	(B)	43	30	27
Improving crop yields	(A)	39	31	29
Using viruses to attack crop pests	(A)	23	26	49
Improving milk yields	(A)	22	30	47
Cloning prize cattle	(A)	7.2	18	72
Changing human physical appearance	(A)	4.5	9.5	84
Producing hybrid animals	(A)	4.5	12	82
Biological warfare	(A)	1.9	2.7	95

efits to consumers and sometimes also to the retailers can be delivered:

(a) Improved and more constant flavour profile of the products, and improved keepability of the organoleptic and physical appearance during retail as well during storage at home. This is important as shown by the acceptance by the consumer of the genetically modified tomato Flavr Savr™. This tomato contains anti-sense polygalacturonase and this results in a slower degradation of pectins in the cell wall of the tomato. Consequently the ripening and decay of the tomato is delayed which was the main reason for consumers to buy these tomatoes, as the keepability at home of the rDNA tomato is increased from 3–5 days to about 10 days.

(b) Better microbiological keepability using natural isolates of lactic acid bacteria that produce inhibitors of other microorganisms (Geis et al., 1983; Van den Berg et al., 1993). Alternatively rDNA lactic acid bacteria can be used to overproduce these bacteriocins or produce bacteriocins that have a wider range of target organisms the activity of bacteriocins (Vandenbergh, 1993). The development of bacteriocins against Gram(-) bacteria will contribute substantially to the keepability of a lot of fermented foods. Especially for products used in developing countries improved microbiological keepability can be of great importance to ensure the food supply for a rapidly growing population.

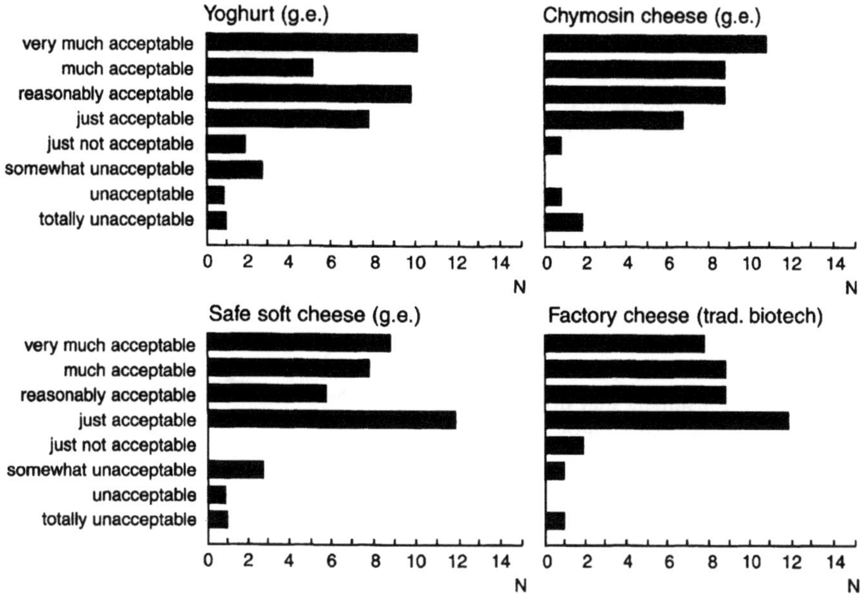

Figure 7. Acceptation of dairy products containing rDNA derived enzymes, GMO's or traditional biotechnological products.

(c) An other interesting option is the use of lactic acid bacteria to produce the right consistency of the product. Especially isolates from meat and olives are able to produce a range of extracellular polysaccharides that improve the viscosity of products (Van den Berg et al., 1993). In this way products with the right consistency can be made without addition of thickeners.

(d) Nutritional aspects of lactic acid bacteria have received a large amount of attention and in the literature various claims are made. Gilliland (1990) and Marteau & Rambaud (1993) summarized the literature on potential beneficial health aspects of lactic acid bacteria (Table 4). In spite of many studies on the nutritional aspects of lactic acid bacteria, no studies have been reported that were designed in such a manner that conclusions on beneficial health aspects could be drawn beyond any reasonable doubt. At the Dutch Institute for Dairy Research (NIZO) and at Unilever Research Vlaardingen quite some attention has been paid on lactic acid bacteria with increased bile salt hydrolase activity. As was proposed by Gilliland (1985), this property would result in a lowering of the blood cholesterol level. We proved that Gilliland's conclusion on the positive action of these lactic acid bacteria was incorrect (Fletcher, personal communication). This confirms findings of Klaver & Van de Meer (1993) on the same topic. Also for the other beneficial aspects given in Table 4, no solid support can be found. It is clear that much better designed experiments are required to change the presently soft, but potentially very important claims into scientifically solid claims about the beneficial health aspects of lactic acid bacteria. If anything can contribute to the acceptance of rDNA lactic acid bacteria in European and American food products, it will be solid health claims.

Clarity and timing of communication to the consumer and environmental organizations and consumer acceptance of rDNA technology

Many studies have been carried out to determine the acceptance of consumers of rDNA technology. Some of the results of these studies are summarized in Table 5, whereas Figure 7 gives specific information on some product containing lactic acid bacteria. In most of these studies it turned out that clarity and timing of communication to consumer- and environmental organizations is a key factor in the acceptance of products made by rDNA technology. SWOKA, the Dutch institute that studied the relation between science and technology and consumer aspects, developed a nice model on the factors that influence consumer acceptance of new products (Figure 8). From another report of SWOKA it is clear that the knowledge of Dutch consumers on biotechnology is quite low as illustrated by the fact that only 58 and 55 % of the consumers know that yoghurt, respectively penicillin, are biotechnological products (Hamstra & Feenstra, 1989). There are several reports that in other countries the situation is similar. In the investigations of Martin & Tait (1992) and Heijs et al. (1993) information was collected on the probability that consumers will buy at a certain moment rDNA-containing products. Table 6 clearly shows that the number of consumers that do not yet know whether they will buy or reject these products is very large. So the information on benefits and perceived or real risks have to be supplied in a clear and structured way to environmental and consumer organizations. Moreover, the timing of supplying information is very important. The development of a new biotechnological product will take somewhere between 4 and 10 years and is characterized in discrete steps of this innovation process (Verrips, 1995). Normally about two years before introduction into the market, sufficient knowledge on the new product is available to communicate the innovation to various organizations. When planned properly, that will not interfere at all with proprietary right on that product, as nearly always patents are filed in an earlier state. Information on the way a product is made will not always result in a more positive attitude of the consumer. Studies of Smink & Hamstra (1995) found that information of production methods for food products (not including transgenic animals) have different effects on different groups of consumers. Consumers with an quite positive attitude are consumers interested in (bio)technology and they like to have information because of this interest. Consumers with a neutral attitude towards biotechnology are not interested in information on the label, whereas consumers with a negative attitude want to have information on the label, because they will use that information to decide whether they will buy the product or not (Figure 9). On the other hand information on production methods including transgenic animals decrease the acceptance. On the question 'I think it is acceptable that this product is made via biotechnology', the two applications of this technology that were rejected by the consumers involved transgenic animals, whereas products like yoghurt, cheese and tomatoes were quite well accepted (Figure 10). This

Table 6. Intentions to buy or reject products made with rDNA technology

	Intention to buy	Intention to protest
Certainly yes	7%	4%
Probably yes	50%	15%
Probably no	34%	46%
Certainly no	9%	36%

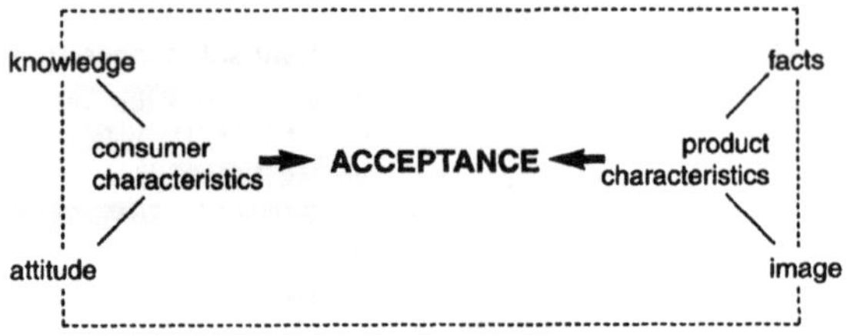

Figure 8. Factors that influence acceptance of new food products by consumers.

illustrates again the importance of a clear division of food products as pointed out in Figure 1.

Conclusions

The introduction of food products made with genetically modified lactic acid bacteria will be accepted most likely in developing countries, Japan and the USA, provided that integrative systems are used and that the benefits to the consumer are clear. In Europe the situation is more complex. If the product does not contain living cells with rDNA, most likely these products with clear consumer benefits will be accepted. For products containing living lactic acid bacteria, which is the majority of fermented food products, the situation is promising provided that consumer benefits are clear. The emphasis of the usage of rDNA technology for lactic acid bacteria should be on the development of clear nutritional and health benefits to the consumers, and on the production of better and more consistent quality of these products. For developing countries, besides the factors mentioned above, also the usage of rDNA technology to extend the microbiological keepability of food products should be investigated. Better and more consumer directed communication is also necessary to get acceptance of the majority of these products by consumers. Finally research should be conducted to quantify the probabilities necessary to carry out proper risk assessments, because as long as

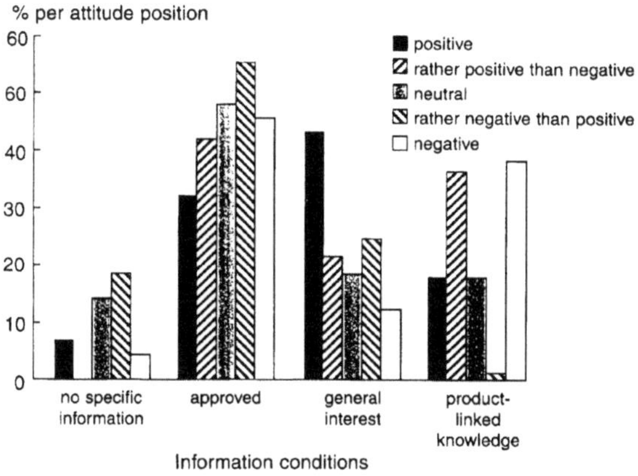

Figure 9. Effect of information and knowledge about food products and acceptance by consumers.

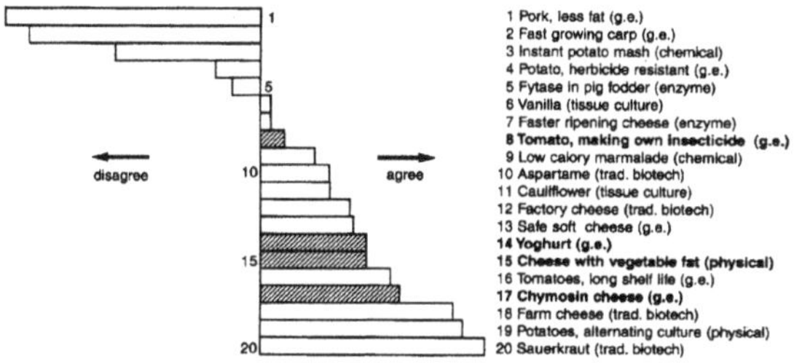

Figure 10. Weighted answers of consumers on the question 'I think it is acceptable that this product is made via biotechnology'.

uncertainties on the safety of certain rDNA products remains these products will not enter the market. Especially the probability of transfer of genetic information via conjugation in the G/I tract *P(c)* should be studied. Until better data are available the use of rDNA lactic acid bacteria containing (conjugative) plasmids is not recommended. Clear, realistic and uniform legislation for the whole European Union is necessary to remain competitive with USA, Japan and some emerging technologically advanced countries.

References

Adams MR & Marteau P (1995) On the safety of lactic acid bacteria from foods. Int. J. Food Microbiol. 27: 263–264

Brockmann E, Jacobsen BL, Hertel C, Ludwig W & Schleifer KH (1996) Monitoring of genetically modified *Lactococcus lactis* in gnotobiotic and conventional rats by using antibiotic resistance markers and specific probe or primer based methods. Syst. Appl. Microbiol. 19 (in press)

Campbell AL (1990) in: Introduction of genetically modified organisms into the Environment (H. A. Mooney and G. Bernardi, Ed.), Scope 44, J. Wiley & Sons, New York

Chan HW, Israel MA, Garon CF, Rowe WP & Martin MA (1979) Molecular cloning of polyoma virus DNA in *Escherichia coli*: Lambda phage vector system. Science 203: 887–892

El Alami, N, Boquien C-Y & Corrieu G (1992) Batch cultures of recombinant *Lactococcus lactis* subsp. *lactis* in stirred fermenter. II Plasmid transfer in mixed cultures. Appl. Microbiol. Biotechnol. 37: 364–368

Gasser F (1994) Safety of lactic acid bacteria and their occurrence in human clinical infections. Bull. Inst. Pasteur 92: 45–67

Geis A, Singh J & Teuber M (1983) Potential of lactic streptococci to produce bacteriocins. Appl. Environ. Microbiol. 45: 205–211

Gilliland SE, Nelson CR & Maxwell C (1985) Assimilation of cholesterol of *Lactobacillus acidophilus*. Appl. Environ. Microbiol. 49: 377–381

Gilliland SE (1990) Health and nutritional benefits for lactic acid bacteria. FEMS Microbiol. Rev. 87: 175–188

Giuseppin MLF, Almkerk JW, Heistek JC & Verrips CT (1993) Comparative study on the production of guar α-galactosidase by *Saccharomyces cerevisiae* SU50B and *Hansenula polymorpha* 8/2 in continuous cultures. Appl. Env. Microbiol. 59: 52–59

Guinee P (1977) Tweede Jaarverslag KNAW Commissie, pg 94–108, KNAW Amsterdam

Hamstra AM & Feenstra MH (1989) SWOKA report, SWOKA, Den Haag

Heijs WJM, Midden CJH & RAJ Drabbe (1993) Biotechnologie, houdingen en achtergronden. Technische Universiteit Eindhoven

Isberg RR & Falkow S (1985) A single genetic locus encoded by *Yersinia pseudotuberculosis* permits invasion of cultured animal cells by *Escherichia coli* K-12. Nature 317: 262–264

Israel MA, Chan HW, Rowe WP & Martin (1979) Molecular cloning of polyoma virus DNA in *Escherichia coli*: Plasmid vector system. Science 203: 883–887

Jett BD, Huycke MM & Gillmore MS (1994) Virulence of enterococci. Clin. Microbiol. Rev. 7: 462–478

Klaver FAM & Van de Meer R (1993) The assumed assimilation of cholesterol by Lactobacilli is due to their bile salt-deconjugating activity. Appl. Environ. Microbiol. 59: 1120–1124

Klein G, Bonaparte C & Reuter G (1992) Laktobazillen als Starterkulturen für die Milchwirtschaft unter dem Gesichtspunkt der Sicheren Biotechnologie. Milchwissenschalt 47: 632–636

Klijn N, Weerkamp AH & de Vos WM (1991) Identification of mesophyllic lactic acid bacteria by using polymerase chain reaction amplified variable regions of 16S rRNA and specific DNA probes. Appl. Environ. Microbiol. 57: 3390–3393

Klijn N, Weerkamp AH & de Vos WM (1995a) Detection and characterization of lactose-utilizing *Lactococcus* spp. in natural ecosystems. Appl. Environ. Microbiol. 61: 788–792

Klijn N, Weerkamp AH & de Vos WM (1995b) Genetic marking of *Lactococcus lactis* shows its survival in the human gastrointestinal tract. Appl. Environ. Microbiol. 61: 2771–2774

Klijn N, Weerkamp, AH & de Vos WM (1995c) Biosafety assessment of the application of genetically modified *Lactococcus lactis* spp. in the production of fermented milk products. System. Appl. Microbiol. 18: 486–492

Langella P, LeLoir Y, Ehrlich SD & Gruss A (1993) Efficient plasmid mobilization by pIP501 in *Lactococcus lactis* subsp. *lactis*. J. Bacteriol. 175: 5806–5813

Leenhouts KJ, Kok J & Venema G (1990) Stability of integrated plasmids in the chromosome of *Lactococcus lactis*. Appl. Environ. Microbiol. 56: 2726–2735

Maat J, Edens L, Ledeboer AM & Verrips CT (1981) Unilever patent application EP-B 0077109

Maat J et al. (1992) Xylanases and their application in Bakery. pp. 349–360. In: Xylans and Xylanases J. Visser et al. ed. Elsevier Science Publishers, Amsterdam 1992

Marteau P & Rambaud J-C (1993) Potential of using lactic acid bacteria for therapy and immuno-modulation in man. FEMS Microbiol. Rev. 12: 207–220

Martin S & Tait J (1992) Attitudes of selected public groups in the UK to biotechnology pg 28–41. In: 'Biotechnology in public: a review of research' (Ed. J Durant), Science Museum for the European Foundation of Biotechnology, London

McKay LL & Baldwin KA (1990) Applications for biotechnology: present and future improvements of lactic acid bacteria. FEMS Microbiol. Rev. 87: 3–14

Overbeeke N, Hughes S & Fellinger A (1989) Unilever Patent WO-A-91/00920

Osinga KA, Bendeker RF, v.d.Plaat JB & de Hollander JA (1988) Gist Brocades patent application EP A 03 06107 A2

Rood JI & Cole ST (1991) Molecular genetics and pathogenesis of *Clostridium perfringens*. Microbiol. Rev. 55: 621–648

Smelt JPPM (1980) Heat resistance of *Clostridium botulinum* in acid ingredients and its signification for the safety of chilled foods. Thesis Utrecht University, The Netherlands

Smink GCJ & Hamstra AM (1995) Research into consumers needs to be informed about the use of biotechnology in foods. SWOKA report 176

Tannock GW (1990) The micro-ecology of lactobacilli in habiting the gastrointestinal tract. In Advances in Microbial Ecology (Marshall KC ed.) 147–171, Plenum Press, New York

Tannock GW, Fuller R, Smith SL & Hall MA (1990) Plasmid profiling of members of the family *Enterobacteriaceae*, lactobacilli and bifidobacteria to study the transmission of bacteria from mothers to infants. J. Clin. Microbiol. 28: 1225–1228

Tannock GW, Luchansky JB, Miller L, Connell H, Thode-Andersen S, Mercer AA & Klaenhammer TR (1994) Molecular characterization of a plasmid-borne (pGT633) erythromycin resistance determinant (ermGT) from *Lactobacillus reuteri* 100–63. Plasmid 31: 60–71

Teuber M (1990) Production and use of chymosin from genetically altered microorganisms. Lebensm. Ind. Milchwirtsch. 35: 1118–1123

Van den Berg DJC, Smits A, Pot B, Ledeboer AM, Kersters K, Verbakel JMA & Verrips CT (1993) Isolation, screening and identification of lactic acid bacteria from traditional food fermentation processes and culture collections. Food Biotechnol. 7: 189–205

Van den Berg JA, van der Laken KJ, van Ooyen AJJ, Renniers TCHM, Rietveld K, Schaap A, Brake AJ, Schultz K, Moyer D, Richman M & Shuster JR (1990) *Kluyveromyces* as a host for heterologous gene expression: expression and secretion of prochymosin. Bio/Technology 8: 135–139

Vandenbergh PA (1993) Lactic acid bacteria, their metabolic products and interference with microbial growth FEMS Microbiol. Rev. 12: 221–238

Verrips CT (1991) Biotechnology for safe and wholesome foods. Food Biotechnol. 5: 347–364

Verrips CT (1995) Structured Risk Assessment of rDNA Products and Consumer Acceptance of These Products. In: Biotechnology (H.J. Rehm and G. Reed Editors), Volume 12 Legal, Economic and ethical dimensions (pp 157–196); VCH, Weinheim.

Vogel RF, Becke-Schmid M, Entgens P, Gaier W & Hammes WP (1992) Plasmid transfer and segregation in *Lactobacillus curvatus* LTH1432 *in vitro* and during sausage fermentation. System. Appl. Microbiol. 15: 129–136

Antonie van Leeuwenhoek **70**: 317–330, 1996.

Lactic acid bacteria as vaccine delivery vehicles

J.M. Wells, K. Robinson, L.M. Chamberlain, K.M Schofield & R.W. F. Le Page
University of Cambridge, Department of Pathology, Cambridge, CB2 1QP, U.K.

Introduction

Every year the U.S. National Institutes of Health publishes a report – called the Jordan Report – which describes progress in the study of infectious diseases and their control by means of vaccination. The most recent report, entitled 'Accelerated Development of Vaccines 1995' neatly summarises the background to recent advances in vaccinology, and emphasises the international perspective of these advances.

Five years ago the 1990 World Summit for Children (endorsing the concepts proposed by the Children's Vaccine Initiative) agreed that every encouragement should be given by international authorities to developing new vaccines and to ensure that they pass into routine use, reaching every child on a sustainable basis.

The scale of this task is large. 500 million health care contacts will be needed each year to immunise a projected global cohort of 125 million children by the year 2000. The new vaccines which are most urgently needed are those that will simplify vaccine distribution and vaccine administration. The new vaccines must reduce dependence on a cold chain, be suitable for oral administration, and be low in cost. Of course, they should also possess the optimal qualities of all good vaccines: be safe, effective and capable of eliciting active immunity in early infancy.

The requirement for mucosal (oral) route administration of new vaccines is particularly challenging to those acquainted with immunology. Although many infectious agents gain access to the body, or damage their hosts by colonising mucosal surfaces, very few of these infections can be effectively prevented by using mucosal route immunisation. Indeed, only in the case of a very limited number of diseases is it known that mucosal route immunisation can elicit protective level immune responses. The classical examples are those which involve the use of polio virus, cholera and typhoid vaccines.

Our general ignorance of how best to develop mucosally active vaccines has led to an enormous increase in research on the physiology and function of the mucosal (local) immune system, and the development of delivery systems for mucosal immunisation. This is the context in which studies have begun in a number of laboratories into the possibility of using recombinant food-grade lactic acid bacteria as antigen-presenting vehicles.

The mucosal immune system

Luminal antigens gain access to the mucosal lymphoid tissues by a sampling process involving the uptake and transport of proteins and particles across the epithelium by specialised transport cells called M cells (Figure 1). M cells are present in the epithelium which overlays sites of organised mucosal lymphoid tissue (O-MALT) where the antigen is encountered and the initial responses induced. The organised mucosal lymphoid tissue of the gut, bronchus, nasal and genital tract is centred around lymphoid follicles (Kraehenbuhl & Neutra, 1992; McGhee et al., 1992; Kato & Owen, 1994).

In humans aggregates of lymphoid follicles form the Peyer's Patches (PP) in the small intestine. The PP are common to many animals and have provided a good model of the O-MALT physiology and function. The dome region of the follicle contains immature B lymphocytes and CD4+ T cells of both Th1 and Th2 subsets as well as dendritic cells and macrophages.

The mucosal immune response is initiated by contact between the antigen and the lymphocytes and antigen presenting cells in the dome of the follicle. Here B cells expressing IgM or IgD on their surface are stimulated to proliferate and differentiate into lymphoblasts expressing IgA on their surface. The follicle macrophages and lymphocytes pass into the efferent lymphatics to the mesenteric lymph nodes and from here can enter the systemic circulation via the thoracic duct (Figure 1). When the lymphoid cells originating at the inductive sites enter distant effector sites (e.g. lamina propria of the intestine, bronchi and urogenital tract & secretory glands) they are selectively maintained by mechanisms not well understood.

These cell migratory pathways are the basis for the observation that mucosal immunisation can sometimes result in the induction of secretory antibody responses at distant mucosal sites. This apparent connectedness between mucosal sites has led to the mucosal immune system being referred to as the common mucosal immune system. At the effector sites B cells clonally expand and mature into IgA plasma cells. Polymeric IgA produced by plasma cells is then secreted across the epithelium into the lumen.

Given the immunological reactivity of the enteric tract, it is not perhaps surprising, that the human intestine should turn out to be the largest immunological organ in the body, containing more than 10^{10} lymphocytes per meter. In humans about 60% of the total immunoglobulin produced daily (several grams) is secreted into the GI tract. It is also apparent that in order to mount an effective response to an invading pathogen the mucosal immune system must be able to respond quickly to an invading pathogen despite the presence of antigens in the diet. Several studies have shown that soluble antigens do not usually induce strong antibody responses following oral immunisation; they can be rendered more antigenic by presenting them to the immune system within inert particles or bacteria (Cox & Taubman, 1984; Wold et al., 1989; Dahlgren et al., 1991). In fact under appropriate conditions systemic tolerance can be induced by oral immunisation with soluble protein antigens (Thomas & Parrot, 1974; Mowat, 1987). This pronounced difference in the immunogenicity between dietary proteins and antigens of bacteria colonising the intestine has presumably evolved to prevent potentially harmful hypersensitivity reactions to food proteins.

Relatively little is known about the regulation of the immune responses to the numerous species of commensal bacteria which colonise the GI tract and the effects that these responses may have on the microbial ecology of the intestine. The difficulties of achieving mucosal immunisation with soluble protein antigens may be due to their degradation in the stomach and intestine, relatively low antigenicity, or limited absorption. To overcome these problems three main strategies have been adopted for the oral delivery of vaccine antigens: (1) the use of inert particles such as biodegradable microparticles, liposomes and cochleates (2) the use of adhesins and lectins and (3) the use of recombinant bacteria (Mestecky, 1991; O' Hagan, 1994).

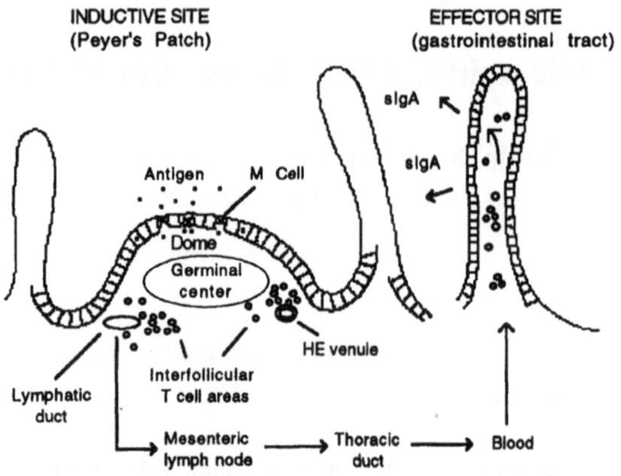

Figure 1. Schematic representation of the uptake of antigens by M cells in the Peyer's patches (PP) of the gastrointestinal tract and the migratory pathways of mucosal lymphocytes. Stimulated lymphocytes from the PP migrate into the submucosal lymphatics and then via the thoracic duct into the bloodstream. The cells then localise in the mucosal effector sites where polymeric IgA produced by plasma cells is transported through the epithelium via its interaction with a receptor (the secretory component) on the basal membrane of epithelial cells. The influx of lymphocytes to the PP occurs across postcapillary high endothelial (HE) venules which have receptors recognised by lymphocyte adhesion molecules. The dome region and germinal centre of the PP lymphoid follicles are indicated.

Lactic acid bacteria as vaccine delivery vehicles

The development of recombinant bacteria as antigen delivery vehicles has so far focused predominantly on the use of live, attenuated strains of pathogenic mycobacteria, salmonella, and clostridia (Aldovini, 1991; Chatfield, 1992; Dougan, 1993; Stover, 1993; Yasutomi et al., 1993). The efficacy of these bacterial vectors as vaccines is believed to depend on their invasiveness, capacity to survive and multiply, and on the occurrence of adequate levels of antigen gene expression *in vivo*. However, it is clear that considerable work is still needed before attenuated strains of pathogenic bacteria that are sufficiently attenuated to pose little or no health risk to paediatric, and partially immunosuppressed recipients are available. It is not yet clear whether such suitably attenuated strains of pathogenic bacteria will in fact retain their ability to colonise and replicate to the extent necessary to elicit an effective immune response. It is the need for improved meth-

Table 1. Examples of model vaccine strains of lactic acid bacteria

Species/Strain	Source	Persistence in GI tract	Persistence in vagina	Efficiency of transformation
Lb. Paracasei LbTGS1.4	Mouse vagina	4–6 days	>21 days	$10^6/\mu g$ DNA
Str. gordonii (Challis)	Human oral cavity	∼ 2 days	>30–42 days	Naturally competent
L. Lactis MG1363	Milk	< 24 hours	NA	$>10^7/\mu g$ DNA
Lb. plantarum NCIB8826	Human	∼ 9 days	∼ 12 days	∼ $10^6/\mu g$ DNA
Staph. xylosus KL117	Human skin	∼ 2 days	NA	10^3–$10^4/\mu g$ DNA
Staph. carnosus TM300	Dry sausage	∼ 3 days	NA	10^5–$10^6/\mu g$ DNA

Studies on the persistence of the model vaccine strains of lactic acid bacteria were done in mice.
NA; data not available.

ods for the oral delivery of antigen at low cost, coupled with concern over the safety of the attenuated live delivery systems which has led to an interest in the use of innocuous bacteria including lactic acid bacteria as vaccine delivery vehicles.

Two main approaches have been adopted for the development of lactic acid bacteria as vaccine delivery vehicles. The first involves the use of lactobacilli which colonise the GI tract and genital tract and *Str. gordonii* which naturally colonises the oral cavity. Here the aim is to implant, at least transiently, a recombinant commensal strain among the endogenous microflora. As mammals are known to develop serum and mucosal antibodies to commensal bacteria (Gold et al., 1978; Warner et al., 1987) it is expected that the transient colonisation by a recombinant lactic acid bacterium will promote an immune response to an expressed antigen. Obviously this approach is dependent on continued antigen expression *in vivo* and on the selection of model strains which colonise the host. The selection of strains of *Lactobacillus* for use as vaccine vectors has depended on their potential to colonise mucosal surfaces, ability to be transformed, capacity to express foreign antigens, and intrinsic adjuvanticity (Table 1).

One model strain of *Lactobacillus* which has been selected for vaccine studies was isolated from the vagina of C57Bl/6xSJL mice and identified as *Lactobacillus casei* spp. *paracasei*, a species also commonly found in the microflora of the human intestine and vagina (Mercenier et al., 1996). This strain is plasmid free and highly transformable (Table 1). The administration of approximately 10^9 cfu of recombinant LbTGS1.4 genetically marked with a plasmid containing the chloramphenicol resistance gene were shown to persist for 4–8 days in the intestine and for at least 21 days in the vagina. Other workers have chosen *Lactobacillus* strains for the development of vaccine vehicles which have an intrinsic capacity to stimulate the immune system in an aspecific manner (Table 1). For example some *Lactobacillus* strains have been selected on the basis of a comparison of the humoral responses to the trinitrophenylated chicken gamma globulin (CGG-TNP) when it is injected intraperitoneally with different strains of *Lactobacillus* or with Specol, an oil based adjuvant (Boersma et al., 1992; Claassen et al., 1995). Interestingly, certain strains of *Lb. casei* and of *Lb. plantarum* have an adjuvant activity which is equivalent to that observed for Specol, while other strains lacked any adjuvanticity. The basis for this adjuvanticity has recently been investigated by using cytochemical methods to follow the production of cytokines in the gut mucosal tissues after oral administration of *Lactobacillus* (Boersma et al., 1995). It was found that while some strains had no effects on cytokine levels in the lamina propria others increased the production of IL–2 and TNFα or IL10, IL–1α and IL–1β. It is not yet known whether *L. lactis*, *Str. gordonii* or other species of lactic acid bacteria have similar effects on the immune system.

Only one strain (Challis) of the oral commensal *Str. gordonii* (formerly *Str. sanguis*) has been exploited as a vaccine delivery vehicle. The derivatives used for vaccine studies carry a mutated allele *str-204* which confers resistance to streptomycin (Pozzi et al., 1990). The Challis strain is a normal inhabitant of the human oral cavity which has also been shown to colonise the oral pharyngeal cavity of mice. In one study mice were inoculated both orally and intranasally with a single combined dose of 1×10^9 c.f.u. of bacteria and monitored over a two month period by taking swabs of the oral cavity and pharynx at regular intervals (Oggioni et al., 1995). The results showed that 75% of the animals were colonised by *Str. gordonii* at each time of sampling and that in 83.3% of mice sample swabs recovered the genetically marked strain of *Str. gordonii*. Importantly there were no differences in the colonizing capacity of the wild type and recombinant strains. All of the recombinant *Str. gordonii* isolated from mice were shown to express the resistance marker and the recombinant antigen of interest on their surface. The genetic stability of these strains *in vivo* is presumably due to the fact that the heterologous antigen genes are expressed in the chromosome rather than on a plasmid which might be segregationally unstable (see below). This strategy assures stable expression of the recombinant antigen *in vivo*, which is obviously an essential feature of a live colonising bacterial vaccine vehicle. In a similar study by these same workers mice were supplied with drinking water containing 5 g/L of streptomycin for two days prior to being given a single inoculum of 8×10^8 c.f.u. of bacteria. These results showed that $> 60\%$ of the 23 mice inoculated were colonised by recombinant *Str. gordonii* during the first three weeks (Medaglini et al., 1995). The numbers of mice exhibiting a positive swab over a two week period then steadily declined to to approximately 30% at nine weeks and 9% at 11 weeks. The main sites of colonisation were the tongue and hard palate. *Str. gordonii* is also able to colonise the vaginal mucosa of mice for up to eight weeks following a single inoculum of only 10^6 bacteria (Mercenier et al., 1996; Medaglini & Pozzi, personal communication).

Our own interest centres on the harmless bacterium *Lactococcus lactis*. Unlike the other lactic acid bacteria being developed as vaccine delivery vehicles *L. lactis* does not colonise the digestive tract of man or other animals and in this respect it is perhaps more analogous to inert microparticles and liposomes than to other recombinant vaccine delivery systems. Although studies in gnotobiotic mice have shown that *L. lactis* is able to colonise the digestive tract, elimination of *L. lactis* occurs when certain commensal bacteria are introduced. In heteroxenic mice colonised with whole human microflora there is only a passive transit of *L. lactis* through the digestive tract (Gruzza et al., 1994). More recently it was shown that after feeding of a genetically marked food-grade strain of *L. lactis* to human volunteers only *L. lactis* cells which passed through the gut within 3 days could be recovered from the faeces (Klijn et al., 1995). However, a specific chromosomal DNA marker present in the genetically marked strain could be extracted and detected in the faeces for up to four days after feeding of live *L. lactis*. In view of the relatively short passage time of *L. lactis* through the gut and the limited capacity for gene expression *in vivo* the effectiveness of the *L. lactis* system is expected to depend primarily on the synthesis of immunogenic quantities of antigen within the bacteria prior to administration. For these reasons it is not considered beneficial to secrete antigens expressed in *L. lactis*. In contrast research groups working on *Lactobacillus* as a vaccine delivery vehicle envisage that proteins are likely to be antigenic if they are secreted during colonisation. It is of considerable interest to determine whether this might be the case as soluble antigens are known to be more susceptible to degradation by acid and proteases than antigens contained with inert particles or bacterial cells. There is also evidence that protein antigens are more immunogenic following oral immunisation when they are contained within inert particles or expressed in recombinant bacteria (Dahlgren et al., 1991; Cox & Taubman, 1984). Of particular concern is the expression of antigens and epitopes as exposed determinants displayed on the surface of bacteria. Genetic constructs which lead to surface display are of considerable interest to most groups working on the development of lactic acid bacteria as vaccine delivery vehicles as it is anticipated that this might enhance their immunogenicty.

Certain species of the genus *Staphylococcus* and *Listeria* fulfill all of the criteria for classification as lactic acid bacteria although these organisms are not generally thought of as members of the LAB group (Holtzapfel & Wood, 1995). *Staphylococcus xylosus*, *Staphylococcus carnosus* and attenuated strains of *Listeria monocytogenes* are being developed as vaccine delivery vehicles but here, attention will only be given to the work with *Staph. xylosus* and *Staph. carnosus* as these species are non-pathogenic like the other LAB mentioned in this review (Table 1).

The strain of *Staph. xylosus* being investigated as a vaccine delivery vehicle was originally isolated from human skin and *Staph. carnosus* is widely used in the ripening process of dry sausages and as a starter culture for the fermentation of meat and fish products. Studies of the pathogenicity of wild type *Staph. xylosus* and *Staph. carnosus* has been assessed in immunocompetent and immunodeficient mice by the inocluation of 10^9 or 10^{10} bacteria by the intraperitoneal, subcutaneous, and oral routes (Stahl et al., 1996). Mortality was only observed when the mice were injected i.p in doses of 10^9 cfu/ml or more for *Staph. carnosus* or in doses of 10^{10} cfu/ml for *Staph. xylosus*. The absence of bacterial growth in all surviving animals indicated that mortality was due to toxicity of the bacterial components rather than to infection. In fact the remarkably low capacity of *Staph. xylosus* and *Staph. carnosus* to cause harm is comparable to the results we have obtained with *L. lactis* in mice. We have not observed harmful effects (such as the formation of an abscess) from the subcutaneous inoculation of mice with live *L. lactis* (up to 5×10^9 bacteria) and live bacteria could not be recovered from the organs of mice one week after subcutaneous, or oral inoculation. *Staph. xylosus* and *Staph. carnosus* do not colonise the GI tract of mice but are able to survive in the intestine for 2–3 days following oral inoculation (Stahl et al., 1996).

Antigen expression in Lactobacillus

The ability to transform and express antigens in different species of *Lactobacillus* has been a major factor in the selection of strains for vaccine studies. The efficiency of transformation even among strains of the same species of *Lactobacillus* seems to be extremely variable. In addition certain intestinal species of lactobacilli are apparently refractory to transformation (Klaenhammer, 1995). Most of the cloning vectors available for use in *Lactobacillus* have been derived from small cryptic plasmids with a rolling circle mechanism of replication (RCR). Several different derivatives of cloning and expression vectors have been based on cryptic plasmids from *Lb. pentosus* and *Lb. plantarum* (Pouwels & Leer, 1993; Leer et al., 1992; Klaenhammer, 1995) Typically, these RCR type plasmids and their derivatives replicate in a wide range of Gram-positive bacteria including a variety of *Lactobacillus* species. A number of expression vectors have been based on the cryptic *Lactobacillus* plasmid p353–2 (Posno et al., 1991a). A typical example is pLPCR2 which contains the promoter of the *Lb. pentosus* xylose operon repressor gene xylR and the xylB gene terminator separated by a multiple cloning site (Posno et al., 1991b). Similarly other derivatives contain regulatory signals from *Lactobacillus* such as promoters from *Lb. amylovorus* α-amylase gene (*amyA*) and *Lb. casei* proteinase gene (*prtP*) (Claassen et al., 1995). In initial studies with these vectors pLPCR2 was used to express a hybrid protein comprising two tandem repeats of a foot-and-mouth disease virus (FMDV) epitope fused to *E. coli* β-galactosidase (Pouwels & Leer, 1993; Claassen et al., 1995). The fusion protein was expressed in *Lb. casei* but at low levels (approx. 0.1% of total cellular protein). Using the *Lb. plantarum* bile acid hydrolase gene (*cbh*) promoter β-galactosidase was expressed at levels of approximately 1–2% of total cell protein in *Lb. plantarum* but not in strains of *Lb. casei* selected for use in vaccine studies (Pouwels & Leer, 1993; Claassen et al., 1995).

The replicons of plasmids isolated from other Gram-positive bacteria such as the lactococci and enterococci have also been exploited in *Lactobacillus*. A *Lactobacillus* expression vector designated pTG2247 has been constructed by cloning a strong *Str. thermophilus* promoter P25 and an *E. coli* terminator into the multiple cloning site of pCK17 (Mercenier pers. comm.) This vector incorporates the replicon of the cryptic lactococcal plasmid pSH71 which has been shown to replicate in *E. coli* and a wide-range of Gram- positive bacteria. Derivatives of pTG2247 have also been constructed which contain translation initiation regions (TIR) from genes isolated from different species of *Lactobacillus*. One of these derivatives containing a TIR derived from *Lb. plantarum* has been used to produce low amounts of the *Vibrio cholerae* toxin B subunit in an insoluble form in the model vaccine strain LbTGS1.4 (Table 1) (Slos and Mercenier pers. comm.). Using pTG2247 in LbTGS1.4 most success has been obtained with the expression of a modified cell-wall anchored M6 protein from *Str. pyogenes* which has been used extensively as an epitope carrier in vaccine strains of *Str. gordonii* (Mercenier et al., 1996). In LbTGS1.4 M6 fusion proteins carrying the HIV–1 V3 loop epitope of gp120 or the gp41 'ELDKWAS' epitope (gp41E) were produced in amounts which were detectable on Western blots but not in Coomassie blue stained gels of protein extracts. Maximum yields of the M6-gp41E fusion protein were estimated to be in range of 0.5% of the total cell protein.

A broad host range expression vector has been recently described which utilises the *Staph. aureus* protein A promoter, TIR and signal leader sequence to direct the expression and secretion of proteins in *Lactobacillus*. Using this vector the variable domain 4 of the chlamydial major outer membrane protein has been expressed and secreted in amounts up to 10 mg/L in several strains of lactobacilli (Rush, 1995).

Recently a novel series of versatile constitutive expression vectors (the pTREX series of *T*heta *R*eplicating *Ex*pression plasmids) have been developed which replicate in a wide range of Gram-positive bacteria (Schofield et al., unpublished). Plasmid pTREX1 contains an expression cassette which incorporates a strong lactococcal promoter, the translation initiation region of *E. coli* bacteriophage T7 *gene10* which has been modified at one nucleotide position to increase complementarity of the Shine Dalgarno (SD) sequence to the ribosomal 16sRNA of *L. lactis* and the bacteriophage T7 RNA polymerase terminator (Figure 2). Unique sites have been positioned between the various sequence elements so that they can be easily manipulated to study gene expression in different species of bacteria. In *L. lactis* pTREX1 has been used to express tetanus toxin fragment C in order to evaluate the effectiveness of the *L. lactis* vaccine delivery system in mice. Plasmid pTREX1-TTFC has also been used to transform three species of *Lactobacillus* including LbTGS1.4. Expression of tetanus toxin fragment C (TTFC) has been detected in *Lb. gasseri, Lb. johnsonii* (Schofield, unpublished data) and *Lb. paracasei* (Mercenier & Kleinpeter, personal communication) which were transformed with pTREX1-TTFC indicating that these vectors may also be useful for the expression of other antigens in vaccine strains of *Lactobacillus.* These results have indicated that it is not necessary to use homologous expression signals to obtain expression in *Lactobacillus.* Although vectors such as pTREX1 which have a theta- mode of replication are inherently more structurally stable than RCR type plasmids they are rapidly lost from *Lactobacillus* in the absence of antibiotic selection. However, the potential exists to increase the segregational stability of pTREX and its derivatives in *Lactobacillus* by incorporating the *orfH* gene of pAMβ1 into these vectors. This gene encodes a resolvase which is now known to facilitate the resolution of plasmid multimers formed during replication (Swinfield et al., 1990; Kiewiet et al., 1993).

The segregational instability of plasmids in *Lactobacillus* is obviously important as the concept behind the use of these bacteria as vaccine delivery vehicles relies on obtaining colonisation and continued gene expression *in vivo.* Most vectors are segregationally unstable in *Lactobacillus* (50–95% loss in 100 generations) when grown in the absence of antibiotic selection with a few notable exceptions (Bringel et al., 1989; Posno et al., 1991b; Shimizu-Kadota et al., 1991). One is plasmid pLPE323 which has been found to be segregationally stable in all but one of a variety of *Lactobacillus* strain tested (Leer et al., 1992). Interestingly there are two examples where the cloning of *E. coli* DNA into segregationally stable plasmids from *Lactobacillus* caused them to become segregationally unstable (Leer et al., 1992).

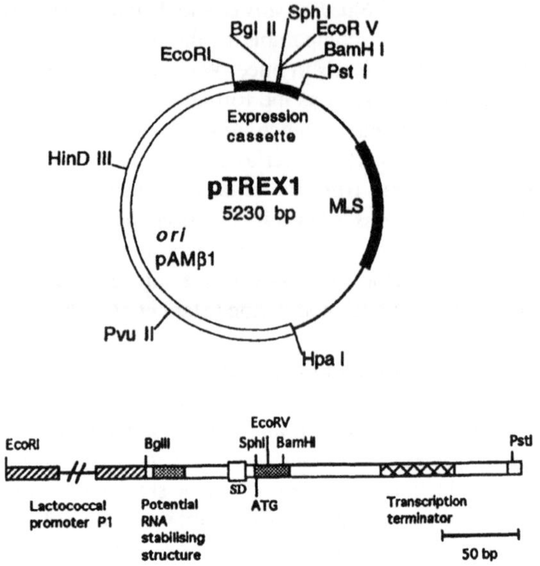

Figure 2. Physical and genetic maps of the broad Gram-positive host range expression plasmid pTREX1. The determinant encoding resistance to macrolides, lincosamides and streptogramin B (MLS) and the origin of replication from the plasmid pAMβ1 (*ori* pAMβ1) are shown as boxes on the plasmid map. The expression cassette is schematically represented; the various sequence elements involved in gene expression and the location of the unique restriction enzyme sites are indicated. SD: Shine Dalgarno sequence, ATG: translation initiation codon.

Immune responses to lactobacilli expressing heterologous antigens

The potential of using *Lactobacillus* as a vaccine delivery vehicle was first demonstrated by immunisation of mice with killed lactobacilli to which the hapten trinitrophenyl (TNP) had been chemically coupled to the surface of the bacteria (Gerritse et al., 1990). Serum antibody responses to TNP were detected in mice after oral priming and intraperitoneal booster immunisations indicating that *Lactobacillus* can provide T-cell help for small epitopes. Similar experiments have recently been reported using live *Lb. plantarum* 80 expressing *E. coli* β-galactosidase (*lacZ*) intracellularly (Claassen et al., 1995). In these experiments doses of 10^6, 10^8 or 10^{10} expressor or non-expressor control strains of bacteria were given orally to mice on days 0, 1 & 2 and boosted after a 4 week interval. No significant antibody responses to β-galactosidase were measured in the intestinal lavages or serum of mice immunised orally with the expressor strains of *Lactobacillus.* In contrast intraperitoneal administration of the expressor bacteria elicited significant anti-β-galactosidase responses in the serum. The reasons for this apparent failure of the *Lactobacillus* delivery system to elicit responses to an expressed β-galactosidase are unclear but may be related to the relatively low levels of β-galactosidase produced in *Lb. plantarum in vitro* and *in vivo*. Similarly, specific immune responses to an α-amylase expressed in LbTGS1.4 were only elicited by intraperitoneal inoculation and not by a mucosal route of immunisation. One possible factor complicating these studies would be that by the oral route of immunisation there is some suppression of immune responses to antigens such as β-galactosidase and α-amylase which are present in numerous other commensal bacteria. At TNO in the Netherlands the continued effort to enhance the levels of antigen expression in lactobacilli has been met with some success recently and it is now possible to express full length antigens such as TTFC and rotavirus proteins 4 and 7 at levels of a few percent of total protein in some strains of *Lactobacillus* (Boersma pers. comm.). Immunological studies are now in progress to determine whether secretory and serum antibody immune responses to these antigens are obtained following mucosal immunisation.

Recently, success has been obtained with the recombinant LbTGS1.4 strains expressing M6-V3 or M6-gp41E fusion proteins (Mercenier et al., 1996). This study has shown that it is possible to elicit systemic immune responses to the M6 protein when BALB/c mice are intra-nasally and intra-gastrically immunised using a regime of two doses of 10^9 bacteria on consecutive days which is then repeated at approximately three weeks and six weeks after the first dose. Although none of the immunisations elicited a specific antibody response to the V3 or gp41E epitopes it is now known that these epitopes are not very immunogenic in mice and that BALB/c mice are relatively poor responders to the gp41E epitope in comparison to other inbred strains of mice. More recent results have indicated that these same constructs produce much higher levels of the M6-fusion proteins in *Lb. plantarum* NCIB 8826 (Table 2). This human isolate is being investigated as a model vaccine strain as it persists in the GI tract and vagina of mice for approximately 9 and 12 days respectively (Aguirre and Mercenier, pers. comm.) (Table 1).

Expression and immune responses to antigens displayed on the surface of *Staph. xylosus* and *Staph. carnosus*

Expression of antigens has been achieved in *Staph. xylosus* using an *E. coli* - *Staphylococcus* shuttle vector containing the expression and targeting signals of the *Staph. aureus* protein A gene. The approach has been to target heterologous antigens and peptides to the bacterial cell envelope by making use of the secretion signal and cell wall anchoring domains (XM) of *Staph. aureus* protein A (Hanson, 1992). A fusion protein comprising the serum albumin binding region of streptococcal protein G, three tandem copies of an epitope from the G glycoprotein of RSV and the C-terminal cell wall anchoring domain of *Staph.* aureus protein A have been expressed on the surface of *Staph. xylosus* (Nguyen et al., 1993). A similar strategy has been employed for the surface display of heterologous proteins on *Staph. carnosus* using the promoter, secretion signal (including the propeptide region) from a lipase gene of *Staphylococcus hyicus* and the cell wall anchoring domains of *Staph. aureus* protein A (Samuelson et al., 1995). Immunoblotting and immunofluorescence was used to demonstrate the surface expression of a chimaeric protein comprising an 80 amino acid polypeptide from a malarial blood stage antigen, the serum albumin binding region of streptococcal protein G and the XM domains *Staph. aureus* protein A in *Staph. carnosus*. The subcutaneous and oral immunisation of mice with live recombinant staphylococci expressing heterologous peptides and proteins on their surface has resulted in the generation of serum antibody responses to the expressed antigens although in the case of the oral immunisations a total of 24 inoculation were given and the responses measured were highly variable (Stahl et al., 1996; Nguyen et al., 1993).

Expression of heterologous antigens on the surface of Str. gordonii

The natural competence of *Str. gordonii* and the high frequency of chromosomal recombination which can occur between homologous DNA sequences which are introduced into these bacteria has been exploited to obtain the expression of heterologous antigen genes integrated in the chromosome. The antigens which have been expressed in *Str. gordonii* have all been M6 gene fusions. The approach taken has been to delete the majority of the surface-exposed segment of a group A streptococcal M protein (M6) and replace it with a

Table 2. Example of antigen genes expressed in lactic acid bacteria

LAB Expression host	Antigen & *Source* (s)	Vector/Exp. system	Location of antigen (amount if known)	Reference
L. lactis	TTFC *Clostridium tetani*	pLET1 T7 system	Intracellular (20% soluble protein)	Wells et al., 1993
L. lactis	TTFC-PrtP anchor *C. tetani/L. lactis*	pLET4 T7 system	Bacterial envelope [a](~ 0.2% total protein)	Wells et al., 1995
L. lactis	TTFC *Clostridium tetani*	pTREX1 P1 promoter	Intracellular [a](1–3% total protein)	Schofield et al., unpublished
L. lactis	TTFC-protein A anchor *C. tetani/S. aureus*	pTREX1–X	Displayed on cell surface	Steidler et al., unpuplished
L. lactis	Sm28 *Schistosoma mansoni*	pLET1 T7 system	Intracellular [a](5–10% total protein)	Chamberlain et al., 1995
L. lactis	TTFC-Sm28 fusion *S. mansoni/C. tetani*	pLET1 T7 system	Intracellular [a](5–10% total protein)	Chamberlain et al., 1995
L. lactis	TTFC-gp120 V3 loop *C. tetani/HIV-1*	pLET1 T7 system	Intracellular	Litt 1996
L. lactis	HIV-1 V3 loop tandem repeats	pLET1/pLET4 T7 system	Intracellular/cell envelope	Litt 1996
L. lactis	TTFC-gp 120 V3 loop *C. tetani/HIV-1*	pLET1 T7 system	Intracellular	Litt 1996
L. lactis	Surface antigen (PAc) *S. mutans*	*PAc* promoter	Intracellular (0.0018% dry wt.)	Iwaki et al., 1990
Staph. xylosus KL117	Antigen Pf155/RESA (80aa) *P. falciparum*	*S. aureus* protein A signals	Displayed on cell surface	Hannon et al., 1992
Staph. xylosus KL117	Variants of 101 aa region of G. protein: RSV G protein	*S. aureus* protein A signals	Displayed on cell surface	Nguyen et al., 1995
Staph. xylosus KL117	G protein epitope- prot. A human RSV- *S. aureus*	*S. aureus* protein A signals	Displayed on cell surface	Nguyen et al., 1993
Staph. carnosus TM300	Antigen Pf155/RESA (80aa) *P. falciparum*	*S. hyicus* lipase signals	Displayed on cell surface	Samuelson et al., 1995
Str. gordonii (Challis)	M6-E7 fusion *S. pyogenes*/HPV16	Promoter in chromosome	Displayed on cell surface	Pozzi et al., 1992
Str. gordonii (Challis)	M6-Ag5.2 fusion *S. pyogenes*/hornet	Promoter in chromosome	Displayed on cell surface	Medaglini et al., 1995
Str. gordonii (Challis)	M6-gp120 loop *S. pyogenes*/HIV-1	Promoter in chromosome	Displayed on cell surface	Pozzi et al., unpublished
Lb. paracasei LbTGS1.4	M6-V3 fusion *S. pyogenes*/HIV-1	pTG292 P25 promoter	Secreted/cellular	Mercenier et al., 1996
Lb. paracasei LbTGS1.4	M6-gp41E fusion *S. pyogenes*/HIV-1	pTG2247 P25 promoter	Secreted/cellular	Mercenier et al., 1996
Lb. plantarum	*β*-galactosidae *E. coli*	*cbh* promoter	Intracellular [a](1–2% of cell protein)	Claassen et al., 1996
Lb. plantarum	VP7- *β* -gal Rotavirus/*E. coli*	*cbh* promoter	Intracellular [a](0.2% of cell protein)	Claassen et al., 1996
Lb. plantarum	*β*-glucuronidase *E. coli*	*lacA* and *cbh* promoter	Intracellular	Claassen et al., 1996

Table 2. continued

LAB Expression host	Antigen & *Source* (s)	Vector/Exp. system	Location of antigen (amount if known)	Reference
Lb. casei ATC323	β -glucuronidase *E. coli*	*lacA* and *cbh* promoter	Intracellular	Claassen et al., 1996
Lb. casei ATC323	FMDV epitope- β -gal FMDV/*E. coli*	pLPCR2 *xylR* promoter	Intracellular [a](0.1% of cell protein)	Claassen et al., 1996
Lb. plantarum NCIB8826	M6-gp41E fusion *S. pyogenes*/HIV-1	Plasmid pGIP	Intracellular (~10% of total protein)	Slos & Mercenier, pers. comm.
Lb. plantarum NCIB8826	M6-gp41E fusion *S. pyogenes*/HIV-1	pTG2247; P25 promoter secreted	Secreted 13/mg/L	Slos & Mercenier, pers. comm.

[a]Estimated by Western blotting.

foreign antigen or epitope (Pozzi et al., 1992; Fischetti et al., 1993; Oggioni and Pozzi, 1996). The constructions are first made on a plasmid in *E. coli* and then used to transform a recipient strain of *Str. gordonii* which has in its chromosome a chloramphenicol resistance gene (*cat*) flanked by DNA sequences also present on the plasmid. The plasmid which is naturally linearised during transformation then recombines with the homologous sequences flanking the *cat* gene resulting in the integration of the hybrid M6-antigen gene and a different selective marker (*ermC*) into the chromosome of *Str. gordonii.* This strategy has the advantage that the genes can be stably maintained in the chromosome without the need for antibiotic selection. Pozzi et al. (1992) were the first to demonstrate the expression of an M6 fusion protein on the surface of *Str. gordonii.* The E7 protein of papilloma virus was shown by immunofluorescence to be located on the surface of intact cells. Using this same strategy a variety of antigens have now been expressed on the surface of *Str. gordonii* including V3 loop sequences of gp120 from HIV–1, a protein antigen from the white faced hornet (Ag5.2), and *E. coli* heat labile toxin subunit B (LTB) (Table 2) (Medaglini et al., 1995; Oggioni et al., 1995; Ricci et al., unpublished).

Immune responses to recombinant proteins expressed on the surface of Str. gordonii

Mice simultaneously inoculated orally and intranasally with recombinant strains of *Str. gordonii* expressing either the E7 protein of human papillomavirus type 16 or white faced hornet venom allergen (Ag5.2) as a fusion with the M6 protein are colonised by the recombinant *Str. gordonii* for up to 11 weeks. Four weeks after inoculation with *Str. gordonii* expressing the E7-M6 fusion protein 61% of the mice had M6-specific serum antibody levels which were significantly different to the mean value of the control mice. Fifty percent of the immunised mice were also positive for the presence of serum antibodies to E7 (Oggioni et al., 1995). Similarly, when mice were inoculated with *Str. gordonii* expressing the M6-Ag5.2 fusion protein significant levels of specific antibody to Ag5.2 were detected in the serum between 4 and 7 weeks after inoculation and continued to increase in level to 11 weeks (Medaglini et al., 1995). The mice immunised with the Ag5.2 expressor strain also showed a significant increase of Ag5.2 specific salivary IgA (> 2% of total IgA) compared to mice colonized with a control strain. However, no significant levels of anti- Ag5.2 IgA were detected in the intestinal lavages of mice inoculated orally and intranasally with the expressor strains suggesting that higher levels of IgA are elicited at the site of colonisation by recombinant *Str. gordonii* than at more distant mucosal surfaces. As expected the injection of the expressor strain subcutaneously with Freund's adjuvant significantly increased the levels of Ag5.2 serum antibody but not the secretory antibody responses to this antigen. Mice inoculated orally and intranasally with a single dose of killed recombinant *Str. gordonii* did not elicit Ag5.2 specific antibody responses in the serum or mucosal secretions.

Antigen expression in L. lactis

As it is expected that *L. lactis* is likely to have only a limited capacity to produce and secrete antigens *in vivo*, in Cambridge we have focused our attention on expressing antigens intracellularly or as fusions to the cell wall anchoring domains of cell surface associated proteins so that the bacteria are pre-loaded with antigen before they are used for immunisation.

In order that immunogenic quantities of antigen can be formed in *L. lactis* a high level inducible expression system which exploits the properties of the *E. coli* T7 bacteriophage RNA polymerase has been developed for use in the lactococci (Wells et al., 1993a). In this expression system the T7 RNA polymerase has been placed under control of the lac promoter in the low-copy-number plasmid pIL227 to generate pILPol. The expression of T7 RNA polymerase is inducible by lactose in strains of *L. lactis* which carry the lactose operon either in the chromosome or on a plasmid. Once formed in the cell the T7 RNA polymerase will transcribe genes cloned downstream of its cognate promoter in the pLET (for *L*actococcal *E*xpression by *T7* RNA polymerase) series of vectors. The vectors in the pLET series each contain a different version of a T7 expression cassette (Figure 3). For example, pLET vectors have been constructed which can be used (a) to produce proteins intracellularly in amounts up to 22% of soluble protein (b) to secrete the protein into the growth medium (Wells et al., 1993b) or (c) to anchor the protein in the bacterial envelope (Wells et al., 1995; Norton et al., 1996).

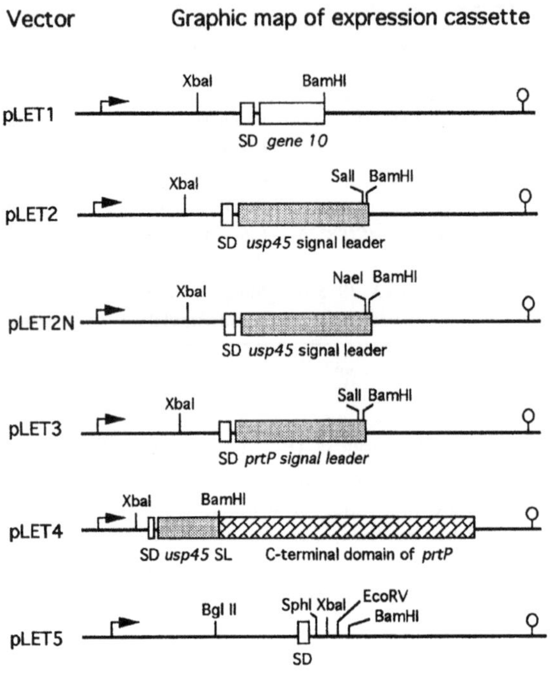

Figure 3. Schematic representation of the expression cassettes present in the pLET series of vectors. All restriction endonuclease sites indicated are unique in the vector. The open reading frames are boxed and labelled accordingly. The position of the T7 promoter → and T7 terminator O is shown. SL: signal leader; SD: Shine Dalgarno motif.

The pLET vectors have been used for the expression of a number of heterologous antigens to high level (2–20% total soluble cell protein) in *L. lactis* e.g. tetanus toxin fragment C (TTFC), diphtheria toxin fragment B, the 28 KDa immunogen of *S. mansoni* (Sm28) and TTFC-HIV-gp120V3 loop fusion proteins [TTFC-V3a = TTFC plus 35 amino acids of the V3 loop; TTFC-V3b = TTFC plus the V3 loop plus 33 flanking amino acids] (Wells & Schofield 1996) (Table 2). Although some antigens are efficiently expressed in *L. lactis* others have only been produced in low or trace amounts. The B subunit antigen of *Vibrio cholerae* toxin (CTB) is an example of an antigen which could not be secreted or formed intracellularly in *L. lactis* using the T7 system despite the fact that abundant levels of mRNA were formed in the cell and that the same expression cassettes produced high levels of CTB in the *E. coli* T7 expression host strains. Overall our results have indicated that the gene expression in *L. lactis* is still largely an empirical process, the levels of expression varying considerably from protein to protein. What is clear however, is that the levels of expression are not solely dependent on the source of the gene, since proteins of bacterial, parasitic and eukaryotic origin have been efficiently expressed in *L. lactis*. It is also evident from the studies of antigen expression in both *Lactobacillus* and *Lactococcus* that the level of expression is greatly influenced by the form and nature of the protein produced. For example, when TTFC is expressed as a fusion to the C-terminal cell-wall spanning and membrane anchoring domain (nt 6518 to 6913) of PrtP in pLET4 (strain UCP1054) the level of protein expressed drops dramatically to about 1/200 of the quantity produced intracellularly (using the pLET1 vector). Immunoblotting with lysostaphin-released cell wall material from UCP1054 ruled out the possibility that this observed lower level of TTFC expression was simply due to its removal from the soluble pool of TTFC by attachment to insoluble peptidoglycan. The restricted amount of antigen formed by strain UCP1054 is presumed to be due either to a toxic effect of TTFC when it is anchored in the membrane as a PrtP fusion protein, or to rapid intracellular degradation of this hybrid protein. Similar approaches have been taken with the cell-wall anchoring domains from *Staph.* aureus protein A to achieve the surface display of heterologous proteins and polypeptides in

L. lactis (Steidler et al., unpublished). In these studies it was shown that heterologous proteins and polypeptides fused to the protein A anchoring domain are firmly attached to the cell wall of *L. lactis* and exposed on the outer surface of the bacteria.

Expression vectors which incorporate constitutively active promoters have also been employed for antigen expression in *L. lactis*. One such vector, designated pTREX1 (Figure 2), is one of a series of medium to low level constitutive expression vectors which has been used to express TTFC, and P28 at levels estimated to be 1–3% of total cell protein (Table 2) (Schofield et al., unpublished). Derivatives of pTREX, designated pTREX1-X, have also been constructed to allow expressed antigens to be fused to the C-terminal region of the cell-wall anchoring domain of protein A in *L. lactis* (Table 2). It is envisaged that these constitutive expression vectors will be more suitable for the expression of antigens which are membrane associated or which show some insolubility or toxicity to bacterial cells when formed at high levels. By using the pTREX series of vectors to express antigens at varying levels in *L. lactis* it should also be possible to investigate the effect of the quantity of antigen present in the inoculum on the magnitude and duration of the immune response.

Immune responses to antigens expressed in L. lactis

The first workers to investigate the capacity of *L. lactis* to act as an antigen carrier immunised groups of 8–9 BALB/c mice orally with killed recombinant *L. lactis* expressing low amounts of a C-terminally truncated 190 kDa surface protein antigen (PAc) from *Streptococcus mutans* (Iwaki et al., 1990). Three consecutive daily doses of 10^9 formalin killed cells were given four weeks apart followed by one single booster inoculation one week later. PAc antigen specific IgA was detected in the saliva of mice immunised with the PAc expressor strain but not with the control strain which lacked the PAc antigen gene. PAc- specific IgG antibody responses were also detected in the serum of mice immunised with the expressor strain. The fragment of the *Str. mutans pac* gene cloned into *L. lactis* included the native promoter but lacked the 3' sequences needed for cell surface anchoring. It was therefore expected that most of the PAc protein produced by *L. lactis* would be secreted into the growth medium. In fact the small amounts of PAc antigen produced by *L. lactis* were only found in the cell extracts and not in the culture supernatant; the reasons for this are unclear. As the *Str. mutants* PAc antigen and related surface proteins of other oral commensals may have been present in the normal microflora of these mice the questions remains as to whether the antibody responses were primary or secondary in nature; it is possible that the *L. lactis* constructs were simply boosting immune responses which had been established earlier by contact with components of their commensal flora.

We have adopted tetanus toxin fragment C (TTFC) as a first model antigen for immunological studies with recombinant *L. lactis* because it is a potent immunogen and the protective significance of the responses generated can be determined by challenging the animals with the holotoxin. Initially, *L. lactis* strains expressing TTFC or TTFC fusion proteins were inoculated subcutaneously into mice in order to determine whether the TTFC formed in *L. lactis* was immunogenic and capable of eliciting immune responses which would protect mice from lethal challenge with tetanus toxin (Wells et al., 1993a, 1995; Norton et al., 1995). These studies have shown that by the subcutaneous route all lactococcal TTFC expressor strains are able to elicit antibodies which protect the mice from lethal toxin challenge (i.e. at least 5–20 × LD_{50} of tetanus toxin). However, the dose of *L. lactis* required to elicit protective antibody responses was dependent on the quantity of antigen produced by the expressor strain and also on the location and form of the antigen. For example it was found that when compared in terms of the dose of expressed TTFC required to elicit protection against lethal challenge a membrane-anchored form of TTFC (see pLET4, Figure 3) was significantly (13–20 fold) more immunogenic than the intracellular form of the protein. However, at least ten fold higher doses of bacteria producing the membrane anchored form of TTFC were required to elicit protective level responses because the amounts of TTFC fusion protein expressed in *L. lactis* were only about 1/200 of that expressed intracellularly (in amounts up to about 20% of total soluble protein) in the lactococcal T7 system. These studies indicated that C57 BL/6 mice were more responsive to immunisation with TTFC than CBA mice and that BALB/c mice were the least responsive.

As the ultimate aim of this work is to use lactococci for the mucosal delivery of vaccines attention has been focused more recently on the immunisation of mice orally and intranasally with recombinant strains of *L. lactis* expressing TTFC. Intranasal inoculation of groups of ten C57 BL/6 mice on days 1, 7 and 29 with either 5 × 10^8 or 5 × 10^9 bacteria of strain UCP1050 expressing TTFC intracellularly (using the lactococcal T7 system) induced a significant (p 0.001) IgG serum antibody response to TTFC (Norton et al., 1995). All of the animals inoculated with the higher dose of the TTFC expressor strain (UCP1059) developed anti TTFC responses which were higher than those seen in animals inoculated with 60 μg of purified TTFC or the control strain UCP1049 lacking the TTFC gene. Oral inoculation of the expressor strain (UCP1050) also significantly elevated the levels of anti-TTFC IgA antibodies detected in the gut secretions (Norton, pers. communication). The mice immunised with the TTFC expressor strain were protected from lethal challenge (at least 20 x LD50) with tetanus toxin.

A more recent study with *L. lactis* strain (UCP1060) expressing TTFC constitutively (using pTREX1) has shown that it is possible to elicit protective level systemic immune responses to TTFC when mice are orally immunised using a regime of three doses on consecutive days which is then repeated four weeks after the first dose (Robinson et al., 1995). The TTFC specific serum antibody levels reached high levels by day 40 giving mean end point titres of approximately 1/10000 compared to mean end point titres of approximately 1/50 for naive mice and mice inoculated with the non expressor control strain of *L. lactis* (Figure 4). The TTFC-specific serum antibody responses were of both the IgG1 and IgG2a isotypes and proved to be protective against lethal challenge with tetanus toxin. Systemic and mucosal immune responses to TTFC were also elicited by immunising mice orally with *L. lactis* strain UCP1050 which utilises the lactococcal T7 expression system to produce TTFC intracellularly.

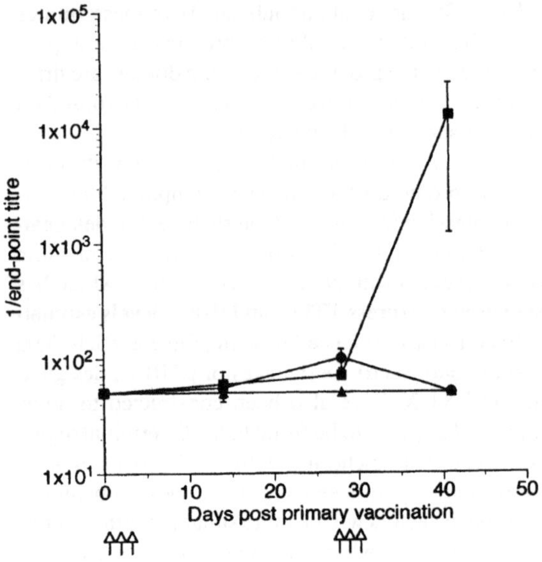

Figure 4. Serum antibody responses to TTFC following oral immunisation of groups of 4 female C57BL/6 mice with 5 x 10^9 c.f.u. of *L. lactis*. The days on which mice were inoculated are indicated with arrows. The TTFC specific serum antibody levels of mice inoculated with the TTFC expressor strain UCP1060 –■– reached high levels by day 40 giving mean end point titres of approximately 1/10000 compared to mean end point titres of approximately 1/50 for naive mice –▲– and mice inoculated with the non expressor control strain of *L. lactis* –●– (vertical bar = mean ± the standard error of the mean).

Concluding remarks

It has now been shown that the oral administration of recombinant *L. lactis*, *Lactobacillus*, and *Str. gordonii* can be used to elicit local IgA and/or serum IgG antibody responses to an expressed antigen. However, it is still not clear to what extent the format in which the antigen expressed by the different recombinant lactic acid bacteria and the amount of the antigen produced affects the magnitude and duration of the responses elicited via mucosal routes of immunisation. It would be of considerable interest to compare the immune responses to a model antigen when it is expressed intracellularly, secreted or displayed on the surface of the different lactic acid bacteria which are currently being developed as vaccine delivery vehicles. This type of study would help to establish the relative importance of colonisation in eliciting an immune response and also the influence of the mucosal site of colonisation on the magnitude and type of immune responses elicited.

Although the idea of the common immune system has been put forward to explain the observation that immunisation at one mucosal site can lead to secretory antibody responses at distal mucosal sites it is often observed that the most vigorous immune responses are elicited at the site of initial priming. In the future we may find that lactic acid bacteria which colonise different mucosal surfaces will be used specifically to vaccinate against pathogens which initiate infection at that particular mucosal site. However, the long term colonisation of any mucosal site by a recombinant lactic acid bacterium may not be desirable as it may result in tolerance to the antigen. In the mucosal colonisation studies which have been described to date there seems to be considerable mouse to mouse variation in both the duration and the extent of colonisation by lactic acid bacteria. This probably reflects the delicate nature of the interaction between microbes and the host defence mechanisms which have evolved to avoid inappropriate responses against beneficial microbes colonising the mucosa. At present our understanding of the mechanism involved in the modulation of the host response to commensal bacteria is limited. However, ongoing studies on the probiotic effects of lactic acid bacteria and their interaction with hosts will ultimately help to identify strains of lactic acid bacteria which are most beneficial and suitable for the different vaccine delivery systems under development.

To date the results are very encouraging as they indicate that lactic acid bacteria are capable of delivering antigen to antigen presenting cells of the mucosal and systemic immune systems following oral immunisation. In the future we can look forward to hearing about the results of oral immunisation with a wider variety of antigens and learning more about the protective significance of the responses in model animal challenge systems.

Acknowledgements

We are very grateful to Annick Mercenier, Gianni Pozzi, Cathy Rush, Stefan Stahl, Wim Boersma, Lothar Steidler and Eric Remaut for their helpful discussion and for providing details of unpublished work. We are especially thankful to Annick Mercenier for her comments and advice on the preparation of the manuscript. Our work on *Lactococcus lactis* is supported by grants from the Biotechnology and Biological Sciences Research Council, The Wellcome Trust, and The Commission of the European Communities (Grants BIO2CT-CT93011 and BIOT/CT94/3055)

References

Aldovini A & Young RA (1991) Humoral and cell-mediated immune responses to live recombinant BCG-HIV vaccines. Nature 351: 479–482

Boersma WJA, Bogaerts WJC, Bianchi ATJ & Claassen E (1992) Adjuvant properties of stable water in oil emulsions: evaluation of the experience with Specol. Res. Immunol. 143: 503–512

Boersma WJA, Glashouwer & Claassen E (1995) Induction of mucosal immune responses using transformed *Lactobacillus* as vaccine carrier. Immunology, 86: supplement 1 No. 4: 21

Bringel F, Frey L & Hubert JC (1989) Characterisation, cloning, curing, and distribution in lactic acid bacteria of pLP1, a plasmid from *Lactobacillus plantarum* CCM 1904 and its use in shuttle vector construction. Plasmid 22: 193–202

Chatfield SN, Charles IG, Makoff AJ, Oxer MD, Dougan G, Pickard D, Slater D & Fairweather NF (1992) Use of the *nirB* promoter to direct the stable expression of heterologous antigens in Salmonella oral vaccine strains: development of a single-dose oral tetanus vaccine. Bio/Technology 10: 888–892

Claassen E, Pouwels PH, Posno M & Boersma W (1995) Development of safe oral vaccines based on *Lactobacillus* as a vector. In: Kurstak, E. (Ed) Recombinant Vaccines: New Vaccinology Int. Comp. Virolog Org., Montreal, (in press)

Cox DS, & Taubman MA (1984) Oral induction of the secretory antibody response by soluble and particulate antigens. Int. Arch. Allergy. Appl. Immun. 74: 249

Dahlgren UIH, Wold AE, Hanson LA & Midtvedt T (1991) Expression of a dietary protein in *E. coli* renders it strongly antigenic to gut lymphoid tissue. Immunology 73: 394

Dougan G, Roberts M, Douce G, Londono P, Hormaeche C, Harrison J & Chatfield S (1993) The Genetics of Salmonella and Vaccine Development.

Fischetti VA, Medaglini D, Oggioni M & Pozzi G (1993) Expression of foreign proteins on Gram-positive commensal bacteria for mucosal vaccine delivery. Cur. Opin. Biotechnol. 4: 603–610

Gerritse K, Posno M, Schellekens MM, Boersma WJA & Claassen E (1990) Oral administration of TNP-*Lactobacillus* conjugates in mice: a model for evaluation of mucosal and systemic immune responses and memory elicited by transformed lactobacilli. Res. Microbiol. 141: 955–962

Gold R, Goldschneider I, Lepow ML, Draper TF & Randolph M (1978) Carriage of *Neisseria meningitidis* and *Neisseria lactamica* in infants and children. J. Infect. Dis. 137: 112–121

Gruzza M, Fons M, Ouriet MF, Duval-Iflah Y, Ducluzeau R (1994) Study of gene transfer *in vitro* and in the digestive tract of gnotobiotic mice from *Lactococcus lactis* strains to various strains belonging to human intestinal flora. Microb. Releases 2, 183–189

Holzapfel WH & Wood BJB (1995) Lactic acid bacteria in contemporary perspective. In: Wood BJB & Holtzapfel WH (Eds.), The Genera of Lactic Acid Bacteria. Vol 2 (pp 1–6). Blackie A & P, London

Iwaki M, Okahashi N, Takahashi I, Kanamoto T, Sugita-Konishi Y, Aibara K & Koga T (1990) Oral immunisation with recombinant *Streptococcus lactis* carrying the *Streptococcus mutants* surface protein antigen gene. Infect. Immun. 58: 2929–2934

Kato T & Owen RL (1994) Structure and function of intestinal mucosal epithelium. In: Ogra PL, Mestecky J, Lamm ME, Strober W, McGhee JR & Bienenstock J. Handbook of Mucosal Immunology (pp 11–26). Academic Press, New York

Kiewiet R, Kok J, Seegers JFML, Venema G, & Bron S (1993). The mode of replication is a major factor in segregational plasmid instability in *Lactococcus lactis*. Appl. Environ. Microbiol. 59: 358–364

Klaenhammer TR (1995) Genetics of intestinal lactobacilli. Int. Dairy Journal 5: 1019–1059

Klijn N, Weerkamp AH & de Vos WM (1995) Genetic marking of *Lactococcus lactis* shows its survival in the human gastrointestinal tract. Appl. Environ. Microbiol. 61, 2771–2774

Kraehenbuhl JP & Neutra MR (1992) Molecular and cellular basis of immune protection of mucosal surfaces. Physiological Rev. 72: 853–79

Leer RJ, Van Luijik N, Poson M & Pouwels PH (1992) Structural and functional analysis of two cryptic plasmids from *Lactobacillus pentosus* MD353 and *Lactobacillus plantarum* ATCC8014. Mol. Gen. Genet. 234: 265–274

Litt D (1996) Immunogenicity of HIV-1 V3 loop epitopes of gp120 expressed in *L. lactis* Ph.D thesis, University of Cambridge, Cambridge, U.K. (in preparation)

Medaglini D, Pozzi G, King TP & Fischetti VA (1995) Mucosal and systemic responses to a recombinant protein expressed on the surface of the oral commensal bacterium *Streptococcus gordonii* after oral colonisation. P.N.A.S. 92: 6868–6872

Mercenier A, Dutot P, Kleinpeter P, Aguirre M, Paris P, Reymund J & Slos P (1993) Development of lactic acid bacteria as live vectors for oral or local vaccines. Advances in Food Science (in press)

McGhee JR, Mestecky J, Dertzbaugh MT, Eldridge JH, Hirasawa M & Kiyono H (1992) The mucosal immune system: from fundamental concepts to vaccine development. Vaccine 10: 75–88

Mowat AMcl (1987) The regulation of immune responses to dietary antigens. Immunol. Today 8: 93–98

Norton PM, Brown HWG, Wells JM, Macpherson AM, Wilson PW & Le Page RWF (1996) Factors affecting the immunogenicity of tetanus toxin fragment C expressed in *Lactococcus lactis*. FEMS Immunol. Med. Microbiol. (in press)

Norton PM, Le Page RWF & Wells JM (1995) Mucosal immunisation with tetanus toxin fragment C expressed in *Lactococcus lactis* generates both enteric IgA and a protective serum antibody response. Folia Microbiologica 40: 225–230

Nguyen TN, Hansson M, Stahl S, BŠchi T, Robert A, Domzig W, Binz H & Uhlén M (1993) Cell-surface display of heterologous epitopes on *Staphylococcus xylosus* as a potential delivery system for oral vaccination. Gene 128: 89–94

Oggioni MR, Manganelli R, Contorni M, Tommasino M & Pozzi G (1995) Immunisation of mice by oral colonization with live recombinant commensal streptococci. Vaccine 13: 775–779

Oggioni MR & Pozzi G (1996) A host-vector system for heterologous gene expression in *Streptococcus gordonii*. Gene (in press)

O'Hagan DT (1994) Novel Delivery Systems for Oral Vaccines. CRC Press, Florida

Posno M, Heuvelmans PTHM, van Giezen MJF, Leer RJ & Pouwels PH (1991a) Complementation of the inability of *Lactobacillus* strains to utilise D-xylose with D-xylose catabolism- encoding genes of *Lactobacillus pentosus*. Appl. Environ. Microbiol. 57: 2764–2766

Posno M, Leer RJ, van Luijk N, van Geizen MJF & Heuvelmans PTHM (1991b) Incompatibility of *Lactobacillus* vectors with replicons derived from small cryptic *Lactobacillus* plasmids and segregational instability of the introduced vectors. Appl. Environ. Microbiol. 57: 1822–1828

Pouwels PH & Leer RJ (1993) Genetics of lactobacilli: plasmids and gene expression. Antonie van Leeuwenhoek 64: 85–107

Pozzi G, Contorni M, Oggioni MR, Manganelli R, Tommasino M, Cavalieri F & Fischetti VA (1992) Delivery and expression of a heterologous antigen on the surface of streptococci. Infect. Immun. 60: 1902–1907

Pozzi G, Musmanno RA, Lievens PMJ, Oggioni MR, Plevani P & Mangenelli R (1990) Method and parameters for genetic transformation of *Streptococcus sanguis*. Res. Microbiol. 141: 659–670

Robinson K, Chamberlain LM, Schofield KM, Wells JM & Le Page RWF (1995) Oral vaccination of mice with recombinant *Lactococcus lactis* expressing tetanus toxin fragment C elicits both secretory and protective high level systemic immune responses. Immunology 86: supplement 1, 27, 139

Rush CM (1995) Antigen delivery to the female reproductive tract using recombinant *Lactobacillus*. Ph.D thesis, Queensland University of Technology, Queensland, Australia

Samuelson P, Hansson M, Ahlborg N, Andréoni C, Götz F, Bächi T, Nguyen TN, Binz H, Uhlén M & Stahl S (1995) Cell surface display of recombinant proteins on *Staphylococcus carnosus*. J. Bact. 177: 1470–1476

Shimizu-Kadota M, Shibahara-Sone H & Ishiwa H (1991) Shuttle plasmid vectors for *Lactobacillus casei* and *Escherishia coli* with a minus origin. Appl. Environ. Microbiol. 57: 3292–3300

Stahl S, Samuelson P, Hansson M, Andréoni C, Goetsch L, Libon C, Liljeqvist S, Gunneriusson E, Binz H, Nguyen N & Uhlén M (1996) Surface display of heterologous antigens on non-pathogenic staphylococci. In: Pozzi G & Wells JM (Eds.), Gram-positive Bacteria as Vaccine Delivery Vehicles for Mucosal Immunisation. R.G. Landes, Biomedical Publishers, Austin, Texas, USA (in press)

Stover CK, Bansal GP, Hanson MS, Burlein JE, Palazynski SR, Young JF, Koenig S, Young DB, Sadziena A & Barbour AG (1993) Protective immunity elicited by recombinant bacille Calmette-Guerin (BCG) expressing outer surface protein A (OspA) lipoprotein: a candidate Lyme disease vaccine. J. Exp. Med. 178: 197–209

Swinfield TJ, Oultram JD, Thompson DE, Brem KJ & Minton NP (1990) Physical characterisation of the replication region of the *Streptococcus faecalis* plasmid pAMβ1. Gene 87: 79–90

Thomas H & Parrot D (1974) The induction of tolerance to a soluble antigen by oral administration. Immunology 27: 631–639

Warner L, Ermak T & Griffiss JM (1987) Mucosal and serum immunity following commensal enteric colonisation. Adv. Exp. Med. Biol. 216: 959–964

Wells JM, Wilson PW, Norton PM, Gasson MJ & Le Page RWF (1993a) Lactococcus lactis: High level expression of tetanus toxin fragment C and protection against lethal challenge. Molec. Microbiol. 8: 1155–1162

Wells JM, Wilson PW, Norton PM & Le Page RWF (1993b) A model system for the investigation of heterologous protein secretion pathways in Lactococcus lactis. Appl. Environ. Microbiol. 59: 3954–3959

Wells JM, Norton PM & Le Page RWF (1995) Progress in the development of mucosal vaccines based on *Lactococcus lactis*. Int. Dairy J. 5: 1071–1079

Wells JM & Schofield KM (1996) Cloning and expression vectors for lactococci. In: Bozoglu F (Ed.), NATO ASI Series Vol. H. Springer Verlag, Heidelberg (in press)

Wold AE, Dahlgren UIH, Hanson LA, Mattsby-Baltzer I & Midvetdt T (1989) Difference between bacterial and food antigens in mucosal immunogenicity. Infect. Immun. 57: 2666–2672

Yasutomi, Y, Koenig S, Haun SS, Stover CK, Jackson RK, Conard P, Conley AS, Emini EA & Furst TR (1993) Immunisation with recombinant BCG-SIV elicits SIV-specific cytotoxic T lymphocytes in rhesus monkeys. J. Immunol. 150: 3101–3107

Antonie van Leeuwenhoek **70:** 331–345, 1996.

Biopreservation by lactic acid bacteria

Michael E. Stiles
Professor of Food Microbiology, Department of Agricultural, Food and Nutritional Science University of Alberta, Edmonton, Alberta, Canada T6G 2P5

Abstract

Biopreservation refers to extended storage life and enhanced safety of foods using the natural microflora and (or) their antibacterial products. Lactic acid bacteria have a major potential for use in biopreservation because they are safe to consume and during storage they naturally dominate the microflora of many foods. In milk, brined vegetables, many cereal products and meats with added carbohydrate, the growth of lactic acid bacteria produces a new food product. In raw meats and fish that are chill stored under vacuum or in an environment with elevated carbon dioxide concentration, the lactic acid bacteria become the dominant population and preserve the meat with a 'hidden' fermentation. The same applies to processed meats provided that the lactic acid bacteria survive the heat treatment or they are inoculated onto the product after heat treatment. This paper reviews the current status and potential for controlled biopreservation of foods.

Abbreviations: LAB – lactic acid bacteria; *C* – *Carnobacterium*; *Lb* – *Lactobacillus; Lc* – *Lactococcus*; *Le* – *Leuconostoc*; *Ls* – *Listeria*; *P* – *Pediococcus*

Historical aspects of food preservation

Preservation of foods in a sound and safe condition has long been, and it remains, an on-going challenge for humans. Drying, salting and fermentation were the traditional methods of preservation. Canning and freezing were relatively recent developments, dating back to Napoleonic times and the 1920s, respectively. In developed societies, food preservation is viewed as a 'convenience' of an efficient food system; in developing societies, food preservation is a key to ensuring the availability of food as a vital benefit. Food fermentations developed by default rather than by design. Foods that spoiled during storage and the product was acceptable for consumption was the most probable basis for the development of fermented foods. Lactic acid bacteria (LAB) play an important role in food fermentations, causing the characteristic flavour changes and exercising a preservative effect on the fermented product. It is estimated that 25% of the European diet and 60% of the diet in many developing countries consists of fermented foods (Holzapfel et al., 1995). The spice trade was the start of the addition of chemical adjuncts to foods. Originally this was done to mask unpleasant flavours of spoiled foods; but it became a way of preserving and imparting flavour and variety to foods. With the industrial revolution and the subsequent development of food industries, food processing moved from kitchen or cottage industries to large scale technological operations with increased need for food preservation. This stimulated the use of food additives, especially those that preserve foods and enhance food quality. In recent years the addition of chemical preservatives has fallen into disfavour with consumers who, it is claimed, are 'seeking foods that are high quality, less severely processed (less intensive heating and minimal freezing damage), less heavily preserved, more natural (freer of artificial additives) and safer' (Goud, 1992). This has resulted in the emergence of a new generation of chill stored, minimally processed foods.

Apart from salt, sugar, and smoke (usually with heat treatment), the food industry uses nitrite, sulphite, the parabens, organic (acetic, lactic and propionic) acids, benzoic and sorbic acids to preserve foods. Reduction of the water activity of foods with added salt and sugar has become less acceptable because of the consumer desire to reduce the amounts of these ingredients in processed foods (Goud, 1992). Many foods provide an environment for microbial growth. Although foods can be contaminated by a variety of microorganisms from food handling, equipment and other sources such as water, air, dust, soil, it is well established that intrinsic factors of the food and extrinsic factors of the storage environment dictate the types of microorganisms that dominate the microflora. Combinations of different physical, chemical and microbial preservation factors have been proposed as the basis of 'hurdle technology' for preservation of foods Leistner & Gorris, 1995). Bacteriocinogenic cultures and (or) their bacteriocins could serve as useful preservation hurdles. Chill storage and modification of the gaseous environment of foods have become important and acceptable methods of food preservation. Chilled storage of raw meats in a vacuum package has a marked effect on the predominant microflora of meat, changing it from putrefactive Gram-negative rod-shaped bacteria to fermentative lactic acid bacteria (LAB). This change is associated with a dramatic extension of their storage life (Dainty & Mackey, 1992).

The preservation of foods by the antagonistic growth of microorganisms was reviewed by Hurst in 1972 (Hurst, 1973). He cited growth of a LAB microflora in milk, sauerkraut and vacuum packaged meats as examples of protective, antagonistic growth. Hurst (Hurst, 1973) also considered the role of 'antibiotics' (bacteriocins) such as nisin in the preservation of foods that support the growth of LAB. In recent times this has been termed 'biopreservation' to differentiate it from the chemical (artificial) preservation of foods. Bacteria preserve foods as a result of competitive growth, products of their metabolism and bacteriocin production. For the purpose of this review, biopreservation refers to extended storage life and enhanced safety of foods using their natural or controlled microflora and (or) their antibacterial products. It may consist of (i) adding bacterial strains that grow rapidly and (or) produce antagonistic substance(s); (ii) adding purified antagonistic substances; (iii) adding the fermentation liquor or concentrate from an antagonistic organism; or (iv) adding mesophilic LAB as a 'fail-safe' protection against temperature abuse. LAB produce lactic acid or lactic and acetic acids, and they may produce other inhibitory substances such as diacetyl, hydrogen peroxide, reuterin (b-hydroxypropionaldehyde) and bacteriocins. Bacteriocins are ribosomally-produced, precursor polypeptides or proteins that, in their mature (active) form, exert an antibacterial effect against a narrow spectrum of closely related bacteria (Jack et al., 1995). While most bacteriocins produced by LAB have a narrow antibacterial spectrum, others are active against closely related species and against *Listeria* and *Enterococcus* species. Nisin is active against a broad spectrum of Gram-positive bacteria, including *Clostridium botulinum* and its spores (Hurst, 1981). LAB that produce broad spectrum bacteriocins offer great potential in biopreservation. LAB also contribute to the nutritive value and healthfulness of foods. The object of this review is to evaluate the status of the research on biopreservation of different food types.

Foods in which lactic acid bacteria are the dominant microflora

LAB are a relatively diffuse group of bacteria that encompass several genera and have associations with many different foods (Table 1). LAB that grow as the adventitious microflora of foods or that are added to foods as cultures are generally considered to be harmless or even an advantage for human health (probiotics). In the United States they are afforded GRAS (Generally Recognized as Safe) status. In comparison, the genus *Streptococcus* includes many human and animal pathogens. Despite this, *Streptococcus thermophilus* is an important nonpathogenic organism that is used in the manufacture of yogurt and several cheese types. The importance of enterococci and their role in foods is difficult to assess because they are associated with human disease, they are commensals of the mammalian intestinal tract, they can dominate the microflora of foods, they are used as starter cultures in some fermented foods and they are used as probiotics (Devriese et al., 1991). The remaining genera associated with foods are considered to be nonpathogenic; but there are reviews of their involvement in human clinical infections (Aguirre & Collins, 1993 and Gasser, 1994). In cases of human clinical infections in which LAB are the only isolates, patients were usually suffering from an underlying disease condition that often involved vancomycin therapy. In general, 'food grade' LAB are an innocuous group of bacteria that have an important ecological role in food preservation. At a European workshop on the safety of lactic acid bacteria from food it was concluded that 'the overall risk of LAB infection is very low, particularly in view of their ubiquity in the environment' (Adams & Marteau, 1995).

The best characterized LAB are those associated with milk fermentations, especially the subspecies of *Lactococcus lactis*. They preserve foods by the low

Table 1. Foods and their associated lactic acid bacteria [a]

Food types	Lactic acid bacteria
Milk and dairy foods	
hard cheeses without eye formation	*Lc. lactis* subsp. *cremoris* and subsp. *lactis*
cottage cheese and cheeses with a few or small eyes (Edam)	*Lc. lactis* subsp. *cremoris* and subsp. *lactis*; and *Le. mesenteroides* subsp. *cremoris*
Cultured butter, buttermilk cheeses with round eyes (Gouda)	*Lc. lactis* subsp. *cremoris* and subsp. *lactis* and var. *diacetylactis*; and *Le. mesenteroides* subsp. *cremoris*
Swiss-type cheeses	*Lb. delbrueckii* subsp. *bulgaricus*; *Lb. helveticus*
Dairy products in general	*Lb. brevis*; *Lb. buchneri*; *Lb. casei*; *Lb. paracasei*; *Lb. fermentum*; *Lb. plantarum*; *Le. mesenteroides* subsp. *cremoris*; *Le. lactis*
Fermented milks	
- yogurt	*Streptococcus thermophilus* and *Lb. delbrueckii* subsp. *bulgaricus*; Lc. lactis subsp. diacetylactis
- acidophilus milk	*Lb. acidophilus*
- kefir	*Lb. kefir*; *Lb. kefiranofaciens*
Meats	
raw	*C. divergens*; *C. piscicola* (*maltaromicus*)
	Lb. sake; *Lb. curvatus*
	Le. carnosum and *Le. gelidum*
semi-preserved	*Lb. viridescens* (spoilage)
	Le. carnosum and *Le. gelidum*
	C. divergens; *C. piscicola* (*maltaromicus*)
fermented meat	*P. acidilactici* and *P. pentosaceus*(inoculated into semi- dry sausages)
	Lb. sake; *Lb. curvatus*
	Lb. farciminis (uninoculated)
Fish	
marinated fish products	*Lb. alimentarius*
	C. piscicola
Fermented vegetables	*P. acidilactici* and *P. pentosaceus*;
	Lb. plantarum; *Lb. sake*; *Lb. buchneri*; *Lb. fermentum*;
cucumbers, sauerkraut	*Le. mesenteroides* (initial fermentation)
	Lb. bavaricus; *Lb. brevis*; *Lb. sake*;
	Lb. plantarum;
olives	*Le. mesenteroides*, *Lb. pentosus*
Soy sauce	*Tetragenococcus* (*Pediococcus*) *halophilus*
Baked goods	
sourdough bread	*Lb. sanfrancisco* (wheat and rye sourdough)
	Lb. farciminis; *Lb. fermentum*;
	Lb. brevis; *Lb. plantarum*;
	Lb. amylovorus; *Lb. reuteri*
	Lb. pontis (rye sourdough)
Wine (malo-lactic fermented)	*Le. oenos*

[a]Sources of information: (Hammes & Tichaczek, 1994); (Wood & Holzapfel, 1995).
[b]*C. Carnobacterium*; *Lb. Lactobacillus*; *Lc. Lactococcus*; *Le. Leuconostoc*; *P. Pediococcus*.

pH and lactic acid that they produce, as well as bacteriocins, in particular nisin that has found widespread application as a food preservative (Delves-Broughton, 1990). *Lc. lactis* is of plant and not animal origin (Sandine et al., 1972). This was confirmed by isolation of nisin- producing strains of *Lc. lactis* from plant material (Harris et al., 1992). Lactobacilli and leuconostocs also have important associations with fermented milk products. In hard cheeses, lactococci are used as the inoculum, but adventitious lactobacilli such as *Lb. casei* play an important role in the ripening process. In contrast, in Swiss-type cheeses, lactobacilli such as *Lb. helveticus* and *Lb. delbrueckii* subsp. *bulgaricus* are added as part of the starter culture. Lactobacilli are an important component of the intestinal microflora, and several species are added to milk for their probiotic effect in the human intestine, in particular *Lb. acidophilus*, *Lb. casei* and *Lb. reuteri*. Developments in the phylogenetic relationships of *Lb. acidophilus* (Fujisawa et al., 1992) make the associations of this 'organism' with intestinal colonization particularly difficult to assess.

Leuconostoc mesenteroides subsp. *cremoris* and *Le. lactis* are important for the production of diacetyl from citrate, but their use for production of this flavour compound in fermented dairy products has largely been replaced by citrate-utilizing strains of *Lc. lactis*, unless the fermentation requires the production of carbon dioxide, for example in Gouda cheese making. The leuconostocs are important in other food fermentations, especially in fermented vegetables where their growth comprises the initial, highly important fermentation (Dellaglio et al., 1995). *Le. oenos* is responsible for the malolactic fermentation in wines, which can be favourable or unfavourable depending on the wine being produced (Henick-Kling, 1995). *Leuconostoc* spp. can be important spoilage organisms in foods. Strains that produce dextrans can cause significant economic loss in sugar processing and in processed meats when sucrose is added in the formulation of the meat. *Le. carnosum* and *Le. gelidum* have been isolated from the LAB microflora of vacuum packaged meat (Shaw & Harding, 1984 and 1989).

The initial microflora of meats consists of a wide range of microorganisms, but refrigeration and packaging affect the microbial ecology of meats and meat products. When raw or cured meats are chill stored and packaged under vacuum or in a modified atmosphere with elevated levels of carbon dioxide, the dominant microflora is psychrotrophic LAB (Shaw & Harding, 1984; 1985 and McMullen & Stiles 1993). The adventitious growth of LAB causes a dramatic extension of the storage life that has been exploited in the marketing of wholesale cuts of red meats (especially beef) and in the retail marketing of sliced semi-preserved (luncheon) meats. LAB also dominate the microflora of modified atmosphere packaged, sliced roast beef; but in cooked hamburgers the development of a LAB microflora was less predictable (McMullen & Stiles, 1989). LAB that grow at refrigeration temperatures are the natural strains that dominate meats. *Carnobacterium* spp. (originally reported as atypical lactobacilli), *Lb. sake* and other homofermentative lactobacilli, and *Le. mesenteroides* subsp. *mesenteroides*, *Le. gelidum* and *Le. carnosum* are important in this regard (Dainty & Mackey, 1992).

LAB and their fermentation products preserve foods, but adventitious growth of LAB in foods could be termed spoilage. In meats, the products of adventitious fermentation are less noticeable, nonetheless at some (unpredictable) time **after** maximum population is achieved the meat will spoil due to the accumulated products of LAB growth. In contrast, if sulphide-producing strains of *Lb. sake* dominate the LAB microflora, overt spoilage of the vacuum packaged meat product will occur (Egan et al., 1989). The normal pH of vacuum packaged meat is 5.5 to 5.6. If muscle glycogen is depleted prior to slaughter, the pH of the meat will not drop to this level. When meat of high pH (>6.0) is vacuum packaged, *Shewanella putrefaciens* and psychrotrophic Enterobacteriaceae grow and produce hydrogen sulphide. Growth of *Brochothrix thermosphacta* is also responsible for off-odours in packaged meats. Growth is generally not affected by pH under aerobic conditions, but under anaerobic conditions growth is inhibited at pH <5.8.

Fermented meat products are well established preserved foods that depend on many different microorganisms for their fermentation, including LAB and micrococci. Important among the LAB used for meat fermentation is *P. acidilactici*. The use of this strain has become widespread in fermented meat production possibly because the predominant strain produces a potent bacteriocin, pediocin PA–1/AcH. Pediococci were isolated and their bacteriocins were studied separately by several laboratories (Marugg et al., 1992; Motlagh et al., 1992; Christensen & Hutkins 1994 and Hoover et al., 1988) but it subsequently transpired that they were studying the same bacteriocin.

Use of LAB and their bacteriocins as biopreservatives also has potential for application in seafood products. LAB isolated from fish intestine, smoked and marinated fish (Pilet et al., 1995) included strains of carnobacteria, lactococci and enterococci. Two bacteriocin-producing carnobacteria were further studied and they show potential for use in fish preservation. Brined shrimp is a highly perishable product that is normally preserved with added sorbic or benzoic acid. Nisin Z, carnocin UI49 or bavaricin A were added to shrimp (Einarsson & Lauzon 1995). Nisin Z was the most effective of these three bacteriocins, extending the storage life from 10 days for control samples without nisin to 31 days for the samples preserved with nisin.

Many vegetables are preserved by brining and (or) fermentation. In the fermentation of cabbage, cucumbers and olives a brine solution is added to favour the growth of LAB. *P. pentosaceus* has been established as an important LAB in vegetable fermentations and strains producing pediocin A have the potential for a preservative role in fermented vegetables (Fleming et al., 1975). The possibility of preserving ready-to-use vegetables with bacteriocin-producing LAB has been investigated (Vescovo et al., 1995). LAB are also important in cereal fermentations, such as African maize, sorghum and millet products. In the Ghanaian fermented maize (corn) product 'Kenkey' there are three environments in which LAB grow: during grain steeping; milling of the steeped grain; and the two day dough fermentation stage. The microflora undergoes a progression towards a uniform population of heterofermentative LAB achieving pH 3.7 in the dough. In a study by Olsen et al., (Olsen et al., 1995) many inhibitory LAB were detected but they concluded that safety and storage stability of fermented maize depends on a mixed population of LAB and that introducing starters as pure cultures may create a microbial risk in the product.

Application of bacteriocins in biopreservation

Nisin is licensed for use as a food additive in over 45 countries (Delves-Broughton, 1990). Other bacteriocins have not been licensed for addition to foods, but studies have shown that there are other bacteriocins that have potential for use as food preservatives, in particular, pediocin A for its antibotulinal effect (Okereke & Montville, 1991 and Okereke & Montville, 1991) and pediocin PA–1/AcH for its anti-*Listeria* activity in dairy products (Pucci et al., 1988) and in meat preservation (Ray, 1992). Many studies on the activity of bacteriocins against target strains were done in laboratory media and not in foods. There are intrinsic factors in foods that could cause reduced activity of a bacteriocin. Class I and II bacteriocins are generally heat resistant, but they can be inactivated by proteolytic enzymes in foods. Most bacteriocins are hydrophobic, hence they can be bound by fats and phospholipids. Nisin activity against *Ls. monocytogenes* is decreased in the presence of increasing fat concentration (Jung et al., 1992), but inactivation of nisin in the presence of fat was decreased with addition of a nonionic emulsifier such as Tween 80, but not by an anionic emulsifier such as lecithin (Jung et al., 1992).

The study of bacteriocins as preservatives in foods can be misleading and confusing unless they have been fully characterized. This was emphasized by the fact that once the amino acid sequence of pediocins PA–1 (Marugg et al., 1992), AcH (Motlagh et al., 1992), JD (Christensen & Hutkins, 1994) and Bac (Hoover et al., 1988), as well as mesenterocin 5 (Daba et al., 1991) was determined, it was realized that they are identical compounds. Bacteriocins produced by leuconostocs of dairy (mesentericin Y105) and meat (leucocin A-UAL187) origin are being studied for their possible use in food preservation (Hastings et al., 1991; Héchard et al., 1992). These two bacteriocins were isolated from unrelated sources yet they differ by only two amino acids; furthermore, leucocin B-Ta11a produced by a strain of *Le. carnosum* isolated from a vacuum packaged, cured meat in South Africa produces a bacteriocin identical to leucocin A, but there are differences in seven residues of their 24-amino-acid N- terminal extensions (Felix et al., 1994). This quite phenomenal distribution of 'leucocin A-like' bacteriocins substantiates the observation with nisins A and Z (Mulders et al., 1993) that minor variants of bacteriocins might be quite widespread in nature. This should encourage studies on site-directed mutagenesis of bacteriocins as a possible means of influencing their antibacterial spectrum. The potential for structural manipulation with a ribosomally synthesized compound is great. This has yet to become a major emphasis of bacteriocin research, but with a gene replacement strategy such as that developed for nisin by Dodd et al., (Dodd et al., 1996), the opportunity to develop genetically engineered variants of nisin is greatly enhanced. It is frequently stated that studies of bacteriocins in foods are lacking. This review summarizes the studies on bacteriocins tested in foods.

Milk and Dairy Products. In 1946, Mattick & Hirsch (Mattick & Hirsch, 1946) showed that a nisin-producing strain of *Lc. lactis* destroyed tubercle bacilli in milk, whereas a nisin-negative strain did not. Subsequently, Hirsch et al., (Hirsch et al., 1951) used nisin-producing starter cultures to preserve Swiss cheese from spoilage by clostridia. These were early examples of biopreservation of foods. Unfortunately, nisin interferes with starter performance and ripening of cheeses so that interest in its use as a biopreservative in cheese making diminished; however, nisin was shown to be an effective inhibitor of clostridial spoilage in pasteurized processed cheese (McClintock et al., 1952). Many other uses for nisin have been listed (Vandenbergh, 1993), including: preservation of canned fruit and vegetables, preservation of nonfermented milk, dairy products, meat and fish, inhibition of diacetyl production by LAB in beer (Ogden & Waites, 1988) and, where necessary, inhibition of the malolactic fermentation in wines (Radler, 1990 and Daeschel et al., 1991). Use of a *Lc. lactis* starter that produces nisin Z was proposed for use in cheeses that are preserved with nitrate (Hugenholtz & de Veer, 1991). Inhibition of the outgrowth of bacterial spores has been nisin's strongest attribute as a biopreservative. To date no other bacteriocin has been reported that competes with nisin for antibotulinal effect, but it has not been able to replace the use of nitrite in processed meats (see below).

The selection criteria for lactic acid bacteria to be used as starter cultures were ably reviewed by Buckenhüskes in 1993 (Buckenhüskes, 1993). Antagonism is cited as a selection criterion for some lactic acid bacteria. The use of bacteriocinogenic cultures in cheese manufacture has many possibilities for inhibition of spoilage and pathogenic bacteria. Unfortunately the use of nisin producing strains of *Lc. lactis* is contraindicated in cheese making because they produce acid at slower rates than commercial starter cultures (Lipinska, 1973). However, a nisin- producing starter that produced acid at rates suitable for Cheddar cheese manufacture was developed with mixed cultures of nisin resistant and nisin producing strains, provided that the strain has a passive mechanism of nisin resistance (Roberts et al., 1992). Nisin successfully inhibits clostridial growth (including *Clostridium botulinum*) in high moisture cheese spreads (Somers & Taylor 1987) and this process has been patented (U.S. Patent 4,597,972). It has also been shown that production of Cheddar cheese for use in processed cheese spreads with a nisin-producing strain of *Lc. lactis* subsp. *cremoris* can significantly reduce spoilage by *Clostridium sporogenes* (Zottola et al., 1994). Use of a nisin-producing strain of *Lc. lactis* to make Camembert cheese from milk inoculated with 10^1, 10^3 or 10^5colony forming units (cfu) of *Ls. monocytogenes* per ml resulted in a marked decrease in the number of *Listeria* in the ripened cheese compared with the nisin- negative control. The effectiveness of the nisin was greatest with low initial levels of *Listeria* in the cheese milk (Maisnier-Patin et al., 1992). Cheddar cheese making with *Lc. lactis* producing a novel broad spectrum bacteriocin, lacticin 3147, gave favourable results for rate of acid production and control of growth of nonstarter lactic acid bacteria (Ryan et al., 1996).

A major focus of bacteriocin research has been their potential to control *Ls. monocytogenes* in foods (Muriana, 1996). With the exception of nisin in white pickled cheese manufacture (Abdalla et al., 1993), use of bacteriocinogenic strains of *Enterococcus faecalis*, *Lb. paracasei* and *Lc. lactis* in Camembert cheese manufacture (Maisnier-Patin et al., 1992 and Sulzer & Busse, 1991) or addition of bacterial ferment containing nisin or pediocin (Zottola et al., 1994 and Pucci et al., 1988) markedly reduced *Listeria* in cheese. A bacteriocin-producing strain of *Enterococcus faecium* was added as a supplementary culture for manufacture of Taleggio cheese, a soft, surface ripened Italian cheese. The bacteriocin was stable during cheese ripening, it did not affect the thermophilic starter bacteria and it has the potential to inhibit *Ls. monocytogenes* (Giraffa et al., 1995). Addition of heat-treated cultures of *Lc. lactis* to Mozzarella cheese inoculated with *Ls. monocytogenes* and stored at 5 °C maintained the *Listeria* count significantly below that of the control sample (Stecchini et al., 1995). Queso Blanco-type cheese made with acidulants and addition of ALTA™2341, a commercial shelf-life extender based on the fermentation product of a lactic acid bacterium (presumably *Pediococcus acidilactici*) resulted in reduced levels of *Ls. monocytogenes* in the experimental cheeses made with ALTA and stored at 4 and 20 °C for 42 and 7 days, respectively (Glass et al., 1995). The potential for use of bacteriocins, their producer strains or their ferments in cheese manufacture is strongly indicated.

Vegetable fermentations. Vegetable fermentations are initiated by addition of salt instead of a brine containing salt, acetic acid and sugar at concentrations that inhibit microbial growth. Sauerkraut, olive and pickle fermentations consist of a complex succession of LAB. It is well established that heterofermentative LAB such as *Le. mesenteroides* dominate the early stages of the fermentations and rapidly die out, and that the final stages of the fermentations are dominated by homofermentative LAB, especially *Lb. plantarum* (Pederson &

Albury, 1961). In the United States, sauerkraut fermentation is done in bulk tanks compared with the European process of packaging and pasteurization after 7 to 9 days of fermentation (Fleming et al., 1988 and Hammes, 1990). The U.S. method is less expensive but it can result in unpredictable product quality. In a study of a commercial sauerkraut fermentation by Harris et al., (Harris et al., 1992), two nisin-producing strains of *Lc. lactis* were isolated. This resulted in a proposed paired starter culture system consisting of nisin resistant *Le. mesenteroides* and nisin-producing *Lc. lactis* (Harris et al., 1992). To determine the survival of the test strains, the study was done in sterile cabbage juice as a model system. Subsequently, Breidt et al., (Breidt et al., 1995) studied the effect of the paired starter system in brined cabbage and compared it with the addition of nisin and a strain of nisin resistant *Le. mesenteroides*. Nisin activity was detected in the fermentation within 24 h, but it was not detected after 72 h of fermentation. Addition of nisin (12,000 IU ml^{-1}) with the nisin resistant strain of *Le. mesenteroides* gave favourable control of the fermentation and delayed the growth of the homofermentative LAB for at least 20 days; whereas, the addition of nisin alone resulted in decreased LAB growth and growth of Gram-negative bacteria. Production of kimchi, a Korean spiced fermented cabbage, fermented at 14 °C and inoculated with sakacin A producing *Lb. sake* or bacteriocinogenic *P. acidilactici* gave dramatically different results for inhibition of *Ls. monocytogenes*. Sakacin A had no effect compared with successful inhibition of *Listeria* by the *Pediococcus* strain during the 16-day fermentation (Choi & Beuchat, 1994).

Spanish-style, green olives undergo a complex, spontaneous fermentation that is associated with the dominant growth of *Lb. plantarum* within one to two weeks of placing the olives in brine. Attempts to produce green olives in a sterilized brine with a starter culture did not yield a high quality product (Durán et al., 1994). Ruiz-Barba et al., (Ruiz-Barba et al., 1994) inoculated brined (6% NaCl) Spanish-style green olives with bacteriocin-producing and isogenic bacteriocin-negative strains of *Lb. plantarum* that were previously isolated from an olive fermentation and were shown to produce bacteriocins, plantaricins S and T (Jiménez-Diaz et al., 1993). The bacteriocin-producing strain inoculated at 10^5 cfu ml^{-1} dominated the LAB microflora during fermentation and produced desired amounts of lactic acid. Bacteriocin production was demonstrated in the brine by ammonium sulphate precipitation.

Vegetables that are packaged and ready-to-use as a convenience product generally have a refrigerated storage life of one week and support the growth of a microbial population that is dominated by pseudomonads and *Enterobacteriaceae* (Huxoll & Bolin, 1989 and Brocklehurst et al., 1987). The possibility of preserving ready-to-use vegetables with bacteriocin-producing LAB has been investigated (Vescovo et al., 1995). Under the conditions of this study it was shown that inoculation of the salads with strains of *Lb. casei* or *P. pentosaceus* resulted in the domination of the vegetables with these bacteria and a dramatic decrease in Enterobacteriaceae that dominated the uninoculated control samples. During the 8-day storage period at 8 °C the inoculated LAB grew and the pH of the salads decreased from about 5.8 to 5.2. The study indicated that inoculation of ready-to-use vegetables with LAB is effective, but no evidence was presented to show that bacteriocin production by the LAB was a factor in this application of LAB as biopreservatives.

Meats and Meat Products. LAB growth on most meats are 'hidden' fermentations because the low carbohydrate content and the strong buffering capacity of meat does not produce a dramatic change in sensory characteristics comparable to the changes in milk and vegetable fermentations. Meat cannot be pasteurized prior to the addition of a LAB starter, as a result a culture for biopreservation of meat must compete with the natural LAB microflora. Chilled storage is essential for the proper preservation of modified atmosphere packaged meat. At 10 °C overt spoilage of meats occurs, as opposed to an extended storage life at 4 °C (McMullen & Stiles, 1993). Decontamination of carcasses with acid washes has been studied without apparent success (Greer & Jones, 1991 and Greer & Dilts, 1995); whereas decontamination of meat surfaces with nisin (Ambicin® at 5000 AU ml^{-1}) spray using a pilot scale washer resulted in a significant reduction of 2 to 3.5 log cfu cm^{-2} on both lean and adipose tissue (Cutter & Siragusa, 1994). In a separate study (Nielsen et al., 1990), bacteriocin produced by a commercial strain of *P. acidilactici* was added to irradiated beef that was subsequently contaminated with *Ls. monocytogenes*. Attachment of *Listeria* was less in meat treated with bacteriocin, and activity of the bacteriocin could be detected on the meat after 28 days of refrigerated storage.

The predominating LAB on vacuum-packaged beef stored at 2 °C were shown to be lactobacilli, carnobacteria, leuconostoc and *Lc. raffinolactis* (Schillinger &

Lücke, 1987). *Lb. sake* and *Lc. raffinolactis* strains were reported to be the most competitive of the LAB microflora. In probably the earliest study on biopreservation of chill stored, vacuum packaged raw meat, Schillinger and Lücke (Schillinger & Lücke, 1987) inoculated fresh beef that had normal pH and low initial microbial load with *Lb. sake* Lb683. The storage life of the inoculated meat was not markedly improved. They concluded that inoculated LAB had only a slight beneficial effect on chill stored vacuum packaged meat with normal pH. In contrast, addition of bacteriocinogenic pediococci or their cell-free extracts to irradiated ground beef was reported to inhibit four target strains that were added to the meat (Skyttä et al., 1991). This was an unconventional study, in that inoculated meat samples were inoculated at 15 °C for 2 days to allow the pediococci to grow and then stored at 6 °C. Inhibition of the target organisms that included *Ls. monocytogenes* and three Gram-negative strains was only observed with the highest inoculum levels. *Lb. sake* was tested as an inoculum in ground pork (German-type fresh Mettwurst) to suppress *Ls. monocytogenes*. *Listeria* did not grow but they survived in normal pH (5.6 to 5.8) pork, but they grew in pork at pH 6.3 (high pH). In high pH pork, bacteriocin-producing *Lb. sake* Lb706 reduced the count and delayed the growth of *Listeria* more effectively than the bacteriocin negative strain (Schillinger & Lücke, 1990 and Schillinger et al., 1991).

Pediocin PA–1/AcH, nisin and Microgard® were added to chill stored, vacuum packaged beef as biopreservatives and compared with the addition of 2% sodium lactate (Rozbeh et al., 1993). There was an immediate inhibitory effect of pediocin and nisin against *Le. mesenteroides* that had been isolated from spoiled beef. Na lactate was the most effective and Microgard was the least effective in controlling the bacterial population during 8 weeks of storage at 3 °C. The effect of four different types of LAB inoculated onto sterile slices of normal pH, lean beef was determined during storage under vacuum at 2 °C for up to 10 weeks and with subsequent aerobic storage at 7 °C for up to 10 days (Leisner et al., 1995). The LAB strains used were *C. piscicola* LV17 and UAL26, *Le. gelidum* UAL187–22 and *Lb. sake* Lb706, all of which produce well characterized bacteriocins, except for the chromosomally-mediated bacteriocin of *C. piscicola* UAL26 that has not been characterized. None of the test strains caused spoilage of the meat during the 10-week storage period under vacuum. Under aerobic conditions at 7 °C, all four strains grew well on the beef samples and *C. piscicola* LV17 and UAL26 and *Lb. sake* Lb706 caused off-odours and discoloration. Deterioration in meat quality during aerobic storage was faster with increased time of storage under vacuum. However, *Le. gelidum* showed good potential to extend the storage life of beef. This was confirmed in a subsequent study in which chill stored, vacuum packaged beef was inoculated with sulfide-producing *Lb. sake* strain 1218 that develops a distinct sulphurous odor in the meat package within three weeks of storage at 2 °C (Leisner et al., 1996). Co-inoculation of the meat with the wild type, bacteriocinogenic strain of *Le. gelidum* UAL187 delayed the spoilage by *Lb. sake* 1218 for up to 8 weeks of storage.

Commercially prepared, vacuum packaged chubs (5 kg packs) of coarse ground beef was inoculated with Bac^+ *Le. gelidum* UAL187 or Bac^+UAL187–13 at 10^5 cfu $g^{-1}$and stored at 2 °C for up to 20 days (Worobo et al., submitted for publication). Both of the inoculated samples had a better storage life based on odour and colour retention in aerobic packages compared with the uninoculated control samples. It is generally considered that refrigerated storage (below 10 °C) is necessary for biopreservation of meats with LAB. A study of surface inoculation of LAB as a means of decontamination of meat under semitropical conditions (25 °C) was reported to give encouraging results (Guerrero et al., 1995). LAB cultures for inoculation of the meat surfaces were obtained from Mexican maize-based beverages. Addition of 10% sucrose to the immersion solution and overwrapping of the meat in plastic film resulted in predominance of *Lb. bulgaricus* and *P. pentosaceus* over the pseudomonad-type bacteria; in fact, the best control of the pseudomonads was achieved when a commercial fermented sausage starter containing of *Lb. plantarum* and *Micrococcus kristinae-varians* was used as the inoculum.

Reports on the use of nisin in cured and fermented meats are equivocal (Vandenbergh, 1993). Although Rayman et al., (Rayman et al., 1981) showed that addition of nisin to pork slurries adjusted to pH 5.5 allowed the level of nitrite necessary for inhibition of outgrowth of *Clostridium botulinum* spores to be reduced to 40 to 60 ppm, it is generally considered that the pH of meat is not suited to nisin solubility or stability. Pediocin PA– 1/AcH is more amenable to use as a preservative for meats (Nielsen et al., 1990; Motlagh et al., 1992; Degnan et al., 1992; Berry et al., 1991; Berry et al., 1990 and Foegeding et al., 1992). *P. acidilactici* is not a psychrotroph, so it is not suited to *in situ* production of pediocin in chill stored meats, but it could be used for 'fail-safe' preservation with temperature abuse. This leaves potential for other bacteriocinogenic LAB that are psychrotrophs to be used for biopreservation of meat, such as sakacin A produced by *Lb. sake* Lb706 (Schillinger & Lücke, 1990 and Schillinger et al., 1991), carnobacteriocins produced by *C. piscicola* LV17 (Worobo et al., 1994 and Quadri et al., 1994), or leucocin A produced by *Le. gelidum* UAL187 (Hastings et al., 1991). These bacteriocins do not have antibacterial spectra equivalent to nisin or pediocin PA–1/AcH, but they are active against *Listeria* spp.

Ls. monocytogenes is a common contaminant of raw and processed meats (Johnson et al., 1990). Several studies have focused on the biopreservation of ready-to-eat meat products with LAB. Using wiener sausage exudates inoculated with strains of *Ls. monocytogenes* it was shown that *P. acidilactici* H producing pediocin PA–1/AcH or the bacteriocin decreased the growth of added *Listeria* strains under refrigerated and abusive temperature conditions (Yousef et al., 1991). Further study in packages of wiener sausages that were surface inoculated with *Ls. monocytogenes* and bacteriocin-producing and isogenic bacteriocin-negative strains of *P. acidilactici* and stored at 4 and 25 °C (Luchansky et al., 1992) revealed that the *Pediococcus* strain did not grow or produce pediocin at 4 °C but that at 25 °C the numbers of *Listeria* were inhibited by the pediocin-producing strain, indicating a protective effect under conditions of temperature abuse. Similar results were obtained with turkey summer sausage fermented with *P. acidilactici* prior to cooking. Residual pediocin was detected during 60 days of storage at 4 °C. Luchansky et al. (Luchansky et al., 1992) demonstrated that the three pediocin-producing strains used in their study are closely related, in fact they are identical. Addition of a bacteriocin-producing or a bacteriocin-negative derivative of *P. acidilactici* JD1–23 (also a producer of pediocin PA–1/AcH) and a five-strain mixture of *Ls. monocytogenes* to cooked, cured sausages during packaging demonstrated an effect of the bacteriocin-producing strain against *Listeria* (Berry et al., 1991). Inhibition of the *Listeria* on the sausages stored at 4 °C required a large inoculum (10^7 cfu g^{-1}) of pediococci. At lower inocula (10^4 cfu g^{-1}) growth of *Listeria* was delayed and an inhibitory effect of the bacteriocin-producing strain was observed.

The presence of *Ls. monocytogenes* in ready-to-eat, meat products has aroused concern for these organisms in dry fermented sausages. Addition of a bacteriocin-producing strain of *P. acidilactici* (strain JD1–23) to semi-dry sausage contaminated with *Ls. monocytogenes* prior to heating to 64.4 °C resulted in 2 log reduction in the *Listeria* but after heating and storage for two weeks, 10% of the sausages were positive for *Listeria* (Berry et al., 1990). Bacteriocin-producing and nonbacteriocinogenic strains of *P. acidilactici* (strain PAC1.0) added to dry fermented sausage inoculated with *Ls. monocytogenes* showed that pediocin production in the sausage provided a safeguard against *Listeria* growth in the absence of sufficient development of acidity (Foegeding et al., 1992). The addition of sakacin K-producing *Lb. sake* to dry fermented sausages was shown to inhibit *Ls. monocytogenes* in a model system and *Ls. innocua* in fermented dry sausage (Hugas et al., 1995). A parallel sensory study of fermented sausage produced with the bacteriocinogenic strain of *Lb. sake* or a nonbacteriocinogenic strain of *Lb. curvatus* revealed that the sausages produced with *Lb. curvatus* had a more acid taste than those produced with *Lb. sake*, otherwise the inoculated sausages were the same, but differed favourably compared with uninoculated sausages. The authors recommend the use of the bacteriocinogenic strain of *Lb. sake* for use as a bioprotective starter in fermented meat products. Subsequently, determination of the composition of sakacin K has shown that it is identical to curvacin A produced by *Lb. curvatus* and sakacin A produced by *Lb. sake* (Aymerich, Nes and Vogel, unpublished data). The addition of bacteriocinogenic and nonbacteriocinogenic strains *Lb. plantarum* MCS to salami prevented growth but not the survival of *Listeria* (Campanini et al., 1993). Reduced numbers but survival of *Listeria* in meats inoculated with bacteriocinogenic LAB is characteristic of most studies.

A bacteriocin-producing strain of *Lb. bavaricus* was added to minimally heat-treated, vacuum packaged beef cubes, with or without added sterile gravy (Winkowski et al., 1993). Depending on the inoculum of *Lb. bavaricus*, added *Listeria* were inhibited or killed. The inhibitory effect of *Lb. bavaricus* was greater in product stored at 4 °C than at 10 °C. Addition of 0.5% glucose to the gravy enhanced the inhibitory effect. The Wisconsin process (addition of carbohydrate and *P. acidilactici* starter culture) has been adapted for use in low-acid refrigerated foods such as chicken salad at pH 5.1 (Hutton et al., 1991). In challenge studies with *Clostridium botulinum* spores during storage at abusive temperatures, the presence of a commercial strain of *P. acidilactici* inhibited the formation of toxin.

State of the art/science for biopreservation of meats

In 1994, Hansen (Hansen, 1994) wrote that 'nisin and other lantibiotics are still in the infancy of their development.' If that can be said for the lantibiotic bacteriocins, then it applies even more strongly for the nonlantibiotic bacteriocins. However, the lantibiotics such as nisin and subtilin with their extensive posttranslational modification require a far more complex system for production than many of the nonlantibiotics that have been characterized to date. It will be interesting to learn the extent to which the structural molecule of nisin can be modified and if improved activity can be achieved without compromising the immunity of the producer organism. For the best opportunity, it would be desirable if 'the leader region of the precursor peptide is the sole arbiter of processing, and that any peptide or protein that is fused to the leader will be conducted through the lantibiotic secretion-modification pathway to undergo dehydration and thioether bridge formation as dictated solely by the locations of serines, threonines and cysteines within the structural sequence.' (Kuipers et al., 1993). When the structural gene for nisin was fused behind the leader peptide of subtilin, failure to produce mature nisin was taken as evidence that the posttranslational events are not independent of the leader region (Kuipers et al., 1993); whereas the accurate processing of subtilin containing mutations at two sites can be encouragingly interpreted (Chakicherla & Hansen, 1995).

Complex genetic systems also exist for the nonlantibiotics including, sakacin A and carnobacteriocins A, BM1 and B2, in which production of the bacteriocin(s) is autoregulated and (or) under the control of an inducer (Axelsson & Holck, 1995 and Saucier et al., 1995). The only apparent posttranslational modifications of these bacteriocins are the cleavage of the leader peptide and in some cases the formation of a disulphide bridge(s). For production of carnobacteriocin BM1 there is a trans-acting reaction between the genetic machinery for carnobacteriocin B2 located on the plasmid and genes for carnobacteriocin BM1 located on the chromosome (Quadri et al., submitted). The genetic control of the class II bacteriocins, pediocin PA–1/AcH (Marugg et al., 1992), leucocin A (Van Belkum & Stiles, 1995) and mesentericin Y105 (Fremaux et al., 1995), is much simpler, requiring approximately 4.5 kb of DNA comprising four or five genes. For leucocin A and mesentericin Y105 the genes include the structural, immunity, ABC transporter and 'accessory' proteins; the fifth gene encoding a small peptide has no identified function. In 1995 there were separate reports of two novel bacteriocins, acidocin B (Leer et al., 1995) and divergicin A (Worobo et al., 1995), that are synthesized as precursors with N-terminal extensions and processing sites characteristic of signal peptides. Acidocin B is produced by a strain of *Lactobacillus acidophilus* and it is active against a relatively broad spectrum of spoilage and potentially pathogenic bacteria, including *Brochothrix thermosphacta*, *Clostridium sporogenes* and *Listeria monocytogenes*. It appears that acidocin B is secreted by the general *sec*-pathway of the cell and that the system may be useful for development of food grade vectors (Van der Vossen et al., 1994). Divergicin A is a narrow spectrum bacteriocin produced by a 3.4-kb plasmid from a strain of *C. divergens*. It was demonstrated that it accesses the general *sec*-pathway of the cell for secretion of the bacteriocin (Worobo et al., 1995). In contrast to the apparent interdependence of the leader sequence and structural peptide for processing of the lantibiotics, the structural portion of divergicin (Worobo et al., 1995) has been fused behind the leader sequences of leucocin A, lactococcin A and colicin V and production of these bacteriocins in homologous and heterologous hosts has been achieved (Van Belkum & Stiles, submitted). This suggests that the leader peptides of lantibiotic class I and nonlantibiotic class II bacteriocins have different functions in bacteriocin secretion. The structural portion of the bacteriocin gene for carnobacteriocin B2 was fused behind the signal sequence of divergicin A and secretion and proper processing of the chimeric peptide was achieved in the absence of the dedicated secretion genes (McCormick et al., submitted). Carnobacteriocin B2 was produced by the wild type strain of *C. divergens* LV13 and in *C. piscicola* LV17C, the nonbacteriocinogenic plasmidless variant of the original carnobacteriocin B2 producer strain. Both of the host strains are sensitive to carnobacteriocin B2 and they both acquired immunity when they were transformed with this construct. The small amount of genetic material required for independent bacteriocin expression has implications for the development of a food-grade multiple bacteriocin expression vector for use in LAB. The structural and immunity genes for leucocin A (Hastings et al., 1991) have also been fused behind the divergicin signal peptide and the active peptide has been produced with only 0.65 kb of genetic material (McCormick & Stiles, unpublished data). This has implications for the development of a food-grade multiple bacteriocin expression vector for use in LAB.

A limiting factor of the class II bacteriocins for use in the preservation of chill stored meats is their relatively narrow antibacterial spectrum compared with that of nisin. Cloning and production of pediocin PA–1/AcH by an appropriate host could provide a good spectrum of antibacterial activity; or bacteriocin gene cassettes for production of multiple bacteriocins by a suitable producer organism could be used as an alternate strategy. By targeting specific spoilage or pathogenic bacteria associated with meats it is possible to obtain bacteriocins suitable for use in these cassettes. *Brochothrix* isolates, including *Brochothrix campestris* ATCC 43754, were screened for bacteriocin production in an effort to obtain bacteriocins active against *Brochothrix thermosphacta,* an important spoilage organism of meats (Gardner, 1981). Production of brochocin-C by *B. campestris* ATCC 43754 was first described by Siragusa and Nettles Cutter (Siragusa & Nettles Cutter, 1993). Brochocin-C was reported to be active against isolates from vacuum packaged beef and pork, including 34 strains of *B. thermosphacta* and 10 strains of *Ls. monocytogenes.* Inhibitory activity was also reported against strains of *Carnobacterium, Enterococcus, Kurthia, Lactobacillus,* and *Pediococcus* but not against Gram-negative species (Siragusa & Nettles Cutter, 1993). In our laboratory we determined that this is a novel, class II bacteriocin with antibotulinal activity (Poon et al., submitted for publication). In fact, brochocin-C has an antibacterial spectrum that appears to be equivalent to that of nisin A. It is stable from pH 2 to 9 when heated at 100 °C for 15 min. If this broad spectrum bacteriocin can be incorporated into gene cassettes for use in suitable LAB, it may also be possible to extend this strategy to target Gram- negative bacteria, for example by incorporating genes for bacteriocins from appropriate Gram-negative bacteria into the gene cassettes.

Our laboratory is positioned to produce bacteriocin gene cassettes. The plasmid pCD3.4 from *C. divergens* LV13 could serve as a food grade vector plasmid. It has many of the characteristics required for a vector and it is a native plasmid that has not been modified with DNA from other sources. Initial interest is in the production of cassettes with leucocin A, brochocin-C and carnobacteriocin B2 to give a broad spectrum of antibacterial activity against Gram-positive bacteria in raw and cured meats, with the express interest in reducing the possibility of *Clostridium botulinum, Ls. monocytogenes* and *Enterococcus* species in cured meats, while extending the storage life of vacuum packaged raw meats. We propose to use *Le. gelidum* UAL187–13 as the carrier for the gene cassette in raw meats and we have characterized a heat tolerant strain of *C. divergens* able to survive heat treatment at 72 °C, that will be studied as the host strain for the gene cassette in cooked meats (McCormick & Stiles, unpublished data).

Developments/needs for application

There is currently a large amount of research on 'natural antimicrobials' for food applications (Goud, 1996), of which bacteriocins comprise one group of compounds that are being studied. Bacteriocins of LAB and other food grade bacteria have the advantage that the organisms generally have GRAS (Generally Recognized As Safe) status with regulatory agencies. Although the purified bacteriocins, except for nisin, have not been licensed for addition to foods, it is clear that bacteriocin residues are currently present in the food supply. Two commercial compounds that have been licensed for addition to foods, Microgard and Alta 2341, are ferments of food grade bacteria that impart antibacterial properties to the foods. It is commonly stated that, except for nisin, applied studies on bacteriocins are lacking. This is understandable because no other bacteriocin has been licensed for addition to foods. Convincing evidence of inhibition of pathogens and spoilage bacteria is required to stimulate commercial interest in bacteriocins as agents for biopreservation. Unfortunately, except for a few bacteriocins, they have a narrow antibacterial spectrum and they are not active against Gram-negative bacteria. Use of nisin with a chelating agent expands the antibacterial spectrum of nisin to include Gram- negative bacteria (U.S. patent 4,980,163) and studies by Stevens et al., (Stevens et al., 1991 and Stevens et al., 1992) demonstrated a marked reduction of enteric bacteria, including *Salmonella* spp. (3 to 7 log cycle reduction), after one hour exposure to 50 μg of nisin and 20 mM EDTA.

The issue of resistance of bacteria to bacteriocins has not been adequately addressed. Studies by Ming & Daeschel (Ming & Daeschel, 1993 and Ming & Daeschel, 1995) confirmed that *Ls. monocytogenes* Scott A develops spontaneous nisin resistance at frequencies of $10^{-6}$to 10^{-8} when they are exposed to nisin at 2 to 8 times the minimum inhibitory concentration. The resistance is a passive, unlike that reported for Bacillus cereus, but the frequency of resistance to nisin (and other bacteriocins) is sufficient to challenge the efficacy of bacteriocins for preservation of foods.

It may not be appropriate to consider bacteriocins as inhibitors by themselves. It would be appropriate to use them in conjunction with other inhibitory factors, such as Leistner's 'hurdle' technology for preservation of foods (Leistner, 1992) in which the combined effects of water activity (a_w), pH, temperature and preservatives are studied in conjunction with added bacteriocinogenic LAB or their bacteriocins. This represents a fertile area for applied food research.

The regulatory aspects of the use of bacteriocins or bacteriocinogenic LAB as biopreservatives was addressed at a recent symposium sponsored by the International Life Sciences Institute (Fields, 1996). The basis for regulation of bacteriocins will depend on the food use and in the United States the use of purified bacteriocins, bacteria producing bacteriocins and genetic expression of bacteriocins in food to produce a preservative effect will fall under the jurisdiction of the Food and Drug Administration for consideration as food ingredients. If the substances are considered GRAS by qualified experts they could be exempt from premarket approval. If the substances are not granted GRAS status, they will require premarket approval as a food additive. In most jurisdictions it is likely that the addition of bacteriocins to meats will also require approval of the Agricultural authorities, in the United States, the Food Safety and Inspection Service of the Department of Agriculture.

The range of minimally processed foods available in the marketplace continues to increase. Furthermore, there is an increased number of foods that are packaged under anaerobic conditions that rely on refrigeration as the sole line of defense against pathogen growth. The studies included in this review were primarily those in which researchers had studied the effect of LAB and (or) their bacteriocins in food matrices. It is expected that various factors in foods will influence the efficacy of bacteriocins and bacteriocinogenic LAB (Daeschel, 1993), such as: adequacy of the food environment for bacteriocin production; loss of bacteriocin-producing capacity; antagonism with other bacteria; inhibition by bacteriophage; development of a bacteriocin-resistant microflora; and inactivation of the bacteriocin by enzymes or binding to fats or proteins. The amount of information on bacteriocins at the basic level and at the applied level for biopreservation of foods is rapidly accumulating. Although there seem to be many constraints to biopreservation of foods with bacteriocinogenic lactic acid bacteria, especially if genetically modified lactic acid bacteria are used [Bucken], the time is approaching when the food industry and government can evaluate the value and the acceptability of these techniques.

References

Abdalla OM, Davidson PM & Christen GL (1993) Survival of selected pathogenic bacteria in white pickled cheese made with lactic acid bacteria or antimicrobials. J. Food Prot. 56: 972–976

Adams MR & Marteau P (1995) On the safety of lactic acid bacteria from food. Int. J. Food Microbiol. 27: 263–264

Aguirre M & Collins MD 1993. Lactic acid bacteria and human clinical infection. J. Appl. Bacteriol. 75: 95–107

Axelsson L & Holck A (1995) A gene involved in production of and immunity to sakacin A, a bacteriocin from *Lactobacillus sake* Lb706. J. Bacteriol. 177: 2125–2137

Aymerich, Nes & Vogel, unpublished data

Berry ED, Hutkins RW & Mandigo RW (1991) The use of bacteriocin-producing *Pediococcus acidilactici* to control post-processing *Listeria monocytogenes* contamination of Frankfurters. J. Food Prot. 540: 681–686

Berry ED, Liewen MB, Mandigo RW & Hutkins RW (1990) Inhibition of *Listeria monocytogenes* by bacteriocin- producing *Pediococcus* during manufacture of fermented semidry sausage. J. Food Prot. 53: 194–197

Breidt F, Crowley KA & Fleming HP (1995) Controlling cabbage fermentations with nisin and nisin-resistant *Leuconostoc mesenteroides*. Food Microbiol. 12: 109–116

Brocklehurst TF, Zaman-Wong CM & Lund BM (1987) A note on the microbiology of retail packs of prepared salad vegetables. J. Appl. Bacteriol. 63: 409–415

Buckenhüskes HJ (1993) Selection criteria for lactic acid bacteria to be used as starter cultures for various food commodities. FEMS Microbiol. Rev. 12: 253–272

Campanini M, Pedrazzoni I, Barbuti S & Baldini P (1993) Behaviour of *Listeria monocytogenes* during the maturation of naturally and artificially contaminated salami: effect of lactic-acid bacteria starter cultures. Int. J. Food Microbiol. 20: 169–175

Chakicherla A & Hansen JN (1995) Role of the leader and structural regions of prelantibiotic peptides as assessed by expressing nisin-subtilin chimeras in *Bacillus subtilis* 168, and characterization of the physical, chemical, and antimicrobial properties. J. Biol. Chem. 270: 23533–23539

Choi SY & Beuchat LR (1994) Growth inhibition of *Listeria monocytogenes* by a bacteriocin of *Pediococcus acidilactici* M during fermentation of kimchi. Food Microbiol. 11: 301–307

Christensen DP & Hutkins RW (1994) Glucose uptake by *Listeria monocytogenes* Scott A and inhibition by pediocin JD. Appl. Environ. Microbiol. 60: 3870–3873

Cutter CN & Siragusa GR (1994) Decontamination of beef carcass tissue with nisin using a pilot scale model carcass washer. Food Microbiol. 11: 481–489

Daba H, Pandian S, Gosselin JF, Simard RE, Huang J & Lacroix C 1991. Detection and activity of a bacteriocin produced by *Leuconostoc mesenteroides*. Appl. Environ. Microbiol. 57: 3450–3455

Daeschel MA (1993) Applications and interactions of bacteriocins from lactic acid bacteria in foods and beverages In: Bacteriocins of Lactic Acid Bacteria (Hoover DG & Steenson LR Eds.), pp. 63–91. Academic Press, New York

Daeschel MA, Jung DS & Watson BT (1991) Controlling wine malolactic fermentation with nisin and nisin-resistant strains of *Leuconostoc oenos*. Appl. Environ. Microbiol. 57: 601–603

Dainty RH & Mackey BM (1992) The relationship between the phenotypic properties of bacteria from chill-stored meat and spoilage processes. J. Appl. Bacteriol. 73 (Supplement): 103S–114S

Degnan AJ, Yousef AE & Luchansky JB (1992) Use of *Pediococcus acidilactici* to control *Listeria monocytogenes* in temperature-abused vacuum-packaged wieners. J. Food Prot. 55: 98–103

Dellaglio F, Dicks LMT & Torriani S (1995) The genus *Leuconostoc*. In: The Lactic Acid Bacteria, Vol. 2. The Genera of Lactic Acid Bacteria (Wood, B.J.B. & Holzapfel, W.H., Eds.), pp. 235–278. Blackie Academic and Professional, Glasgow

Delves-Broughton J (1990) Nisin and its uses as a food preservative. Food Technol. 44(11): 100, 102, 104, 106, 108, 111, 112, 117

Devriese LA, Collins MD & Wirth R (1991) The genus *Enterococcus*. In: The Prokaryotes, Vol. II, 2nd Edn. (Balows A, Trüper HP, Dworkin M, Harder W & Schleifer KH, Eds.), pp. 1465–1481. Springer-Verlag, New York

Dodd HM, Horn N, Giffard CJ & Gasson MJ (1996) A gene replacement strategy for engineering nisin. Microbiology 142: 47–55

Durán MC, Garcia P, Brenes M & Garrido A (1994) Induced lactic acid fermentation during the preservation stage of ripe olives from Hojiblanca cultivar. J. Appl. Bacteriol. 76: 377–382

Egan AF, Shay BJ & Rogers PJ (1989) Factors affecting the production of hydrogen sulphide by *Lactobacillus sake* L13 growing on vacuum packaged beef. J. Appl. Bacteriol. 67: 255–262

Einarsson H & Lauzon HL (1995) Biopreservation of brined shrimp (*Pandalus borealis*) by bacteriocins from lactic acid bacteria. Appl. Environ. Microbiol. 61: 669–676

Felix JV, Papathanasopoulos MA, Smith AA, von Holy A & Hastings JW (1994) Characterization of leucocin B-Ta11a: a bacteriocin from *Leuconostoc carnosum* Ta11a isolated from meat. Curr. Microbiol. 29: 207–212

Fields FO (1996) Use of bacteriocins in food: regulatory considerations. J. Food Prot. 1996 (Supplement): 82–86

Fleming HP, Etchells JL & Costilow RN (1975) Microbial inhibition by an isolate of *Pediococcus* from cucumber brines. Appl. Microbiol. 30: 1040–1042

Fleming HP, McFeeters RF & Humphries EG (1988) A fermentor for study of sauerkraut fermentation. Biotechnol. Bioeng. 31: 189–197

Foegeding PM, Thomas AB, Pilkington DH & Klaenhammer TR (1992) Enhanced control of *Listeria monocytogenes* by in situ-produced pediocin during dry fermented sausage production. Appl. Environ. Microbiol. 58: 884–890

Fremaux C, Héchard Y & Cenatiempo Y (1995) Mesentericin Y105 gene clusters in *Leuconostoc mesenteroides* Y105. Microbiology (Reading) 141: 1637–1645

Fujisawa T, Benno Y, Yaeshima T & Mitsuoka T (1992) Taxonomic study of the *Lactobacillus acidophilus* group, with recognition of *Lactobacillus gallinarum* sp. nov. and *Lactobacillus johnsonii* sp. nov. and synonymy of *Lactobacillus acidophilus* group A3 (Johnson et al., 1980) with the type strain of *Lactobacillus amylovorus* (Nakamura 1981). Int. J. Syst. Bacteriol. 42: 487–491

Gardner GA (1981) *Brochothrix thermosphacta* (*Microbacterium thermosphactum*) in the spoilage of meats: a review. In: Psychrotrophic microorganisms in spoilage and pathogenicity (Roberts TA, Hobbs G, Christian JHB & Skovgaard N, Eds.), p. 139–173. Academic Press, New York

Gasser F 1994. Safety of lactic acid bacteria and their occurrence in human clinical infections. Bull. Inst. Pasteur 92: 45–67

Giraffa G, Picchioni N, Neviani E & Carminati D (1995) Production and stability of an *Enterococcus faecium* bacteriocin during Taleggio cheesemaking and ripening. Food Microbiol. 12: 301–307

Glass KA, Prasad BB, Schlyter JH, Uljas HE, Farkye NY & Luchansky JB (1995) Effects of acid type and Alta™2341 on *Listeria monocytogenes* in a Queso Blanco type cheese. J. Food Prot. 58: 737–741

Gould GW (1992) Ecosystem approach to food preservation. J. Appl. Bacteriol. 73 (Supplement): 58S–68S

— (1996) Industry perspectives on the use of natural antimicrobials and inhibitors for food applications. J. Food Prot. 1996 (Supplement): 82–86

Greer GG & Jones SDM (1991) Effects of lactic acid & vacuum packaging on beef processed in a research abattoir. Can. Inst. Food Sci. Technol. J. 24: 161–168

Greer GG & Dilts BD 1995. Lactic acid inhibition of the growth of spoilage bacteria and cold tolerant pathogens on pork. Int. J. Food Microbiol. 25: 141–151

Guerrero I, Mendiolea R, Ponce E & Prado A (1995) Inoculation of lactic acid bacteria on meat surfaces as a means of decontamination in semitropical conditions. Meat Sci. 40: 397–411

Hammes WP (1990) Bacterial starter cultures in food production. Food Biotechnol. 4: 383–397

Hammes WP & Tichaczek PS (1994) The potential of lactic acid bacteria for the production of safe and wholesome food. Z. Lebensm. Unters. Forsch. 198: 193–201

Hansen JN (1994) Nisin as a model food preservative. Crit. Rev. Food Sci. Nutr. 34: 69–93

Harris LJ, Fleming HP & Klaenhammer TR (1992) Characterization of two nisin-producing *Lactococcus lactis* subsp. *lactis* strains isolated from a commercial sauerkraut fermentation. Appl. Environ. Microbiol. 58: 1477–1483

— (1992) Novel paired starter culture system for sauerkraut, consisting of a nisin- resistant *Leuconostoc mesenteroides* strain and a nisin-producing *Lactococcus lactis* strain. Appl. Environ. Microbiol. 58: 1484–1489

Hastings JW, Sailer M, Johnson K, Roy KL, Vederas JC & Stiles ME (1991) Characterization of leucocin A-UAL 187 and cloning of the bacteriocin gene from *Leuconostoc gelidum*. J. Bacteriol. 173: 7491–7500

Héchard Y, Dérijard B, Letellier F & Y (1992) Characterization and purification of mesentericin Y105, an anti- *Listeria* bacteriocin from *Leuconostoc mesenteroides*. J. Gen. Microbiol. 138: 2725–2731

Henick-Kling T (1995) Control of malo-lactic fermentation in wine: energetics, flavour modification and methods of starter culture preparation. J. Appl. Bacteriol. 79 (Supplement): 29S–37S

Hirsch A, Grinsted E, Chapman HR & Mattick ATR (1951) A note on the inhibition of an anaerobic sporeformer in Swiss- type cheese by a nisin-producing *Streptococcus*. J. Dairy Res. 18: 205–206

Holzapfel WH, Geisen R & Schillinger U (1995) Biological preservation of foods with reference to protective cultures, bacteriocins and food-grade enzymes. Int. J. Food Microbiol. 24: 343–362

Hoover DG, Walsh PM, Kolaetis KM & Daly MM (1988) A bacteriocin produced by *Pediococcus* species associated with a 5.5 MDal plasmid. J. Food Prot. 51: 29–31

Hugas M, Garriga M, Aymerich MT & Monfort JM 1995. Inhibition of *Listeria* in dry fermented sausages by the bacteriocinogenic *Lactobacillus sake* CTC494. J. Appl. Bacteriol. 79: 322–330

Hugenholtz J & de Veer GJCM (1991) Application of nisin A and nisin Z in dairy technology. In: Nisin and Novel Lantibiotics (Jung G & Sahl HG, Eds.), pp. 440–447. ESCOM, Leiden, The Netherlands

Hurst A (1973) Microbial antagonism in foods. Can. Inst. Food Sci. Technol. J. 6: 80–90

— (1981) Nisin. Adv. Appl. Microbiol. 27: 85–123

Hutton MT, Chehak PA & Hanlin JH (1991) Inhibition of botulinum toxin production by *Pediococcus acidilactici* in temperature abused refrigerated foods. J. Food Safety 11: 255–267

Huxoll CC & Bolin HR (1989) Processing and distribution alternatives for minimally processed fruits and vegetables. Food Technol. 43(2): 124–128

Jack RW, Tagg JR & Ray B 1995. Bacteriocins of Gram-positive bacteria. Microbiol. Rev. 59: 171–200

Jiménez-Diaz R, Rios-Sánchez RM, Desmazeaud M, Ruiz-Barba JL & Piard JC (1993) Plantaricins S and T, two new bacteriocins produced by *Lactobacillus plantarum* LPCO 10 isolated from a green olive fermentation. Appl. Environ. Microbiol. 59: 1416–1424

Johnson JL, Doyle MP & Cassens RG (1990) *Listeria monocytogenes* and other *Listeria* spp. in meat and meat products. J. Food Prot. 53: 81–91

Jung DS, Bodyfelt FW & Daeschel MA (1992) Influence of fat and emulsifiers on the efficacy of nisin in inhibiting *Listeria monocytogenes* in fluid milk. J. Dairy Sci. 75: 387–393

Kuipers OP, Rollema HS, de Vos WM & Siezen RJ (1993) Biosynthesis and secretion of a precursor of nisin Z by *Lactococcus lactis*, directed by the leader peptide of the homologous lantibiotic subtilin from *Bacillus subtilis*. FEBS Lett. 330: 23–27

Leer RJ, van der Vossen JMBM, van Giezen M, van Noort JM & Pouwels PH (1995) Genetic analysis of acidocin B, a novel bacteriocin produced by *Lactobacillus acidophilus*. Microbiology (Reading) 141: 1629–1635

Leisner JJ, Greer GG, Dilts BD & Stiles ME (1995) Effect of growth of selected lactic acid bacteria on storage life of beef stored under vacuum and in air. Int. J. Food Microbiol. 26: 231–243

Leisner JJ, Greer GG & Stiles ME (1996) Control of spoilage of beef by a sulfide-producing *Lactobacillus sake* with bacteriocinogenic *Leuconostoc gelidum* UAL187 during anaerobic storage at 2 °C. Appl. Environ. Microbiol. (in press)

Leistner L (1992) Food preservation by combined methods. Food Res. Int. 25: 151–158

Leistner L & Gorris GM (1995) Food preservation by hurdle technology. Trends Food Sci. Technol. 6: 41–46

Lipinska E (1973) Use of nisin-producing lactic streptococci in cheese making. Annu. Bull. Int. Dairy Fed. 73: 1–24

Luchansky JB, Glass KA, Harsono KD, Degnan AJ, Faith NG, Cauvin B, Baccus-Taylor G, Arihara K, Bater B, Maurer AJ & Cassens RG (1992) Genomic analysis of *Pediococcus* starter cultures used to control *Listeria monocytogenes* in turkey summer sausage. Appl. Environ. Microbiol. 58: 3053–3059

Maisnier-Patin S, Deschamps N, Tatini SR & Richard J (1992) Inhibition of *Listeria monocytogenes* in Camembert cheese made with a nisin-producing starter. Lait 72: 249–263

Marugg JD, Gonzalez CF, Kunka BS, Ledeboer AM, Pucci MJ, Toonen MY, Walker SA, Zoetmulder LCM & Vandenbergh PA 1992. Cloning, expression, and nucleotide sequence of genes involved in production of pediocin PA–1, a bacteriocin from *P. acidilactici* PAC1.0. Appl. Environ. Microbiol. 58: 2360–2367

Mattick ATR & Hirsch A (1946) Sour milk and the tubercle bacillus. Lancet (i): 417–418

McClintock M, Serres L, Marzolf JJ, Hirsch A & Mocquot G (1952) Action inhibitrice des streptocoques producteurs de nisine sur le développement des sporulés anaérobies dans le fromage de Gruyère fondu. J. Dairy Res. 19: 187–193

McCormick JK, Worobo RW & Stiles ME Expression of the antimicrobial peptide carnobacteriocin B2 by a signal peptide, *Sec*-dependent pathway. Appl. Environ. Microbiol. submitted for publication

McCormick & Stiles, unpublished data

McMullen LM & Stiles ME (1993) Microbial ecology of fresh pork stored under modified atmosphere at –1, 4.4 and 10 °C. Int. J. Food Microbiol. 18: 1–14

— (1989) Storage life of selected meat sandwiches at 4 °C in modified gas atmospheres. J. Food Prot. 52: 792–798

Ming X & Daeschel MA (1993) Nisin resistance of foodborne bacteria and the specific resistance responses of *Listeria monocytogenes* Scott A. J. Food Prot. 56: 944–948

— (1995) Correlation of cellular phospholipid content with nisin resistance of *Listeria monocytogenes* Scott A. J. Food Prot. 56: 944–948

Motlagh AM, Bhunia AK, Szostek F, Hansen TR, Johnson MC & Ray B (1992) Nucleotide and amino acid sequence of *pap*-gene (pediocin AcH production) in *Pediococcus acidilactici* H. Lett. Appl. Microbiol. 15: 45–48

Motlagh AM, Holla S, Johnson MC, Ray B & Field RA (1992) Inhibition of *Listeria* spp. in sterile food systems by pediocin AcH, a bacteriocin produced by *Pediococcus acidilactici* H. J. Food Prot. 55, 337–343

Mulders JWM, Boerrigter IJ, Rollema HS, Siezen RJ & de Vos WM (1993) Identification and characterization of the lantibiotic nisin Z, a natural nisin variant. Eur. J. Biochem. 201: 581–584

Muriana PM (1996) Bacteriocins for control of Listeria spp. in foods. J. Food Prot. 1996 (Supplement): 54–63

Nielsen JW, Dickson JS & Crouse JD (1990) Use of a bacteriocin produced by *Pediococcus acidilactici* to inhibit *Listeria monocytogenes* associated with fresh meat. Appl. Environ. Microbiol. 56: 2142–2145

Ogden K & Waites MJ (1988) The action of nisin on beer spoilage lactic acid bacteria. J. Inst. Brew. 92: 463–467

Okereke A & Montville TJ (1991) Bacteriocin inhibition of *Clostridium botulinum* spores by lactic acid bacteria. J. Food Prot. 54: 349–356

— (1991) Bacteriocin-mediated inhibition of *Clostridium botulinum* spores by lactic acid bacteria at refrigeration and abuse temperatures. Appl. Environ. Microbiol. 57: 3423–3428

Olsen A, Halm M & Jakobsen M 1995. The antimicrobial activity of lactic acid bacteria from fermented maize (kenkey) and their interactions during fermentation. J. Appl. Bacteriol. 79: 506–512

Pederson MS & Albury MN 1961. The effect of pure culture inoculation on fermentation of cucumbers. Food Technol. 15: 351–354

Pilet MF, Dousset X, Barré R, Novel G, Desmazeaud M & Piard JC (1995) Evidence for two bacteriocins produced by *Carnobacterium piscicola* and *Carnobacterium divergens* isolated from fish and active against *Listeria monocytogenes*. J. Food Prot. 58: 256–262

Poon A, Sailer M, van Belkum MJ, Roy KL, Vederas JC & Stiles ME Biochemical and genetic characterization of brochocin C, an antibotulinal class II bacteriocin produced by *Brochothrix campestris* ATCC 43754. (Submitted for publication)

Pucci MJ, Vedamuthu ER, Kunka BS & Vandenbergh PA (1988) Inhibition of *Listeria monocytogenes* by using bacteriocin PA–1 produced by *Pediococcus acidilactici* PAC1.0. Appl. Environ. Microbiol. 54: 2349–2353

Pucci MJ, Vedamethu ER, Kunka BS & Vandenbergh PA (1988) Inhibition of *Listeria monocytogenes* by using pediocin PA–1 produced by *Pediococcus acidilactici* PAC1.0. Appl. Environ. Microbiol. 54: 2349–2353

Quadri LEN, Sailer M, Roy KL, Vederas JC & Stiles ME (1994) Chemical and genetic characterization of bacteriocins produced by *Carnobacterium piscicola* LV17B. J. Biol. Chem. 269: 12204–12211

Quadri LEN, Roy KL, Vederas JC & Stiles ME Inactivation of four genes from *Carnobacterium piscicola* LV17B affects Production of antimicrobial peptides and immunity. Submitted

Radler F (1990) Possible use of nisin in winemaking. II. Experiments to control lactic acid bacteria in winemaking. Am. J. Enol. Vitic. 41: 7–11

Ray B (1992) Bacteriocins of starter culture bacteria as food biopreservatives: an overview. In: Food Biopreservatives of Microbial Origin (Ray, B. & Daeschel, M., Eds.), pp. 177–205. CRC Press, Boca Raton, Florida

Rayman MK, Aris B & Hurst A (1981) Nisin: a possible alternative or adjunct to nitrite in the preservation of meats. Appl. Environ. Microbiol. 41: 375–380

Roberts RF, Zottola EA & McKay LL (1992) Use of a nisin-producing starter culture suitable for cheddar cheese manufacture. J. Dairy Sci. 75: 2353–2363

Rozbeh M, Kalchayanand N, Field RA, Johnson MC & Ray B (1993) The influence of biopreservatives on the bacterial level of refrigerated vacuum packaged beef. J. Food Safety 13: 99–111

Ruiz-Barba JL, Cathcart DP, Warner PJ & Jiménez-Díaz R (1994) Use of *Lactobacillus plantarum* LPCO10, a bacteriocin producer, as a starter culture in Spanish-style green olive fermentations. Appl. Environ. Microbiol. 60: 2059–2064

Ryan MP, Rea MC, Hill C & Ross P (1996) An application in Cheddar cheese manufacture for a strain of *Lactococcus lactis* producing a novel broad-spectrum bacteriocin, lacticin 3147. Appl. Environ. Microbiol. 62: 612–619

Sandine WE, Radich PC & Elliker PR (1972) Ecology of the lactic streptococci. A review. J. Milk Food Technol. 35: 176–185

Saucier L, Poon A & Stiles ME (1995) Induction of bacteriocin in *Carnobacterium piscicola* LV17. J. Appl. Bacteriol. 78: 684–690

Schillinger U & Lücke FK 1987. Lactic acid bacteria on vacuum-packaged meat and their influence on shelf life. Fleischwirtsch. 67: 1244–1248

— Lactic acid bacteria as protective cultures in meat products. Fleischwirtsch. 70: 1296–1299

Schillinger U, M. Kaya & FK Lücke 1991. Behaviour of *Listeria monocytogenes* in meat and its control by a bacteriocin- producing strain of *Lactobacillus sake*. J. Appl. Bacteriol. 70: 473–478

Shaw BG & Harding CD (1984) A numerical taxonomic study of lactic acid bacteria from vacuum-packed beef, pork, lamb and bacon. J. Appl. Bacteriol. 56: 25–40

— (1989) *Leuconostoc gelidum* sp. nov. and *Leuconostoc carnosum* sp. nov. from chill-stored meats. Int. J. Syst. Bacteriol. 39: 217–223

— (1985) Atypical lactobacilli from vacuum-packaged meats: comparison by DNA hybridization, cell composition and biochemical tests with a description of *Lactobacillus carnis* sp. nov. Syst. Appl. Microbiol. 6: 291–297

Siragusa GR & Nettles Cutter C (1993) Brochocin-C, a new bacteriocin produced by *Brochothrix campestris*. Appl. Environ. Microbiol. 59: 2326–2328

Skyttä E, Hereijgers W & Mattila-Sandholm T (1991) Broad spectrum antibacterial activity of *Pediococcus damnosus* and *Pediococcus pentosaceus* in minced meat. Food Microbiol. 8: 231–237

Somers EB & Taylor SL (1987) Antibotulinal effectiveness of nisin in pasteurized processed cheese spreads. J. Food Prot. 50: 842–848

Stecchini ML, Aquili V & Sarais I (1995) Behavior of *Listeria monocytogenes* in Mozzarella cheese in presence of *Lactococcus lactis*. Int. J. Food Microbiol. 25: 301–310

Stevens KA, Sheldon BW, Klapes NA & Klaenhammer TR (1991) Nisin treatment for inactivation of *Salmonella* species and other Gram-negative bacteria. Appl. Environ. Microbiol. 57: 3613–3615

— (1992) Effect of treatment conditions on nisin inactivation of Gram-negative bacteria. J. Food Prot. 55: 763–766

Sulzer G & Busse M (1991) Growth inhibition of *Listeria* spp. on Camembert cheese by bacteria producing inhibitory substances. Int. J. Food. Microbiol. 14: 287–296

Van Belkum MJ & Stiles ME (1995) Molecular characterization of genes involved in the production of the bacteriocin leucocin A from *Leuconostoc gelidum*. Appl. Environ. Microbiol. 61: 3573–3579

Van Belkum MJ & Stiles ME The double-glycine-type leader peptides direct the secretion of bacteriocins by the ABC transporter proteins. Submitted

Vandenbergh PA (1993) Lactic acid bacteria, their metabolic products and interference with microbial growth. FEMS Microbiol. Rev. 12: 221–238

Van der Vossen JMBM, van Herwijnen MHM, Leer RJ, ten Brink B, Pouwels PH & Huis in 't Veld JHJ (1994) Production of acidocin B, a bacteriocin of *Lactobacillus acidophilus* M46 is a plasmid-encoded trait: Plasmid curing, genetic marking by in vivo plasmid integration, and gene transfer. FEMS Microbiol. Lett. 116: 333–340

Vescovo M, Orsi C, Scolari G & Torriani S (1995) Inhibitory effect of selected lactic acid bacteria on microflora associated with ready-to-use vegetables. Lett. Appl. Microbiol. 21: 121–125

Winkowski K, AD Crandall & TJ Montville 1993. Inhibition of *Listeria monocytogenes* by *Lb. bavaricus* MN in beef systems at refrigeration temperatures. Appl. Environ. Microbiol. 59: 2552–2557

Wood BJB & Holzapfel WH (1995) The Lactic Acid Bacteria, Vol. 2. The Genera of Lactic Acid Bacteria. Blackie Academic and Professional, Glasgow

Worobo RW, Henkel T, Sailer M, Roy KL, Vederas JC & Stiles ME (1994) Characteristics and genetic determinant of a hydrophobic peptide bacteriocin, carnobacteriocin A, produced by *Carnobacterium piscicola* LV17A. Microbiology (Reading) 140: 517–526

Worobo RW, van Belkum MJ, Sailer M, Roy KL, Vederas JC & Stiles ME (1995) A signal peptide secretion- dependent bacteriocin from *Carnobacterium divergens*. J. Bacteriol. 177: 3143–3149

Worobo RJ, Greer GG, Stiles ME & McMullen LM submitted for publication

Yousef AE, Luchansky JB, Degnan AJ & Doyle MP (1991) Behavior of *Listeria monocytogenes* in wiener exudates in the presence of *Pediococcus acidilactici* H or pediocin AcH during storage at 4 or 25 °C. Appl. Environ. Microbiol. 57: 1461–1467

Zottola EA, Yezzi TL, Ajao DB & Roberts RF (1994) Utilization of cheddar cheese containing nisin as an antimicrobial agent in other foods. Int. J. Food Microbiol. 24: 227–238

Antonie van Leeuwenhoek **70:** 347–358, 1996.

Clinical uses of probiotics for stabilizing the gut mucosal barrier: successful strains and future challenges

S. Salminen*[,1], E. Isolauri[2] & E. Salminen[3]
[1]*Department of Biochemistry and Food Chemistry, University of Turku, Finland and Key Centre for Appl.ied and Nutr.itional Toxicology, RMIT, Melbourne, Australia;* [2]*Medical School, University of Tampere, Finland;* [3]*Department of Oncology and Radiotherapy, Turku University Hospital, Turku, Finland. (* author for correspondence)*

Abstract

Probiotic bacteria are used to treat disturbed intestinal microflora and increased gut permeability which are characteristic to many intestinal disorders. Examples include children with acute rotavirus diarrhoea, subjects with food allergy, subjects with colonic disorders and patients undergoing pelvic radiotherapy and sometimes changes associated with colon cancer development. In all such disease states altered intestinal microflora, impaired gut barrier and different types of intestinal inflammation are present. Successful probiotic bacteria are able to survive gastric conditions and colonize the intestine, at least temporarily, by adhering to the intestinal epithelium. Such probiotic microorganisms appear to be promising candidates for the treatment of clinical conditions with abnormal gut microflora and altered gut mucosal barrier functions. They are also promising ingredients to future functional foods and clinical foods for specific disease states provided that basic requirements for strains and clinical studies are carefully followed.

Introduction

Probiotics are viable bacteria that influence the health of the host in a beneficial manner. The bulk of evidence on probiotic cultures and foods is based on anecdotal reports and poorly controlled studies making the work inconclusive and recommendations difficult. However, evidence is currently accumulating from well designed, randomized and placebo- controlled double blinded studies indicating that a few well characterized lactic acid bacteria strains have documented probiotic health promoting effects when defined doses are administered (Lee & Salminen, 1995). Increasing numbers of colonization and dose-response studies defining the required doses have been published (Saxelin et al., 1995; Wolf et al., 1995).

The mechanisms behind the specific benefits include a few mechanisms of action which are discussed in relation to the strengthening of the gut mucosal barrier: gut microflora modification, adherence to intestinal mucosa with capacity to prevent pathogen adherence or pathogen activation, modification of dietary proteins by intestinal microflora, modification of bacterial enzyme activity, and influence on gut mucosal permeability (Table 1).

The intact intestinal epithelium with the normal intestinal microflora represent a barrier to the movement of pathogenic bacteria, antigens and other noxious substances from the gut lumen. In healthy subjects this barrier is stable protecting the host and providing normal intestinal function (Figure 1). When either the normal microflora or the epithelial cells are disturbed by triggers such as dietary antigens, pathogens, chemicals or radiation, defects in the barrier mechanisms become evident (Figure 1). Altered permeability further facilitates the invasion of pathogens, foreign antigens and other harmful substances (Isolauri, 1995). Disturbed intestinal microflora may lead to diarrhoea, mucosal inflammation or activation of harmful drugs and carcinogens in intestinal contents (Salminen et al., 1996). For future research and the development of new food and clinical applications of probiotic bacteria a thorough understanding of the mechanisms of this barrier system is essential.

Table 1. Desirable properties of probiotic bacteria

Probiotic Strain Characteristics	Functional and technological properties
Human origin	Species dependend health effects and maintained viability; applicability to fermented foods
Acid and bile stability	Survival into the intestine, maintaining adhesiveness; maintenance of flavour and aroma profiles during processing and storage
Adherence to human intestinal cells	Immune modulation, competitive exclusion of pathogens: maintenance of mild acidity throughout storage time, good adicity profile
Colonization of the human intestinal tract	Multiplication in the intestinal tract at least temporarily, immune modulation; maintenance of this ability throughout processing and storage
Production of antimicrobial substances	Pathogen inactivation in the intestine, normalization of gut flora; good storage stability and shelf-life in functional food products
Antagonism agains cariogenic and pathogenic bacteria	Prevention of dental decay and pathogen exclusion, prevention of pathogen adhesion, normalization of gut flora, ability to keep the properties after freeze-drying, drying and other processing methods
Safety in food and clinical use	Accurate strain identification (genus, species), documented safety
Clinically validated and documented health effects	Dose-response data for minimum effective dosage in different products

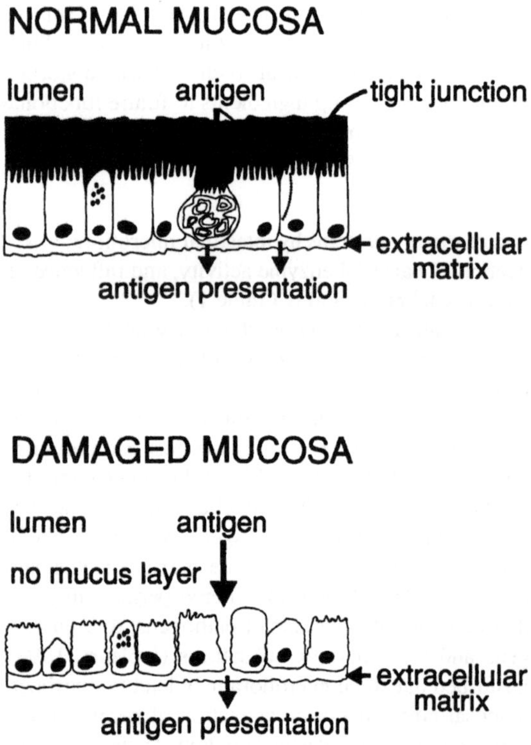

Figure 1. The structure of normal intestinal epithelium and changes caused by outside factors such as diarrhoea, pathogens, radiation and inflammation.

Mucosal and microflora defects and disease

Intestinal barrier

The intestinal mucosa is an important organ of defence providing a barrier against the antigens encountered by the enteric route, and most foreign antigens are excluded by the intestine's mucosal barrier (Sanderson & Walker, 1993). Apart from the barrier function, the intestinal mucosa is efficient in assimilating antigens. For this purpose, there are specialized antigen transport mechanisms in the villous epithelium and particularly in Peyer's patches, essential for evoking specific immune responses (Heyman et al., 1982).

Even in physiological conditions, a quantitatively unimportant but immunologically important fraction of antigens bypasses the defence barrier. They are absorbed across the epithelial layer by transcytosis along two functional pathways (Heyman et al., 1982). The main degradative pathway entails lysosomal processing of the protein to smaller peptide fragments which reduces immunogenicity of the protein and is important in host-defence in diminishing the antigen load. More than 90% of the protein internalized passes in this way. A minor pathway allows the transport of intact proteins which results in antigen- specific immune responses. In health paracellular leakage of macromolecules is not allowed due to intact intercellular tight junctions maintaining the macromolecular barrier. The integrity of the defence barrier is necessary to prevent inappropriate and uncontrolled antigen transport.

Intestinal antigen handling determines subsequent immune responses to the antigen. These include immune exclusion of antigens encountered by the enteric route by interfering with the adherence of antigens, immune elimination of substances that have penetrated the mucosa, and immune regulation of the systemic immune response to antigen-specific systemic hyporesponsiveness (Strobel, 1991). There is evidence that during the absorption process across the intestinal mucosa, antigens are altered into tolerogenic form (Bruce & Ferguson, 1986).

Immature gut defence barrier

The barrier functions are incompletely developed in infancy and early childhood. Intestinal permeability can be transiently increased postnatally, particularly in premature infants (Beach et al., 1982; Axelsson et al., 1989). The binding of antigens to immature gut microvillus membrane is increased compared to the mature mucosa, which has been shown to correlate with the increased uptake of intact macromolecules (Stern et al., 1984). An increased antigen load may evoke aberrant immune responses and lead to sensitization (Holt et al., 1990; Isolauri et al., 1995).

Intestinal inflammation

As a result of local intestinal inflammation, a greater amount of antigens may traverse the mucosal barrier and the routes of transport are altered (Heyman & Desjeaux, 1992). Aberrant antigen transport results in overriding the normal tolerogenic signal into an immunogenic stimulus favouring allergic reactions (Holt, 1994; Fargeas et al., 1995). Foreign antigens such as viruses, bacteria or dietary antigens can induce local inflammation in the intestinal mucosa and, in worst cases, systemic infections and inflammation. Noxious substances in the intestinal contents may also pass the abnormal mucosal barrier. Prior to transport they may be modified by the intestinal microbes and their enzymes (Gorbach, 1990). These functions have in part similar influences on intestinal microflora, intestinal permeability and related factors offering a rational for successful use of lactic acid bacteria for the treatment and prevention of such changes (Figure 2).

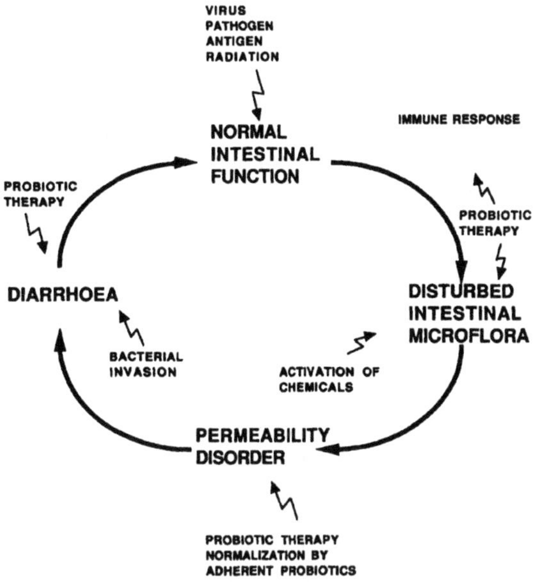

Figure 2. Changes during intestinal disorders and potential targets of treatment and prevention.

Acute gastroenteritis. Rotavirus is the most common cause of acute childhood diarrhoea world-wide (Claeson & Merson, 1990). Rotaviruses invade the highly differentiated absorptive columnar cells of the small intestinal epithelium, where they replicate. Partial disruption of the intestinal mucosa ensues with loss of microvilli and decrease in the villus/crypt ratio. Rotavirus infection has been shown to be associated with increased intestinal permeability (Jalonen et al., 1991). Moreover, the levels of immune complexes containing dietary b-lactoglobulin in sera were significantly higher in patients with rotavirus diarrhoea than in nondiarrhoeal patients. Macromolecular absorption has also been shown to be increased in rotavirus gastoenteritis (Jalonen et al., 1991; Heyman et al., 1987; Uhnoo et al., 1990; Isolauri et al., 1993). The intestinal microflora affects gut permeability, so that in the absence of intestinal microflora, disturbance in intestinal absorption of macromolecules is more severe than in its presence.

Food allergy. Food allergy is defined as an immunologically mediated adverse reaction against dietary antigens. The immaturity of the immune system and the gastrointestinal barrier may explain the peak prevalence of food allergies in infancy (Isolauri & Turjamaa, 1995). In food allergy, intestinal inflammation (Majamaa et al., 1995) and disturbances in intestinal permeability (Jalonen, 1991) and antigen transfer (Heyman et al., 1988) occur when an allergen comes into contact with the intestinal mucosa. During dietary elimination of the antigen, the barrier and transfer functions of the mucosa are normal (Majamaa et al., 1995; Jalonen, 1991; Heyman et al., 1988). It has therefore been concluded that impairment of the intestine's function is secondary to an abnormal intestinal immune response to the offending antigens.

Atopic dermatitis. Atopic dermatitis is a common and complex, chronically relapsing skin disorder of infancy and childhood. Hereditary predisposition is an important denominator of atopic dermatitis, and hypersensitivity reactions contribute the expression of this predisposition (Sampson & MacCaskill, 1985). The relationship between environmental allergens and exacerbation of atopic dermatitis is particularly apparent in infancy so that dietary antigens predominate and allergic reactions to foods are common (Isolauri & Turjanmaa, 1995). In a recent study (Majamaa & Isolauri, 1995), macromolecular absorption across the intestinal mucosa was assessed in children (aged 0.5–8 years) with atopic dermatitis. In these patients, the offending foods were identified and eliminated, and the intestinal mucosa was not challenged in vitro nor in vivo. Significantly increased absorption of protein, in intact and degraded form, was found in the atopic dermatitis patients compared to controls. The result may reflect a primarily altered antigen transfer in atopic dermatitis. Aberrant antigen absorption could partly explain why patients with atopic dermatitis frequently show hightened immune responses to common environmental antigens, including dietary antigens.

Crohn's disease. Crohn's disease is a chronic and idiopathic inflammation of the gastrointestinal tract with characteristic patchy transmural lesions containing granulomas. The outbreak of Crohn's disease is thought to require genetic predisposition, immunologic disturbance and the influence of intraluminal triggering agent(s), for example bacteria or viruses. Crohn's disease is associated with impairment of the barrier function. In a recent in vitro study (Isolauri, 1995), a rise in macromolecular absorption in uninvolved parts of the intestine was detected in patients with clinically moderate or severe Crohn's disease. An interplay between the immune effector cells and the intestinal vascular endothelium has been suggested to result in disrupted vasculature, cell-mediated immunity with lymphokine production and a vigorous IgG response and finally dysfunction of the mucosa (Podolsky, 1991; MacDonald, 1993).

Rheumatoid arthritis. Dysfunction of the gut mucosal barrier as a consequence of abnormal intestinal microflora, intestinal inflammation and/ or secondary to diet or medication may be the key connecting the gut and joint in rheumatoid arthritis (Midtvedt, 1987). Evidence for this suggestion is afforded by observations that patients with rheumatoid arthiritis gain relief of clinical symptoms by fasting or specific dietary regimens. In a recent study (Malin et al., 1996), the enzyme activities in faeces were investigated in patients with juvenile chronic arthritis (JCA) and controls. Specifically the activity of urease but not the activities of b-glucosidase and b-glucuronidase in faeces was increased in JCA patients. Urease catalyzes the hydrolysis of urea to yield ammonia. Urease and ammonia may contribute to tissue damage, and elevated urease activity may indicate disturbance in the population of anaerobic bacteria in JCA.

Pelvic radiotherapy. Radiotherapy has a profound effect on the intestinal microflora and mucosa. Radiation alters the intestinal microflora, vascular permeability of the mucosa and intestinal motility (Friberg, 1980; Silva, 1993). The villi are shortened and flattened and there is decreased mitosis in intestinal crypts with necrosis Heavy infiltration of lamina propria with plasma cells is seen with polymorphonuclear leukocytosis (Friberg, 1980; Silva, 1993). The result is a flat surface covered by thin columnar epithelial cells which may also be lost leading to ulcerated surface and appearance of radiation induced fibroblasts. Widening of the tight junctions between cells is common (Silva, 1993). Within three to ten days the intestinal epithelium may be completely denuded and the villous surface is replaced by a layer of exudate in which masses of bacteria are present. Overgrowth of pathogenic microorganisms has been proposed as a factor enhancing the severity of radiation enteritis (Danielsson et al., 1991). Bacteria can penetrate the damaged villi leading to bacteremia in extreme cases (Friberg, 1980; Silva, 1993). The changes leading to radiation enteropathy in man include both damage to intestinal mucosa, changes in the intestinal microflora and impaired immune response (Figure 2).

Clinically, the primary reactions start during the first and second weeks of treatment giving such symptoms as nausea, vomiting and diarrhoea with peritonitis resulting from the necrotizing effects of radiation. The late secondary reaction, e.g. fibrosis and obstruction of the intestine, may give clinical symptoms years after the treatment. The relationship between early and late reactions is not clear, although some studies have indicated that the severe early reactions precede serious late effects (Silva, 1993; Dahl et al., 1994). *Lactobacillus* supplementation in lethally irradiated mice has been reported to prolong their survival (Dong et al., 1987).

Intestinal inflammation and chemical exposure. Long term inflammatory changes, changes in gut microflora, alterations in drug and carcinogen metabolism, nutrition of intestinal epithelial cells, and genetic factors may alter the ability of intestinal tract to handle drugs and carcinogens. It has been reported in several studies, reviewed by Rogers & Nauss (1985), that nutritional factors such as increasing fat intake, can alter the incidence and number of chemically induced tumours in animals. These changes have been suggested to be related to intestinal microflora, intestinal microbial enzyme activities, intestinal immune response and intestinal mutagen production (Goldin et al., 1996). Subjects with intestinal inflammatory conditions, such as ulcerative colitis and Crohn's disease, may have an increased risk for colon cancer with the risk increasing with the duration and extent of disease (Tytgat et al., 1995). In a study conducted in the United Kingdom, cancer risk was compared in two cohorts of patients with extensive ulcerative colitis and equally extensive colonic Crohn's disease. Both the relative risk and the cumulative incidences of cancer were the same (Gillen et al., 1994).

In the colitic mucosa there may be a spectrum of changes occurring with cycles of inflammation that may contribute to harmful alterations in the mucosa (Tytgat et al., 1995). Inflammatory alterations may be potentiated by the disturbed microflora and its metabolic activities. Intestinal microflora influences the handling of drugs and carcinogens in a healthy subject by assisting the removal of drugs and carcinogens as well as products of enterohepatic circulation (Goldin & Gorbach, 1984; Goldin et al., 1992). These changes may be counteracted by oral administration of suitable lactic acid bacteria either in fermented milks or in other foods.

Probiotic treatment of human intestinal inflammation

Acute gastroenteritis. *Lactobacillus* GG has been proven effective in the treatment of rotavirus diarrhoea. It repeatedly reduces the duration of diarrhoea to about half in children with rotavirus diarrhoea. It has also been proven effective in watery diarrhoea in several studies in Asia (Raza et al., 1995). One such study has been reported for *Lactobacillus casei* Shirota strain and one prevention study for *Bifidobacterium bifidum* (Sugita & Togawa, 1994; Saavedra et al., 1995).

Different lactic acid bacteria were compared for their effects on the immune response to rotavirus in children with acute rotavirus gastroenteritis (Kaila et al., 1992; Majamaa et al., 1995). Serum antibodies to rotavirus, total number of immunoglobulin-secreting cells (ISC) and specific antibody-secreting cells (sASC) to rotavirus were measured at the acute stage and at convalescence. The treatment with *Lactobacillus* GG was associated with an enhancement of IgA sASC to rotavirus and serum IgA antibody level at convalescence. It was therefore suggested that certain strains of lactic acid bacteria, particularly *Lactobacillus* GG, promote systemic and local immune response to rotavirus, which may be of importance for protective immunity against reinfections (Kaila et al., 1992). Next, a study was made to compare the immunological effects of viable and heat inactivated lactic acid bacteria (Majamaa et al., 1995). *Lactobacillus* GG administered as a viable preparation during acute rotavirus gastroenteritis resulted in a significant rotavirus specific IgA response at convalescence. The heat inactivated *Lactobacillus* GG was clinically as efficient, but the IgA response was not detected. This result suggests that viability of the strain is critical in determining the capacity of lactic acid bacteria to induce immune stimulation. Also, in a study with different preparations of lactic acid bacteria in the treatment of rotavirus diarrhoea it was shown that *Lactobacillus* GG was most effective whilst a preparation containing *Streptococcus thermophilus* and *Lactobacillus bulgaricus* or a *Lactobacillus rhamnosus* did not have any effect on the duration of diarrhoea (Majamaa et al., 1995). In a Japanese study, $1{,}5 \times 10^{9-10}$ cfu of Biolactis powder (*Lactobacillus casei* Shirota) was administered to children with rotavirus diarrhoea (n = 17) along with the normal Japanese treatment using lactase and albumin tannate. The controls (n = 15) received lactase and albumin tannate only (Sugita & Togawa 1994). The calculated days until improvement were 3.8 in the *Lactobacillus* group and 5.3 in the control group ($p < 0.05$). While the result is positive, further studies are required to assess the efficacy using standard biomarkers such as the duration of diarrhoea.

Food allergy. The mechanisms of the immune enhancing effect of *Lactobacillus* GG are not entirely understood, these may relate to the antigen transport in the intestinal mucosa. Therefore, the effect of *Lactobacillus* GG on the gut mucosal barrier was investigated in a suckling rat model (Isolauri et al., 1993). Rat pups were divided into three experimental feeding groups to receive a daily gavage of cow milk, or *Lactobacillus* GG with cow milk, while controls were gavaged with water. At 21 days, the absorption of horseradish peroxidase across patch-free jejunal segments and segments containing Peyer's patches was studied in Ussing chambers. Gut. immune response was indirectly monitored by the ELISPOT method of sASC to b- lactoglobulin. Prolonged cow milk challenge increased macromolecular absorption, whereas *Lactobacillus* GG stabilized the mucosal barrier with a concomitant enhancement of antigen-specific immune defense and proportional transport across Peyer's patches. These results indicate that there exists a link between stabilization of non-specific antigen absorption and enhancement of the antigen-specific immune response. They further suggest that the route of antigen absorption is an important determinant of the subsequent immune response to the antigen (Isolauri et al., 1993).

Atopic dermatitis. The capacity of *Lactobacillus* GG to degrade food antigens, and thereby modify their immunoactivity was investigated (Sütas et al., 1996). For this purpose, the immunoactivity of caseins and *Lactobacillus* GG-degraded caseins was assessed in lymphocyte transformation tests in healthy adults. Casein, as1-casein, b-casein suppressed the lymphocyte proliferation capacity, whereas k-casein induced it. Degradation of casein, as1-casein and b-casein by *Lactobacillus* GG enhanced the suppressive effect of these caseins on the lymphocyte proliferation capacity. Degradation of k-casein by *Lactobacillus* GG reversed its inducive effect to profound suppression. These results indicate that *Lactobacillus* GG can degrade food antigens such as bovine casein, and down-regulate T-cell responses. The results were identical in healthy children and in children with atopic dermatitis (Sütas et al., 1996). Moreover, in cultures of atopic children, casein was found to increase the IL–4 production capacity while *Lactobacillus* GG- degraded casein was found to reduce the IL–4 generation. These results suggest that probiotic bacteria may have the capacity to release tolerogens from allergens.

Crohn's disease. The effect of a ten days' oral bacteriotherapy with *Lactobacillus* GG in Crohn's disease was investigated (Malin et al., 1996b). Irrespective of the activity of Crohn's disease, an increase in IgA sASC to dietary b-lactoglobulin and casein was detected. The result indicates that probiotic bacteria may have the potential to increase gut IgA and thereby promote the gut immunological barrier.

Rheumatoid arthritis. In JCA patients, a ten days' oral bacteriotherapy reduced elevated urease activity in faeces suggesting an effect to counteract imbalanced microflora in JCA (Malin et al., 1996a).

Pelvic radiotherapy. In a randomised study, it was reported that patients receiving pelvic radiotherapy and a fermented milk containing viable *Lactobacillus acidophilus* NCFB 1748 had a significant decrease in diarrhoea (Salminen et al., 1988). In a five-year follow up study there was trend to less late serious intestinal complications (Salminen et al., 1995). It has been shown that *Lactobacillus acidophilus* (NCFB 1748) has ability to colonise human colon mucosa *in vitro* eventhough adhesion to Caco–2 cells is relatively small. In a Swedish study (Henriksson et al., 1995) fermented milk intake decreased the severity of late effects caused by pelvic radiotherapy. This indicates an important role for lactic acid bacteria in the intestinal tract following radiotherapy. Also earlier studies have shown that some *Lactobacillus acidophilus* and *Bifidobacterium* preparations may have potential in controlling diarrhoea and intestinal side effects of radiation (Haller & Kräubig, 1960; Mettler et al., 1973). However, the strains utilized are not described in an up-to-date manner and further studies are required in this area.

Intestinal inflammation, chemical exposure and colon cancer related parameters. There is evidence that lactic acid bacteria influence the mutagenicity of intestinal contents and the levels of faecal microbial enzymes, such as b-glucuronidase, b-glucosidase, nitroreductrase and urease. *Lactobacillus acidophilus* (NCFB 1748) has been shown to significantly decrease faecal mutagenicity and urinary mutagenicity in healthy volunteers consuming fried ground beef. The same strain decreased faecal *E. coli* levels in colon

Table 2. Important studies for the safety assessment of probiotic lactic acid bacteria and other bacteria

Type of property studied	Safety factor to be assessed
Intrinsic properties of lactic acid bacteria	Adhesion factors, antibiotic resistance, existence of plasmids and plasmid transfer potential, harmful enzyme profile
Metabolic products	Concentrations, safety and other effects
Toxicity	Acute and subacute effects of ingestion of large amounts of tested bacteria
Mucosal effects	Adhesion, invasion potential, intestinal mucus degradation, infectivity in immunocompromised animals (e.g. following lethal irradiation)
Dose-response effects	Dose-response studies by oral administration in volunteers
Clinical assessment	Potential for side-effects, careful evaluation in healthy volunteers and disease specific studies
Epidemiological studies	Surveillance of large populations following introduction of new strains and products

cancer patients and reduced the faecal b-glucuronidase levels (Lidbeck et al., 1991). Similar results have been reported for *Lactobacillus* GG, *Lactobacillus acidophilus*, *Lactobacillus casei* Shirota strain and other strains (Ling et al., 1994; Goldin & Gorbach, 1984; Goldin et al., 1992; Morotomi, 1996).

It has also been demonstrated with the *Lactobacillus casei* Shirota strain that there is potential in the area of 'dietary prevention' of cancer. Parenteral administration of the strain has had antitumour and immunostimulating activities on experimentally implanted tumours (Morotomi, 1996). Similar effects have been verivied by oral administration. In a recent study *Lactobacillus* GG was shown to cause a dramatic reduction in DMH induced tumour formation in rats. It was reported that *Lactobacillus* GG inhibited the initiation and early phase promotion on the tumorigenesis process (Goldin et al., 1996). However, it was reported that the organisms was not able to prevent the growth of tumours once they had been established (Goldin et al., 1996). The fact that *Lactobacillus* GG is an organisms of human origin capable of surviving in the gastrointestinal tract and that it inhibits tumour initiation or early promotion in the colon provides a basis for further studies of these factors in humans. In a similar manner, *Lactobacillus casei* Shirota strain has been shown to have inhibitory properties on chemically induced tumours in animals (Morotomi, 1996; Kato et al., 1994). In a clinical study, prophylactic effects of oral administration of *Lactobacillus casei* Shirota on the recurrence of superficial bladder cancer have been reported in two Japanese studies (Aso et al., 1993; Aso et al., 1995). Lactic acid bacteria have also been indicated in colonic fermentation producing butyrate or butyric acid in the colon. This alteration in intestinal metabolism may be due to changes in the intestinal microecology following probiotic intake or the direct metabolism of slowly absorbable components by orally administered colonising lactic acid bacteria. Butyrate has been shown in in vitro studies to slow the growth of cultured colon cancer cells. Young (1996) has proposed that butyrate may be a diet-regulated, natural anti-tumour compound at least partly responsible for the antitumour effect of dietary components and also dietary probiotics. When these studies are related to animal data on several *Lactobacillus* strains and colon cancer related parameters (Table 3), it is important to assess the potential of probiotic lactobacilli for cancer chemoprevention.

Important properties for probiotic bacteria

Adhesion and colonization

Adhesion of lactic acid bacteria to intestinal cells is the first step of colonisation or temporary colonisation. Adhesion can be non-specific, based on physicochemical factors, or specific involving adhesin molecules on the surfaces of adherent bacteria and receptor molecules on epithelial cells. Variability of adhesion properties tested using human intestinal cells has become a standard procedure for selecting new probiotic strains. The adhesion properties of a strain are considered often species specific, but often there appears to be adhesion on non-host species. In general, common dairy strains are not among the most adhesive ones, but probiotic strains appear to have strong adhesion. Based on adhesion, some lactic acid bacteria of human origin and some of dairy origin show moderate to good adhesion properties in human cell lines. *Lactobacillus acidophilus LB*, *Lactobacillus* GG (ATCC 53103),

Table 3. Successful probiotic bacteria and their reported effects

Strain	Reported effects in clinical studies	Selected References
Lactobacillus acidophilus LA1	Immune enhancer, adjuvant, adherent to human intestinal cells, balances intestinal microflora	Link-Am.ster et al., 1994; Bernet et al., 1994; Bernet et al., 1993; Schiffrin et al., 1996
Lactobacillus acidophilus NCFB 1748	Lowering faecal enzyme activity, decreased faecal mutagenicity, prevention of radiotherapy related diarrhoea, treatment of constipation	Lidbeck et al., 1991; Salminen et al., 1988; Sarem-Damerdji et al., 1995; Salminen et al., 1995
Lactobacillus GG (ATCC 53013)	Prevention of antibiotic associated diarrhoea, treatment and prevention of rotavirus diarrhoea, treatment of relapsing *Clostridium difficile* diarrhoea, prevention of acute diarrhoea, Crohn's disease, antagonistic against cariogenic bacteria, vaccine adjuvant	Kaila et al., 1992; Siitonen et al., 1991; Isolauri et al., 1992; Salminen et al., 1993; Majamaa et al., 1995; Raza et al., 1995; Kaila et al., 1995; Meurman et al., 1994
Lactobacillus casei Shirota	Prevention of intestinal disturbances, treatment of rotavirus diarrhoea, balancing intestinal bacteria, lowering faecal enzyme activities, positive effects in the treatment of superficial bladder cancer, immune enhancer in early colon cancer, immune enhancement	Asa et al., 1995; Tanaka & Ohwaki, 1994; Salminen et al., 1993; Sugita & Togawa, 1994; Sawamura et al., 1994; Kato et al., 1994; Okawa et al., 1989
Streptococcus thermophilus; Lactobacillus bulgaricus	No effect on rotavirus diarrhoea, no immune enhancing effect during rotavirus diarrhoea, no effect on faecal enzymes	Majamaa et al., 1995; Goldin et al., 1992
Bifidobacterium bifidum	Treatment of rotavirus diarrhoea, balancing intestinal microflora, treatment of viral diarrhoea	Saavedra et al., 1994; Marteau et al., 1990
Lactobacillus gasseri (ADH)	Fecal enzyme reduction, survival in the intestinal tract	Pedrosa et al., 1995
Lactobacillus reuteri	Colonizing the intestinal tract, Mainly animal studies so far, possibly an emerging human probiotic	Casas et al., 1996; Molin et al., 1993

Lactobacillus acidophilus NCFB 1748, *Lactobacillus reuteri, Lactobacillus rhamnosus* (LA750) and *Lactobacillus acidophilus* BG2FO4 have been shown to be adherent in Caco–2 cells or in other systems (Elo et al., 1991; Chauviere et al., 1992). Temporary colonisation has been reported to *Lactobacillus* GG (ATCC 53103), *Lactobacillus reuteri, Lactobacillus gasseri* ADH and *Lactobacillus acidophilus* LA1 (Goldin et al., 1992; Saxelin et al., 1991; Saxelin et al., 1995).

In clinical studies, adherence and colonisation are thought to be important for health effects. It has been shown in a number of studies that adherent lactic acid bacteria are good candidates for the treatment of acute diarrhoea, radiation gastroenteritis and intestinal inflammation (Salminen et al., 1988; Isolauri et al., 1991). Apart from that, adherent lactic acid bacteria are also thought to be responsible for changes in intestinal microflora, intestinal bacterial enzyme activities and stabilising of intestinal permeability. Thus, adherence studies and colonisation studies form a basis for future probiotic research and adherent strains are the key candidates for succesfull probiotic therapy.

Stability. The most important properties for future probiotics include the acid and bile tolerance, adherence to human intestinal mucosa, temporary colonization of the human gastrointestinal tract, production of antimicrobial substances and inhibition of pathogen growth (Lee & Salminen, 1995). It is also important that the strains used are of human origin since many of the properties may be species dependent (Table 1).

Safety. The safety of lactic acid bacteria used in clinical and functional food is of great importance. In general, lactic acid bacteria have a good record of safety, and no major problems have occurred. Cases of infection have been reported with several strains, most commonly with the ones that are naturally most abundant in the human intestinal mucosa (Gasser, 1994; Aguirre & Collins, 1993). Studies on safety have been documented on dairy strains (Saxelin et al., 1996a; Saxelin et al., 1996b) and a review of current knowledge on safety of probiotic bacteria has been reported by Donohue and Salminen (1996). It is most important for future probiotic lactic acid bacteria that their safety has been assured and that they confirm to all regulations. A proposed scheme for safety assessment is presented in Table 2. (Donohue & Salminen, 1996). Most stringent studies have to be completed for genetically modified strains intended to human consumption.

Successful probiotic strains

Probiotic bacteria with these properties and documented clinical effects include *Lactobacillus acidophilus* (NCFB 1478), *Lactobacillus casei* Shirota strain, *Lactobacillus* GG (ATCC 53103) and *Lactobacillus acidophilus* LA1 (Lee & Salminen, 1995). A large number of published studies exists on each preparation documenting their health effects. All of these are currently further tested for different intestinal problems and they offer alternatives for dietary treatment of intestinal disorders (Table 3). In future, it is likely that we shall see more specific clinical targets for probiotic therapy and then the above mentioned strains are likely to play an important role in new products. New strains emerge and are likely to be included in our diet.

Table 4. Requirements for good clinical studies of demonstration of probiotic properties for functional and clinical use

Each strain documented and tested independently, on its own merit
Extrapolation of data from closely related strains not acceptable
Well-defined probiotic strains, well-defined study preparations
Double-blind, placebo controlled human studies
Randomized human studies
Results confirmed by several independent research groups
Publication in peer-reviewed journals

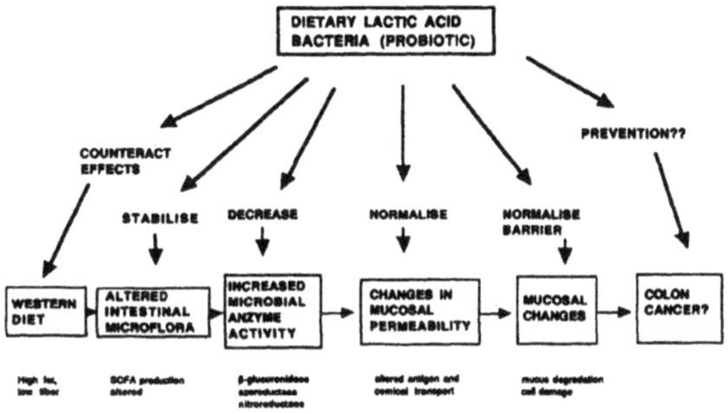

Figure 3. Possible mechanisms of lactic acid bacteria influencing colon cancer related parameters.

Food and clinical applications of probiotics

Probiotic bacteria (e.g., *Bifidobacterium bifidum* and *Lactobacillus* GG, *Lactobacillus casei* Shirota) have beneficial effects on the clinical course of rotavirus diarrhoea, including prevention and treatment, and the enhanced immune response thereafter (Isolauri et al., 1991; Kaila et al., 1992; Saavedra et al., 1994). In a similar manner, *Lactobacillus acidophilus* preparations and *Lactobacillus casei* (powder prepared from heat killed *Lactobacillus casei* Shirota) preparations have been beneficial in the prevention of radiation enterophathy (Salminen et al., 1988; Salminen et al., 1995; Okawa et al., 1993). *Lactobacillus acidophilus* and *Lactobacillus* GG have enhancing effects on the immune system in connection with oral vaccines and *Lactobacillus reuteri* on the duration of acute diarrhoea in children (Casas et al., 1995). Am.ong the possible mechanisms responsible for the favourable clinical response is promotion of the immunologic and nonimmunologic defence barrier in the gut.

Oral introduction of *Lactobacillus acidophilus* LA1 and *Lactobacillus GG* have been associated with alleviation of intestinal inflammation and *Lactobacillus GG* with normalization of increased intestinal permeability (Isolauri et al., 1993) and gut microflora (Isolauri et al., 1994). Such studies have not been reported for the other successful probiotic strains. Another explanation for the gut stabilizing effect of *Lactobacillus GG* could be improvement of the intestine's immunologic barrier, particularly intestinal IgA responses (Isolauri et al., 1993).

The above mentioned health effects and known mechanisms offer a basis for both applications of these strains in functional food products and, in specific disease states, also in clinical foods and medical foods. Due to differences between even closely related strains all strains shouls be tested on their own merit with good clinical study protocols as outlined in Table 4.

Types of clinical food applications

Management in food allergy includes exclusion of responsible food from the diet. In attempts to eliminate cow milk allergens from the diet, the most common approach is predigestion of bovine casein or whey to create formulae that provide nitrogen as mixtures of peptides and amino acids (Isolauri et al., 1995). The data taken together suggest that probiotic bacteria down-regulate hypersensitivity reactions and promote endogenous barrier mechanisms. The use of probiotic bacteria supplemented formula can further reduce antigenicity of substitute formulae and consequently have an important role in the development of specific therapy for patients who have food allergy.

Management of acute diarrhoea requires infant formulae with efficient lactic acid bacteria. Another alternative is to supply oral rehydration solutions with similar bacteria. In clinical nutrition, elemental diets are used for the treatment of intestinal inflammation, radiation enteritis and inflammatory bowel disease. Probiotic lactic acid bacteria do have a role in alleviating these disturbances and thus, the introduction of such strains into clinical foods and special dietary foods is likely. A new area is opened with the possibility of influencing food allergy and a variety of products could be developed for this area using the specific strains.

Another promising area for future research is the prevention and adjuvant treatment of cancer of the colon or bladder. Human studies using *Lactobacillus casei* Shirota strain appear promising in both areas. Other strains, such as *Lactobacillus* GG and *Lactobacillus acidophilus* NFCB 1748 have been beneficial in animal studies with colon carcinogens and in human studies using faecal bacterial enzymes as biomarkers.

Conclusion

The applications described in this paper indicate that there are some common background characteristics for many intestinal disturbances which facilitate effective use of probiotic bacteria. It is clear that probiotic bacteria have potential in the treatment of clinical conditions with altered gut mucosal barrier functions. Probiotic bacteria offer new dietary alternatives for the stabilization of the intestinal microflora. They can be used for immunotherapy to counteract local immunological dysfunctions and to stabilize the natural gut mucosal barrier mechanisms. It is important that the probiotic properties of each strain are demonstrated in carefully planned and controlled human studies.

References

Adams MR & Marteau P (1995) On the safety of lactic acid bacteria from food. Int. J. Food Microbiol. 27: 262–264

Aguirre M & Collins MD (1993) Lactic acid bacteria and human clinical infection. J. Appl. Bact. 75: 95–107

Aso Y & Akazan H (1992) Prophylactic effect of a *Lactobacillus casei* preparation on the recurrence of superficial bladder cancer. Urol. Int. 49: 125–129

Aso Y, Akaza H, Kotake T, Tsukamoto T, Imai K, Naito S (1995) Preventive effect of a *Lactobacillus casei* preparation on the recurrence of superficial bladder cancer in a double-blind trial. Eur. Urol. 27: 104–109

Axelsson I, Jakobsson I, Lindberg T, Polberger S, Benediktsson B & Räihä N. (1989) Macromolecular absorption in preterm and term infants. Acta. Paediatr. Scand. 78: 532–537

Beach RC, Menzies IS, Clayden GS & Scopes JW (1982) Gastrointestinal permeability changes in the preterm neonate. Arch. Dis.. Child 57: 141–145

Bernet M F, Brassart D, Neeser J R & Servin A (1994) *Lactobacillus acidophilus* LA1 binds to cultured human intestinal cell lines and inhibits cell attachment and cell invasion by enterovirulent bacteria. Gut. 35: 483–489

Bernet M F, Brassart D, Neeser J-R & Servin A (1993) Adhesion of human bifidobacterial strains to cultured human intestinal epithelial cells and inhibition of enteropathogen-cell interactions. Appl. Environm. Microbiol. 59: 4121–4128

Bruce MG & Ferguson A (1986) The influence of intestinal processing on the immunogenicity and molecular size of absorbed, circulating ovalbumin in mice. Immunol.ogy 59: 295–300

Casas I, Shornikova A, Isolauri E, Salminen S, Vesikari T (1996) Colonization of *Lactobacillus reuteri* in the gastrointestinal tract of children. Abstracts of the SOMED Meeting, Puerto Rico 1995. Micr. Ecol. Health Dis.. (in press)

Claeson M & Merson MH (1990) Global progress in the control of diarrheal disease. Pediatr. Infect. Dis.. 9: 345–355

Dahl O, Horn A, Mella O (1994) Do acute side effects during radiotherapy predict tumour response in rectal carcinoma. Acta. Oncol. 33: 409–413

Danielsson å, Nyhlin H, Persson H, Stendah U, Stenling R & Suhr O (1991) Chronic diarrhoea after radiotherapy for gynaecological cancer: occurrence and aetiology. Gut. 32: 1180–1187

Dong M-Y, Chang T-W & Gorbach SL (1987) Effects of feeding *Lactobacillus* GG on lethal irradiation in mice. Diagn. Microbiol. Infect. Dis.. 7: 1–7

Donohue D & Salminen S (1996) Safety of probiotic bacteria. Asia Pacific J. Clin. Nutr. 5: 25–28

Elo S, Saxelin M & Salminen S (1991) Attachment of *Lactobacillus casei* strain GG to human colon carcinoma cell line Caco-2: comparison with other dairy strains. Lett. Appl. Microbiol. 13: 154–156

Fargeas MJ, Theodorou V, More J, Wal JM, Fioramonti J & Bueno L (1995) Boosted systemic immune and local responsiveness after intestinal inflammation in orally sensitized guinea pigs. Gastroenterology 109: 53–62

Friberg L-G (1980) Effects of irradiation on the small intestine of the rat. A SEM study. Ph.D. Thesis, University of Lund, Lund, Sweden

Gasser F (1994) Safety of lactic acid bacteria and their occurrence in human clinical infections. Bull Inst. Pasteur. 92: 45–67

Gillen CD, Prior P, Andrews HA & Allan RN (1994) Ulcerative colitis and Crohn's disease:comparison of colorectal cancer risk in extensive colitis. Gut. 35: 1590–1592

Goldin B & Gorbach SL (1984) The effect of milk and *Lactobacillus* feeding on human intestinal bacterial enzyme activity. Am. J. Clin. Nutr. 33: 15–18

Goldin B, Gorbach SL, Saxelin M, Barakat S, Gualtieri L & Salminen S (1992) Survival of *Lactobacillus* species (strain GG) in human gastrointestinal tract. Dig. Dis. Sci. 37: 121–128

Goldin B, Gualtieri L & Moore R (1996) The effect of *Lactobacillus* GG on the initiation and promotion of dimethylhydrazine-induced intestinal tumors in the rat. Nutr. Cancer 25: 197–204

Gorbach SL (1990) Lactic acid bacteria and human health. Ann Med. 22: 37–41

Henriksson R, Franzen L, Sandström K, Nordin A, Arevarn M & Grahn E (1995) The effects of active addition of bacterial cultures in fermented milk to patients with chronic bowel discomfort following irradiation. Support Care Cancer 3: 81–83

Heyman M & Desjeux JF (1992) Significance of intestinal food protein transport. J. Pediatr. Gastroenterol. Nutr. 15: 48–57

Heyman M, Ducroc R, Desjeux JF & Morgat JL (1982) Horseradish peroxidase transport across adult rabbit jejunum in vitro. Am. J. Physiol. 242: G558–G564

Heyman M, Gorthier G, Petit A, Meslin J-C, Moreau C & Desjeux J-F (1987) Intestinal absorption of macromolecules during viral enteritis: an experimental study on rotavirus-infected conventional and germ-free mice. Pediatr. Res. 22: 72–78

Heyman M, Grasset E, Ducroc R & Desjeux JF (1988) Antigen absorption by the jejunal epithelium of children with cow's milk allergy. Pediatr. Res. 24: 197–202

Holt PG, McMenamin C & Nelson D (1990) Primary sensitisation to inhalant allergens during infancy. Pediatr. Allergy Immunol.. 1: 3–13

Holt PG (1994) Immunophophylaxis of atopy: light at the end of the tunnel. Immunol. Today 15: 484–489

Isolauri E & Turjanmaa K (1996) Combined skin prick and patch testing enhances identification of food allergy in infants with atopic dermatitis. J. Allergy Clin. Immunol. 97: 9–15

Isolauri E, Joensuu J, Suomalainen H, Luomala M, Vesikari T. (1995). Improved immunogenicity of oral DxRRV reassortant rotavirus vaccine by *Lactobacillus casei* GG. Vaccine 13: 310–312

Isolauri E, Juntunen M, Rautanen T, Sillanaukee P & Koivula T (1991) A human *Lactobacillus* strain (*Lactobacillus* GG) promotes recovery from acute diarrhea in children. Pediatrics 88: 90–7

Isolauri E, Kaila M, Arvola T, Majamaa H, Rantala I, Virtanen E & Arvilommi H (1993) Diet during rotavirus enteritis affects jejunal permeability to macromolecules in suckling rats. Pediatr. Res. 33: 548–53

Isolauri E, Kaila M, Mykkänen H, Ling WH & Salminen S (1994) Oral bacteriotherapy for viral gastroenteritis. Dig. Dis. Sci. 39: 2595–2600

Isolauri E, Majamaa H, Arvola T, Rantala I, Virtanen E & Arvilommi H (1993) *Lactobacillus casei* strain GG reverses increased intestinal permeability induced by cow milk in suckling rats. Gastroenterology 105: 1643–50

Isolauri E (1995) Intestinal Integrity and IBD. Kluwer Academic Publishers, Lancaster, UK, 85: 553–5

Isolauri E (1995) The treatment of cow's milk allergy. Eur J. Clin. Nutr. 49 (Suppl 1): S49-S55

Isolauri E, Sütas Y, Mäkinen-Kiljunen S, Oja SS, Isosomppi R & Turjanmaa K (1995) Efficacy and safety of hydrolysed cow milk and amino acid-derived formulas in infants with cow milk allergy. J. Pediatr. 127: 550–7

Jalonen T, Isolauri E, Heyman M, Crain-Denoyelle A-M, Sillanaukee P & Koivula T (1991) Increased ß-lactoglobulin absorption during rotavirus enteritis in infants: relationship to sugar permeability. Pediatr. Res. 30: 290–293

Jalonen T (1991) Identical intestinal permeability changes in children with different clinical manifestations of cow's milk allergy. J. Allergy Clin. Immunol. 88: 737–742

Kaila M, Isolauri E, Soppi E, Virtanen E, Laine S & Arvilommi H (1992) Enhancement of the circulating antibody secreting cell response in human diarrhea by a human *Lactobacillus* strain. Pediatr. Res. 32: 141–144

Kato I, Endo K, Yokokura T. (1994) Effects of oral administration of Lactobacillus casei on antitumor responses induced by tumor resection in mice. Int. J. Immunopharmac. 16: 29–36

Kato I, Yokokura T, Mutai M (1988) Correlation between inncrease in Ia-bearing macrophages and induction of T-cell dependent antitumor activity by *Lactobacillus casei* in mice. Cancer Immun. Immunother. 26: 215–221

Lee Y-K & Salminen S (1995) The coming of age of probiotics. Trends Food Sci. Technol. 6: 241–245

Ling WH, Korpela R, Mykkänen H, Salminen S & Hänninen O (1994) strain GG supplementation decreases colonic hydrolytic and reductive enzyme activities in healthy female adults. J. Nutr. 124: 18–23

Link-Am.ster H, Rochat F, Saudan KY, Mignot O & Aeschlimann JM. Modulation of a specific humoral immune response and changes in intestinal flora mediated through fermented milk intake. FEMS Immunol. Med. Micr 10: 55–64

MacDonald TT (1993) Aetiology of Crohn's disease. Arch Dis. Child 68: 623–625

Majamaa H & Isolauri E (1996) Evaluation of the gut mucosal barrier: evidence for increased antigen transfer in children with atopic eczema. J. Allergy Clin. Immunol. (in press)

Majamaa H, Isolauri E, Saxelin M & Vesikari T (1995) Lactic acid bacteria in the treatment of acute rotavirus gastroenteritis. J. Ped. Gastroenterol. Nutr. 20: 333–338

Majamaa H, Miettinen A, Laine S & Isolauri E (1996) Intestinal inflammation in children with atopic eczema: faecal eosinophil cationic protein and tumour necrosis factor-a as noninvasive indicators of food allergy. Clin. Exp. Allergy. 26: 181–187

Malin M, Verronen P, Mykkänen H, Salminen S & Isolauri E (1996a) Increased bacterial urease activity in faeces in juvenile chronic arthritis: evidence of altered intestinal microflora? Br. J. Rheumatol. (in press)

Malin M, Suomalainen H, Saxelin M & Isolauri E (1996b) Promotion of IgA immune response in petients with Crohn's disease by oral bacteriotherapy with *Lactobacillus* GG. Ann Nutr. Met. (in press)

Meurman J, Antila H, Korhonen A, Salminen S (1995) Effect of *Lactobacillus rhamnosus* strain GG (ATCC 53103) on the growth of *Streptococcus sorbinus* in vitro. Eur. J. Oral. Sci. 103: 253–258

Midtvedt T (1987) Intestinal bacteria and rheumatoid disease. Scand. J. Rheumatol. 64(suppl): 49–54

Molin G, Jeppsson B, Johansson ML, Ahrne S, Nobaek S, Ståhl M & Bengmark S (1993) Numerical taxonomy of Lactobacillus spp. associated with healthy and diseased mucosa of the human intestines. J. Appl. Bact. 74: 314–323

Morotomi M. (1996) Properties of *Lactobacillus casei* Shirota strain as probiotics. Asia Pacific J. Clin. Nutr. 1996; 5: 29–30.

Okawa T, Niibe H, Arai T, Sekiba K, Noda K, Takeuchi S, Hashimoto S & Ogawa N (1993) Effect of LC9018 combined with radiation therapy on carcinoma of the uterine cervix. Cancer 72: 1949–1954

Okawa T, Kita M, Arai T, Iida K, Dokiya T, Tagekava Y, Hirokawa Y, Yamazaki K, Hashimoto S. Phase II randomized clinical trial of LC 9018 concurrently used with radiation in the treatment of carcinoma of the uterine cervix; its effects on tumor reduction and histology. Cancer 69: 1769–1776

Pedrosa MC, Golner BB, Goldin BR, Barakat S, Dallal G & Russel R (1995) Survival of yogurt-containing organisms and Lactobacillus gasseri (ADH) and their effect on bacterial enzyme activity in the gastrointestinal tract of healthy and hypochlorhydric elderly subjects. Am. J. Clin. Nutr. 61: 353–358

Podolsky DK (1991) Inflammatory bowel disease. (First of two parts). New Engl. J. Med. 325: 928–937

Raza S, Graham SM, Allen SJ, Sultana S, Cuevas L & Hart CA (1995) *Lactobacillus* GG promotes recovery from acute non-bloody diarrhea in Pakistan. Pediatr. Infect. Dis. J 14: 107–111

Saavedra JM, Bauman N, Oung I, Perman J & Yolken R (1994) Feeding of *Bifidobacterium bifidum* and *Streptococcus thermophilus* to infants in hospital for prevention of diarrhoea and shedding of rotavirus. Lancet 344: 1046–1049

Saavedra JM (1995) Microbes to fight microbes: a not so novel approach to controlling diarrheal diseases. J. Ped. Gastroenterol. Nutr. 21: 125–129

Salminen E, Elomaa I, Minkkinen J, Vapaatalo H & Salminen S (1988) Preservation of intestinal integrity during radiotherapy using live *Lactobacillus acidophilus* cultures. Clin. Radiol. 39: 435–437

Salminen E, Salminen S, Vapaatalo H & Holsti LR (1995) Adverse effects of pelvic radiotherapy. International Meeting on Progress in Radio-oncology, HD Kogelnik (ed.,) (pp 501–504) Monduzzi Editore, Italy

Salminen S, Deighton M, Gorbach SL (1993) Lactic acid bacteria in health and disease. In: Salminen S & von Wright A (Eds) Lactic Acid Bacteria (pp 199–225). Marcel Dekker Inc, New York

Salminen S, Isolauri E & Salminen E (1996) Probiotics and stabilization of the gut mucosal barrier. Asia Pac. J. Clin. Nutr. 5: 53–56

Sampson HA & McCaskill CC (1985) Food hypersensitivity and atopic dermatitis: evaluation of 113 patients. J. Pediatr. 107: 669–675

Sanderson IR & Walker WA (1993) Uptake and transport of macromolecules by the intestine: possible role in clinical disorders (an update). Gastroenterology 104: 622–639

Sarem-Damerdji, L-O, Sarem F, Marchal L & Nicolas J-P (1995) In vitro colonization ability of human colon mucosa by exogenous *Lactobacillus* strains. FEMS Microbiol. Lett. 131: 133–137

Sawamura A, Yamaguchi Y, Toge T, Nagata N, Ikeda H, Nakanishi K, Asakura A (1994). The enhancing effect of oral *Lactobacillus casei* on the immunologic acitivty of colon cancer patients. Biotherapy 8: 1567–1572

Saxelin M, Elo S, Salminen S & Vapaatalo H (1991) Dose response colonisation of faeces after oral administration of *Lactobacillus casei* strain GG. Micr. Ecol. Health Dis. 4: 209–214

Saxelin M, Pessi T & Salminen S (1995) Fecal recovery following oral administration of *Lactobacillus* strain GG (ATCC 53103) in gelatine capsules to healthy volunteers. Int. J. Food Microbiol. 25: 199–203

Saxelin M, Rautelin H, Chassy B, Gorbach SL, Salminen S & Mäkelä H (1996a) Lactobacilli and septic infections in Southern Finland. Clin. Infect. Dis. 22: 564–566

Saxelin M, Rautelin H, Salminen S & Mäkelä, H (1996b) The safety of commercial products with vieable *Lactobacillus* strains. Infect. Dis. Clin. Practice (in press)

Schiffrin EJ, Brassart D, Servin AL, Rochat F & Donnet-Hughes A (1996) Immune modulation of blood leukocytes in man by lactic acid bacteria. Criteria for strain selection. Am. J. Clin. Nutr. (in press)

Siitonen S, Vapaatalo H, Salminen S, Gordin A, Saxelin M, Wikberg R & Kirkkola AL (1990) Effect of *Lactobacillus* GG yoghurt in prevention of antibiotic associated diarrhoea. Ann Med. 22: 57–59

Silva C (1993) The use of elemental diets in radiation enteritis. In: Bounous G (Ed)., Uses of elemental diets in clinical situations (pp 122–134). CRC Press, Boca Raton, FL, USA

Stern M, Pang KY & Walker WA (1984) Food proteins and gut mucosal barrier. II. Differential interaction of cow's milk proteins with the mucous coat and the surface membrane of adult and immature rat jejunum. Pediatr. Res. 18: 1252–1257

Strobel S (1991) Mechanisms of gastrointestinal immunoregulation and food induced injury. Eur J. Clin. Nutr. 45: 1–9

Sugita T, Togawa M. (1994). Efficacy of *Lactobacillus* preparation Biolactis powder in children with rotavirus enteritis. Japan J. Pediatr. 47: 2755–2762

Sūtas Y, Soppi E, Korhonen H, Syväoja EL, Saxelin M, Rokka T & Isolauri E (1996) Suppression of lymphocyte proliferation in vitro by bovine caseins hydrolysed with *Lactobacillus* GG-derived enzymes. J. Allergy Clin. Immunol. (in press)

Sūtas Y, Hurme M & Isolauri E (1996) Differential regulation of cytokine production by bovine caseins: evidence for generation of tolerogens from allergens by intestinal bacteria. Scand. J. Immunol. (in press)

Tanaka R & Ohwaki M (1994) A controlled study on the ingestion of *Lactobacillus casei* fermented milk on the intestinal microflora, its microbiology and immune system in healthy adults. Proceedings of XII Riken Symposium on Intestinal Flora, Tokyo, Japan, 85–104

Tytgat GNJ, Dhir V & Gopinath N (1995) Endoscopic appearance of dysplasia and cancer in inflammatory bowel disease. Eur. J. Cancer 31: 1174–1177

Uhnoo IS, Freihorst J, Riepenhoff-Talty M, Fisher JE & Ogra PL (1990) Effect of rotavirus infection and malnutrition on uptake of a dietary antigen in the intestine. Pediatr. Res. 27: 153–160

Wolf BW, Garleb DG, Casas I (1995) Safety and tolerance of *Lactobacillus reuteri* in healthy adult male subjects. Microb. Ecol. Health Dis. 8: 41–50

Young G. (1996) Prevention of colon cancer: role of short chain fatty acids produced by intestinal flora. Asia Pacific J. Clin. Nutr. 5: 44–47

MIX
Papier aus verantwortungsvollen Quellen
Paper from responsible sources
FSC® C105338

If you have any concerns about our products,
you can contact us on
ProductSafety@springernature.com

In case Publisher is established outside the EU,
the EU authorized representative is:
Springer Nature Customer Service Center GmbH
Europaplatz 3, 69115 Heidelberg, Germany

Printed by Libri Plureos GmbH
in Hamburg, Germany